Computational Intelligence for Green Cloud Computing and Digital Waste Management

K. Dinesh Kumar
Amrita Vishwa Vidyapeetham, India

Vijayakumar Varadarajan
The University of New South Wales, Australia

Nidal Nasser
College of Engineering, Alfaisal University, Saudi Arabia

Ravi Kumar Poluru
Institute of Aeronautical Engineering, India

A volume in the Advances in Computational
Intelligence and Robotics (ACIR) Book Series

Published in the United States of America by
    IGI Global
    Engineering Science Reference (an imprint of IGI Global)
    701 E. Chocolate Avenue
    Hershey PA, USA 17033
    Tel: 717-533-8845
    Fax: 717-533-8661
    E-mail: cust@igi-global.com
    Web site: http://www.igi-global.com

Library of Congress Cataloging-in-Publication Data

CIP PENDING

TITLE: Computational Intelligence for Green Cloud Computing and Digitial Waste Management
 2024 Engineering Science Reference

ISBN: 9798369315521
eISBN: 9798369315538

This book is published in the IGI Global book series Advances in Computational Intelligence and Robotics (ACIR) (ISSN: 2327-0411; eISSN: 2327-042X)

British Cataloguing in Publication Data
A Cataloguing in Publication record for this book is available from the British Library.

All work contributed to this book is new, previously-unpublished material. The views expressed in this book are those of the authors, but not necessarily of the publisher.

For electronic access to this publication, please contact: eresources@igi-global.com.

Advances in Computational Intelligence and Robotics (ACIR) Book Series

Ivan Giannoccaro
University of Salento, Italy

ISSN:2327-0411
EISSN:2327-042X

MISSION

While intelligence is traditionally a term applied to humans and human cognition, technology has progressed in such a way to allow for the development of intelligent systems able to simulate many human traits. With this new era of simulated and artificial intelligence, much research is needed in order to continue to advance the field and also to evaluate the ethical and societal concerns of the existence of artificial life and machine learning.

The **Advances in Computational Intelligence and Robotics (ACIR) Book Series** encourages scholarly discourse on all topics pertaining to evolutionary computing, artificial life, computational intelligence, machine learning, and robotics. ACIR presents the latest research being conducted on diverse topics in intelligence technologies with the goal of advancing knowledge and applications in this rapidly evolving field.

COVERAGE

- Intelligent Control
- Artificial Life
- Pattern Recognition
- Computer Vision
- Artificial Intelligence
- Fuzzy Systems
- Heuristics
- Cognitive Informatics
- Robotics
- Adaptive and Complex Systems

IGI Global is currently accepting manuscripts for publication within this series. To submit a proposal for a volume in this series, please contact our Acquisition Editors at Acquisitions@igi-global.com or visit: http://www.igi-global.com/publish/.

Titles in this Series

For a list of additional titles in this series, please visit:
www.igi-global.com/book-series/advances-computational-intelligence-robotics/73674

AIoT and Smart Sensing Technologies for Smart Devices
Fadi Al-Turjman (AI and Robotics Institute, Near East University, Nicosia, Turkey & Faculty of Engineering, University of Kyrenia, Kyrenia, Trkey)
Engineering Science Reference • © 2024 • 250pp • H/C (ISBN: 9798369307861) • US $300.00

Industrial Applications of Big Data, AI, and Blockchain
Mahmoud El Samad (Lebanese International University, Lebanon) Ghalia Nassreddine (Rafik Hariri University, Lebanon) Hani El-Chaarani (Beirut Arab University, Lebanon) and Sam El Nemar (AZM University, Leanon)
Engineering Science Reference • © 2024 • 348pp • H/C (ISBN: 9798369310465) • US $275.00

Principles and Applications of Adaptive Artificial Intelligence
Zhihan Lv (Uppsala University, Sweden)
Engineering Science Reference • © 2024 • 316pp • H/C (ISBN: 9798369302309) • US $325.00

AI Tools and Applications for Women's Safety
Sivaram Ponnusamy (Sandip University, Nashik, India) Vibha Bora (G.H. Raisoni College of Engineering, Nagpur, India) Prema M. Daigavane (G.H. Raisoni College of Engineering, Nagpur, India) and Sampada S. Wazalwar (G.H. Raisoni College of Engineering, Nagpur, India)
Engineering Science Reference • © 2024 • 362pp • H/C (ISBN: 9798369314357) • US $300.00

Advances in Explainable AI Applications for Smart Cities
Mangesh M. Ghonge (Sandip Institute of Technology and Research Centre, India) Nijalingappa Pradeep (Bapuji Institute of Engineering and Technology, India) Noor Zaman Jhanjhi (School of Computer Science, Faculty of Innovation and Technology, Taylor's University, Malaysia) and Praveen M. Kulkarni (Karnatak Law Society's Institute of Management Education and Research (KLS IMER), Belagavi, India)
Engineering Science Reference • © 2024 • 506pp • H/C (ISBN: 9781668463611) • US $270.00

Applications of Machine Learning in UAV Networks
Jahan Hassan (Central Queensland University, Australia) and Saeed Alsamhi (Insight Centre for Data Analytics, University of Galway, Ireland)
Engineering Science Reference • © 2024 • 406pp • H/C (ISBN: 9798369305782) • US $285.00

701 East Chocolate Avenue, Hershey, PA 17033, USA
Tel: 717-533-8845 x100 • Fax: 717-533-8661
E-Mail: cust@igi-global.com • www.igi-global.com

Table of Contents

Detailed Table of Contents

Chapter 1

Rajashri Roy Choudhury, Brainware University, India
Piyal Roy, Brainware University, India
Shivnath Ghosh, Brainware University, India
Ayan Ghosh, Brainware University, India

Green computing is an innovative approach to making computer systems environmentally friendly, energy-efficient, and low in carbon emissions. It uses advanced techniques from machine learning and deep learning to optimize real-time resource allocation, reducing energy consumption. This approach enhances workload patterns and uses methods like convolutional and recurrent neural networks to enhance architectural efficiency. The integration of ML and DL techniques allows for accurate temperature forecasting and alternative cooling strategies. Despite challenges, the synergistic fusion of ML and DL algorithmic software with green computing holds great promise for reducing energy consumption and enhancing environmental sustainability.

Chapter 2

Gudivada Lokesh, Jawaharlal Nehru Technological University, India
B. Rupa Devi, Annamacharya Institute of Technology and Sciences, India
N. Badrinath, Vellore Institute of Technology, India
L. N. C. Prakash K., CVR College of Engineering, India
Pole Anjaiah, Institute of Aeronautical Engineering (Autonomous), India
T. Ravi Kumar, Aditya Institute of Technology and Management, India

The internet of things (IoT) connects devices of all sizes to the internet, providing seamless communication and ease. However, this technical improvement has prompted environmental concerns. Sustainable development has grown as we work to offset the effects of technology on our planet, economy, and consumerism. Innovative companies are exploring technology-based sustainability solutions. Green IT, a developing idea, uses sustainable design to reduce or eliminate IoT operations' environmental implications. This chapter evaluates IoT problems, defines Green IT, explores Green IT design methodologies, and describes how to implement these green designs as sustainable environmental solutions. It also thoroughly analyses numerous author proposals to find the best sustainable IoT architectures. Green IT solutions are essential to a sustainable and ecologically responsible future in a world where information technology is central to our lives, businesses, and marketing tactics.

Chapter 3

    J. Jeyaranjani, SRM Madurai College for Engineering and Technology, India

    K. Rangaswamy, Rajeev Gandhi Memorial College of Engineering and Technology (Autonomous), India

    A. Ashwitha, Manipal Institute of Technology Bengaluru, Manipal Academy of Higher Education, India

    Ramakrishna Gandi, Madanapalle Institute of Technology and Science, India

    R. Roopa, Madanapalle Institute of Technology and Science, India

    P. Anjaiah, Institute of Aeronautical Engineering (Autonomous), India

The adoption of green computing is crucial due to cost reduction and environmental responsibility. However, challenges hinder progress. This research explores solutions for balancing industrial growth with sustainability and addresses green computing obstacles. Recognizing industry's importance while mitigating its environmental impact is vital. Strategies to reduce waste and energy use, like cloud computing and virtualization, can help. Implementing a circular economy approach makes products regenerative and less wasteful. This research provides real-world case studies and insights, aiding businesses in adopting greener, more sustainable computing practices for a cost-effective, environmentally responsible tech landscape.

Chapter 4

    Dhanabalan Thangam, Presidency Business School, Presidency College, Bengaluru, India

    Haritha Muniraju, Triveni Institute of Commerce and Management, Bengaluru, India

    R. Ramesh, Department of Management Studies, Knowledge Institute of Technology, Salem, India

    Ramakrishna Narasimhaiah, Department of Economics, Jain University, Bengaluru, India

    N. Muddasir Ahamed Khan, Department of Management, Acharya Institute of Graduate Studies, Bengaluru, India

    Shabista Booshan, ISBR College, India

    Bharath Booshan, Department of Management, Acharya Institute of Graduate Studies, Bengaluru, India

    Thirupathi Manickam, Christ University, India

    R. Sankar Ganesh, Vel Tech Rangarajan Dr. Sagunthala R&D Institute of Science and Technology, India

The data-driven economy is transforming with data centers becoming a crucial business infrastructure. However, the increasing reliance on data centers is posing a threat to the environment. Climate change activists are focusing on reducing emissions from sectors like automotive, aviation, and energy. Data centers consume more electricity than the UK, accounting for 3% of global electricity supply and 2% of total greenhouse gas emissions. By 2040, digital data storage is projected to contribute to 14% of the world's emissions. The number of data centers worldwide has surged from 500,000 in 2012 to over 8 million, with energy consumption doubling every four years. The rise in internet penetration rates and the introduction of 5G technologies and IoT devices will further exacerbate the issue, increasing the demand for data processing.

    Dhaarini K. N. Hathwar, REVA University, India
    Srinidhi R. Bharadwaj, REVA University, India
    Syed Muzamil Basha, REVA University, India

This chapter describes the central role that virtualization technologies play in promoting sustainable computing practices. The authors thoroughly explore the complexities of green data center and server operations and highlight the importance of server virtualization in collaborative integration efforts. Essential technologies such as dynamic voltage and frequency scaling (DVFS) will be examined for their potential to reduce energy consumption. Additionally, they introduce a new approach called communication-aware request stacking to optimize energy efficiency. By advocating best practices in network design, they are committed to embracing green networks and leveraging energy-efficient resources and nodes. The proposed framework integrates network virtualization and adaptive link rate, promising improved network performance and a greener operational paradigm. This chapter provides rich insights for practitioners, researchers, educators, and policy makers working to promote environmental sustainability in computing and networking.

    K. Gopi, M.G.R. Arts and Science College, Hosur, India
    Anil Sharma, Parul Institute of Management and Research, India
    M. R. Jhansi Rani, ISBR Business School, Bangalore, India
    K. Praveen Kamath, ISBR Business School, Bangalore, India
    Thirupathi Manickam, Christ University, India
    Dhanabalan Thangam, Presidency Business School, Presidency College, Bengaluru, India
    K. Ravindran, Presidency Business School, Presidency College, Bengaluru, India
    Chandan Chavadi, Presidency Business School, Presidency College, Bengaluru, India
    Naveen Pol, ISBR College, India

Data centers and transmission networks are crucial in the digital age, with the market expected to grow from $50 billion in 2021 to $120 billion by 2030. However, their extensive computing infrastructure and continuous operation generate significant heat, necessitating energy-intensive cooling systems. The IEA report revealed that data center power consumption surged by over 60% between 2015 and 2021, with transmission networks experiencing a 60% usage increase. Addressing these growing energy demands poses significant challenges for the industry, with some countries considering restrictions on new data center licenses due to environmental concerns. To mitigate the climate impact, the industry must prioritize the procurement of low-carbon or carbon-free electricity to reduce Scope 2 emissions related to electricity, heating, and cooling. Tech giants like AWS, Google, and Meta/Facebook have already adopted ambitious public targets, either running on carbon-free electricity or investing in global projects for cost-effective and large-scale emissions reduction.

Chapter 7

Sana Dahmani, Independent Researcher, Germany

Intelligent resource management across fog, edge, and cloud computing entails dynamic allocation, optimal utilization, and robust security measures. Cloud computing adopts a centralized approach to provision resources, whereas edge and fog computing allocate resources at the periphery, strategically minimizing latency. The incorporation of AI/ML algorithms has a pivotal role, enabling resource prediction for anticipating demand, detecting anomalies, and optimizing allocation efficiently. The self-organizing management aspect facilitates autonomous adaptation. Resource virtualization abstracts physical resources into flexible virtual counterparts, complemented by meticulous accounting that tracks consumption and costs. The inclusion of security-aware measures ensures protection against unauthorized access. This comprehensive approach not only enhances performance, scalability, and security but also promotes adaptive scaling and proactive decision-making. Additionally, the implementation of green IT practices optimizes resource utilization, effectively reducing environmental impact.

Chapter 8

*V. Vijayalakshmi, Department of DSBS, SRM Institute of Science and Technology,
Kattankulathur, India*

R. Radha, Department of DSBS, SRM Institute of Science and Technology, Kattankulathur, India

*S. Sharanya, Department of DSBS, SRM Institute of Science and Technology,
Kattankulathur, India*

As the rapid pace of technological advancement continues to propel society into the digital age, the surge in electronic waste (e-waste) poses significant challenges to environmental sustainability. This research explores modern technological innovations in digital waste management that contribute to the reduction, recycling, and responsible disposal of electronic devices. Special attention is given to advancements in recycling methods, the application of artificial intelligence (AI), machine learning (ML), deep learning (DL), robotics, and IoT in automated e-waste processing. The research investigates the utilization of exploring how materials can facilitate easier recycling and reduce the environmental impact of electronic devices. The research also explores the role of extended producer responsibility (EPR) in adapting to sustainable practices in product design, disposal, and recycling. This research contributes by offering an understanding of the tools, strategies, and policies that can contribute to a more sustainable volume of e-waste in our increasingly digitized world.

Chapter 9

S. Sharanya, Data Science and Business Systems, SRM Institute of Science and Technology, India

*V. Vijayalakshmi, Data Science and Business Systems, SRM Institute of Science and
Technology, India*

R. Radha, Data Science and Business Systems, SRM Institute of Science and Technology, India

The accelerated growth in artificial intelligence, internet of devices, machine learning (ML), and deep learning at breakneck speed has attracted the attention of researchers in developing novel green solutions for reclaiming the green society. The intersection of these technologies with green sustainability will greatly impact the deployment of cutting-edge technologies with green solutions. Leveraging ML technologies to improve engineering techniques to reduce the toxins released in the environment in various forms is discussed in this work. The predominant area of focus is applying is developing green AI-based solutions with sustainability measures and metric in mind. The primary contribution of this work is the holistic analysis of the employment of green ML and deep learning techniques in fostering a sustainable environment. The potential scope of this research is to benefit the research community in developing novel ML and deep learning technologies for improving green sustainability.

Chapter 10

    Diya Biswas, Brainware University, India
    Anuska Dutta, Brainware University, India
    Shivnath Ghosh, Brainware University, India
    Piyal Roy, Brainware University, India

Cloud providers place a high value on reducing energy consumption in cloud computing since it reduces operating costs and improves service sustainability. Cloud services are frequently replicated across providers to ensure high availability and dependability, which increases provider resource utilization and overhead. Finding the right balance between service replication and consolidation to lower energy usage and boost service uptime can be challenging for cloud resource management decision-makers. This chapter addresses this problem by presenting a ground-breaking technique known as "CRUZE," which is based on cuckoo optimization and considers energy efficiency, dependability, and comprehensive resource management in cloud computing, encompassing cooling systems, servers, networks, and storage. Using cloud resources, effectively illuminating and executing a range of jobs, CRUZE has significantly reduced energy usage by 20.1% while improving dependability and CPU utilization by effectively illustrating and executing a variety of workloads on cloud resources that have been allocated.

Chapter 11

    N. Manikandan, SRM Institute of Science and Technology, India
    D. Vinod, SRM Institute of Science and Technology, India
    R. Anto Arockia Rosaline, SRM Institute of Science and Technology, India
    P. Nancy, SRM Institute of Science and Technology, India
    G. Premalatha, SRM Institute of Science and Technology, India

The term "green computing" describes the efficient use of resources in computing and IT/IS infrastructure. This study suggests a unique method for dispersed fog data centres' work scheduling and resource allocation based on digital waste management. Here, the bandwidth differential preemption evolution moving average method (BDPEMA) is used to control the network's digital waste while allocating resources. Reinforcement adversarial hierarchical group multi-objective cuckoo optimisation (RAHMCO) is used to schedule network tasks. In terms of resource sharing rate, energy efficiency, reaction time, quality of service, and makespan, experimental study is conducted. The proposed approaches have been evaluated in a simulated cloud environment. The proposed method outperformed the current rules when QoS features were considered. The proposed technique attained QoS of 66%, energy efficiency of 96%, resource sharing of 88%, response time of 45%, and makespan of 61%.

Chapter 12

B. Manjunatha, New Horizon College of Engineering, India

K. Dinesh Kumar, Amrita Vishwa Vidyapeetham, India

Sam Goundar, RMIT University, India

Balasubramanian Prabhu Kavin, SRM Institute of Science and Technology, India

Gan Hong Seng, XJTLU Entrepreneur College, Xi'an Jiaotong-Liverpool University, China

E-waste is an invisible, indirect waste that contaminates natural resources like the air, water, and soil, endangering the ecosystem, people, and animals. Long-term waste accumulation and contamination can harm the resources found in the environment. Since traditional waste management systems are very inefficient and the number of people living in urban areas is increasing, waste management systems in these areas face challenges. However, by combining a variety of sensors with deep learning (DL) models, waste resources can be used effectively. For this chapter, firstly, the Trashnet dataset with 2527 images in six classes and the VN-trash dataset, which comprises three classes and 5904 images, are collected. Then the collected images are preprocessed using truncated gaussian filter. After that, pre-trained convolutional neural network (CNN) models (Resnet20 and VGG19) are applied to the images in order to extract features. In order to enhance the predictive performance, this study then creates a MobileNetV2 model for trash classification (TC) called MNetV2-TC.

Chapter 13

Sana Dahmani, Independent Researcher, Germany

The intersection of environmental protection, Sustainable Development Goals, and the role of information technology (IT) is to foster it. Artificial intelligence addresses environmental challenges, offering solutions such as emissions reduction, cost savings, legal compliance, HR attraction, optimized investments, and waste management. The "three Rs" of green IT emphasize sustainable hardware management through reuse, refurbishment, and recycling. Green IT aligns with broad sustainability and ESG standards and is fostered with innovative AI solutions to integrate and optimize AI-based green computing algorithms which address environmental impacts and data utilization strategies, optimizing energy consumption, and mitigating digital waste and carbon footprint. The conclusion advocates increased use of AI-built technologies for the adoption of renewable energy sources, energy-efficient hardware, hardware optimization, and exploration of external cloud solutions for a more sustainable future.

Chapter 14

B. Santosh Kumar, New Horizon College of Engineering, India

K. A. Jayasheel Kumar, New Horizon College of Engineering, India

Balasubramanian Prabhu Kavin, SRM Institute of Science and Technology, India

Gan Hong Seng, XJTLU Entrepreneur College, Xi'an Jiaotong-Liverpool University, China

One of the hottest new technologies that allows users to handle a broad range of resources and massive amounts of data in the cloud is green computing resource management. One of the biggest obstacles is task scheduling, and poor management leads to a decrease in productivity. The task must be efficiently scheduled to ensure optimal resource utilisation and minimal execution time. Given this, this study suggests a fresh method for efficient task scheduling in a green computing environment that also offers improved security. There is a development of an enhanced spotted hyena optimizer (ISHO). Finding the ideal or almost ideal subset with a straightforward structure to minimise the specified fitness function is a strong point of the SHO. It enhances a switch strategy in the spotted hyena's position updating mechanism and generates random positions in place of the violated spotted hyenas in order to support the proposed ISHO's exploration characteristics. Second, the data is encrypted using the ChaCha20–Poly1305 authenticated encryption algorithm, ensuring secure data transmission.

Chapter 15

A. V. Kalpana, SRM Institute of Science and Technology, India

S. Suchitra, SRM Institute of Science and Technology, India

Ram Prasath, SRM Instiute of Science and Technology, India

K. Arthi, SRM Institute of Science and Technology, India

J. Shobana, SRM Institute of Science and Technology, India

T. Nadana Ravishankar, SRM Institute of Science and Technology, India

Efficient waste management is crucial in today's environmental landscape, necessitating comprehensive approaches involving recycling, landfill practices, and cutting-edge technological integration. The proposed approach introduces a sophisticated waste management system, harnessing dual or twofold convolutional neural networks (D-CNN or TF-CNN) and a histogram density segmentation (HDS) algorithm. This intelligent system equips users with the means to enact essential safety protocols while handling waste materials. Notably, this research presents groundbreaking contributions: Firstly, a geometrically designed smart trash box, incorporating ultrasonic and load measurement sensors controlled by a microcontroller, aimed at optimizing waste containment and collection. Secondly, an intelligent method leverages deep learning for the precise classification of digestible and indigestible waste through image processing. Lastly, a cutting-edge real-time waste monitoring system, employing short-range Bluetooth and long-range IoT technology through a dedicated Android application was proposed.

Chapter 16

Rajalakshmi Shenbaga Moorthy, Sri Ramachandra Institute of Higher Education and
Research, India

K. S. Arikumar, VIT-AP University, India

Sahaya Beni Prathiba, Vellore Institute of Technology, Chennai, India

P. Pabitha, Anna University, India

The increasing population rate plays a vital role in bringing challenges in the provisioning of health care. The data was initially collected and kept in the cloud, where the machine learning algorithm was run on the data and decisions were then transmitted back to the client device. This incurs a significant delay for transferring the data and getting back the result. Thus, in this chapter, fog layer is introduced between device layer and cloud layer for processing the sensor data. The introduction of fog layer tends to minimize the delay incurred by the cloud, as analyzing the health data is close to the device that generates the data. For conducting the best analytics on health data received from sensors, the grey wolf optimization (GWO)-based k-nearest neighbor (K-NN) is proposed. GWO K-NN is integrated in the fog nodes, which is close to the device generating the health data, thereby providing timely decisions.

The proposed GWO K-NN works on the fitness of accuracy and misclassification rate of K-NN, and it models the hunting behavior of wolves.

Shashi, Department of Computer Application, CCS University Meerut, Meerut, India

M. Dhanalakshmi, Department of Computer Science and Engineering, New Horizon College of Engineering, Bangalore, India

K. Tamilarasi, School of Computer Science and Engineering, Vellore Institute of Technology, Chennai, India & VIT University, India

S. Saravanan, Department of Mechanical Engineering, Bannari Amman Institute of Technology, Sathyamangalam, India

G. Sujatha, Department of Networking and Communications, SRM Institute of Science and Technology, Chennai, India

Sampath Boopathi, Department of Mechanical Engineering, Muthayammal Engineering College, Namakkal, India

The automotive industry is increasingly focusing on autonomous vehicles, leading to a need for intelligent systems that enable safe and efficient self-driving. Fog computing is a promising paradigm for real-time data processing and communication in autonomous vehicles. This chapter presents a comprehensive framework and solutions for integrating fog computing into intelligent vehicle systems, enabling autonomous features, low-latency data processing, reliable communication, and enhanced decision-making capabilities. By offloading computational tasks to nearby fog nodes, this framework optimizes resource utilization, reduces network congestion, and enhances vehicle autonomy. The chapter discusses various use cases, architectures, communication protocols, and security considerations within fog computing, ultimately contributing to the evolution of intelligent and autonomous vehicles.

Preface

In a cloud environment, servers' compute instances and workload are described as physical or virtual sets of computing resources such as CPU, memory, storage, network, etc. Service providers dynamically assign computing resources to the user applications based on their requirements. A service provider can provide computing resources from lightweight web applications to high computational cloud applications. Due to the escalation of compute instances and workloads, the installation of data centers is rising exponentially yearly. Additionally, the growing requirement for advanced technologies like Data Science, Big-data analytics, and Internet-of-Things has increased the requirement for cloud datacenters to provide effective and smart services. These advanced technologies need high-performance computing resources for processing the data with a high throughput rate.

On the other side, the primary issue with the exponential growth of datacenters is the power consumption of datacenters is increasing rapidly and producing high CO_2 into the environment. According to the survey reports, datacenters alone producing nearly 80 million metric tons of CO_2 worldwide, and the power consumption datacenters will increase to 8000 TWh by 2030 if we do not utilize computing resources effectively. Now, the high power consumption of datacenters has become one of the major reasons for global warming issues. The main reason behind this issue is inefficient resource management in a cloud environment. To avoid SLA violations and poor QoS, cloud service providers allocate extra resources to the applications, which may lead to over-provisioning of computing resources to the applications, and finally, it leads to under-utilization of resources and high power consumption datacenters. Therefore, efficient and intelligent resource management is necessary to achieve a green computing solution in a cloud environment while solving global warming issues with minimizing the high power consumption of datacenters.

Green computing is defined as an eco-friendly computing environment. In detail, green computing mainly focuses on global warming issues due to the high power consumption in the IT industry and the effective usage of computing resources of different technologies and smart applications. Green computing solutions mainly focus on researching the design of computational intelligence resource management to provide effective resource management while minimizing the high power consumption of datacenters. Green computing solutions are necessary for all technical environments, from lightweight computing infrastructure to hyper-scale datacenters.

This book aims to provide computational intelligence resource management in several computing environments like Cloud/Fog/Edge/Distributed computing with in-depth research and results to provide significant solutions and accelerate future research scope in the green computing environment. The major research areas of interest include machine learning and deep learning methods to offer intelligence resource management in computing environments. Also, machine learning and deep learning solutions for minimizing high power consumption of datacenters, green virtualization technologies, resource monitoring systems, challenges of green computing, intelligence-based green computing, green parallel and distributed computing, and energy efficiency challenges. The major motivation of this book is to encourage the research communities to contribute significant and innovative solutions, groundbreaking research, and in-depth case studies towards green computing.

The first chapter is titled "Machine Learning and Deep Learning Algorithms for Green Computing." Green computing is an innovative approach to making computer systems environmentally friendly, energy-efficient, and low in carbon emissions. It uses advanced techniques from Machine Learning and Deep Learning to optimize real-time resource allocation, reducing energy consumption. This approach enhances workload patterns and uses methods like Convolutional and Recurrent Neural Networks to enhance architectural efficiency.

The second chapter is titled "Green Computing and the Quest for Sustainable Solutions." Green IT, a developing idea, uses sustainable design to reduce or eliminate IoT operations' environmental implications. This article evaluates IoT problems, defines Green IT, explores Green IT design methodologies, and describes how to implement these green designs as sustainable environmental solutions. It also thoroughly analyses numerous author proposals to find the best sustainable IoT architectures.

The third chapter is titled "Navigating Green Computing Challenges and Strategies for Sustainable Solutions." This research explores solutions for balancing industrial growth with sustainability and addresses green computing obstacles. Recognizing industry's importance while mitigating its environmental impact is vital. Implementing a circular economy approach makes products regenerative and less wasteful. This research provides real-world case studies and insights, aiding businesses in adopting greener, more sustainable computing practices for a cost-effective, environmentally responsible tech landscape.

The fourth chapter is titled "Impact of Data Centers on Power Consumption, Climate Change, and Sustainability." The data-driven economy is transforming, with data centers becoming a crucial business infrastructure. However, the increasing reliance on data centers is posing a threat to the environment. Climate change activists are focusing on reducing emissions from sectors like automotive, aviation, and energy. Data centers consume more electricity than the UK, accounting for 3% of global electricity supply and 2% of total greenhouse gas emissions.

The fifth chapter is titled "Power-Aware Virtualization- Dynamic Voltage Frequency Scaling Insights and Communication-Aware Request Stacking." This chapter describes the central role that virtualization technologies play in promoting sustainable computing practices. We thoroughly explore the complexities of green data center and server operations and highlight the importance of server virtualization in collaborative integration efforts. Essential technologies such as dynamic voltage and frequency scaling (DVFS) will be examined for their potential to reduce energy consumption. Additionally, introduces a new approach called communication-aware request stacking to optimize energy efficiency.

The sixth chapter is titled "Strategies to Achieve Carbon Neutrality and Foster Sustainability in Data Centers." Data centers and transmission networks are crucial in the digital age, with the market expected to grow from $50 billion in 2021 to $120 billion by 2030. However, their extensive computing infrastructure and continuous operation generate significant heat, necessitating energy-intensive cooling systems. The IEA report revealed that data center power consumption surged by over 60% between 2015 and 2021, with transmission networks experiencing a 60% usage increase. Addressing these growing energy demands poses significant challenges for the industry, with some countries considering restrictions on new data center licenses due to environmental concerns.

The seventh chapter is titled "Computational Intelligence for Green Cloud Computing and Digital Waste Management: Intelligence Computing Resource Management in Cloud/Fog/Edge Distributed Computing." Intelligent resource management across fog, edge, and cloud computing entails dynamic allocation, optimal utilization, and robust security measures. Cloud computing adopts a centralized approach to provision resources, whereas edge and fog computing allocate resources at the periphery, strategically minimizing latency. The incorporation of AI/ML algorithms has a pivotal role, enabling resource prediction for anticipating demand, detecting anomalies, and optimizing allocation efficiently.

The eight chapter is titled "Modern Technological Innovations in Digital Waste Management: E-Waste Management." As the rapid pace of technological advancement continues to propel society into the digital age, the surge in Electronic Waste (E-Waste) poses significant challenges to environmental sustainability. This research explores modern technological innovations in digital waste management that contribute to the reduction, recycling, and responsible disposal of electronic devices. Special attention is given to advancements in recycling methods, the application of Artificial Intelligence (AI), Machine Learning (ML), Deep Learning (DL), Robotics and IoT in automated e-waste processing.

The ninth chapter is titled "Achieving Green Sustainability in Computing Devices in Machine Learning and Deep Learning Techniques." The accelerated growth in Artificial Intelligence, Internet of Devices, Machine Learning (ML), and Deep Learning at breakneck speed has attracted the attention of researchers in developing novel green solutions for reclaiming the Green society. The intersection of these technologies with green sustainability will greatly impact the deployment of cutting-edge technologies with green solutions.

The tenth chapter is titled "Future Trends and Significant Solutions for Intelligent Computing Resource Management." Cloud providers place a high value on reducing energy consumption in cloud computing since it reduces operating costs and improves service sustainability. Cloud services are frequently replicated across providers to ensure high availability and dependability, which increases provider resource utilization and overhead. Finding the right balance between service replication and consolidation to lower energy usage and boost service uptime can be challenging for cloud resource management decision-makers. This article addresses this problem by presenting a ground-breaking technique known as "CRUZE," which is based on cuckoo optimization and considers energy efficiency, dependability, and comprehensive resource management in cloud computing, encompassing cooling systems, servers, networks, and storage.

The eleventh chapter is titled "Green Computing-Based Digital Waste Management and Resource Allocation for Distributed Fog Data Centers." The term "green computing" describes the efficient use of resources in computing and IT/IS infrastructure. This study suggests a unique method for dispersed fog data centres' work scheduling and resource allocation based on digital waste management. Here, the bandwidth differential preemption evolution moving average method (BDPEMA) is used to control the network's digital waste while allocating resources. Reinforcement adversarial hierarchical group multi-objective cuckoo optimization (RAHMCO) is used to schedule network tasks.

The twelfth chapter is titled "Sustainable Waste Management OOA Enhanced MobileNetV2-TC Model for Trash Image Classification." A waste is an invisible, indirect waste that contaminates natural resources like the air, water, and soil, endangering the ecosystem, people, and animals. Long-term waste accumulation and contamination can harm the resources found in the environment. Since traditional waste management systems are very inefficient and the number of people living in urban areas is increasing, waste management systems in these areas face challenges. However, by combining a variety of sensors with deep learning (DL) models, waste resources can be used effectively.

The thirteenth chapter is titled "Computational Intelligence for Green Cloud Computing and Digital Waste Management: AI Solutions for Green Computing and Digital Waste Management." The intersection of environmental protection, Sustainable Development Goals, and the role of information technology (IT) is to foster it. Artificial Intelligence addresses environmental challenges, offering solutions such as emissions reduction, cost savings, legal compliance, HR attraction, and optimized investments, waste management. The "Three Rs" of Green IT emphasize sustainable hardware management through reuse, refurbishment, and recycling. Green IT aligns with board sustainability and ESG standards and is fostered with innovative AI solutions to integrate and optimize AI-based green computing algorithms which address environmental impacts and data utilization strategies, optimizing energy consumption; and mitigating digital waste and carbon footprint.

The fourteenth chapter is titled "Resource Management in Green Computing based on ISHOA Task Scheduling With Secure ChaCha20-Poly1305 Authenticated Encryption-Based Data Transmission." One of the hottest new technologies that allows users to handle a broad range of resources and massive amounts of data in the cloud is green computing resource management. One of the biggest obstacles is task scheduling, and poor management leads to a decrease in productivity. The task must be efficiently scheduled to ensure optimal resource utilisation and minimal execution time. Given this, this study suggests a fresh method for efficient task scheduling in a green computing environment that also offers improved security. There is a development of an enhanced spotted hyena optimizer (ISHO).

The fifteenth chapter is titled "Dual-CNN-Based Waste Classification System Using IoT and HDS Algorithm." Efficient waste management is crucial in today's environmental landscape, necessitating comprehensive approaches involving recycling, landfill practices, and cutting-edge technological integration. The proposed approach introduces a sophisticated waste management system, harnessing Dual or Twofold Convolutional Neural Networks (D-CNN or TF-CNN) and a Histogram Density Segmentation (HDS) algorithm. This intelligent system equips users with the means to enact essential safety protocols while handling waste materials.

The sixteenth chapter is titled "Intelligent Healthcare Provisioning in Fog Using Grey Wolf Optimization." The increasing population rate plays vital role in bringing challenges in the provisioning of health care. The data was initially collected and kept in the cloud, where the machine learning algorithm was run on the data and decisions were then transmitted back to the client device. This incurs a significant delay for transferring the data and getting back the result. Thus, in this paper, Fog layer is introduced between device layer and cloud layer for processing the sensor data. The introduction of Fog layer tends to minimize the delay incurred by the cloud, as analyzing the health data is close to the device that generates the data.

The seventeenth chapter is titled "Fog Computing-Based Framework and Solutions for Intelligent Systems: Enabling Autonomy in Vehicles." The automotive industry is increasingly focusing on autonomous vehicles, leading to a need for intelligent systems that enable safe and efficient self-driving. Fog computing is a promising paradigm for real-time data processing and communication in autonomous vehicles. This chapter presents a comprehensive framework and solutions for integrating fog computing into intelligent vehicle systems, enabling autonomous features, low-latency data processing, reliable communication, and enhanced decision-making capabilities.

K. Dinesh Kumar
Amrita Vishwa Vidyapeetham, India

Vijayakumar Varadarajan
The University of New South Wales, Australia

Nidal Nasser
College of Engineering, Alfaisal University, Saudi Arabia

Ravi Kumar Poluru
Institute of Aeronautical Engineering, India

Chapter 1
Machine Learning and Deep Learning Algorithms for Green Computing

Rajashri Roy Choudhury
Brainware University, India

Shivnath Ghosh
Brainware University, India

Piyal Roy
Brainware University, India

Ayan Ghosh
Brainware University, India

ABSTRACT

Green computing is an innovative approach to making computer systems environmentally friendly, energy-efficient, and low in carbon emissions. It uses advanced techniques from machine learning and deep learning to optimize real-time resource allocation, reducing energy consumption. This approach enhances workload patterns and uses methods like convolutional and recurrent neural networks to enhance architectural efficiency. The integration of ML and DL techniques allows for accurate temperature forecasting and alternative cooling strategies. Despite challenges, the synergistic fusion of ML and DL algorithmic software with green computing holds great promise for reducing energy consumption and enhancing environmental sustainability.

1. INTRODUCTION

Green computing, sometimes referred to as ecological computing, is a method that makes use of semiconductors, computer systems, and software to maximize energy efficiency. It's a sustainable approach that minimizes negative environmental effects by minimizing the need for more computers. This approach extends across the supply chain, from the raw materials used to recycling systems. Green computers must give the greatest work for the least energy, commonly measured by performance per watt. Optimizing the energy efficiency of hardware and minimizing system temperature is also enhanced by DL approaches such as compression neural networks or recurrent neural networks.

DOI: 10.4018/979-8-3693-1552-1.ch001

For millennia, intelligent people have contemplated building a device capable of reproducing the functions of the human brain. Originally postulated by Aristotle in 300 B.C., "associationism" is where deep learning originated. Because it is essential that scientists understand how effectively human recognition systems work, this hypothesis served as the impetus for people to become more interested in learning about the brain. The McCulloch-Pitts model, which is commonly referred to as the origin of artificial neural models, was initially proposed in 1943, marking the beginning of the current deep learning era. They designed a computer model based on neural networks that functionally replicate the neocortex in human brains. Reduced energy use, electrical waste reduction, and decreased operating expenses are only a few advantages of green computing Deep compression (Han. S et al, 2015). Data centres, networks, hardware, software, and other components of computer systems are all included in the domain of "green computing." Green computing techniques include improving software efficiency, virtualizing servers, lowering power usage, using renewable energy sources, and managing e-waste.

2. ENERGY-EFFICIENT COMPUTING USING MACHINE LEARNING

Using the scheduling mechanism, the power consumed can be decreased in the software approach. The scheduling approach allocates service requests from clients to virtual machines (VMs) that can fulfil them within the parameters of the service level agreement (SLA). The scheduling approach chooses virtual machines (VMs) running on physical machines (servers) that have the potential to use less power in order to meet the goal of lowering the consumption level of power. Numerous suggested techniques rely on the scheduling or software approach (A. Hameed et al, 2016). Energy efficiency is highly valued in modern cloud computing since it reduces operating costs and adheres to green computing ideals. Resource management in the cloud encompasses many different aspects, such as workload consolidation, job scheduling, virtual machine deployment, and more. Researchers work to establish the best policies for this management. In these attempts, machine learning is essential. In this research, we conduct a comprehensive assessment of the literature on machine learning (ML) in recent works to provide recommendations for energy conservation in cloud computing systems. Large-scale data centres are outfitted with mechanical and electrical gear and sensors that throw millions of data points a day. A machine learning method called "neural network framework" analyses these data points to assess how effectively energy is used. The strategy was tested in Google's data centre. Results indicated that it could result in energy savings (Hatzivasilis, G et al., 2008) (Hassan, M. B et al., 2022). The most important greenhouse gas is carbon dioxide (CO2). According to recent research, five automobiles' worth of on release of carbon dioxide. over course of a Natural Language Processing (NLP) model's development utilising deep neural networks "Deep Learning" (Ning, Z et al., 2019) (Bharany, S et al., 2022) (Luo, T et al., 2023). The mathematical model's initially unknowable parameters are estimated using machine learning. Specifically, we need to ascertain the task performance level. Predict a priori utilisation of resources (e.g., CPU consumption) by different activities under existing workloads and agreements (e.g., reaction periods) given burden characteristics, host features, and competition between jobs on the same host (Raja, S. P., 2021) (Gholipour, N et al., 2021). When paired with precise or approximate schedulers, algorithms based on machine learning can precisely predict system behaviour and assign jobs to hosts in a way that balances power consumption, income, and service quality (Yu, P et al., 2020).

2.1 Energy Consumption Modelling

Role of HPC in training ML algorithms significantly impacts their advancements, but increased energy consumption due to numerous changes and training runs raises concerns about making AI ecologically feasible. Finding solutions that enable energy-efficient AI requires evaluating the performance of ML algorithms and the relationship between their energy usage in various HPC environments. The majority of ML algorithms were not implemented with a focus on Green AI because they were created decades ago before HPC and AI convergence was even considered. These models aid in the error and problem detection for the hardware, software, and algorithm implementations (Ahmed, K. I et al., 2019). By creating accurate energy consumption models, researchers and consumers can safely choose approaches, configure hardware, and use optimisation techniques (Luo, T et al., 2023). This finding paves the way for the development of computer systems with fewer operating costs and negative environmental effects. A vital first step towards a better future is to make ML and DL algorithms more energy-efficient because doing so makes these technologies more publicly available and environmentally benign. As computer technology develops, energy consumption modelling will become increasingly important in the endeavour to support green computing projects (Ning, Z et al., 2019).

2.2 Dynamic Voltage and Frequency Scaling

Due to the enormous amount of CO_2 emissions, current computing paradigms and information and communication technologies, combined with their broad deployment, Particularly, the usage of cloud computing has a big effect on the ecosystem. The study area of "green" and low power network has seen a tremendous increase in the last ten years. The energy consumption of GPUs is a difficult problem given the most recent advancements in deep learning and other computationally demanding machine learning frameworks. Dynamic Voltage Frequency Scaling (DVFS) has generated a lot of research interest and produced positive results in each of these situations. For the purpose of consuming less energy, data centres employ Dynamic Voltage and Frequency Scaling (DVFS) without sacrificing Quality of Service. However, processing various spatio-temporal queries demands time- and resource-intensive algorithms and techniques due to the magnitude and complexity of the spatio-temporal information. The key enablers of energy-efficient management are the correlations between energy, DVFS, consolidation, and performance in all aspects. In this chapter, we investigate various DVFS methods currently in use as well as issues that arise when analysing and processing spatiotemporal queries in a cloud-based paradigm. For illustrating energy usage and the value of dynamic voltage and frequency scaling, movement datasets have been taken into consideration (Gholipour, N et al., 2021) (Ghosh, S., & Das, J., 2022).

2.3 Power Management in Data Centres

Data centre hosting is becoming more and more popular as a means of meeting the computational and storage demands of cloud services platforms and information technology (IT), which in turn require more electricity to operate the IT equipment and meet data centre cooling requirements. The increasing demand of information centre infrastructure has made it more difficult to optimise power utilisation and ensure that data facility energy safety remains unchanged (Mehlin, V et al., 2023). Consequently, various machine learning-based optimisation techniques have been suggested for raising total power effectiveness. In order to assess the methods that scholars are employing to optimise data centre energy usage through the use of algorithms that utilise machine learning, this study endeavours to ascertain and scrutinize the principal research projects conducted between 2015 and 2021, respectively (Kumar, R et al., 2022). With the growth of data and the accompanying applications (such as machine learning, cloud storage, the Internet of Things, and so on), the data centre market is expanding quickly. The shift in internet activity during the COVID-19 epidemic has recently enhanced growth. There are many obstacles to lowering data centre energy use, which are made worse by Singapore's tropical climate with its high humidity and warmth. The current compartmentalized strategy of operating the facility systems and information technology (IT) separately has led to unnecessary over-provisioning (Zhou, X et al., 202) (Ali, Z et al., 2019). The infrastructure of the data centre, the jobs that must be completed, and In order to accomplish efficient job scheduling, the desired profit are expressed using a mathematical programming framework that can be solved in a variety of methods. An exact solver for mixed linear programming is used to demonstrate the idea. But we also demonstrate that approximation solutions can be used to obtain nearly optimal schedules, as this is a problem with NP completeness. The parameters of the mathematical model that were initially unknown are estimated via machine learning. Given the characteristics of the workload, the host, and the competition between tasks on the same host, in order to estimate task commitments to service levels (like response time) and predict resource utilization (like CPU usage) by multiple tasks under the current workloads, a priori predictions are required. (Anley, M. B., & Awgichew, R. B., 2022,) (Berral, J. L et al., 2011).

2.4 Energy-Aware Task Scheduling

Energy consumption has been a major factor in the rapid expansion of cloud data centres. It not only raises the cost for electrical power for suppliers of services but also contributes significantly to emissions of greenhouse gases and contamination of the environment, as well as adversely affecting network performance and trustworthiness. Since work done at cloud data centres are the basis for many parallel scheduling applications and indicators of energy efficiency and consumption have grown in importance. In this study, we provide best heuristic scheduling (BHS), a two-phase scheduling method for directing acyclic graph (DAG) scheduled on cloud-based information centre processors. The programme divides up resources into tasks in the first phase by sorting them using four heuristic techniques and a grasshopper algorithm. Then, it selects the most suitable method to complete each work based on the important factor determined by the end consumer or the service supplier to accomplish a solution stated at the right time. In the second phase, BHS minimises the duration and energy usage while taking into account the virtual machines' start, setup, finish, and energy profiles in addition to the importance factor selected by the service provider or end user. Next, utilising a test dataset, the suggested BHS method is contrasted with the metaheuristic allocation of resources algorithm (MHRA) .Results indicate that compared to the MHRA algorithm, the suggested algorithm enables 19.71% more energy storage. Furthermore, in heterogeneous situations, the make span is decreased by 56.12% (Peng, Z et al., 2020). In order to minimise delay and energy consumption, resource management—which entails making the best use of various resources, including as bandwidth, computing power, and energy—is crucial for Internet of Things transportation. Latency reduction is essential to IoT transportation because it ensures efficient and timely data delivery. This is especially crucial for applications involving autonomous vehicles since passengers' and other drivers' safety depends on real-time data. IoT transportation systems may reduce latency by managing resources like as bandwidth and processing power, which can speed up data transmission. Energy consumption is a significant component of IoT mobility. Many IoT devices are powered by batteries, which have a limited lifespan and need to be replaced or recharged on a regular basis (Kumar, M. R et al., 2023). IoT transportation systems can reduce the amount of energy required for data transfer by making the most use of resources like computing power and wireless connection. Increased lifespan of batteries, fewer replacements needed, and lower operational costs might be the outcome of this. In conclusion, reducing delay and energy consumption in IoT transportation requires effective resource management. By making the most of a variety of resources, IoT transportation systems can transfer data quickly, efficiently, and with minimum energy use. This has the potential to improve the security, reliability, and performance of IoT systems for transportation (Atiq, H. U et al., 2023).

3. RESOURCE ALLOCATION AND OPTIMIZATION

The key components of ML and DL algorithms in the context of green computing are resource allocation and optimization. The availability of these algorithms is vital to increase computer systems' efficiency and sustainability. ML and DL algorithms make it feasible to deploy resources more effectively, by examining the real-time data trends and distribution of workloads. Dynamic power management and workload predictions that lower energy consumption and operating expense make it viable to optimize allocation (Ahmed, K. I. et al., 2019) (Mehlin, V et al., 2023). In order to ensure optimum resource consumption, these algorithms are also dealing with load imbalance. Examples of DL approaches that help in the creation of hardware with energy efficiency include convolutional and repeating neural networks. They find architectural configurations that maximize system designs by balancing performance and power efficiency (Luo, T. et al., 2023). Green computing is a crucial instrument to battle climate change, the grave danger of our time. In the past century, there has been an approximate 1.2°C increase in global temperatures. Because of this, ice caps have begun to melt, raising sea levels by about 20 centimetres and intensifying and increasing the frequency of catastrophic storms. Increasing power use is one factor contributing to global warming. Although data centres only make about 1% of the total power used annually or 200 terawatt-hours, they are a developing factor that needs to be taken into consideration. Strong, low-power computer systems are a component of the solution. They are improving human well-being and science, particularly our ability to understand and adjust to changes in the environment (Yu, P. et al., 2020) (Raja, S. P., 2021). Even when a significant amount of IoT devices are connected and initiating data-driven conversations with distant applications, network congestion is typical. Furthermore, cloud data centres have increased computational strain. IoT-enabled CPSs find that the services they need cannot be provided by the stored in the cloud application execution model. In the Internet of Things (IoT) context, massive amounts of data are continuously generated by sensors, systems for communication, handheld devices, and social networks, resulting in a new kind of network architecture. The distributed, demanding, and dynamic nature of the Internet of Things ecosystem poses a number of technological difficulties. Among the technological challenges are latency, connection, expertise, cost, power, adaptability, and reliability. Using the computing capability of fog devices, the Fog computational environment is a state-of-the-art processing model (Kumar, M. R. et al., 2023). Imagine an IoT-based transportation network in which the surroundings are composed of different automobiles, gadgets, and sensors. Sensitive equipment such as speed sensors, cameras, sensors for temperature, GPS sensors, and vehicle tracking sensors generate data. Applications can help us make informed judgements in this situation. This includes offering advice on how to save gas or delivering up-to-date traffic information in emergency situations, including approaching a rescue vehicle or a pedestrian. Therefore, in order to build an IoT-based transportation environment, numerous apps need to compute in real-time. A number of criteria are required to achieve a certain application objective, such as speed, distance travelled, road condition, driving patterns, traffic sensors, temperature, and other variables. It is not feasible to rely only on the cloud to handle processing due to the constantly changing nature of the application's environment and the requirement for low latency. In order to lower latency, an intermediate processing and storage node—also known as a "Fog device"—is needed. Utilising a large number of Fog devices allows for the execution of near to consumer programming with little latency. This also reduces network traffic and the strain on the cloud's application processing infrastructure. The dynamic and intricate nature of the Fog environment makes allocating resources to user applications challenging. Not all applications are limited to fog devices. Communication via wireless networks in a fog environment, device autonomy, and decentralised administration can all lead to resource and connection problems as well as delay (Songhorabadi, M. et al., 2023).

3.1 Virtual Machine Placement

The potential of increased efficiency in cloud data centres is made possible by virtualization, which improves isolation between application resource utilisation, allocation, and management. Live migration is among the most fascinating features of virtualization. Virtualization consequently gains attractiveness. Live migration makes it simple to move running virtual machines (VMs) between physical hosts. Many users can access virtual application resources simultaneously thanks to virtual machines (VMs). Any operating system within an organization can function as if it would a stand-alone computer with a desktop application window thanks to virtual machines (VMs). The server's use of past data affects how VM functions. The main benefits of VM are its flexibility and affordability. By sending software to computers that are only partially or moderately used, virtualization technology helps to balance the load on the server. In order to reduce the burden on the physical device, this process involves offsetting many virtual computers. There are several elements that need to be carefully considered and adjusted, including hosting, energy consumption, service level agreements (SLA), and the quantity of migrations (Balicki, J., 2021). Any discrepancy between the measurements offered can negatively impact the data centre's services performance and quality of service. This means that it should be simple to tell the difference between overused and underused equipment (Han, S., Mao, H., & Dally, W. J., 2015). Investigating actual and virtual machines requires a lot of time in this regard. For this reason, machine learning architecture makes sense. To address the critical elements that affect service quality, artificial intelligence and swarm intelligence have been merged (Talwani, S. et al., 2022).

3.2 Load Balancing Techniques

As a key component of cloud computing, load balancing ensures that resources deployed on physical servers, such as computer systems, network interfaces, hard discs for storage, and virtual machines (VMs) are all subject to the same amount of load. In order to achieve quality of service (QoS) objectives in cloud computing, such as improved efficiency of resources and throughput, deep learning (DL) techniques can be applied. In order to increase system resilience, these strategies can also reduce delay, response times, and expenses while distributing the workload among processors. As a result of DL's accurate and efficient decision-making, the most suitable resource is chosen to satisfy the incoming requests. The term "load" can be used to describe network load, CPU load, memory load, or storage load (Kaur, A. et al.,

2020). The goal is to evenly spread the workload across the computers to enhance the efficiency of job execution across cloud resources (memory, storage, network, and computing power). When computers are used below their capacity, load-balancing solutions oversee assigning resources with the least amount of resource waste and preventing overloading and underloading of virtual machines. Consequently, the workload placed on the virtual or physical system exceeds its capabilities (Agarwal, R., & Sharma, D. K., 2021). The applications respond more slowly as a result. This problem is resolved by shifting tasks from a highly loaded virtual machine to one that is less laden. The process of moving load from heavily laden virtual machines (VMs) towards correspondingly lightly loaded VMs is known as load balancing in the context of clouds. Deep Learning (DL) is a method that learns from training samples to recognize and remember items. The intricate arrangements of brain neurons and their connections serve as the inspiration for this method. Hybrid approaches can be used to address the workflow-based scheduling issue in heterogeneous environments like the cloud, where applications with workflow inputs have become more sophisticated. Another scheduling method that uses genetic algorithms to load balance virtual machines on hosts. A load balancing technique called LB3M is based on a mixture of the algorithms Max and Max-Min. This algorithm divides up the work among the nodes based on how much processing power each one has.

3.3 Adaptive Resource Provisioning

On-demand resource allocation is facilitated by many applications using cloud computing, which offers services and resources through the Internet. These dynamic networks distribute the required resources according to the needs of the users, and thus need an appropriate resource allocation plan. If resources are not allocated properly, users will use a variety of resources, which will cause a load imbalance in the system. Utilising devices connected to the internet for storage and computation involves not just sharing cloud resources but also connecting the hardware to the network using various protocols. These changes turn the network into a dense, heterogeneous, and complicated system. It is recommended to utilise a comprehensive reinforcement learning framework to give network users an efficient way to allocate resources. The dimensionality issue cannot be resolved by the conventional Q-learning model when the state space expands exponentially. Deep reinforcement learning and fair resource allocation are integrated in the proposed approach to produce better allocation schemes than the traditional methodology (Karthiban, K., & Raj, J. S., 2020).

4. DEEP LEARNING FOR ENERGY EFFICIENT SYSTEM

Deep Learning has made numerous strides in the development of machine learning applications during the last few years. However, there are increasing concerns as to the impact on the environment of Deep Learning algorithms that already use a great deal of energy for computations. In the last few years, researchers have given much attention to energy efficiency deep learning and it has made a considerable amount of progress. The purpose of this study is to compile information about these developments from the research literature and demonstrate how and where energy consumption can be decreased over the Deep Learning lifecycle (IT-Infrastructure, Information, especially Modelling, Instruction, Deployment, Evaluation) (Ghosh, S., & Das, J., 2022). Deep Learning is of paramount importance in Machine Learning and Algorithms for Green Computer, where energy efficiency systems are concerned. The inventive strategies for lowering computer systems' energy usage can be demonstrated by convolutional and recurrent neural nets, which are examples of deep learning techniques (Gholipour, N. et al., 2021). These methods allow system data to be analyzed in detail, thus making the use of prediction models for forecasting workload and allocating resources more efficient. Deep learning is used to identify patterns of usage that reduce energy waste, in support of dynamic power management and load balance. It also helps to find good architectures for balancing the performance of computing and energy efficiency, through deep learning, which aids hardware designs that are energy efficient (Paul, S. G. et al., 2021).

4.1 Neural Architecture Search for Efficiency

Neural Architecture Search (NAS) provides a powerful technique in machine learning and deep learning that tries to automate the creation of neural network designs. NAS can be quite useful in maximizing the computational efficiency of models when it comes to green computing or energy-efficient machine learning. Deep learning's ability to automate the feature engineering process—instead of manually designing hierarchical feature extractors—is partly responsible for its success in perceptual tasks (Elsken, T. et al., 2021). This accomplishment has raised the need for architecture engineering because entails manually creating ever-more complex neural structures. Neural Architecture Search (NAS), an automated approach for architecture engineering, is hence a natural progression from machine learning automation. On a number of jobs, including semantic segmentation, object detection, and image classification, NAS approaches have already surpassed manually developed systems. They also have a substantial amount in common with hyperparameter optimization. Sorting NAS techniques has been done based on the three criteria considering search space, search techniques, and techniques for performance estimate.

Figure 1. A visual summary of techniques employed in neural architecture search

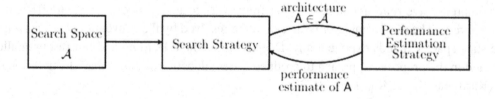

4.1.1 Search Space

The set of potential neural network topologies that can be defined by the field of search in NAS considered during the search process. It essentially determines what architectural building blocks can be combined to create a neural network. By applying past information about typical features of concepts well-suited for a task, the region to be searched can be reduced in size and the process of searching made simpler. But this also brings in a human bias, which could prevent new architectural building blocks from being discovered that go beyond current human knowledge (Zhang, T. et al., 2021). As Figure 2 (left) illustrates, the field containing chain-structured neural network structures is a rather simple search space. A chain-structured artificial neural network design A has n layers altogether, with layer 1 providing the input for the i'th layer Li. Layer I + 1 uses the result, A = Ln... L1 L0, as its input. Next, the following variables are used to parametrize the search space: (i) The number of layers (n), which can be as high as possible; (ii) the computation that each layer carries out, such pooling, convolution, data or more intricate operations like dilated convolutions or depth-wise separable convolutions; and (iii) Hyper-parameters associated with the operation. Recent NAS research has incorporated cutting-edge design aspects from manually built structures, like skip connections, which enable the construction of intricate, multi-branch networks, as shown on the right side of Figure 2. Formally speaking, the input from layer I in this instance can be defined as an algorithm gi(L out i−1,..., Lout 0) that combines the outputs of earlier layers. When utilising this function, the degrees of freedom increase significantly. One can find two special examples of these multi-branch architectures: (i) the chain-structured networks, where the previous layer outputs are added together (gi(L out i−1,..., Lout 0) = L out i−1 + L out j, j < i − 1) or additionally (iii) here the outputs of the preceding layers are combined into one (gi(L out i−1,..., Lout 0) = concat(L out i−1,..., Lout 0)) (Lu, Z. et al., 2019).

Figure 2. Example of many architectural spaces

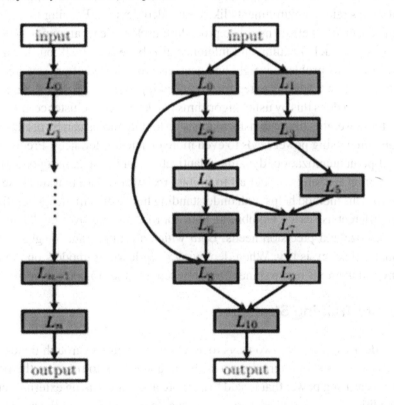

4.1.2 Search Strategy

The optimal method for sorting across the search space, that can be sometimes exponentially large or even unbounded, is explained by the search strategy. It involves the classic exploration-exploitation trade-off because, while it is desirable to uncover efficient structures quickly, early convergence to an area of subpar topologies should be avoided.

4.1.3 Performance Estimation Strategy

A common goal of NAS is to find architectures with outstanding forecasting performance on unseen data. The method of figuring out this performance is called performance estimate. The simplest approach is to use data for training and validating the architecture using standard methods; however, this is computationally costly and limits the variety of topologies that may be explored. As a result, developing techniques to reduce the expenditure of these performance estimates is a major area of interest for current research (Lu, Z. et al., 2019).

4.2 Pruning and Quantization Techniques

Pruning: Network pruning is a technique developed in the 1990s to reduce memory and bandwidth in neural networks. It involves reducing a large network into a smaller one without retraining, making it suitable for constrained environments like embedded devices. Pruning is one of the techniques utilised to raise a model's efficiency. The procedure uses the least amount of energy while reducing the size of the model. Training and inference also benefit as well, but it can lead to a loss in accuracy. To solve this problem, weight and neurons are adjusted after they are cut.

Quantization: Quantization takes this a step further by taking continuous signals and turning them into strings of integers. It does this by using algorithms like k-means to cluster certain weights states. To decompress these weights linear transformations or lookup tables can be used. Reducing memory utilization and processing power leads to even more efficient inferences. Precision is also reduced through fixed-point quantization, dynamic quantization, and post-training quantization. Synthetic data can take on many forms — but all to imitate real-world data that machine learning models would run into. This method helps with understanding how well a model generalizes problems and performs on different use cases. Combining both approaches improves model performance further by reducing its size and precision needs. Both which are important for green computing where computational efficiency is key. When deploying deep learning models on devices with limited resources, hybrid methods really shine through when it comes to performance.

4.3 Energy-Aware Training Strategies

For heterogeneous devices, there are two ways to assess power: either through the use of external sensors or internal sensors. Utilising internal sensors, Ferro and others to measure the power directly. An indirect method of measuring power that is said to be more accurate is using external sensors the amount of energy used. Additionally, external sensors are utilized to determine both the accuracy and the initial learning for the evaluation of a specific power model's parameters. Readings have been taken at regular intervals to determine the amount of electricity consumed over a given time period in order to obtain energy estimations. The energy is subsequently ascertained by computing the integral over the required time from the power-versus-time graph.

4.3.1 Power Measurement Sensors

In general, AC power is measured using external power metres, like universal inline metres, and DC power is measured using metres that link the test instrument to the power supply. In addition to the additional hardware, components have built-in sensors. They enable quick querying and tracking of the power data by academics and users. Linux monitoring tools, such as nvml-tools and lm_sensors for CPUs are the most used sensors that may offer hardware information (Cheema, S., & Khan, G. N., 2020).

4.3.2 Power Related Diagnostic Tools

The requirement to comprehend the GPU's power consumption has increased because it is thought that GPUs are crucial to the high-performance and scientific computing communities. Two types of power-related diagnostic tools have been discussed that employed in our research.

- **Powertop:** This Linux tool is utilised to analyse and process power usage. For the CPU's C and P states, detailed panels that display the device and software execution processes are accessible. As with every previous version, the CPU, together with other hardware and software, is diagnosed, and the approximate power usage is displayed, using the Advanced Configuration and Power Management, or ACPI, consumption information.
- **Psensor**: Psensor is a graphical application that may be used with Linux must periodically read the internal sensors to monitor the temperature [36]. Using the Linux lm_sensors, it can keep an eye on the CPU and motherboard temperatures. Utilizing XNVCtrl, the temperature of the NVidia GPUs is tracked.

4.4 Transfer Learning for Green Computing

This section presents a new deep learning transfer learning approach to classifying judicial data has been introduced that combines the simultaneous learning of shared inter-domain representation and domain-specific networks through numerous correlation mapping hidden layers.

- **Problem Formulation:** Finding a representation space for common features in which to transfer labelled source data for target instance annotation is a key challenge in deep transfer learning. Co-occurrence data, a group of instances that exist across domains but are unlabeled, should be gathered to address it. Websites or social media platforms can offer adequate amounts of this additional information (Liu, G. et al., 2019). For text-to-image transfer learning, for instance, image annotations or documents and the accompanying images can be used as co-occurrence pairings. They are thus useful for domain-agnostic shared feature space tilting (Kashyap, P. K. et al., 2019).
- **Deep Transfer Learning via Semantic Mapping Mechanism:** Combination of CCA with deep neural networks to create a multi-layer correlation matching model that can identify a deep semantic mapping representation and learn the domain-specific networks for source and target domains. This approach is motivated by the ability of CCA to maximise correlations between domains by deriving the projected common subspace as a joint representation. A multi-layer source and target domain-specific deep learning network is joined by CCA for the deep semantic mapping space learning in the proposed deep semantic mapping model, as shown in Figure 3. Domain network

matching is guided by CCA to learn the corresponding semantic space. The hidden features of co-occurrence data are first simultaneously learned by the source and target domain networks in forward propagation. The canonical connection follows. The produced hidden features are analysed to determine the correlation coefficients between heterogeneous features, which can be utilised to fine-tune the network parameters in the future. The high-level semantic mapping space and domain-specific networks are created by corrective matching between domains for numerous layers (Liu, G. et al., 2019) (Kashyap, P. K. et al., 2019).

Figure 3. Deep semantic mapping model

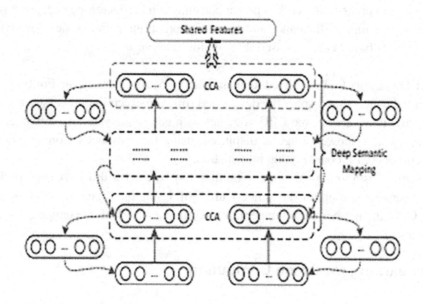

5. SUSTAINABLE PRACTICE AND ENVIRONMENTAL IMPACT

The adoption of sustainable practices and consideration of the environment's effects are important elements in relation to machine learning algorithms for green computing, such as Microsoft ML and Deep Learning (Phiri, M. et al., 2023). The use of these technologies enables alternative ways to lessen the damaging effects that computer systems have on the environment. Predictive modelling for job management and resource allocation that uses less energy is made possible by ML and DL algorithms. By analyzing past data and reducing energy use and carbon emissions, these tools allow timely decisions to be made (Coelho Jr, C. N. et al., 2021). To ensure optimal utilisation of available resources and minimize operating waste, these algorithms take account of load distribution (Ahmed, K. I. et al., 2019). The development of hardware with reduced energy use is largely dependent on DL, including convolutional and recurrent neural networks. It lists systems to support sustainability objectives through maximization of performance while minimizing energy use (Bharany, S. et al., 2022).

5.1 E-Waste Reduction through Predictive Maintenance

An IoT sensor-based alarm system has been added to a Waste Management System (WMS), which is now entirely manual. It can now notify the appropriate personnel when waste is overflowing so they may take urgent action. In terms of both time and money, it was inefficient. Additionally, WMS is not concerned with recycling or removing electronic debris (E-waste). The use of electronic gadgets has grown significantly along with information technology. Laptops, displays, mobile phones, headphones, tablets, and other electronic gadgets deteriorate in value and usage over time, increasing the amount of e-waste produced (Khatiwada, B. et al., 2023). Even while this increases the amount of e-waste that poses a hazard to the environment, the possibility of recovering valuable, recovered minerals is a plus. For the best use and management of e-waste, numerous e-waste management systems have been developed. Surveys are still looking into the idea of applying image processing to recycle e-waste. Several deep-learning techniques are used for the picture processing, improving categorization and prediction accuracy. A CNN algorithm is used in a deep learning model to extract features, and an RBM model is then used to analyse the features for improved accuracy. The suggested system is tested using an open-source dataset, and the findings are contrasted with those of other current methods. The comparison demonstrates that the suggested method offers greater e-waste forecast accuracy (96%).

The Method: Niblack's algorithm is used to preprocess the e-waste photos, and the sliding window is used for prediction. The advent of deep learning algorithms has improved the accuracy of photo segmentation and prediction., and in this research, the CNN technique is utilized for prediction. In image analysis and prediction, the CNN algorithm is frequently employed. But in this model, the CNN algorithm is used to process and enhance the image quality before making predictions. This increases the e-waste prediction's accuracy. The predictability approach is combined with a review of the devices' ability to be reused.

Figure 4. Workflow of the proposed work

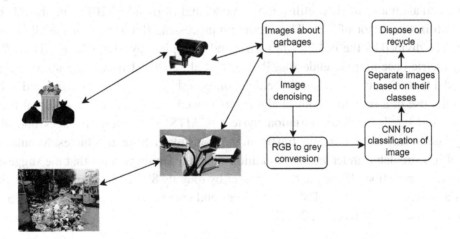

5.2 Environmental Monitoring With IoT and ML

A system known as the IoT is composed of a network of various sensors, applications, and other technology, that nevertheless allows for human work to remain flexible and practical. An embedded system is used in this scenario as a framework for a variety of sensors and software to carry out one or more activities simultaneously. An intelligent network of internet-connected devices called the Internet of Things, or IoT, is growing quickly. Because superior performance and a sophisticated environment, linked IoT devices require a significant amount of energy. The G-IoT aims to connect smart devices and sensors, automate workflows, and promote energy-conserving behaviors (Songhorabadi, M. et al., 2023). The energy and financial conservation of There are numerous ways to implement IoT systems, including using artificial neural networks, Bayesian models, C-means, and K-means. The requirement and framework of the system affect energy consumption, thus implementing green wireless sensor networks and green computing can moderate energy usage (Nalla, K., & Pothabathula, S. V., 2021).

5.3 Carbon Footprint Estimation and Mitigation

Tracking energy and carbon consumption has been revolutionized by a fresh technology in machine learning, with a dashboard dedicated exclusively to energy-effective reinforcement learning algorithms. Not just a valuable model for other machine learning fields, this development encourages ethical study. Simplified accounting promotes energy-efficient algorithm testing and upholds the long-term expansion of machine learning research. With Experiment-impact-tracker, there is the energy, computation, and carbon footprint of machine learning (ML) systems may now be reported in detail and uniformly using a lightweight framework. As a result, this alleviates the strain on energy grid emissions. When merged, this information has the ability to estimate carbon and energy measures through rough calculations. Consider the combination of the employed graphics processing units' thermal design power (W) with the experiment time (h), for example. As a result, a Watt-hour energy metric is produced (Anthony, L. F. W. et al., 2020) (Henderson, P., Hu, J. et al., 2020). To calculate the quantity of CO_2 emitted, multiply this by the carbon intensity of the regional electricity system. This estimating technique ignores CPU consumption and assumes full GPU utilisation. Amodei and Hernandez (2018), on the other hand, employ a GPU utilisation factor of 33%. The maximum processing throughput of the GPU (measured in PFLOPs) can be divided by the PFLOPs-hr metric and multiplied by TDP (Watts). This offers a Watt-hour energy metric once more (Henderson, P., Hu, J. et al., 2020). Through the use of a three-layered learning model that takes into account on-vehicle, on-co-vehicle, as well as on-fog-and-vehicle learning through the use of a squadron control algorithm as well as federated instruction at the Fog level, the Computerised Intelligent Transportation System (CAITS) effectively manages both the services provided by intelligent vehicles and the traffic on the roads by traditional vehicles. Simulations are run on the CloudSim simulator under various conditions, and the findings show that the suggested scheme improves context prediction efficacy at the Fog layer by roughly 8%–24% compared to existing models, which translates into a reduction in EV service times and energy consumption as well as a decrease in ambient CO_2 (Reddy, K. H. K. et al., 2023).

5.4 Circular Economy Approaches in Computing

A circular economy (CE) is built like an ecosystem that regenerates itself. Current items' useful lives are extended; however, the use of fresh materials is constrained. When resources are used repeatedly by several parties, closed economic loops are created. (i) Durable design; (ii) maintenance; (iii) repair; (iv) reuse; (v) remanufacturing; (vi) refurbishing; and (vii) recycling are, in general, the fundamental components of CE models (Hatzivasilis, G. et al., 2021). The digitalization of many tasks and procedures, such as smart manufacturing, preventive maintenance, and water distribution, is made possible by the IoT, and it has the potential to advance the circular economy and create a more sustainable society. Paradoxically, IoT technologies and concepts like edge computing, while holding great potential for the digital revolution towards sustainability, are not yet assisting in the sustainable evolution of the IoT sector itself (Atiq, H. U. et al., 2023). Green IoT and AI are emerging as potential solutions to businesses' carbon footprint issues, addressing finite natural resources and energy consumption. However, the sustainable vision of G-IoT is at odds with Edge AI's increased energy usage (Afolabi, H. A., & Aburas, A., 2021), The smart city concept can also be explained from this. The term "smart city" has rapidly emerged, referring to the fast adoption of technologies such as context awareness, decentralised computation at the edge of the network, and the IoT. There are new issues in managing the infrastructure of ICT (information and communication technology) in such a dynamic environment at scale. For instantaneous fashion vertical and cross-vertical services, context-sharing and migration are part of the current fog layer resource management system. On the other hand, poor context migration may have a detrimental effect on performance. The authors of this work have envisioned using artificial intelligence and context-aware computing to enhance the Quality of Service (QoS) for connected transportation while reducing the amount of data that must be sent in real-time between intelligent cars and fog nodes (Reddy, K. H. K. et al., 2023).

6. CHALLENGES AND FUTURE SCOPE

MLDL systems face challenges in balancing model complexity and energy efficiency due to high computing costs. To address this, focus on improving model topology and developing less power-hungry alternatives. Address accessibility and data quality issues by exploring representative data sets and using transfer learning strategies. Scaling up energy-efficient models requires adapting smaller algorithms for larger systems. Establish standardized evaluation criteria for fair assessments. Addressing these challenges can lead to more robust computer systems.

6.1 Data and Privacy Concerns and Security

Security ensures data integrity, availability, and confidentiality, while privacy regulates access to sensitive information on a device. Figure 5 depicts risks related to security and privacy that could impact any CoT device (Hatzivasilis, G. et al, 2021).

Figure 5. Taxonomy of threats in IoT

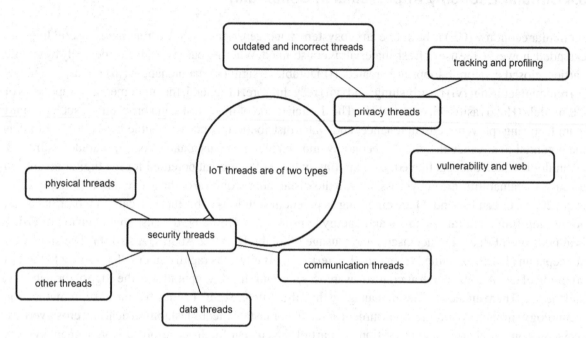

6.1.1 Security Threats

This category compromises an IoT system's security, includes risks related to data, service delivery, physical, communication, and other areas. The dangers indicated are explored below-

- **Communication Threats:** An attacker may launch several threats in a communication threat by using a communication channel. Attacks that cause a denial of service (DoS) are one common example. Software defects, resource depletion, hardware malfunctions, and the illegal streaming of high-energy signals are all possible outcomes of denial-of-service (DoS) attacks. These kinds of assaults will severely impair or maybe entirely stop a network from performing its intended function. Due to its hardware limitations, CoT is susceptible to DoS attacks.
- **Physical Threats:** There are several of these threats. They are situations that have the potential to physically harm or destroy connected Internet of Things gadgets. These risks can come from the inside (fire, unstable power supplies, etc.), the outside (lightning, earthquakes, etc.), or the people (theft, arson, careless or careless mistakes, etc.).
- **Data Breaches:** One of the most prevalent types of cybersecurity risks is data breaches. Plagiarism, turning off security features, and stealing and/or damaging data fall under this category. Threats that rely on CoT data include data leaks, bogus data injection, and data destruction and leakage. Malicious insiders, shared infrastructure flaws, misuse of cloud servers, and other serious threats that are unrelated to either of the examples above include.

6.1.2 Privacy Threats

Since users of a system are the ones who are most concerned about privacy, privacy threats are more concerned with breaching and stealing the personal information of CoT system consumers. A CoT environment may be subject to a variety of attacks that undermine the privacy of some users.

6.2 Generalization to Diverse Computing Environments

Implementing deep learning and machine learning techniques for green computing requires generalization to various computing environments. To ensure energy efficiency and sustainability across a variety of computing systems, these algorithms should be adaptable enough to accept different hardware and software configurations. It can be explained it by an example such as deep learning techniques Lateral Control and Reinforcement Learning for Autonomous Driving. Convolutional neural network technology (CNN) designed for multi-task learning (MTL) has been successfully used in various sectors including harmful item identification and is able to address the perception difficulty by applying the common knowledge of various jobs. The control module, which aims to generate the optimal or almost optimal control command to keep the vehicle on course, is another essential element as determined by the planning module. Popular control techniques include model predictive control (MPC), fuzzy logic, and linear quadratic regulator (LQR). According to the RL paradigm, the decision-making problem for vehicle lateral control is an ongoing condition and continuous action problem. Thus, to solve this problem, apply the policy gradient strategy. To forecast the important track features within the vision-based lateral management system, the driver-view picture is supplied into the perceptual module depending on the MTL neural network (Li, D. et al., 2019). The vehicle's trajectory is then maintained by an RL control module that links these track elements to a steering operation. The MTL-RL controller is the name given to it. The RL agent tests and demonstrates its control strategy through an iterative process that often requires several samples to converge. Thus, training an agent on a real car would be risky and expensive. A big sample set is also necessary for CNN or other deep neural network training. An environment for autonomous driving that combines agent learning and image processing at the same time is created for the aim of testing and training the perception and control algorithms (Afolabi, H. A., & Aburas, A., 2021). The VTORCS can be configured in a variety of ways, including with track curvatures, vehicle numbers, and varied lane numbers among other things, to create the ideal simulation environment with various degrees of complexity. The separation of perception and control allows for the option of using state-of-the-art algorithms for every module. The agent can obtain the visual or tangible information from the VTORCS at each time step and make a decision based on the underlying policy. May 2019 issue of IEEE Artificial Intelligence Magazine Issue 85 Back at the VTORCS, the action completes the one-step simulation.

VTORCS Environment: The environment is made up of four components, as depicted in Fig Six blocks: a controller interface block and a sensor block, a networking interface block, and a game engine block. The original TORCS provides the yellow block in Figure 6, whereas the green bricks are our own creations. The C++ language is used to implement the game engine settings section according to the original TORCS. Both Python and C++ are used to construct additional components. The environment offers a Python interface for the implementation of the RL algorithm and deep learning (DL). Two signal streams are part of the working flow. The sensors employ a protocol for communication to transfer all of the data into a shared memory after first obtaining measurement information from the game engine. After reading this data and compiling the observation, the control interface transfers it to the external perception/control algorithm. The control interface receives the control command from the external algorithm, computes it, and sends it to shared memory via the signal that controls stream. Once the game engine has read the command via the communication protocol, the a single-step simulation is finished.

Figure 6. VTORCS environment

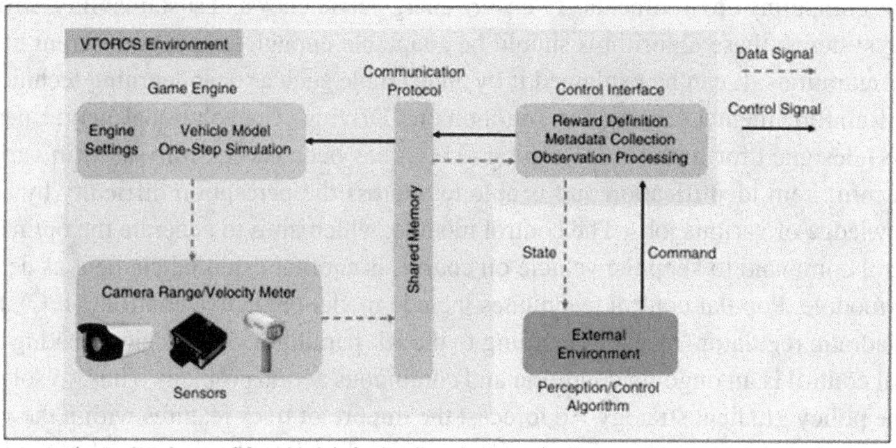

7. CONCLUSION

The integration of ML and DL algorithms in the environment of green computing holds an enormous pledge for adding sustainability in the information age. These algorithms break significant problems of resource allocation, environmental impact and energy effectiveness in colourful operations. ML/ DL enables dynamic power operation, workload soothsaying, and intelligent resource allocation, which significantly reduces energy use and carbon emigration. Their significance in green computing systems is further emphasized by their function in optimizing tackle design, cargo balancing, and cooling systems. still, prostrating obstacles similar to model complexity, data quality, and scalability is necessary for successful perpetration. Research must concentrate on creating durable datasets, optimizing algorithms for energy effectiveness, and assuring scalability across colourful computing settings.

REFERENCES

Afolabi, H. A., & Aburas, A. (2021). An evaluation of machine learning classifiers for prediction of attacks to secure green IoT infrastructure. *International Journal (Toronto, Ont.)*, 9(5).

Agarwal, R., & Sharma, D. K. (2021, February). Machine learning & Deep learning based Load Balancing Algorithms techniques in Cloud Computing. In *2021 International Conference on Innovative Practices in Technology and Management (ICIPTM)* (pp. 249-254). IEEE. 10.1109/ICIPTM52218.2021.9388349

Ahmed, K. I., Tabassum, H., & Hossain, E. (2019). Deep learning for radio resource allocation in multi-cell networks. *IEEE Network*, *33*(6), 188–195. doi:10.1109/MNET.2019.1900029

Ali, Z., Jiao, L., Baker, T., Abbas, G., Abbas, Z. H., & Khaf, S. (2019). A deep learning approach for energy efficient computational offloading in mobile edge computing. *IEEE Access : Practical Innovations, Open Solutions*, *7*, 149623–149633. doi:10.1109/ACCESS.2019.2947053

Anley, M. B., & Awgichew, R. B. (2022, January). Machine Learning Approach for Green Usage of Computing Devices. *Proceeding of the 2 nd Deep Learning Indaba-X Ethiopia Conference 2021.*

Anthony, L. F. W., Kanding, B., & Selvan, R. (2020). Carbontracker: Tracking and predicting the carbon footprint of training deep learning models. *arXiv preprint arXiv:2007.03051.*

Atiq, H. U., Ahmad, Z., Uz Zaman, S. K., Khan, M. A., Shaikh, A. A., & Al-Rasheed, A. (2023). Reliable resource allocation and management for IoT transportation using fog computing. *Electronics (Basel)*, *12*(6), 1452. doi:10.3390/electronics12061452

Balicki, J. (2021). Many-objective quantum-inspired particle swarm optimization algorithm for placement of virtual machines in smart computing cloud. *Entropy (Basel, Switzerland)*, *24*(1), 58. doi:10.3390/e24010058 PMID:35052084

Berral, J. L., Gavalda, R., & Torres, J. (2011, September). Adaptive scheduling on power-aware managed data-centers using machine learning. In *2011 IEEE/ACM 12th International Conference on Grid Computing* (pp. 66-73). IEEE. 10.1109/Grid.2011.18

Bharany, S., Badotra, S., Sharma, S., Rani, S., Alazab, M., Jhaveri, R. H., & Gadekallu, T. R. (2022). Energy efficient fault tolerance techniques in green cloud computing: A systematic survey and taxonomy. *Sustainable Energy Technologies and Assessments*, *53*, 102613. doi:10.1016/j.seta.2022.102613

Cheema, S., & Khan, G. N. (2020, October). Power and Performance Analysis of Deep Neural Networks for Energy-aware Heterogeneous Systems. In *2020 IEEE International Conference on Systems, Man, and Cybernetics (SMC)* (pp. 2184-2189). IEEE. 10.1109/SMC42975.2020.9283092

Coelho, C. N. Jr, Kuusela, A., Li, S., Zhuang, H., Ngadiuba, J., Aarrestad, T. K., Loncar, V., Pierini, M., Pol, A. A., & Summers, S. (2021). Automatic heterogeneous quantization of deep neural networks for low-latency inference on the edge for particle detectors. *Nature Machine Intelligence*, *3*(8), 675–686. doi:10.1038/s42256-021-00356-5

Elsken, T., Metzen, J. H., & Hutter, F. (2019). Neural architecture search: A survey. *Journal of Machine Learning Research*, *20*(1), 1997–2017.

Fraga-Lamas, P., Lopes, S. I., & Fernández-Caramés, T. M. (2021). Green IoT and edge AI as key technological enablers for a sustainable digital transition towards a smart circular economy: An industry 5.0 use case. *Sensors (Basel)*, *21*(17), 5745. doi:10.3390/s21175745 PMID:34502637

Gallego, F., Martín, C., Díaz, M., & Garrido, D. (2023). Maintaining flexibility in smart grid consumption through deep learning and deep reinforcement learning. *Energy and AI*, *13*, 100241. doi:10.1016/j.egyai.2023.100241

Gholipour, N., Shoeibi, N., & Arianyan, E. (2021). An energy-aware dynamic resource management technique using deep q-learning algorithm and joint VM and container consolidation approach for green computing in cloud data centers. In *Distributed Computing and Artificial Intelligence, Special Sessions, 17th International Conference* (pp. 227-233). Springer International Publishing. 10.1007/978-3-030-53829-3_26

Ghosh, S., & Das, J. (2022). Dynamic Voltage and Frequency Scaling Approach for Processing Spatio-Temporal Queries in Mobile Environment. In *Green Mobile Cloud Computing* (pp. 185–199). Springer International Publishing. doi:10.1007/978-3-031-08038-8_9

Hameed, A., Khoshkbarforoushha, A., Ranjan, R., Jayaraman, P. P., Kolodziej, J., Balaji, P., Zeadally, S., Malluhi, Q. M., Tziritas, N., Vishnu, A., Khan, S. U., & Zomaya, A. (2016, July). A survey and taxonomy on energy efficient resource allocation techniques for cloud computing systems. *Computing*, *98*(7), 751–774. doi:10.1007/s00607-014-0407-8

Han, S., Mao, H., & Dally, W. J. (2015). Deep compression: Compressing deep neural networks with pruning, trained quantization and huffman coding. *arXiv preprint arXiv:1510.00149*.

Hassan, M. B., Saeed, R. A., Khalifa, O., Ali, E. S., Mokhtar, R. A., & Hashim, A. A. (2022, May). Green Machine Learning for Green Cloud Energy Efficiency. In *2022 IEEE 2nd International Maghreb Meeting of the Conference on Sciences and Techniques of Automatic Control and Computer Engineering (MI-STA)* (pp. 288-294). IEEE. 10.1109/MI-STA54861.2022.9837531

Hatzivasilis, G., Ioannidis, S., Fysarakis, K., Spanoudakis, G., & Papadakis, N. (2021). The green blockchains of circular economy. *Electronics (Basel)*, *10*(16), 2008. doi:10.3390/electronics10162008

Henderson, P., Hu, J., Romoff, J., Brunskill, E., Jurafsky, D., & Pineau, J. (2020). Towards the systematic reporting of the energy and carbon footprints of machine learning. *Journal of Machine Learning Research*, *21*(1), 10039–10081.

Irtija, N., Anagnostopoulos, I., Zervakis, G., Tsiropoulou, E. E., Amrouch, H., & Henkel, J. (2021). Energy efficient edge computing enabled by satisfaction games and approximate computing. *IEEE Transactions on Green Communications and Networking*, *6*(1), 281–294. doi:10.1109/TGCN.2021.3122911

Karthiban, K., & Raj, J. S. (2020). An efficient green computing fair resource allocation in cloud computing using modified deep reinforcement learning algorithm. *Soft Computing*, *24*(19), 14933–14942. doi:10.1007/s00500-020-04846-3

Kashyap, P. K., Kumar, S., & Jaiswal, A. (2019, November). Deep learning based offloading scheme for IoT networks towards green computing. In *2019 IEEE International Conference on Industrial Internet (ICII)* (pp. 22-27). IEEE. 10.1109/ICII.2019.00015

Kaur, A., Kaur, B., Singh, P., Devgan, M. S., & Toor, H. K. (2020). Load balancing optimization based on deep learning approach in cloud environment. *International Journal of Information Technology and Computer Science*, *12*(3), 8–18. doi:10.5815/ijitcs.2020.03.02

Khatiwada, B., Jariyaboon, R., & Techato, K. (2023). E-waste management in Nepal: A case study overcoming challenges and opportunities. e-Prime-Advances in Electrical Engineering. *Electronics and Energy*, *4*, 100155.

Kumar, M. R., Devi, B. R., Rangaswamy, K., Sangeetha, M., & Kumar, K. V. R. (2023, April). IoT-Edge Computing for Efficient and Effective Information Process on Industrial Automation. In *2023 International Conference on Networking and Communications (ICNWC)* (pp. 1-6). IEEE. 10.1109/ICNWC57852.2023.10127492

Kumar, R., Khatri, S. K., & Diván, M. J. (2022). Optimization of power consumption in data centers using machine learning based approaches: A review. *Iranian Journal of Electrical and Computer Engineering*, *12*(3), 3192. doi:10.11591/ijece.v12i3.pp3192-3203

Li, D., Zhao, D., Zhang, Q., & Chen, Y. (2019). Reinforcement learning and deep learning based lateral control for autonomous driving. *IEEE Computational Intelligence Magazine*, *14*(2), 83–98. doi:10.1109/MCI.2019.2901089

Liang, T., Glossner, J., Wang, L., Shi, S., & Zhang, X. (2021). Pruning and quantization for deep neural network acceleration: A survey. *Neurocomputing*, *461*, 370–403. doi:10.1016/j.neucom.2021.07.045

Liu, G., Ying, Z., Zhao, L., Yuan, X., & Chen, Z. (2018, July). A New Deep Transfer Learning Model for Judicial Data Classification. In *2018 IEEE International Conference on Internet of Things (iThings) and IEEE Green Computing and Communications (GreenCom) and IEEE Cyber, Physical and Social Computing (CPSCom) and IEEE Smart Data (SmartData)* (pp. 126-131). IEEE. 10.1109/Cybermatics_2018.2018.00053

Lu, Z., Whalen, I., Boddeti, V., Dhebar, Y., Deb, K., Goodman, E., & Banzhaf, W. (2019, July). Nsga-net: neural architecture search using multi-objective genetic algorithm. In *Proceedings of the genetic and evolutionary computation conference* (pp. 419-427). 10.1145/3321707.3321729

Luo, T., Wong, W. F., Goh, R. S. M., Do, A. T., Chen, Z., Li, H., Jiang, W., & Yau, W. (2023). Achieving Green AI with Energy-Efficient Deep Learning Using Neuromorphic Computing. *Communications of the ACM*, *66*(7), 52–57. doi:10.1145/3588591

Mehlin, V., Schacht, S., & Lanquillon, C. (2023). Towards energy-efficient Deep Learning: An overview of energy-efficient approaches along the Deep Learning Lifecycle. *arXiv preprint arXiv:2303.01980*.

Nalla, K., & Pothabathula, S. V. (2021). Green IoT and Machine Learning for Agricultural Applications. *Green Internet of Things and Machine Learning: Towards a Smart Sustainable World*, 189-214.

Ning, Z., Dong, P., Wang, X., Guo, L., Rodrigues, J. J., Kong, X., Huang, J., & Kwok, R. Y. (2019). Deep reinforcement learning for intelligent internet of vehicles: An energy-efficient computational offloading scheme. *IEEE Transactions on Cognitive Communications and Networking*, *5*(4), 1060–1072. doi:10.1109/TCCN.2019.2930521

Ounifi, H. A., Gherbi, A., & Kara, N. (2022). Deep machine learning-based power usage effectiveness prediction for sustainable cloud infrastructures. *Sustainable Energy Technologies and Assessments*, *52*, 101967. doi:10.1016/j.seta.2022.101967

Paul, S. G., Saha, A., Arefin, M. S., Bhuiyan, T., Biswas, A. A., Reza, A. W., ... Moni, M. A. (2023). A Comprehensive Review of Green Computing: Past, Present, and Future Research. *IEEE Access*.

Peng, Z., Barzegar, B., Yarahmadi, M., Motameni, H., & Pirouzmand, P. (2020). Energy-aware scheduling of workflow using a heuristic method on green cloud. *Scientific Programming*, *2020*, 1–14. doi:10.1155/2020/8898059

Phiri, M., Mulenga, M., Zimba, A., & Eke, C. I. (2023). Deep learning techniques for solar tracking systems: A systematic literature review, research challenges, and open research directions. *Solar Energy*, *262*, 111803. doi:10.1016/j.solener.2023.111803

Raja, S. P. (2021). Green computing and carbon footprint management in the IT sectors. *IEEE Transactions on Computational Social Systems*, *8*(5), 1172–1177. doi:10.1109/TCSS.2021.3076461

Rao, H. (n.d.). *Machine Learning endorsing "Green Computing" by optimizing energy efficiency of data centers*. https://www.linkedin.com/pulse/machine-learning-endorsing-green-computing-optimizing-himanshu-rao

Reddy, K. H. K., Goswami, R. S., & Roy, D. S. (2023). A futuristic green service computing approach for smart city: A fog layered intelligent service management model for smart transport system. *Computer Communications*, *212*, 151–160. doi:10.1016/j.comcom.2023.08.001

Senthil Kumar, R., Saravanan, S., Pandiyan, P., Suresh, K. P., & Leninpugalhanthi, P. (2022). Green Energy Using Machine and Deep Learning. *Machine Learning Algorithms for Signal and Image Processing*, 429-444.

Songhorabadi, M., Rahimi, M., MoghadamFarid, A. M., & Haghi Kashani, M. (2023). Fog computing approaches in IoT-enabled smart cities. *Journal of Network and Computer Applications*, *211*, 103557. doi:10.1016/j.jnca.2022.103557

Talwani, S., Singla, J., Mathur, G., Malik, N., Jhanjhi, N. Z., Masud, M., & Aljahdali, S. (2022). Machine-Learning-Based Approach for Virtual Machine Allocation and Migration. *Electronics (Basel)*, *11*(19), 3249. doi:10.3390/electronics11193249

Tuli, S., Gill, S. S., Xu, M., Garraghan, P., Bahsoon, R., Dustdar, S., Sakellariou, R., Rana, O., Buyya, R., Casale, G., & Jennings, N. R. (2022). HUNTER: AI based holistic resource management for sustainable cloud computing. *Journal of Systems and Software*, *184*, 111124. doi:10.1016/j.jss.2021.111124

Wu, C. J., Raghavendra, R., Gupta, U., Acun, B., Ardalani, N., Maeng, K., ... Hazelwood, K. (2022). Sustainable ai: Environmental implications, challenges and opportunities. *Proceedings of Machine Learning and Systems*, *4*, 795–813.

Yu, P., Zhou, F., Zhang, X., Qiu, X., Kadoch, M., & Cheriet, M. (2020). Deep learning-based resource allocation for 5G broadband TV service. *IEEE Transactions on Broadcasting*, *66*(4), 800–813. doi:10.1109/TBC.2020.2968730

Zhang, T., Lei, C., Zhang, Z., Meng, X. B., & Chen, C. P. (2021). AS-NAS: Adaptive scalable neural architecture search with reinforced evolutionary algorithm for deep learning. *IEEE Transactions on Evolutionary Computation*, *25*(5), 830–841. doi:10.1109/TEVC.2021.3061466

Zhou, H., Jiang, K., Liu, X., Li, X., & Leung, V. C. (2021). Deep reinforcement learning for energy-efficient computation offloading in mobile-edge computing. *IEEE Internet of Things Journal*, *9*(2), 1517–1530. doi:10.1109/JIOT.2021.3091142

Zhou, X., Wang, R., Wen, Y., & Tan, R. (2021). Joint IT-facility optimization for green data centers via deep reinforcement learning. *IEEE Network*, *35*(6), 255–262. doi:10.1109/MNET.011.2100101

Chapter 2
Green Computing and the Quest for Sustainable Solutions

Gudivada Lokesh
https://orcid.org/0009-0005-0669-0393
Jawaharlal Nehru Technological University, India

B. Rupa Devi
https://orcid.org/0009-0005-1298-737X
Annamacharya Institute of Technology and Sciences, India

N. Badrinath
https://orcid.org/0000-0003-4058-7136
Vellore Institute of Technology, India

L. N. C. Prakash K.
CVR College of Engineering, India

Pole Anjaiah
Institute of Aeronautical Engineering (Autonomous), India

T. Ravi Kumar
Aditya Institute of Technology and Management, India

ABSTRACT

The internet of things (IoT) connects devices of all sizes to the internet, providing seamless communication and ease. However, this technical improvement has prompted environmental concerns. Sustainable development has grown as we work to offset the effects of technology on our planet, economy, and consumerism. Innovative companies are exploring technology-based sustainability solutions. Green IT, a developing idea, uses sustainable design to reduce or eliminate IoT operations' environmental implications. This chapter evaluates IoT problems, defines Green IT, explores Green IT design methodologies, and describes how to implement these green designs as sustainable environmental solutions. It also thoroughly analyses numerous author proposals to find the best sustainable IoT architectures. Green IT solutions are essential to a sustainable and ecologically responsible future in a world where information technology is central to our lives, businesses, and marketing tactics.

DOI: 10.4018/979-8-3693-1552-1.ch002

INTRODUCTION

Green Computing (Podder, 2022) is a crucial paradigm shift in the world of technology, and it emphasizes the need for environmentally conscious Information and Communications Technology (ICT) (Mory-Alvarado, 2023) practices and sustainable solutions to mitigate the unfavorable effects of technology on our planet. The IoT systems comprise various components critical to their operation, including identification, monitoring, locating, sensing, tracking, communication, and computation services (Tewari, 2020). The IoT is an evolution of internet-based networks that transcend traditional human-to-human interactions. It brings about a new era where communication extends its reach to include the dynamic interplay between humans and things and the interconnectedness between things themselves (Ahmad, 2019). The general usage of IoT devices has led to a concerning growth in energy consumption, resulting in a growing amount of waste, excessive heat, and carbon emissions that are poisonous to the environment (Al-Turjman, 2019). Additionally, the production and disposal of IT hardware have been a significant contributor to environmental issues, with complex processes that have the potential to disrupt the delicate balance of our ecosystem (Murugesan, 2008). According to contemporary studies, IoT is transforming rapidly, with 24 billion out of 34 billion interconnected devices projected to join the IoT realm by 2020 (Ahmad, 2019), and this highlights significant growth opportunities and the importance of staying ahead of the curve.

Green Computing encompasses many practices, such as using energy-efficient hardware, eco-friendly data centers, sustainable software design, and adopting sustainable strategies in designing, producing, operating, and disposing of ICT equipment and services (Naim, 2021). The significance of green computing lies in its potential to make technology a catalyst for environmental conservation rather than its adversary.

Green Computing presents a promising solution for a sustainable future by striving to minimize the environmental impact of computing systems. It achieves this by reducing the harmful materials used in computers, increasing energy efficiency, and promoting waste recyclability (Kaur, 2015). A primary area of focus for Green Computing is data centers that consume vast amounts of power. To address this, it uses various strategies like virtualization, power management techniques, and innovative cooling systems to curtail power consumption. Virtualization is a pivotal component of Green Computing that reduces the need for numerous servers, leading to significant energy savings. Energy-efficient cooling systems help dissipate the heat data centers generate, lowering power consumption and minimizing environmental impact (Zhang, 2021). Several tips, plans, tools, and technologies are available for pursuing Green Computing, ranging from choosing energy-efficient hardware components to implementing environmentally responsible disposal methods for obsolete equipment. Adopting sustainable practices like recycling and refurbishing computer components promotes a circular economy, reducing electronic waste and conserving valuable resources.

This research explores the multifaceted approaches and technologies that facilitate environmentally friendly IT practices. It will investigate the integration of Green Computing principles into the realm of IoT and other emerging technologies to mitigate the potential environmental impact. The aim is to inspire individuals, businesses, and policymakers to embrace a sustainable approach and work together to build a more environmentally friendly society and economy.

GREEN COMPUTING AND SUSTAINABILITY

Green Computing is a holistic approach to utilizing computer systems and related technology in an environmentally responsible manner to minimize the negative impact of digital technology on the environment while maximizing efficiency and sustainability. The principles of Green Computing encompass various aspects:

Figure 1. Green computing concept

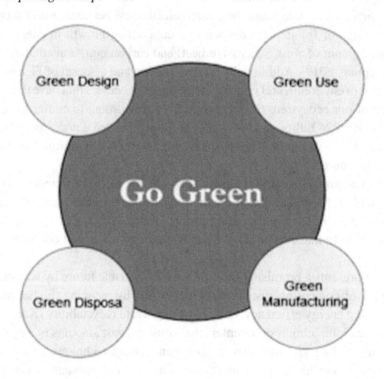

1. **Green Use:** This principle highlights the efficient use of computer systems and peripheral devices to reduce energy consumption and promote eco-friendly practices. It concerns optimizing software and hardware configurations to reduce energy usage during operation (Al-Zamil, 2020). To minimize the harmful effects on the environment and ensure sustainability, it is essential to use computer systems and peripheral devices with care. To optimize device performance and reduce their power consumption, appropriate adjustments to both software and hardware parameters are necessary. To promote eco-friendly practices, consider implementing the following:
    Use power-saving modes such as sleep, hibernation, and standby when the devices are not in use or inactive for a specific period.
    Dim the brightness and quality of display panels and turn off additional features like Bluetooth, Wi-Fi, and GPS when not in use.
    Employ energy-efficient software like virtualization, cloud computing, and green algorithms to reduce the computational complexity and resource requirements of applications.

2. **Green Disposal**: Proper management of e-waste, which refers to electronic devices and components that are discarded or obsolete, is crucial for the environment and human health. E-waste contains hazardous materials such as lead, mercury, cadmium, and brominated flame retardants, which can cause various diseases and disorders by contaminating the soil, water, and air. Disposing of non-recyclable e-waste safely while minimizing the threats to human health and the environment is essential. Landfilling, incineration, and exporting are acceptable methods of disposal only if they comply with environmental laws and regulations such as the Basel Convention and the RoHS Directive (Peri, 2022).

3. **Green Design**: Green Design (Ojo, 2019) is a design approach that aims to develop energy-efficient digital devices with a reduced carbon footprint. It emphasizes creating technology that consumes less energy, has prolonged lifespans, and emits fewer greenhouse gases during its lifecycle. Designers and manufacturers can adopt several strategies to achieve these objectives, including using eco-friendly materials such as biodegradable plastics, organic fibers, and recycled metals. Incorporating low-power processors, sensors, and switches can also help regulate energy consumption, optimizing device performance in alignment with user preferences and needs. Moreover, modular, upgradeable, and repairable components can enhance device durability and compatibility, extending their lifespan and minimizing replacements and disposals. Applying recognized certifications and standards such as Energy Star, EPEAT, and LEED can ensure adherence to green design principles, enhancing the environmental performance and quality of digital devices and contributing to a more sustainable and eco-friendly future.

4. **Green Manufacturing**: Green manufacturing is an approach to developing computer systems, servers, and other digital equipment while minimizing their adverse impact on the environment (Lavanya, 2020). The process involves utilizing resources that are recyclable, biodegradable, or renewable. The ultimate goal of green manufacturing is to reduce environmental harm, waste generation, and pollution arising from both the manufacturing processes and the resultant goods. Adopting green manufacturing practices not only enhances businesses' environmental performance but also contributes to their social responsibility, economic competitiveness, and customer satisfaction. By prioritizing materials that are recyclable or biodegradable, manufacturers contribute to a more sustainable production cycle, reducing the long-term ecological footprint of their products. Simultaneously, by minimizing energy usage and greenhouse gas emissions during production, companies actively work towards lessening their overall environmental impact. It is not just a strategic move for environmental conservation but also a means to meet consumer expectations for environmentally friendly and socially responsible products.

Green Computing Benefits

Green computing is an approach to computing that aims to minimize the environmental impact of technology. There are several benefits of green computing, including creating an eco-friendly workplace, preventing pollution to protect the environment and increase agriculture throughput, using IT resources in an energy-efficient and cost-effective manner, gaining a competitive advantage for environmental and agriculture sustainability, decreasing IT equipment emissions to protect the rate of oxygen in the air, and minimizing energy consumption to save costs.

Sustainability Perspective on Environmental Concerns in IoT: IoT has become an integral part of our digital ecosystem, but it also suggests environmental challenges:

1. **Energy Consumption**: A pivotal challenge in IoT design revolves around ensuring prolonged device operation without excessive energy consumption. To address this hurdle, sustainable IoT design integrates low-power components, encompassing sensors, microcontrollers, and batteries. Additionally, the adoption of reliable data transmission protocols like MQTT, CoAP, and 6LoWPAN facilitates efficient connectivity among IoT devices and the cloud, minimizing bandwidth and resource utilization. The incorporation of these techniques in sustainable IoT design significantly diminishes operational costs and mitigates the environmental impact of IoT systems.

2. **Resource Efficiency**: Resource efficiency stands as a crucial factor in the production of IoT devices. Components like sensors, communication modules, processors, and memory chips necessitate various resources, such as energy and materials. The escalating demand for these devices raises concerns about resource shortages, particularly rare earth metals. Addressing these challenges, sustainable IoT solutions adopt a lifecycle approach, assessing the environmental impact of devices from production to disposal. This approach involves utilizing recyclable, biodegradable, or renewable materials and embracing safe, effective, and clean production methods. Furthermore, it emphasizes the design of durable, upgradeable, and easily maintainable devices, accompanied by appropriate recycling or disposal procedures at the end of their lifecycle.

   ○ **Data Management**: The inherent capacity of IoT to collect and analyze data from various sources poses challenges to data centers responsible for storage, transfer, and analysis, often consuming substantial energy and resources. Sustainable IoT systems employ various techniques to curtail energy usage and mitigate environmental impacts associated with data management. These include:

   ○ **Data optimization:** Reduced generation, transmission, and storage of data by IoT systems and devices is necessary to achieve this. Data encryption, deduplication, filtering, aggregation, and compression are a few methods that may be used to maximize data.

   ○ **Edge computing:** This entails shifting a portion of the data processing workload from the cloud to the Internet of Things devices' location at the network's edge. When data is sent and analyzed, this can lower latency, bandwidth, and energy usage while also improving data security and privacy.

   ○ **Efficient cloud infrastructure:** Using energy-efficient cooling and ventilation technologies, such as heat recovery, liquid cooling, and free cooling, as well as developing data centers that employ renewable energy sources, including solar, wind, and hydropower, are all part of this challenge.

3. **E-Waste**: When electronic equipment like computers, cellphones, and TVs are not disposed of correctly, the materials and components they contain may represent a risk to human health and the environment. Over time, these gadgets may break down or become outdated, producing e-waste that is frequently disposed of in landfills or burned, releasing harmful materials into the ground, water, and atmosphere, including lead, mercury, and cadmium. Recovering the valuable elements from the e-waste, such as glass, metals, and plastics, and using them to create new goods is crucial for preventing this. Recycling has the potential to decrease greenhouse gas emissions and pollutants related to the management of e-waste, as well as the need for energy and raw materials.

4. **Carbon Footprint**: Data centers and IT operations require massive amounts of power which are primarily produced from fossil fuels like coal, oil, and gas. However, the energy produced from these fuels releases greenhouse gases such as carbon dioxide, methane, and nitrous oxide into the atmosphere, which trap heat and contribute to global warming. According to research, data centers were responsible for 0.3% of global carbon emissions and approximately 1% of the world's power demand in 2018.

5. **Resource Usage**: A variety of components and materials used in computer hardware, including memory, storage, and CPUs, come from non-renewable, limited resources including metals and minerals. Certain resources are in great demand and limited availability, such as gold, silver, copper, cobalt, and rare earth elements. Their extraction and processing can result in negative social and environmental effects like deforestation, pollution, and breaches of human rights. Green manufacturing principles are a collection of rules and procedures that, by using several tactics, seek to reduce the amount of resources used and the environmental effect of the creation of computer hardware.

In today's digital age, Green Computing and sustainability involve a set of principles and practices that aim to minimize the environmental impact of technology to ensure a more sustainable future, and it is essential to prioritize eco-friendly design, responsible disposal, and efficient resource management while developing and using digital devices and systems. By doing so, we can reduce the harmful effects of technology on our environment and contribute to a healthier planet.

CLOUD COMPUTING

The intersection of cloud computing and artificial intelligence (AI) has given rise to a new era of intelligent applications and services. Cloud-based AI services, such as machine learning platforms and natural language processing APIs, provide developers with powerful tools to integrate AI capabilities into their applications without the need for extensive expertise in AI algorithms. Cloud computing promises continued innovation, efficiency, and empowerment on a global scale, permeating industries and shaping the way we process information, develop applications, and harness the power of data. The concept of cloud computing, which has completely changed the information technology environment, is a complex and dynamic ecosystem that goes much beyond simple data storage. Fundamentally, cloud computing is the Internet-based distribution of computer resources, such as storage, processing power, and applications. Unlike conventional on-premise computing architectures, this novel method offers unmatched scalability, flexibility, and cost-efficiency. A deeper exploration of cloud computing reveals that this technology has developed into a complex catalyst for digital transformation, impacting how people engage with technology, how organizations run, and how societies use data. The ability of cloud computing to enable universal access to a shared pool of computer resources is one of its core principles. Networks, servers, storage, apps, and services are all included in this resource pool and given to end users as needed. Because cloud computing is on-demand, customers may adjust their consumption to meet unique needs and can also guarantee that resources are always accessible. Businesses can now quickly adjust to shifting workloads and market conditions without having to make significant upfront expenditures in physical infrastructure because of this flexibility, which is a game-changer. Three unique service models that address different aspects of computing demands are infrastructure as a Service (IaaS), Platform as a Service (PaaS), and Software as a Service (SaaS) in the vast world of cloud computing.

IaaS offers pay-as-you-go services like virtual machines and storage over an Internet-based virtualized computing infrastructure. PaaS, on the other hand, takes things a step further by offering a platform that frees developers from the burdens of managing infrastructure so they can create, launch, and maintain apps. The most user-centric approach, SaaS, distributes software via the Internet, saving customers from having to update, manage, or install it on their devices.

The functionality of cloud computing is also heavily dependent on its architectural foundation. Three general categories which categorize cloud architectures: public, private, and hybrid. Private clouds are devoted to a particular business and allow more control and customization, whereas third-party service providers run public clouds and give resources to the broader public. As the name implies, hybrid clouds combine aspects of public and private clouds to provide a more adaptable solution that meets a range of business requirements. Security considerations loom large in the cloud computing landscape, and providers have responded with a robust array of measures to safeguard data and ensure privacy. Encryption, identity and access management, and regular security audits are integral components of cloud security strategies. However, the shared responsibility model prevails, emphasizing the collaborative effort required between cloud service providers and users to maintain a secure computing environment. Users must be vigilant in configuring and managing security settings, while providers play a crucial role in fortifying the underlying infrastructure. The advent of edge computing, a paradigm that brings computational resources closer to the data source, intrinsically links to the evolution of cloud computing. Enabling data processing to occur closer to where it is generated, edge computing embodies an architectural approach that decentralizes computing power. This proximity reduces latency, enhances real-time processing capabilities, and minimizes the need for large-scale data transfers to centralized cloud servers. As the Internet of Things (IoT) continues to proliferate, the symbiotic relationship between cloud computing and edge computing becomes increasingly apparent, offering a seamless and efficient solution for managing the deluge of data generated by interconnected devices.

ENERGY CONSUMPTION MODEL FOR SMART CITY

The energy consumption model for a smart city involves various components that contribute to energy consumption. Some of these components include energy consumption by residential, commercial, and manufacturing sectors, intelligent transportation, the creation of green energy, waste-to-energy, local cooling and heating, public infrastructure and services, data centers, and the infrastructure supporting information and communication technology.

Usage of Residential Energy: The energy we consume in our homes is of utmost importance. It is the energy utilized by various housing structures such as homes, apartments, and smart houses. This consumption encompasses electricity, natural gas, and other fuels such as kerosene, propane, and fuel oil. Energy is mostly used in homes for air conditioning, space heating, water heating, lights, appliances, and electronics. It's noteworthy to notice that homeowners are increasingly using power to charge their vehicles as electric cars become more and more popular. The location, climate, and size of the household are some of the variables that affect how much energy a house uses. Data from surveys, meters, sensors, and models are used to estimate and analyze home energy consumption trends and evaluate their effects on the environment and the economy.

Industrial and Commercials: Energy use varies throughout businesses, factories, and other industrial and commercial institutions. In order to maximize energy efficiency, it is important to monitor the energy use of various sectors, considering factors such as lights, HVAC systems, machinery, and other equipment (Tuysuz, 2020). It is essential to maximize energy efficiency in factories, enterprises, and other commercial and industrial establishments. It goes without saying that these industries have a significant impact on the world's energy consumption and greenhouse gas emissions. It is essential to comprehend the distinct energy consumption patterns of every business, including manufacturing, retail, hospitality, and transportation. Opportunities to raise energy efficiency, lower costs, and improve environmental performance can be found by closely observing and evaluating the energy use in various industries. This knowledge is necessary for efficient energy management, which allows for the improvement of energy efficiency, the use of energy-saving strategies, and the uptake of clean energy technologies such as microgrids, energy storage, and renewable energy production.

Smart Transportation: More than ever, we need transportation options that are affordable, environmentally friendly, and easy to use. The solution lies in intelligent mobility. It includes mass transit, electric vehicles (EVs), and sophisticated traffic management. Because they run on electricity, EVs have lower fuel and maintenance costs and produce less air, noise, and greenhouse gas pollution. The essential components of the charging infrastructure required for the widespread adoption of EVs are wireless charging systems, home chargers, and public stations. The emphasis on intelligent traffic management, public transit, and electric vehicles (EVs) by the smart transportation movement presents a comprehensive plan for addressing problems with energy utilization in modern cities. Let's work toward a more ecologically friendly future by implementing astute transportation techniques. Electric vehicles, or EVs, are autos that run on electricity rather than fossil fuels. Apart from reducing emissions of air, noise, and greenhouse gases, they can also help owners save money on fuel and upkeep. However, the proper infrastructure—which can include wireless charging systems, home chargers, and public stations—is required in order to charge them. Furthermore, the availability and dependability of energy are determined by supply and demand. Mass transit refers to public transportation modes such as buses, trains, subways, or trams that can transport large numbers of people quickly and cheaply. Mass transit can improve accessibility and equity for urban residents and reduce commuting times, traffic congestion, and environmental impacts. However, it also demands significant investment in operations, vehicles, and infrastructure. Moreover, it has to adapt to changing travel preferences while ensuring staff and passenger safety. Intelligent traffic management leverages data, sensors, and communication technology to monitor and optimize traffic flow, speed, and safety. It can enhance customer satisfaction, reduce emissions and accidents, and optimize traffic efficiency. However, implementing intelligent traffic management poses institutional and technological challenges, including data collection and analysis, integration and interoperability, privacy and security, governance, and regulation.

Waste Management: Smart cities provide heating and cooling services to buildings by using district heating and cooling and waste-to-energy technology. The former system involves the distribution of hot or cold water through underground pipelines to multiple buildings while the latter burns municipal solid waste to generate electricity and heat. By utilizing renewable energy sources and waste heat sources, these systems can significantly reduce greenhouse gas emissions while improving energy efficiency. According to (Guo, 2019), it's vital to monitor and optimize the energy efficiency and environmental impact of these systems..

Creating Renewable Energy: Integrating renewable energy (Li, 2018) resources into urban grids, such as solar panels and wind turbines, is a common focus of smart cities. Information about the power produced by these sources must be included in the model. Lastly, because smart cities significantly rely on data processing and communication technology, data centers and ICT infrastructure should have their energy usage tracked.

Figure 2. Energy consumption model (smart city)

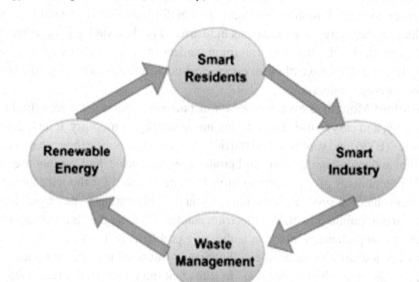

IMPLEMENTATION OF IOT

This section will thoroughly examine the decentralized edge cloud computing (DECC) idea. We will examine its design, underlying ideas, and significance for cloud computing and the Internet of Things. Massive amounts of data are being created from sensors, gadgets, and machines due to the expansion of IoT. Numerous industries, including healthcare, transportation, manufacturing, and smart cities, can benefit from this data. While the conventional cloud architecture can manage this data, it cannot fulfill the IoT's needs for real-time processing. Especially for applications that need instantaneous response, the time required to transport data from the edge to centralized cloud data centers might be a substantial barrier. DECC is an innovative solution that processes and analyzes IoT data in real-time while optimizing bandwidth usage (Ahmed, 2021).

DECC acts as a bridge between the worlds of IoT and cloud computing. By fusing cloud and edge computing, it combines the benefits of both (Hassan, 2018). With cloud-like services integrated into its edge nodes, distributed computing environments can ensure real-time processing while reducing latency. DECC achieves the best of both worlds by combining edge computing's low latency and real-time processing capabilities with the cloud's scalability, storage, and resource optimization (Gezer, 2021).

The Architecture of DECC

1. IoT Devices

The core technology driving DECC's operational efficiency is the Internet of Things (IoT). This network comprises devices equipped with sensors, cameras, and data-collection components, enabling seamless communication among themselves and with remote servers for processing and analysis. Leveraging IoT, DECC effectively monitors and manages environmental factors such as lighting, temperature, security, and energy consumption. Furthermore, IoT empowers DECC to deliver enhanced services and experiences to its guests and clients.

2. Edge Nodes

Situated near data sources on the network periphery, edge nodes are pivotal in minimizing the necessity for data transfer to a central server or cloud. This localization facilitates swifter and more efficient data processing and analysis. These nodes are capable of executing applications that interact with IoT devices, sensors, cameras, and other peripheral devices. Equipped with requisite hardware and software components, edge nodes encompass:

- **Virtualization layer:** An integral element of DECC's infrastructure is the virtualization layer. This component dynamically supports resource allocation, ensuring equitable distribution of CPU, memory, and storage for every program or task. The choice of virtualization technology varies based on the specific requirements of each edge node, potentially incorporating hypervisors or containerization.
- **Resource allocation algorithm:** The virtualization layer depends on the resource distribution method. It determines how resources are distributed across various workloads or apps on the edge node. The key goal is to use resources as efficiently as possible while considering the need for IoT apps to operate in real time. This method may be improved with machine learning and AI approaches, which can also be used to decide how to allocate resources based on workload requirements dynamically.
- **Central Cloud:** DECC operates mostly peripherally but can also be linked to a central cloud infrastructure for long-term archiving and additional processing power. The central cloud is valuable because it provides extra storage, backup, and resources for programs that do not require real-time processing. It connects the conventional cloud with the edge, ensuring safe data archiving that can be retrieved when necessary for batch processing or historical analysis.
- **Security measures:** Securing edge nodes, which manage data in close proximity to IoT devices, requires robust precautions. Encryption is vital to protect data both in transit and at rest. The growing deployment of edge nodes brings an array of security risks, including denial-of-service attacks, malicious injections, and data breaches. To counter these threats, establishing comprehensive policies is crucial for safeguarding both edge nodes and the associated data. Among the most effective security measures is the implementation of encryption methods. These methods ensure the confidentiality and integrity of data stored on edge nodes and during transmission to other devices or the cloud. Another critical aspect is authenticating objects and entities communicating with edge nodes, allowing access only to the necessary resources. Secure bootstrapping

techniques, like token or certificate usage, foster trust between edge nodes and connected devices. Furthermore, employing access control techniques, such as attribute-based or role-based access control, helps prevent unauthorized access to sensitive data and resources on the edge node. These techniques also facilitate the enforcement of policies and regulations governing access and usage of the edge node.

Figure 3. Edge computing

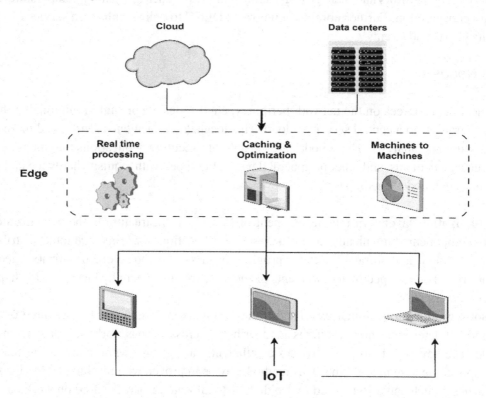

IMPLEMENTING GREEN COMPUTING STRATEGIES

Transitioning to green computing has gained significant importance in recent years, serving not only to optimize data center efficiency and cut costs but also to address the environmental impact of computing. Various measures have been implemented, encompassing new power and cooling methods, energy-efficient hardware utilization, virtualization, and software enhancements, along with the deployment of power and workload management strategies. These efforts collectively contribute to a sustainable and eco-friendly future, as outlined by (Mmeah, 2018).

- **Data center infrastructure:** There is a wide range of infrastructure equipment required to run a data center, including chillers, power supplies, storage devices, switches, pumps, fans, and network equipment. However, many data centers are now over ten years old, and their infrastructure equipment is reaching the end of its useful life. As a result, they tend to be more power-hungry and less efficient, consuming 2 or 3 times more energy overall than the IT equipment, mostly due to cooling. The diverse range of infrastructure equipment required to operate a data center, including chillers, power supplies, storage devices, switches, pumps, fans, and network equipment, often faces obsolescence in aging data centers, leading to increased power consumption, particularly in cooling. To tackle this challenge, many data centers are either investing in new energy-efficient facilities or retrofitting existing ones.

- **Power and workload management:** Research indicates that the implementation of power and workload management software can result in substantial cost savings, ranging from $25 to $75 per desktop per month, with even higher savings for servers. This software dynamically adjusts processor power states (P-states) to match workload requirements, efficiently utilizing processor power during peak demands and conserving energy during lighter workloads. Some companies are also making the transition from desktops to laptops due to the latter's superior power-management capabilities.

- **Thermal load management:** Advancements in technology have led to more compact data centers with higher power density, necessitating improved heat dissipation. The power usage of ventilation and cooling systems now rivals that of the servers themselves. Thermal load management strategies include variable cooling delivery, airflow management, and raised floor designs to ensure optimal airflow. Data centers are also adopting more efficient air conditioning equipment, leveraging ambient air, liquid heat removal systems, heat recovery systems, and smart thermostats to enhance thermal efficiency.

- **Product design:** To reduce energy consumption in data centers, we can shift to multiple-core microprocessors that run at slower speeds and lower voltages. Dynamic frequency and voltage scaling features enable microprocessor performance to ramp up or down based on the workload, further reducing energy usage. Additionally, energy-proportional computing designs servers that consume energy in proportion to the work performed. By prioritizing these approaches, we can double server efficiency and preserve a considerable amount of energy.

- **Virtualization:** Virtualization has become a popular solution to cope with the increasing computing needs of businesses. It is an effective way to optimize IT operations, reduce energy consumption, and cut costs. By improving the utilization of existing IT infrastructure, virtualization can help companies reduce energy use, capital investment, and human resource costs (Ryder, 2008). Data center virtualization impacts four major areas: server hardware and operating systems, storage, networks, and application infrastructure. For example, server virtualization allows companies to increase server utilization by consolidating applications on fewer servers. Through virtualization, data centers can deploy new applications while consuming less power, physical space, and labor.

SUSTAINABLE IT SERVICES

In today's dynamic business landscape, the adoption of sustainable IT services is no longer a choice but a strategic imperative for organizations aiming to thrive in the contemporary world. As the imperative to embrace sustainable business practices intensifies, it becomes imperative to recognize that sustainable IT extends beyond mere energy efficiency or carbon footprint reduction. Its scope encompasses the enduring impact of IT on the organization, its clientele, and society at large. Sustainable IT plays a pivotal role in delivering unparalleled value to customers, attaining a formidable market position, and safeguarding the longevity of the IT organization.

By aligning IT initiatives with overarching business strategies, organizations can achieve not just success but market-leading excellence while concurrently fostering societal and environmental benefits. The era of sustainable IT services presents a unique opportunity for businesses to stay ahead and maintain their competitive edge. Seize the chance to prioritize sustainable IT, and the rewards outlined in the seminal work by (Butner, 2008) become attainable.

In the contemporary business landscape, organizations heavily rely on IT services to operate efficiently and meet customer demands. However, ensuring the sustainability of these services is imperative to avert disruptions and maintain seamless business continuity. The sustainability of IT services can be examined from various perspectives, including service, temporal, cost, organizational, and environmental sustainability.

Service sustainability: The backbone of IT services lies in service sustainability. This entails the implementation of effective and reliable processes for service delivery, performance management, continuous security, system recovery planning, and software currency. Such measures guarantee that IT services run seamlessly, providing essential support to organizational operations.

Temporal sustainability: Sustaining IT services over time is at the core of temporal sustainability. A clear understanding of the value created by these services, a robust business case, and adaptability to changing business conditions are essential. This not only creates value for customers, society, and the business but also enables organizations to remain competitive and responsive to customer needs.

Cost sustainability: Another critical facet in ensuring the sustainability of IT services is cost sustainability. By managing acquisition and operating costs effectively, organizations can reduce the total cost of ownership. This involves selecting cost-effective hardware and software with attributes such as low power consumption and enhanced resource utilization. Additionally, considerations like life cycle management and replacement costs contribute to overall cost sustainability.

Organizational sustainability: Managing change within the organization is crucial for organizational sustainability. Given the inevitability of organizational change, IT services must be adaptable and innovative, irrespective of ongoing changes. Well-documented systems with comprehensive training mechanisms are better equipped to handle change, ensuring the continuous smooth operation of services.

Environmental sustainability: Environmental sustainability is about ensuring that IT services provide customer and business value while using the Earth's resources at a rate that ensures replenishment. Organizations must ensure that their IT services are designed to be environmentally sustainable, meeting the existing needs without compromising the ability of future generations (Senge, 2008).

In today's world, sustainability is a critical aspect that must be considered in every business practice, and IT services are no exception. Sustainable IT services must be evaluated not only from a business perspective but also from a societal value perspective. The value of these services goes beyond economic gains to include environmental and social responsibility requirements that benefit society as a whole. By systematically integrating and aligning individual IT service components, we can create services that deliver superior societal value, thus ensuring that we meet our sustainability goals. Therefore, all aspects of IT services must align with the sustainability goals of society (Savitz, 2013).

Despite the pressing need for strategies to ensure the environmental sustainability of IT services, there is no established literature on effective practices. Companies have taken an incremental approach to sustainability through "greener IT". Sustainable IT services have a broader focus on the role of IT in society, driven by corporate social responsibility (CSR) . Prioritizing sustainable IT helps companies fulfill their CSR goals while actively contributing to preserving the planet. Sustainable IT has been a crucial focus for IT organizations in the past decade as the cost of power for data centers has soared rapidly. The first wave of sustainable IT initiatives has primarily concentrated on boosting data center efficiency, with infrastructure, power and workload management, thermal management, product design, virtualization, and cloud computing strategies at the forefront of both strategic and tactical focus. The era of sustainable IT services presents not only a challenge but a strategic opportunity for organizations to fortify their positions in the market while making meaningful contributions to society and the environment. Embrace sustainability, align IT with business strategy, and unlock the full spectrum of benefits that (Butner, 2008) envisions for forward-thinking businesses.

SUSTAINABLE SOLUTION

Adopting a multifaceted approach is vital for promoting eco-friendly and sustainable practices. Three key strategies for achieving meaningful change include industry collaborations and standards, green certification programs, and public awareness and education campaigns.

1. **Industry Collaborations and Standards:** Collaboration and coordination among industries are essential to address the complex and urgent sustainability challenges. Collaborative efforts and coordinated actions among industries are crucial to addressing sustainability challenges effectively. Businesses should establish and adhere to industry-specific sustainability standards, focusing on issues like energy efficiency, waste reduction, and ethical resource procurement. By sharing innovative ideas and best practices, businesses can learn from one another, improving sustainability performance and minimizing environmental impact.

2. **Green Certification Programs:** In today's world, going green is more important than ever. Green certification programs offer businesses a structured framework to implement sustainable practices and demonstrate their commitment to sustainability. These programs guide companies in adopting eco-friendly measures, such as water conservation, energy use, waste management, and carbon footprint reduction, resulting in operational efficiency and cost savings. Green certification programs offer a structured framework for businesses to implement sustainable practices. These programs guide companies in adopting eco-friendly measures, symbolizing a commitment to sustainability and facilitating cost savings through operational efficiency (Butt, 2020). Certification covers various aspects, including water conservation, energy use, waste management, and carbon footprint reduction. Beyond showcasing commitment, certified companies enhance operational efficiency, reduce costs, and gain a competitive edge by attracting clients and investors valuing environmentally conscious initiatives.

3. **Public Awareness and Education Campaigns:** Creating a sustainable future requires raising awareness and familiarity with environmental challenges and their possible solutions. To inspire action for the environment, we can employ various educational and communication strategies such as media campaigns, community workshops, and school curriculums. By gaining knowledge about

the origins, effects, and innovations of environmental challenges, as well as the possibilities and innovations for sustainability, people can make more informed and ethical decisions in their everyday lives. This, in turn, can encourage support for companies and laws that share environmental values and promote sustainability.

FUTURE DIRECTION

Pursuing sustainable computing has led to the development of innovative AI-powered solutions expected to revolutionize the IT industry. AI-enhanced circular IT ecosystems help create equipment that can be updated, repaired, and recycled using predictive algorithms, hence fostering circular economies that reduce electronic waste. Eco-blockchain for Sustainable Computers combines the transparency and immutability of blockchain technology with AI forecasts of the environmental effects of computer components, permitting customers to make knowledgeable and environmentally responsible purchase decisions. AI-optimized energy harvesting encourages using renewable energy sources, and autonomous energy management in small-scale computer technologies reduces reliance on non-renewable electricity. Virtual Green Computing Assistants provide users with real-time guidance on energy-efficient practices. Eco-Friendly IT Procurement AI predicts a company's most environmentally friendly IT choices, considering factors such as energy efficiency, recyclability, and supplier environmental policies. These AI-driven technologies represent a shift towards environmentally conscious computing practices that embrace environmental responsibility throughout the IT lifecycle.

CONCLUSION

Pursuing sustainable computing has led to the development of innovative AI-powered solutions expected to revolutionize the IT industry. AI-enhanced circular IT ecosystems help create equipment that can be updated, repaired, and recycled using predictive algorithms, hence fostering circular economies that reduce electronic waste. Eco-blockchain for Sustainable Computers combines the transparency and immutability of blockchain technology with AI forecasts of the environmental effects of computer components, permitting customers to make knowledgeable and environmentally responsible purchase decisions. AI-optimized energy harvesting encourages using renewable energy sources, and autonomous energy management in small-scale computer technologies reduces reliance on non-renewable electricity. Virtual Green Computing Assistants provide users with real-time guidance on energy-efficient practices. Eco-Friendly IT Procurement AI predicts a company's most environmentally friendly IT choices, considering factors such as energy efficiency, recyclability, and supplier environmental policies. These AI-driven technologies represent a shift towards environmentally conscious computing practices that embrace environmental responsibility throughout the IT lifecycle. The introduction to green computing and the design of sustainable solutions underscore the critical need for environmentally conscious practices in the field of technology. Green computing is essential to promoting environmentally conscious practices in technology. It emphasizes designing and implementing sustainable solutions that reduce energy consumption, minimize electronic waste, and promote overall environmental responsibility. By incorporating eco-design principles, life cycle assessment, and resource efficiency, we can create technologies that contribute to a sustainable and resilient future. Implementing energy-efficient technologies and practices not only reduces the carbon footprint but also contributes to cost savings for businesses. Sustainable design methodologies provide a framework for the development and deployment of environmentally friendly IoT solutions that encompass the entire lifecycle of IoT devices.

Putting these green ideas into practice requires a concerted effort from various stakeholders, including businesses, policymakers, and consumers. Companies at the forefront of technological innovation are actively seeking and implementing sustainable solutions, including energy-efficient data centers, renewable energy sources, and IoT devices designed with a focus on environmental responsibility. Policymakers play a crucial role in incentivizing green practices through regulations and providing a supportive framework for sustainable technology development. The search for Green IT solutions is paramount in ensuring a sustainable and ecologically responsible future. As we navigate the complexities of a connected world, embracing and advancing Green IT is not just a choice but a necessity for a harmonious coexistence of technology and the environment. By implementing sustainable practices, we can mitigate the negative impacts of IoT and create a more sustainable future for generations to come. This paper will delve into the challenges posed by IoT, define the characteristics of Green IT, explore sustainable design methodologies, and elucidate the process of implementing these green principles as long-term solutions to pressing environmental problems. As part of its effort to determine the most successful sustainable IoT concepts, it also provides a thorough and lucid analysis of numerous suggestions from various writers. The search for Green IT solutions is critical in ensuring a sustainable and ecologically responsible future in a world where information technology is a vital part of our lives, businesses, and marketing plans.

Integrating green computing principles into technological innovation requires collaboration among researchers, engineers, policymakers, and industry leaders. By doing so, we can create a sustainable and eco-friendly digital landscape that meets present needs without compromising future generations' ability to meet their own.

REFERENCES

Ahmad, R. a. (2019). *Green IoT—issues and challenges.* Academic Press.

Ahmed, M. B. (2021). A Decentralised Mechanism for Secure Cloud Computing Transactions. IEEE, 1-5.

Al-Turjman, F. S. (2019). *The green internet of things (g-iot).* Hindawi. doi:10.1155/2019/6059343

Al-Zamil, A. S., & Saudagar, A. K. J. (2020). Drivers and challenges of applying green computing for sustainable agriculture: A case study. *Sustainable Computing : Informatics and Systems, 28,* 100–264. doi:10.1016/j.suscom.2018.07.008

Almalki, F. A. (2021). Green IoT for eco-friendly and sustainable smart cities: Future directions and opportunities. *Mobile Networks and Applications,* 1–25.

Butner, K. a. (2008). Mastering carbon management: Balancing trade-offs to optimize supply chain efficiencies. IBM Global Business Services. *IBM Institute for Business Value.*

Butt, S. A. (2020). Green Computing: Sustainable Design and Technologies. IEEE, 1-7.

Gezer, V. W. (2021). Real-Time Edge Framework (RTEF): Decentralized Decision Making for Offloading. IEEE, 1-6.

Guo, L. X., Xu, Liu, & Wang. (2019). Understanding firm performance on green sustainable practices through managers' ascribed responsibility and waste management: Green self-efficacy as moderator. *Sustainability (Basel), 11*(18), 49–76. doi:10.3390/su11184976

Hassan, N. G., Gillani, S., Ahmed, E., Yaqoob, I., & Imran, M. (2018). The role of edge computing in internet of things. *IEEE Communications Magazine, 56*(11), 110–115. doi:10.1109/MCOM.2018.1700906

Kaur, S. a. (2015). *Green Computing-Saving the environment with Intelligent use of computing.* Know Your CSI.

Lavanya, R. (2020). Green Scrum Model: Implementation of Scrum in Green and Sustainable Software Engineering. *International Research Journal of Engineering and Technology*, 1583-1587.

Li, W. Y., Yang, T., Delicato, F. C., Pires, P. F., Tari, Z., Khan, S. U., & Zomaya, A. Y. (2018). On enabling sustainable edge computing with renewable energy resources. *IEEE Communications Magazine, 56*(5), 94–101. doi:10.1109/MCOM.2018.1700888

Mmeah, S. a. (2018). Assessing the Influence of Green Computing Practices on Sustainable IT Services. *International Journal of Computer Applications Technology and Research*, 390-397.

Mory-Alvarado, A. M. (2023). *Green IT in small and medium-sized enterprises: A systematic literature review.* Elsevier.

Murugesan, S. (2008). *Harnessing green IT: Principles and practices.* IEEE.

Naim, A. (2021). Green Information Technologies in Business Operations. *Periodica Journal of Modern Philosophy, Social Sciences and Humanities*, 36-49.

Ojo, A. O., Raman, M., & Downe, A. G. (2019). Toward green computing practices: A Malaysian study of green belief and attitude among Information Technology professionals. *Journal of Cleaner Production, 224*, 246–255. doi:10.1016/j.jclepro.2019.03.237

Peri, G. L., Licciardi, G. R., Matera, N., Mazzeo, D., Cirrincione, L., & Scaccianoce, G. (2022). Disposal of green roofs: A contribution to identifying an "Allowed by legislation" end—of—life scenario and facilitating their environmental analysis. *Building and Environment, 226*, 109–739. doi:10.1016/j.buildenv.2022.109739

Podder, S. K. (2022). *Green computing practice in ICT-based methods: innovation in web-based learning and teaching technologies.* IGI Global.

Ryder, C. (2008). *Improving energy efficiency through application of infrastructure virtualization: introducing IBM WebSphere Virtual Enterprise.* The Sageza Group.

Savitz, A. (2013). *The triple bottom line: how today's best-run companies are achieving economic, social and environmental success-and how you can too.* John Wiley & Sons.

Senge, P. (2008). The necessary revolution: How individuals and organisations are working together to create a sustainable world. *Management Today*, 54–57.

Tewari, A. a. (2020). *Security, privacy and trust of different layers in Internet-of-Things (IoTs) framework.* Elsevier. doi:10.1016/j.future.2018.04.027

Tuysuz, M. F., & Trestian, R. (2020). From serendipity to sustainable green IoT: Technical, industrial and political perspective. *Computer Networks*, *182*, 107–469. doi:10.1016/j.comnet.2020.107469

Zhang, Q., Meng, Z., Hong, X., Zhan, Y., Liu, J., Dong, J., Bai, T., Niu, J., & Deen, M. J. (2021). A survey on data center cooling systems: Technology, power consumption modeling and control strategy optimization. *Journal of Systems Architecture*, *119*, 102253. doi:10.1016/j.sysarc.2021.102253

Chapter 3
Navigating Green Computing Challenges and Strategies for Sustainable Solutions

J. Jeyaranjani

SRM Madurai College for Engineering and Technology, India

K. Rangaswamy

Rajeev Gandhi Memorial College of Engineering and Technology (Autonomous), India

A. Ashwitha

(iD) https://orcid.org/0009-0002-9264-2328

Manipal Institute of Technology Bengaluru, Manipal Academy of Higher Education, India

Ramakrishna Gandi

Madanapalle Institute of Technology and Science, India

R. Roopa

(iD) https://orcid.org/0009-0002-3640-2645

Madanapalle Institute of Technology and Science, India

P. Anjaiah

Institute of Aeronautical Engineering (Autonomous), India

ABSTRACT

The adoption of green computing is crucial due to cost reduction and environmental responsibility. However, challenges hinder progress. This research explores solutions for balancing industrial growth with sustainability and addresses green computing obstacles. Recognizing industry's importance while mitigating its environmental impact is vital. Strategies to reduce waste and energy use, like cloud computing and virtualization, can help. Implementing a circular economy approach makes products regenerative and less wasteful. This research provides real-world case studies and insights, aiding businesses in adopting greener, more sustainable computing practices for a cost-effective, environmentally responsible tech landscape.

DOI: 10.4018/979-8-3693-1552-1.ch003

INTRODUCTION

Green computing and Green IT have become more well-known in recent years. They describe the environmentally responsible design and use of computer hardware, information technology, and communication technologies. This includes adopting energy-efficient servers, peripherals, and central processing units, as well as encouraging resource conservation and appropriate e-waste disposal. Modern technology and gadgets are all around us as if they were essential components of our daily existence, and their effects on the environment are something we cannot overlook. The scope of IoT is broad, ranging from everyday appliances to industrial machinery and environmental sensors.

On the other hand, green computing is a practice that focuses on environmentally conscious methods within the Information Technology realm. Its relevance has increased due to the rapid expansion of cloud computing and the associated energy costs. By utilizing energy-efficient hardware, optimizing data centers, and promoting efficient renewable energy adoption to power I.T. infrastructure, Green computing seeks to make I.T. systems and operations more energy-efficient and eco-friendly (Alsharif, 2023). Promoting an ecologically conscious and sustainable attitude towards technology is crucial. This can be achieved by reducing energy usage and minimizing technological waste. Cloud computing has significantly increased the cost of operating I.T. infrastructure. In response to this problem, green computing was created. Today, there is growing concern about the relationship between energy use and carbon emissions and the need to reduce both. The expansion of data centers has led to a significant increase in energy consumption, which has had detrimental effects on the environment. Enterprise data centers are responsible for more than 50% of a company's energy costs and roughly half its carbon footprint (Foley, 2007; Kumar, 2014).

Growing environmental concerns have brought considerable attention to green computing and sustainable design approaches in modern years (Jänick, 2012). This study explores the connection between these two concepts, highlighting the challenges and possibilities they provide for creating a more sustainable future. With the industry's rapid growth, energy consumption and electronic waste have become critical environmental concerns (Xu M. a., 2019). We need to find innovative strategies to mitigate the environmental impacts of information technology. This study examines ways to maximize energy efficiency, reduce electronic waste, and promote responsible technology usage. It also scrutinizes the tenets of sustainable design, emphasizing the need for environmentally friendly products that consider the entire life cycle. To help organizations, politicians, and consumers make well-informed decisions regarding technology usage and design, this study addresses these issues and promotes sustainable design practices. The journey towards a more sustainable and eco-friendly digital future begins with an understanding of these challenges and a commitment to responsible technology usage and design.

Using some of the most well-known academic research as a starting point, this article provides an extensive overview of the key ideas behind green computing. It is challenging to implement all of the suggested solutions at once since many of the ones made in the published literature to address the problems with IoT are similar. The publication urges researchers to reassess previous research and pinpoint problems and gaps that might be resolved in the next proposals. Further research is needed on Green implementation methods for corporations to manage Green Use effectively.

OVERVIEW OF GREEN COMPUTING

The concept of green computing was presented by the U.S. Environmental Protection Agency in 1992 by implementing an Energy Star rating for monitors and other electronic equipment. Green computing has become increasingly important, and there is a need for more eco-friendly computing solutions. According to the Greenpeace Guide to Greener Electronics published in May 2010, only Nokia and Sony Ericsson obtained a reasonably green rating from 18 P.C. and other electronics manufacturers listed. The heavy reliance on Information Technology in virtually every commercial organization of any size has resulted in a rise in technology-related power consumption. While individuals can take steps such as ensuring their P.C.s are not unnecessarily switched on, organizations have a massive scope for affecting energy use, recycling, and profit by adopting a green approach to I.T. With rising electricity prices, businesses are increasingly keen to make energy and cost savings.

Furthermore, public and corporate opinion is shifting favor of environmental responsibility, especially with climate change becoming a political and regulatory reality. Implementing relatively straightforward practical measures with existing hardware can result in real environmental savings in computing. For instance, powering down P.C.s while they are not in use can lead to savings of 1500-4500 Rs. per P.C., and if each of the 100 companies in a city has an average of 100 computers, they can save 1,50,000 – 4,50,0000 per year. I.T. management needs to understand the basic concepts of green computing to make investments that can be improved through this approach. The solution falls into three general categories:

- Enhancing energy efficiency by decreasing carbon footprint
- Reducing e-waste
- Helping lifestyle changes that lower effect on the environment

Figure 1. The cycle of green manufacturing

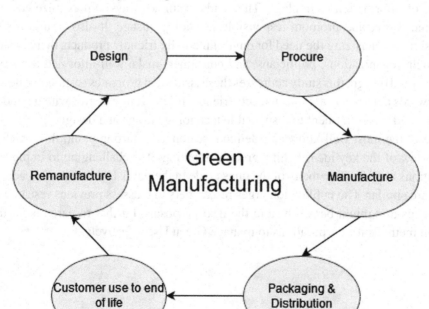

Green computing is an essential aspect of environmental science that enables cost-effective solutions for conserving natural resources and the environment. Its key components include designing, manufacturing, and using computers and their sources more efficiently to minimize environmental impact (Sagar, 2021). The immediate goal of green computing is to control power and energy efficiency, using eco-friendly hardware and software devices and recyclable materials to enhance product life.

Nowadays, energy-efficient technologies such as star management are employed to reduce energy consumption waste. To achieve green computing, various steps need to be taken into consideration. These steps include designing, manufacturing, usage, and disposal of components of computers, which have both hardware and software parts. Green usage involves minimizing the energy consumption of computers and information systems and using them in an environmentally friendly manner. Green disposal involves reusing old computers and recycling rejected computers and electronic devices.

Green design involves designing sound components that are more energy-efficient and environment-friendly, such as computers, servers, cooling equipment, and data centers. Green manufacturing involves developing electronic components, computers, and other related subsystems with minimum environmental impact. The main practice in green computing is the thorough use of computing and I.T. resources more efficiently. It is the responsibility of human beings to protect the environment and save energy costs. For the sustainability of computing and information technology, green computing and green information technology are used for analysis. Green computing requires reducing the use of hazardous devices, increasing energy efficiency, and promoting the reusability of computing devices and information technology waste. Green computing is the only way to implement constructive policies for the future, including environmental sustainability, energy efficiency commercialization, and total cost estimation of disposal and recycling of computing and I.T. devices.

ENVIRONMENTAL AND INDUSTRIAL ISSUES

Industrial and environmental solutions consist of green computing, and those terms and issues are discussed.

Environmental Impact

E-waste improperly disposed of has the potential to cause severe environmental damage. Lead, cadmium, mercury, and brominated are among the hazardous materials commonly detected in electronic garbage. These substances can pollute soil and water, posing serious threats to human health and the environment.

Global Concerns

The global consumption of electronic devices has led to increased electronic waste (e-waste) production. This trend is particularly noticeable in developed countries, which often export electronic waste to underdeveloped nations. Unfortunately, this practice has far-reaching consequences, exposing workers in informal recycling operations within these recipient nations to significant health risks. The disposal and processing of electronic waste in such environments can lead to many adverse health effects, creating a pressing need for comprehensive solutions to address the global implications of this burgeoning issue.

Figure 2. E-waste in environment

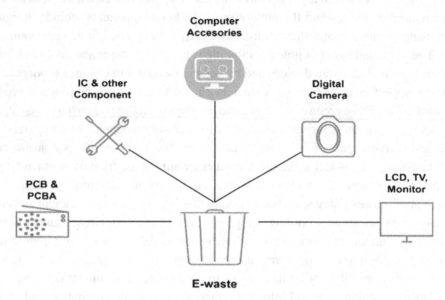

Resource Depletion

Valuable resources like copper, silver and precious metals like gold are used to create electrical equipment. Unfortunately, these priceless resources are lost when electronic garbage, or "e-waste," is not properly recycled. This adds greatly to the depletion of resources and the waste of precious materials. This further strains the finite supply of natural resources by necessitating the continuous extraction and refinement of new minerals. One effective and sustainable solution to this problem is the intentional use of recycling processes. Recycling in the context of e-waste is the deliberate removal and retrieval of functional components from outdated electronic equipment.

There is less demand for newly mined resources when reclaimed materials may be used again to create new goods. It is essential to properly and legally dispose of e-waste by local laws when recycling is not an option. This entails abiding by the guidelines established for properly disposing of electronic waste. Individuals can use approved e-waste disposal businesses or approved e-waste collection locations. By forming these responsible behaviors, society may contribute to conserving finite resources and developing a more robust electronic environment.

Waste Volume

Since technology is always developing and our reliance on electronic gadgets in our everyday lives is rising, the issue of electronic rubbish, or "e-waste," has become increasingly more concerning. Due to inadequate control over the ever-increasing volume of electronic trash that is being dumped, this global issue has sparked worries for both public health and the environment. The amazing 53.6 million metric tons of e-waste created worldwide in 2019 were reported by the Global E-waste Monitor 2020. With just 17.4% of this enormous volume of e-waste being collected and processed through appropriate recycling procedures, the problem is made worse by the stark variations in waste management policies.

The remainder is frequently disposed of incorrectly, ending up in landfills with negative environmental repercussions. Regretfully, if corrective action is delayed, it is predicted that the annual production of e-waste might reach a frightening 74 million metric tons by 2030 (Singh, 2020). This growing tendency emphasizes the necessity of all-encompassing policies that address every stage of the lifetime of electronic devices, from design and production to end-of-life disposal. For the present and future generations, a coordinated worldwide effort is required to lessen the mounting risks posed by the expanding amount of e-waste, safeguard the environment, and encourage sustainable practices.

Industrial Impact

Data centers are responsible for a significant amount of carbon emissions generated by service companies. This poses a significant environmental challenge for information-intensive organizations that rely heavily on data centers, as they are responsible for more than half of their carbon footprint. The Green Computing project aims to address and reduce the environmental impact of data center operations, which have negative effects and are prompted by rising energy prices. This initiative aims to reduce the energy requirements and carbon footprint of data centers.

Energy Expenses: Data centers functioning largely depend on the infrastructure supporting electricity and cooling. However, the associated costs of powering and cooling the data centers often surpass that of the computing equipment itself, significantly impacting the total cost of ownership. Assuming an electricity cost of $0.08 per kilowatt-hour, a $4,000 server with a 500-watt power rating would use approximately $4,000 for power and cooling over three years (Baldwin, 2007). Energy consumption expenses in Japan are considerably higher, almost twice that of other countries. The proportion of power and cooling expenses, compared to equipment costs, has been increasing steadily. From a ratio of 0.1 to 1 in 2000, it reached a 1-to-1 ratio in 2007 (Mmeah, 2018). As the number of data centers and servers increases, the financial burden associated with energy consumption for power and cooling is projected to continue upward, compounded by the introduction of carbon cap-and-trade schemes.

Lower Server Utilization: Data center operations need to improve their efficiency, namely low server utilization rates, which results in significant energy consumption challenges. Studies have shown that many large data centers achieve only a 5% to 10% server utilization rate, which indicates that a vast amount of computing infrastructure still needs to be utilized (Foley, Google in Oregon: Mother Nature meets the data center, 2007). Low server utilization has two consequences. It first results in financial inefficiencies as businesses spend more on electricity, upkeep, and operational support, even if they only utilize a small portion of their server capacity. Secondly, it leads to unnecessary energy consumption from an environmental perspective, contributing to a larger carbon footprint.

Sustainable Materials: Ensuring environmental responsibility in developing I.T. equipment through incorporating sustainable materials is a significant challenge. Because of their complexity and the need to be environmentally friendly, these gadgets require careful investigation of substitutes for traditional, sometimes dangerous materials. In this context, sustainability entails a thorough search for materials that satisfy I.T. components' performance and durability requirements and strict environmental norms.

Figure 3. Green industry

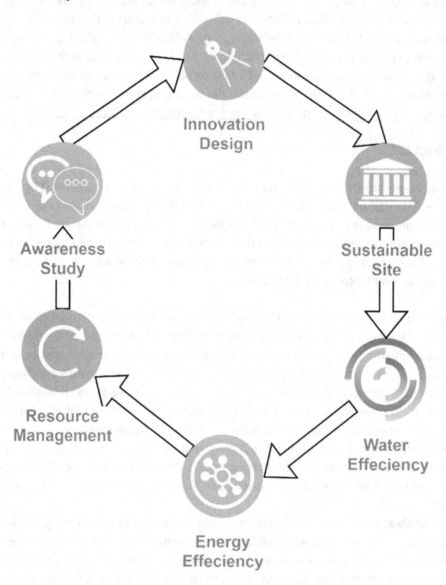

The goal is to create a closed-loop system that enables appropriate disposal and reuse of computing equipment at the end of their lives, in addition to minimizing the environmental effect of these devices while they are in use. A sophisticated strategy is needed to search for sustainable materials for I.T. devices, considering aspects like extraction, manufacturing procedures, and component end-of-life concerns. This entails exploring the complexities of waste management, supply chain logistics, and material science. Furthermore, industry cooperation is required to create guidelines and procedures that support the widespread use of sustainable materials. The difficulty grows as the need for I.T. equipment rises worldwide. It becomes critical to balance cost-effectiveness, technical innovation, and environmental stewardship. Businesses that manufacture I.T. equipment must constantly navigate this complex environment in search of creative answers that satisfy consumers' increasing need for cutting-edge technology while upholding sustainability standards.

Regulatory and Policy Hurdles

Numerous legislative and policy barriers might prevent the broad acceptance and use of green computing. These difficulties result from the technological sector's complexity, environmental issues, and the need to balance sustainability with development. The following are some of the major policy and regulatory barriers:

Absence of Established Regulations: There is a rule gap that may direct and encourage industries and enterprises to adopt eco-friendly computing methods. Through establishing strong legal frameworks, policymakers may offer a well-defined plan for implementing green computing concepts. These frameworks can include a broad range of initiatives, such as energy-efficiency requirements, rules for responsibly disposing of electronic trash, and financial incentives for creating and applying sustainable technology. Including green computing in regulatory frameworks may do more than ensure compliance; it can spur innovation and change the industry. Legislative frameworks can accelerate the development of energy-efficient technology and environmentally friendly computer solutions. They may also level the playing field by ensuring that companies that follow green computing guidelines do not have an unfair edge over others trailing behind in sustainability initiatives.

The need for defined rules in green computing is a demand for a controlled and encouraging atmosphere that encourages a group commitment to environmental responsibility. Regulatory frameworks may guide the business toward a more environmentally sound future by setting clear objectives and providing incentives for sustainable behavior. As a result, a green computing regulatory environment puts sustainable technology practices at the forefront of industrial development and innovation while addressing the present lack of rules.

Industry Resistance: The implementation of green computing regulations faces significant obstacles due to industrial reluctance, a complex problem based on possible interruptions, increased prices, and technology limitations. Persuading industrial stakeholders of the need for environmentally friendly practices is a complex process beyond simple economic concerns. The fear of operational disruptions is a common source of resistance to adopting green computing policies since they can potentially upset established workflows and procedures. This anxiety calls for a thorough plan that tackles the pressing issues and provides a smooth transition schedule to minimize business interruptions. Cost factors can contribute to industry opposition since they may need larger initial financial outlays when adopting environmentally sustainable methods. It is also crucial to present a strong business case that goes beyond the short-term expenses and emphasizes the long-term advantages of green computing: reduced resource consumption, enhanced operational resilience, and enhanced brand perception. Technological limitations add another difficulty level since they may need significant changes to current systems and infrastructures to comply with green computing guidelines. A comprehensive strategy that considers compatibility concerns, technology advancements, and potential limits in present capabilities is necessary to persuade stakeholders. Providing a plan for the progressive integration of green technology without sacrificing essential operational features is part of this. Overcoming industrial opposition requires a communication strategy beyond operational and financial concerns. It is important to raise stakeholders' awareness of the wider environmental and social consequences of their activities, highlighting the industry's role in promoting sustainable practices and the ecological well-being of the world.

INDUSTRIAL IOT

Industry 4.0 represents a significant shift away from traditional hierarchical corporate organizational structures. It involves the integration of industrial technologies with information and communication technologies (ICT) to enable a seamless transition between the virtual and physical worlds. The combination of embedded systems, the Internet of Things (IoT), and cyber-physical systems (CPS) forms the basis of Industry 4.0. It enables "smart" production, the paradigm change of which Tabaa (2020) speaks. One of the key features of Industry 4.0 is smart factories, which prioritize efficient use of manufacturing resources to deliver customized goods faster to market, lower costs per unit, and boost total profitability. An age of digital and industrial technology optimization through process optimization has begun with this, ushering in a new phase of industrial change. A global introduction of the concept of Industry 4.0 took place during the 2011 Hanover Industrial Technology Fair. Within the framework of its "High-Tech Strategy 2020" action plan, Germany was a leader in formally adopting Industry 4.0 in 2013 (Feng, 2020). The German project's four main goals are to create a Cyber-Physical Systems network, address topics related to intelligent manufacturing and plants, and accomplish three different forms of integration: point-to-point, vertical, and horizontal. This approach has gained significant traction in the German sector, and businesses are actively engaged in this revolutionary period. As the concept of Industry 4.0 gained acceptance, other nations began implementing comparable policies.

The strategy aims to improve China's industrial status globally by integrating industrial and digital technology. The idea of Industry 4.0 has been adopted by several other countries, such as the United States and France, indicating a worldwide trend toward a more technologically advanced and networked industrial environment. Implementing Industry 4.0 offers several advantages for businesses embracing this revolutionary movement. Intelligent procedures powered by machines that can analyze data make predictive maintenance scheduling possible, maximizing equipment performance and uptime. Industry 4.0 also revolutionizes supply chain management by improving inventory control, logistical monitoring, and traceability. This optimization reduces inefficiencies and increases overall productivity by ensuring a more responsive and efficient supply chain. Industry 4.0's intrinsic digitization seeks to eliminate paper-based procedures. Businesses may lower mistakes, boost operational effectiveness, and enhance compliance by digitizing mill activities and coordinating them in real time with scheduled processes.

GIIoT: The smart grid is a complex system often described as a network of subsystems due to the diverse range of infrastructure it encompasses. This infrastructure combines energy and communication technologies, forming a heterogeneous network. The smart grid's intricate architecture comprises several components, including energy control, optimization systems, and distributed power generation, which adds to its complexity. The smart grid's nodes are highly compatible with the IoT ecosystem's tenets. This makes it ideal for energy management and the ubiquitous impact of ICT applications. As a result, these components come together to form the concept of the Internet of Green Devices, in which networked gadgets work together to promote ecologically responsible energy use (Narciso, 2020) (Saleem, 2019) (Fentis, 2019).

The Industrial Internet is a concept that General Electric first presented. It links sophisticated physical machinery with networked sensors and Software. This concept combines machine learning, big data, IoT, and machine-to-machine (M2M) communication to gather and handle machine data. It then leverages this data to improve operational changes (Zhu, 2015) (Huang, 2014). The Industrial Internet Consortium (IIC) defines the Industrial Internet as a network that connects human workers, machines, and intelligent gadgets to enable sophisticated analytics for business results that can significantly revolutionize industries. Unlike the consumer-centric orientation of the Internet of Things, the Industrial Internet employs "Internet thinking" in industrial settings to address the nonconsumer aspects of the IoT (Dharfizi, 2018). It centers on three key components that capture the spirit of the idea:

RENEWABLE ENERGY ADAPTION

The integration and utilization of renewable energy sources, including geothermal, biomass, hydroelectric, solar, and wind power, to satisfy energy demands is known as renewable energy adaption. A broad acceptance and use of appropriate technology and behaviors are required to use these clean energy sources effectively and sustainably.

Utilizing computer resources sensibly and effectively to reduce technology's negative effects on the environment is known as "green computing." It entails the ecologically sustainable design, production, usage, and disposal of computers and other electronic equipment. Reducing the carbon footprint of the I.T. business requires integrating renewable energy into the field of green computing.

The goals of green computing and renewable energy adoption are similar in that they seek to lessen the negative effects of information technology and the energy sources that fuel it on the environment. The paper proposes (Xu M. T., 2020) to minimize carbon emissions. The article suggests a self-adaptive method for controlling resource consumption in cloud data centers and optimizing the use of renewable energy. The suggested method has been implemented in a working prototype system and assessed using real-world web services. The findings indicate a 10% increase in the use of renewable energy sources and a 21% decrease in the use of brown energy.

ZERO-WASTE AND THE CIRCULAR ECONOMY

In green computing, circular economy and e-waste management are crucial in promoting sustainable development and safeguarding the environment.

- **Circular Economy**

The circular economy is an economic model that prioritizes resource sustainability via waste reduction and increased material and product reuse and recycling. A circular economy approach to green computing promotes the long-term and robust design of I.T. systems and electrical gadgets. This increases a product's lifespan and lessens the need for frequent replacement. To ensure that parts and devices may be utilized for longer periods and reduce the development of electronic waste, circular economy concepts also encourage the repair and refurbishing of electronic equipment. (Demestichas, 2020)

Table 1. Renewable energy strategy

1. Energy efficient hardware	• With no moving parts, SSDs utilize less energy than conventional hard disc drives (HDDs). They help increase system responsiveness overall, speed up startup times, and produce less heat and electricity. • Power supplies are necessary to transform A.C. electricity into the D.C. power that computer components need. Computer power consumption may be decreased overall, and energy waste can be minimized using a high power factor correction (PFC) efficient power supply. • Through the integration of these energy-efficient hardware components into computer systems, people and institutions may minimize their energy usage, cut down on power expenses, and help promote a more environmentally friendly computing approach.
2. On-Site Renewable Energy	• Solar energy is crucial to green computing. This method produces relatively little heat or other gas emissions, particularly carbon Dioxide. Large panels in solar computing employ cells. After installation, solar cells require extremely minimal maintenance during their lifetime. For several years, there have been no more expenses. There are a ton of really dependable, silent, efficient, and completely non-polluting solar-powered gadgets on the market (Anwar, 2013). On-site renewable energy provides long-term financial savings and energy security in addition to lessening the environmental effect of data centers. Encouraging green computing practices and attaining sustainability goals need this measure
3. Biomass Energy	Biomass energy is a sustainable source that may create biofuels, heat, and electricity from organic materials. A vast variety of materials may be included in biomass, including wood, crop waste, algae, agricultural wastes, and even municipal solid trash. Because these organic components can be constantly created or regrown, they are regarded as renewable, making biomass a sustainable energy source. • Organic waste, sewage, and manure may all be broken down anaerobically to create biogas. A renewable natural gas supply for heating, power generation, or transportation fuel may be obtained by capturing the methane gas released during this operation.
4. Virtualization and Cloud Computing	Virtualization and cloud computing are effective instruments in the quest for green computing because they provide diverse approaches to maximize resource usage, reduce energy use, and lessen environmental effects. • Newly developed energy-efficient storage can replace the cloud's current storage. Since a data center's lifespan is just nine years, engineers can utilize energy-efficient memory, such as solid-state storage, while rebuilding an old data center. Unlike hard disc drives, solid-state storage has no moving mechanical parts. As a result, a solid-state needs less cooling than a hard disc drive, hence using less energy for cooling. (Jain, 2013) Firms may adjust their computer resources to real demand thanks to the scalability and flexibility of virtualization and cloud computing technologies, which lowers energy waste and encourages green computing practices. Businesses that care about the environment may find cloud providers appealing since they also aggressively try to match their services with sustainability objectives.

- **E-Waste Management**

When electronic or electrical equipment is discarded, it is called "e-waste" or "electronic waste." Due to the rapid development of technology and the short lifespan of electronic devices, e-waste has become a significant environmental issue. If not managed properly, e-waste can have harmful effects on both the environment and public health. Thus, it is crucial to ensure appropriate disposal and treatment to minimize the impact of e-waste. Careful management is necessary to eliminate the detrimental effects of e-waste on the environment and public health. The management techniques involve recycling, repairing, reusing, and disposing of electronic equipment by local laws. International projects and legislation have been established to address the issue of e-waste management and promote sustainable habits. The increasing use of electronic devices and the rapid development of technology have made the volume of e-waste a significant concern globally. The improper management of the resulting large volume of e-waste has created environmental and human health hazards. According to the Global E-waste Monitor 2020, around 53.6 million metric tons of e-waste were produced worldwide in 2019, but only 17.4% was collected and recycled. The research predicts that if proper measures are not taken, the amount of e-waste generated will reach 74 million metric tons by 2030. (Debnath, 2016), (Mukta, Review on E-waste management strategies for implementing Green Computing, 2020) (Mukta, E-Waste Management Strategies for Implementing Green Computing, 2021)

- **Efficiency of Resources:** Green computing's circular economy techniques reduce the need to extract raw materials by reusing and recycling electronic components, which helps conserve significant resources.
- **Cost Reductions:** Less energy and material usage and lower waste disposal costs might result from using circular economy concepts.
- **Ecological Innovation:** As producers look for methods to reduce waste and boost resource efficiency, the emphasis on the circular economy and e-waste management spurs creativity in designing more environmentally friendly and sustainable electronic devices.
- **Observance of Rules:** Manufacturers must follow proper disposal and recycling techniques due to the several nations and areas that have implemented rules about the management of e-waste.

IMPLEMENTATION OF GREEN INTERNET OF THINGS

Green IT is essential in our efforts to protect the environment. Green IT seeks to protect resources, lessen carbon impact, and promote energy-efficient technology use. (Alsamhi, 2019) et al. have proposed implementing "Green IoT" strategies to reduce carbon emissions and improve energy efficiency. The term "green" IoT describes environmentally friendly technologies that collect, store, access and regulate data through services and storage. These sustainable methods will help us design, produce, and dispose of computing devices, servers, and related subsystems more effectively and with less adverse environmental effects. (Alsamhi, 2019) et al. have discussed several technologies, including Green Radio-Frequency Identification (GRFID), Green Machine-to-Machine (GM2M), Green Wireless Sensor Network (GWSN), Green Data Center (GDC), and Green Cloud Computing (GCC), which minimize environmental stress. IoT objects may all be uniquely identified thanks to the complimentary technologies of WSN and RFID. Applications that monitor people, things, or products in real-time employ RFID, an adaptable and affordable technology. According to (Al-Turjman, 2019), this wireless technology is rapidly expanding and offers several advantages because of its capabilities. Using novel materials to manufacture biodegradable RFID tags is a green approach that eliminates negative environmental impacts. Tracking and gathering environmental data, weather, rainfall, emission levels, farming conditions, and identifying fire falls to wireless sensor networks or WLANs. After that, a substantial quantity of electricity and power are used to transmit this data to the Base Station (B.S.). Adopting Green WSN practices, such as shifting sensor nodes into sleeping mode to save energy, using ultra-low power data communication, and harvesting energy from natural sources, will reduce energy consumption pollution and enhance the lifespan of sensor node performance.

It is imperative to move toward a greener future for the sake of the ecosystem and long-term advantages, including lower energy use, pollution, and non-renewable and raw materials. (Maksimovic, 2017) emphasizes the role of ICT in addressing environmental issues. Developing a "green and sustainable place for living" depends heavily on green IoT. By connecting billions of gadgets, automobiles, and infrastructure, we can create a system that will lower energy, water, and carbon emissions. Combining a planned device architecture with big data analytics allows the Internet of Things (IoT) to be implemented with greater safety, efficiency, and human well-being. Green IoT must be utilized to develop and manufacture eco-friendly products and technologies that reduce pollution and carbon emissions while increasing energy efficiency (Maksimovic, 2017).

The G-IoT aims to integrate the most important and advanced Green ICTs, including biometrics, M2M communication, cognitive radio (C.R.), cellular networks, RFID, WSNs, and energy harvesting devices. These technologies must be integrated into green designs to support green communication, processing, consumption, and disposal. To realize the potential and significance of the Internet of Things (Maksimovic, 2017) delineates the subsequent essential actions that must be taken:

1) Using bioproducts and environmentally friendly designs while producing G-IoT components.
2) Using designs to save operating expenses and energy use.
3) Turning off or enabling sleep scheduling algorithms to improve data center, cooling, and power supply efficiency.
4) Using green, renewable energy sources such as geothermal, wind, solar, and oxygen.
5) Sending information only when required.
6) Reducing the wireless data channel's size and the data path's length.
7) Modulating communication-related processing (compressive sensing).
8) Lastly, using cutting-edge communication strategies like MIMO.

Figure 4. Green internet of things

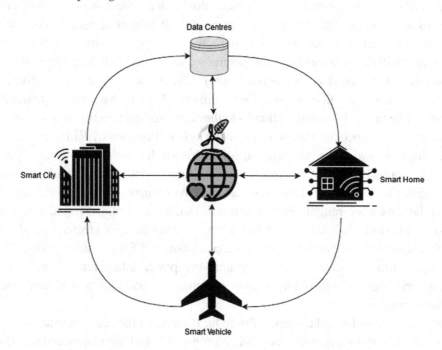

These characteristics provide efficient QoS and advantages of G-IoT over IoT. Reduced waste and greenhouse gas emissions, increased energy efficiency, and little to no negative effects on human health or the environment characterize the benefits of G-IoT. We can reduce our carbon footprint by adopting G-IoT and contribute to a sustainable future. So, let's embrace the potential of G-IoT and work towards a greener tomorrow.

Green IoT is a solution to address IoT's energy consumption issues. It offers significant benefits to the sustainability of environments, with energy efficiency being one of the most important factors to consider. The field of green computing and communications is concentrated on designing, manufacturing, disposing, and operating computers and associated components and subsystems such as monitors, printers, storage devices, and networking systems in a way that is both efficient and effective while having a minimal or no impact on the environment.

SUSTAINABLE IT SERVICE

Information technology (I.T.) services delivered environmentally conscious and resource-efficiently are collectively referred to as sustainable I.T. services or green computing. The objectives of sustainable I.T. services are to reduce the environmental effect of I.T. operations, conserve resources, and advance a more environmentally conscious and sustainable approach to technology. (Harmon, 2009), (Pazowski, 2015)

- **Sustainable Energy:** For I.T. services to be sustainable, switching to renewable energy sources like wind or solar power to power data centers and I.T. operations is essential. Energy consumption optimization is the main goal of sustainable I.T. services. It focuses on designing and utilizing energy-efficient hardware and Software. Power management techniques, low-power components, and energy-efficient CPU development are examples. This shift towards green energy would help reduce carbon emissions and promote a healthier environment. By adopting these renewable energy options, I.T. services can be a key player in fostering long-term environmental sustainability.
- **Develop Software in an Ecological Way:** Ecologically producing Software requires a thoughtful approach to program design and deployment, primarily aiming to reduce environmental effects. I.T. systems can reduce their computational needs by creating and utilizing energy-efficient software applications and optimizing code. Organizations can support a more ecologically conscious and environmentally friendly approach to technology by using sustainable software development standards, which helps to reduce the carbon footprint associated with I.T. systems.
- **Monitoring and Controlling Energy:** Energy consumption tracking and improvement possibilities can be found using energy monitoring and management systems. A crucial component of sustainable I.T. services is a reliable system for monitoring and managing energy usage. This entails implementing cutting-edge procedures and technologies to constantly track I.T. infrastructure energy use. Organizations can discover patterns in energy usage, pinpoint inefficient regions, and take proactive measures to resolve problems using real-time monitoring. By reducing waste and guaranteeing responsible resource usage, this systematic approach to energy management supports the larger objective of attaining sustainability in I.T. operations.
- **Training and Awareness for Employees:** Workers are encouraged to embrace green I.T. practices and are trained in energy-efficient computing techniques. These programs provide staff members with the knowledge and skills necessary to adopt green I.T. practices and incorporate

energy-efficient computing methods into their regular workdays. Companies can cultivate a culture of environmental responsibility through extensive training programs and promote awareness of sustainable I.T. principles. This, in turn, can lead to increased engagement in efforts to reduce the ecological footprint of information technology.

- **Optimizing Performance:** To assure efficiency and lower energy usage, I.T. systems and resources should be routinely assessed and optimized. This entails a systematic and continuous examination of hardware and software components to find possible bottlenecks, improve workflow, and boost overall efficiency. Performance optimization includes many things, such as optimizing hardware resources, enhancing data storage and retrieval systems, and fine-tuning algorithms. This continuous cycle of analysis and enhancement is crucial for sustaining an I.T. environment that is both high-performing and environmentally responsible, in line with the overarching objectives of operational efficiency.

Sustainable I.T. services may save businesses money and increase their operational effectiveness, in addition to helping the environment by lowering carbon emissions and resource usage. Using green I.T. practices has become a top focus for many companies and organizations as people become more conscious of the value of sustainability.

CONCLUSION

The escalating power expenses of data centers have caused I.T. businesses to focus more and more on. Sustainable I.T. during the last ten years. The definition of I.T.'s place in an organization's Corporate Social Responsibility (CSR) plan is at the center of the sustainable I.T. paradigm. Developing a thorough roadmap, setting baseline measurements, reengineering business processes, encouraging involvement, and radically changing company culture to accept new paradigms are all necessary for successfully navigating this terrain. Therefore, to successfully respond to the dynamic environment of emerging sustainability efforts, this shift towards sustainability necessitates considerable changes in I.T. governance and decision-making processes.

An insightful summary of the current discussion around sustainable I.T. is given in this article, which also emphasizes important elements to take into account while creating a sustainable strategy. The significance of further study delving deeply into the complex relationships between customer, corporate, and societal value is emphasized. A detailed understanding of the wider effect requires an understanding of how the use of sustainable I.T. solutions resonates across different dimensions. The study also emphasizes the necessity of investigating sustainable services that balance profit value and cost-effectiveness and are suited to the manufacturing and consumer sectors. Finding these sustainable service dynamics is essential to adjusting company plans to changing customer demands, making the business model more robust and sustainable in the long run. Creating a thorough model that directs the creation and use of sustainable solutions is crucial. This model should delicately weave together to match I.T. sustainability inside the company with more comprehensive enterprise-positioning initiatives. This comprehensive approach should address the critical issue of aligning I.T. practices with the broader sustainability ambitions across the commercial ecosystem. By integrating these aspects, businesses may effectively promote Sustainable I.T. while cultivating a comprehensive strategy that penetrates their whole operational and strategic framework, resulting in favorable environmental and societal effects.

REFERENCES

Ahmad, R. a. (2019). *Green IoT—issues and challenges.* Academic Press.

Al-Turjman, F. S. (2019). *The green internet of things (g-iot).* Hindawi. doi:10.1155/2019/6059343

Alsamhi, S. H., Ma, O., Ansari, M. S., & Meng, Q. (2019). Greening internet of things for greener and smarter cities: A survey and future prospects. *Telecommunication Systems, 72*(4), 609–632. doi:10.1007/s11235-019-00597-1

Alsharif, M. H., Jahid, A., Kelechi, A. H., & Kannadasan, R. (2023). Green IoT: A review and future research directions. *Symmetry, 15*(3), 757. doi:10.3390/sym15030757

Anwar, M. (2013). Green computing and energy consumption issues in the modern age. *IOSR Journal of Computer Engineering, 12*(6), 91–98. doi:10.9790/0661-1269198

Baldwin, E. a. (2007). *Managing it innovation for business value: Practical strategies for it and business managers.* Academic Press.

Debnath, B., Roychoudhuri, R., & Ghosh, S. K. (2016). E-waste management—A potential route to green computing. *Procedia Environmental Sciences, 35*, 669–675. doi:10.1016/j.proenv.2016.07.063

Demestichas, K., & Daskalakis, E. (2020). Information and communication technology solutions for the circular economy. *Sustainability (Basel), 12*(18), 7272. doi:10.3390/su12187272

Dharfizi, A. D. (2018). The Energy Sector and the Internet of Things Sustainable Consumption and Enhanced Security through Industrial Revolution 4.0. *Journal of International Students*, 99–117.

Feng, B., Sun, K., Chen, M., & Gao, T. (2020). The impact of core technological capabilities of high-tech industry on sustainable competitive advantage. *Sustainability (Basel), 12*(7), 2980. doi:10.3390/su12072980

Fentis, A. a. (2019). *Short-term nonlinear autoregressive photovoltaic power forecasting using statistical learning approaches and in-situ observations.* Academic Press.

Foley, J. (2007). *Google in Oregon: Mother Nature meets the data center.* InformationWeek's Google Weblog.

Harmon, R. R. (2009). Sustainable I.T. services: Assessing the impact of green computing practices. IEEE, 1707-1717.

Huang, J., Meng, Y., Gong, X., Liu, Y., & Duan, Q. (2014). A novel deployment scheme for green internet of things. *IEEE Internet of Things Journal, 1*(2), 196–205. doi:10.1109/JIOT.2014.2301819

Jain, A. a. (2013). Energy efficient computing-green cloud computing. IEEE, 978-982.

Jänick, M. (2012). Green growth: From a growing eco-industry to economic sustainability. *Energy Policy, 48*, 13–21. doi:10.1016/j.enpol.2012.04.045

Kaur, S. a. (2015). *Green Computing-Saving the environment with Intelligent use of computing*. Know Your CSI.

Kumar, T. V. (2014). Green computing-an eco friendly approach for energy efficiency and minimizing e-waste. *International Journal of Engineering Research*, 356-359.

Maksimovic, M. (2017). The role of green internet of things (G-IoT) and big data in making cities smarter, safer and more sustainable. *International Journal of Computing and Digital Systems*, 175-184.

Mmeah, S. a. (2018). Assessing the Influence of Green Computing Practices on Sustainable I.T. Services. *International Journal of Computer Applications Technology and Research*, 390-397.

Mory-Alvarado, A. M. (2023). *Green IT in small and medium-sized enterprises: A systematic literature review*. Elsevier.

Mukta, T. A. (2020). Review on E-waste management strategies for implementing green computing. *Int. J. Comput. Appl*, 45–52.

Mukta, T. A. (2021). *E-Waste Management Strategies for Implementing Green Computing*. Academic Press.

Murugesan, S. (2008). *Harnessing green I.T.: Principles and practices*. IEEE.

Naim, A. (2021). Green Information Technologies in Business Operations. *Periodica Journal of Modern Philosophy, Social Sciences and Humanities*, 36-49.

Narciso, D. A., & Martins, F. G. (2020). Application of machine learning tools for energy efficiency in industry: A review. *Energy Reports*, *6*, 1181–1199. doi:10.1016/j.egyr.2020.04.035

Pazowski, P. a. (2015). *Green computing: latest practices and technologies for ICT sustainability*. Academic Press.

Podder, S. K. (2022). *Green computing practice in ICT-based methods: innovation in web-based learning and teaching technologies*. IGI Global.

Sagar, S. a. (2021). A review: Recent trends in green computing. *Green Computing in Smart Cities: Simulation and Techniques*, 19-34.

Saleem, Y., Crespi, N., Rehmani, M. H., & Copeland, R. (2019). Internet of things-aided smart grid: Technologies, architectures, applications, prototypes, and future research directions. *IEEE Access : Practical Innovations, Open Solutions*, *7*, 62962–63003. doi:10.1109/ACCESS.2019.2913984

Singh, N., Duan, H., & Tang, Y. (2020). Toxicity evaluation of E-waste plastics and potential repercussions for human health. *Environment International*, *137*, 105–559. doi:10.1016/j.envint.2020.105559 PMID:32062437

Tabaa, M., Monteiro, F., Bensag, H., & Dandache, A. (2020). Green Industrial Internet of Things from a smart industry perspectives. *Energy Reports*, *6*, 430–446. doi:10.1016/j.egyr.2020.09.022

Tewari, A. a. (2020). *Security, privacy and trust of different layers in Internet-of-Things (IoTs) framework.* Elsevier. doi:10.1016/j.future.2018.04.027

Xu, M. a. (2019). Optimized renewable energy use in green cloud data centers. Springer, 314330.

Xu, M. T. (2020). A self-adaptive approach for managing applications and harnessing renewable energy for sustainable cloud computing. *IEEE Transactions on Sustainable Computing*, 544–558.

Zhang, Q. a. (2021). A survey on data center cooling systems: Technology, power consumption modeling and control strategy optimization. *Journal of Systems Architecture*, 102253.

Chapter 4
Impact of Data Centers on Power Consumption, Climate Change, and Sustainability

Dhanabalan Thangam
https://orcid.org/0000-0003-1253-3587
Presidency Business School, Presidency College, Bengaluru, India

Haritha Muniraju
Triveni Institute of Commerce and Management, Bengaluru, India

R. Ramesh
Department of Management Studies, Knowledge Institute of Technology, Salem, India

Ramakrishna Narasimhaiah
https://orcid.org/0000-0002-4973-5775
Department of Economics, Jain University, Bengaluru, India

N. Muddasir Ahamed Khan
Department of Management, Acharya Institute of Graduate Studies, Bengaluru, India

Shabista Booshan
ISBR College, India

Bharath Booshan
Department of Management, Acharya Institute of Graduate Studies, Bengaluru, India

Thirupathi Manickam
https://orcid.org/0000-0001-7976-6073
Christ University, India

R. Sankar Ganesh
https://orcid.org/0000-0003-0708-8327
Vel Tech Rangarajan Dr. Sagunthala R&D Institute of Science and Technology, India

ABSTRACT

The data-driven economy is transforming with data centers becoming a crucial business infrastructure. However, the increasing reliance on data centers is posing a threat to the environment. Climate change activists are focusing on reducing emissions from sectors like automotive, aviation, and energy. Data centers consume more electricity than the UK, accounting for 3% of global electricity supply and 2% of

DOI: 10.4018/979-8-3693-1552-1.ch004

total greenhouse gas emissions. By 2040, digital data storage is projected to contribute to 14% of the world's emissions. The number of data centers worldwide has surged from 500,000 in 2012 to over 8 million, with energy consumption doubling every four years. The rise in internet penetration rates and the introduction of 5G technologies and IoT devices will further exacerbate the issue, increasing the demand for data processing.

INTRODUCTION

The exponential growth of technology in the digital age has resulted in an astounding volume of digital content. A vast amount of digital content is created by every text message, email, picture, video, document, presentation, and spreadsheet. Some of this content is saved for later use, while less significant pieces remain unaltered.Data is produced by almosteachcompany and human being on the planet. The expansion of digital content is likely to persist despite effective regulation. The COVID-19 pandemic, which prompted an increased reliance on remote work, learning, and entertainment, resulted in a substantial 56% growth in digital content from 2019 to 2020. This growth surpassed the rate observed from 2018 to 2019, more than doubling it. Projections indicate that data creation is predictable to exhibit a annual growth rate of around 19%, reaching over 180 zettabytes (ZB) by the year 2025. The Global Data Sphere, according to IDC, measures and examines the volume of data generated, collected, and duplicated globally in any given year. This figure incorporates data from academic institutions, governments, corporations, and the worldwide consumer network. A significant portion of this data finds its home in digital repositories known as data centers, with the larger ones being referred to as hyperscalers. In the current digital era, where connectivity plays a crucial role and information serves as the central axis around which the public and countries revolve, datacenters thus become the essential pillars sustaining the immense load of storage and data traffic (Birke et al., 2012). Yet, the question remains: what precisely are these data centers, and what factors contribute to their heightened importance in modern society?

Fundamentally, data centres are extremely specialised spaces created to handle, distribute, and store enormous volumes of data. These include straightforward emails that are transmitted quickly over the world, intricate financial transactions, and even high-definition video streaming that is now a common occurrence for many people (Wilson et al., 2023). One cannot overestimate the importance of data centres to the global digital network. Like its pulsing heart, they are. A lot of the scientific comforts that have ingrained themselves so easily hooked on our everyday lives would not work without them (Guoet al., 2021). All things considered, these centres have been crucial to the digital revolution over the past few decades. The capacity of modern data centres is correlated with the development of artificial intelligence, the proliferation of online-connected systems (Liu et al., 2020), the augment of cloud compute technique, and the increasing reliance on big data analytics (Matsveichuk&Sotskov, 2023; Hashem et al., 2015). These breakthroughs have fueled technological progress as well as the democratisation of knowledge and the globalisation of economies (Skare& Soriano, 2021).

However, increased reliance on digital hubs, powerful as they may be, brings about a corresponding responsibility. The heightened use of these extensive facilities, filled with servers and infrastructure, necessitates a continuous and substantial energy supply (Townend et al., 2019). Briscar (2017) mentioned that, the yearly energy usage of data centres is comparable to the productivity of 334 coal-fired power plants or the energy required to run every residence in New York City for a period of two years. Katal et al. (2023) forecast that by 2030, data centre power consumption will have increased from 200 TWh in 2016 to about 2967 TWh. Maintaining ideal temperatures to avoid overheating and malfunctions is crucial for critical components like cooling systems (Park &Seo, 2018). Continuous energy consumption, frequently from non-renewable sources, results in expanding carbon quantity and observable ecological effects (Manganelli et al., 2021). The necessity to minimise digital waste is highlighted by the fact that the storage industry's environmental effect is growing along with our reliance on digital storage. 'Digital waste' refers to the environmental effects of creating, utilising, and discarding digital devices and data storage systems. This includes life cycle appraisal, utilization of energy,emission of carbon, and electronic waste (e-waste) from abandoned hardware, such as computers, smartphones, and storage devices, which can be hazardous to the environment because of their toxic contents.

A number of topics need to be addressed in connection with sustainability of digital material. Power efficiency in storage systems is critical, particularly in hyperscalers and data centres. There can be a big impact from data centres switching to renewable energy, natural air circulation, and efficient water use. Sustainability is facilitated by virtualization and optimisation of data centres, which reduce energy consumption, optimise resource utilisation, combine data processing and storage, and reduce the need for separate physical storage devices. With calls for sustainable practices and emerald solutions diagonally all sectors, the urgency of climate change has emerged as a defining challenge (Chen et al., 2022; Manganelli, et al., 2021). Since unchecked carbon discharge from datacenters could competitor those of entire country or major worldwide industries, the ecological footprint of datacenters has drawn considerable attention from ecologists, policymakers, and the stakeholders of various industries (Guitart, 2017).

A never-before-seen era of connectedness and technical innovation has begun with the rise of data centres, the foundation of our digital infrastructure. But along with this boom comes a turning point that necessitates a close look at how data centres affect the environment in terms of electricity usage, global warming, and sustainability in general. This chapter was inspired by the realisation of the significant influence data centres have on the environment. With these centres becoming more and more essential to our digital lives, it is critical to fully comprehend their environmental impact. Deep-seated concerns about the rising power consumption of data centres, their role in contributing to climate change, and the wider consequences for global sustainability are the driving forces behind our effort. Through a close examination of these interrelated aspects, our paper hopes to clarify the environmental issues raised by data centres and, more significantly, act as a spur for innovative solutions. The conviction that a thorough investigation of these effects is necessary to educate industry executives, legislators, and stakeholders about the significance of incorporating sustainable practices into the data centre environment is the driving force for this endeavour. In addition to bringing attention to the environmental issues, our goal is to stimulate group efforts towards a more sustainable and responsible integration of data centres into our digital future through in-depth study and thought-provoking ideas. This chapter aims to serve as a call to action, imploring the international community to confront the environmental consequences of data centres and set out on a course that balances technological advancement with the necessity of preserving our planet for future generations.

The chapter "Impact of Data Centres on Power Consumption, Climate Change, and Sustainability" is extremely pertinent today, as society makes its way through the digital era, which is characterised by an unparalleled dependence on data-driven technology. Evaluating the environmental impact of data centres is vital given their increasing number in response to the ever-increasing needs of a digitalized society. Given the substantial quantity of electricity that data centres consume, it is critical to comprehend how they affect energy supplies and how this affects climate change. Considering the severity of the world's sustainability concerns and the need to reduce carbon emissions and mitigate the consequences of climate change, the research becomes especially pertinent. Furthermore, this study offers pertinent insights into how data centres might adjust and favourably support sustainable development goals at a time when governments and industry struggle to build rules and practices that support these goals. In the current context, where sustainable practices are vital to constructing a resilient and eco-friendly future, the research examines the intersection of technical advancement, environmental responsibility, and social well-being.

As far as the methodology is concerned, this chapter was created by reading scientific publications from key sources, including Google Scholar, Science Direct, Wiley Online Library, and Springer Link. It is entirely based on secondary data sources. In addition to these sources, the keywords "data centres," "power consumption," "climate change," "sustainable data centres," and "greenhouse gas emissions" have also been utilised in online sources. The most recent data on the chapter has been gathered via the search. The gathered data was further filtered to reveal the potential impact of data centres on greenhouse gas emissions, global warming, and climate change. As a result, the current chapter is formatted.

AN OVERVIEW OF GLOBAL DATA CENTERS AND THEIR POWER CONSUMPTION

Hinton et al. (2011) reported that the ICT industry in industrialized nations accounted for around 6% of total electricity utilization in 2011. The Internet was responsible for 1% of all power usage, but this number was expected to rise quickly due to rising data admittance speeds. This trend may become unmanageable, requiring lower data center energy utilization during periods of low demand, enhancing core router energy efficiency, and implementing more energy-efficient admittance network technology. Van Heddeghem et al. (2014) conducted a global analysis of ICT electricity usage, estimating that infrastructure and digital gadgets are consuming more power than the world's total electricity consumption. They found that data centres, personal computers, and communication networks have grown annually at rates of 10, 5, and 4 percent, respectively. This development is more than the 3% global increase in power usage during the same period. Belkhir and Elmeligi's (2018) study estimates that greenhouse gas emissions (GHGE) from ICT component manufacturing could increase from 1-1.6% in 2007 to over 14% by 2040, accounting for over half of the current contribution of the transportation sector. The study also highlights the significant contribution of smart phones, with their footprint surpassing desktops, laptops, and displays by 2020. Morley et al. (2018) addressed the rising requirement for electricity from data centers, focusing on the everyday peak requirement for data transmission for video torrenting and video interaction. They suggested restraining the increase in transfer to keep up with competence enhancements, which goes beyond simply relying on technical competence. Andrae and Edler's (2015) work projected an absolute increase in power consumption for the whole communications technology industry from 2,200 TWh in 2010 to 7,800 TWh in 2030. However, their projections, updated in 2019 to 950 TWh for data centers in 2030, are still questionable. According to Andrae and Edler (2015), if wireless admittance arrangements and permanent access systems/data centres do not achieve sufficient improvements in their power effectiveness, the use and manufacture of user devices, communication networks, and data centers could account for as much as 51% of the world's power consumption in 2030.

There are currently 70,000 lakhs Internet-connected gadgets in use, and this number is continually rising. Large percentages of these devices produce large amounts of data, which means procedures for gathering, sending, storing, analysing, and retrieving information are required. With the introduction of Industry 4.0 and the Internet of Things (IoT), producers are increasingly using are using data analytics to maximize the effectiveness, output, safekeeping, and cost-efficiency of their process.But internal data management is getting harder, more time-consuming, and more expensive (gray.com, 2023). In an effort to save money on infrastructure and save energy,even major corporations such as Cisco are contemplating the closure of a few of their internal datacenters.As a result, an increasing number of companies are contracting with specialised third-party data centre management firms to handle their data operations. In particular, colocation data centres are becoming more and more well-liked since they offer physical space, electricity, server cooling systems, and links to local communication networks. Rentableroom up to 100,000 feet or more is offered by large colocation providers such as QTS, Digital Realty, Equinix,, Compass Datacenters, and Cologix (Angus Loten, 2019).

Figure 1. Explains the Number of data centers worldwide in 2023

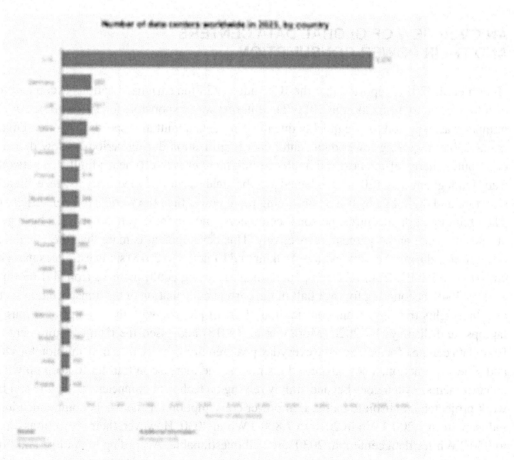

According to a recent investigation and market analysis, the worldwide datacenter market is expected to expand at a multiple yearly growth speed of more than two percent between 2020 and 2025. By 2024, the datacenter industry in the US is anticipated to generate more than $70 billion in revenue. Hyperscale data centres, which are owned and run by the corporations they service, like Apple or Google, and colocation service providers are the main sources of these reserves in datacenters.The US regions with the quickest growth are the Southeast and Pacific Northwest (Research and Markets, 2020). Figure 1 explains the Number of data centers worldwide in 2023.The United States has the highest number of data centres globally as of September 2023, with 5,375 recorded locations. Furthermore, 522 of them were located in Germany, while 517 were in the United Kingdom.

A growing number of datacenters, vary from Tier one (less complicated and safe) to Tier four (very intricate with strict IT/safekeeping standards), are being built to meet the growing need for data. These centres are often hyperscalecentres. Robust communications infrastructure and minimal energy expenses are characteristics of optimal sites. Over 300 colocation projects, including both new construction and extensions, have been completed in modern days, demonstrating the mounting demand for colocation and managed services (Research and Markets, 2020). Significant hyperscale initiatives are also taking shape in Australia, Hong Kong, India, China, and Hong Kong. Colocation service providers may purchase these strategically positioned facilities with first-rate supporting infrastructure and affordable energy costs as enterprises shift away from possessing and managing own datacenters, frequently leaving behind unoccupied space. For example, Equinix, a well-known California-based supplier of data centre colocation, recently paid $3.6 billion to Verizon Communications for the acquisition of 24 data centre locations (Austen Hufford and Drew FitzGerald, 2016).

DATA CENTERS AND THEIR POWER CONSUMPTION

Globally, there are millions of data centers, each power-intensive and capable of housing tens of thousands of servers, consuming environmental resources on a scale surpassing entire countries. The collective environmental impact of these data centers has become a growing concern for governments and the general public, prompting the need for action from data center operators. A significant development in recent years is the rapid emergence of massive "hyperscale" data centers. Within a short timeframe, these colossal centers, some as large as multiple football fields, have doubled their energy consumption, approaching an annual total of 100 terawatt-hours. Operating 24/7, 365 days a year, these data centers are augmented by the rise of new "edge" data centers, further contributing to the already high power consumption rates (grcooling.com, 2020).

As the essential processing centres of the Internet era, data centres support society's growing reliance on social networks, cloud computing, online banking, and an expanding range of devices. But there is a negative aspect to this reliance on data centers a previously unnoticed environmental cost. One wonders how much data centre electricity use will increase from its current 1% worldwide share. By 2030, will it grow forty times, five times more moderately, or less? In an effort to answer this question, Sean Ratka and Francisco Boshell of IRENA look at technological advancements that lower emissions and power costs.The datacentercalculatingproductivity increased six fold from 2010 to 2018, yet their poweruse increased by six percent. This is encouraging evidence. Big digital firms like Amazon, Google, Facebook, and Microsoft are embracing energy-efficient strategies like centralising cloud computing, using AI to balance usage prototype, placing servers in chillyweather, and making investments in renewable energy sources as their prices come down. Sector coupling and demand-side management provide further efficiency advantages. The authors urge other industries that are having difficulty reducing their carbon footprint to carefully consider these various options (Sean Ratka and Francisco Boshell, 2020).

Although data centre energy consumption is a growing concern, especially with the rapid expansion of global computing capability, a different viewpoint indicates that datacenters could present a chance to expedite the shift to sustainable energy sources. Datacenters are perceived as being at the crossroads of powercompetence, renewable power, and the expanding data market made possible by digitalization, rather than as a danger to the sustainable powerconversion specified in the Paris Agreement and larger climate goals. Data centres are setting the standard for other power-intensive industries to follow by incorporating cutting-edge technologies and taking advantage of the more and more advantageous economics combined with the effectiveness made probable by AI (The Global Renewables Outlook, 2020; irena.org, 2019).

PRESENT AND FUTURE PROJECTED POWER USE FROM DATA CENTRES

The necessitate of datacentres is only going to increase due to the expansion in collection and use of data. The power cooling infrastructure, backups, storage devices, and servers in these data centres all need a lot of electricity. A solitary internet search employs around 0.0003 kWh, or 1 KJ of power, according to Google Administration (2009). To place things in viewpoint, this much energy could run a 60W lightbulb for about 17 seconds. Even while they may appear little individually, when we take into account the presentenormous flow of data in our increasingly digitalized society, the cumulative effect on power consumption and carbon discharge from datacentres become significant (Sean Ratka and Francisco Boshell, 2020).

In the year 2018 Datacenters used 205 TWh, or about one percent of the globalsumof electricity usage (Masanet et al., 2020; Pearce, 2018). Emissions of carbon from this level of use are comparable to those from the commercial aviation sector. According to projections, data centre power consumption might, in the worst-case scenario, exceed 8,000 TWh by 2030, while in the best-case scenario, it could only reach 1,100 TWh (Sean Ratka and Francisco Boshell, 2020).

ENVIRONMENTAL BRUNT OF DATACENTERS' POWER CONSUMPTION

The establishment of mega datacenters have proven immensely advantageous for businesses and the global economy, facilitating the daily activities of billions of individuals, it comes at a considerable environmental cost. The operation of millions of servers, coupled with the need to manage the substantial heat generated by power-intensive processors through cooling systems, demands vast amounts of electricity. Remarkably, the Environmental Protection Agency (EPA) of USA estimates around 62 percent of the global electrical delivery happening from fossil-fuels, primarily natural gas and coal. As a result, the generation of electrical energy is the second-largest contributor to conservatory gas discharge. In certain instances, up to half of the electricity consumed by data centers are directed towards outdated air-cooling systems rather than powering the servers directly. Transitioning from air-cooled systems to liquid cooling solutions presents an opportunity to reduce this waste and mitigate the adverse environmental impacts (grcooling.com, 2020).

Figure 2.

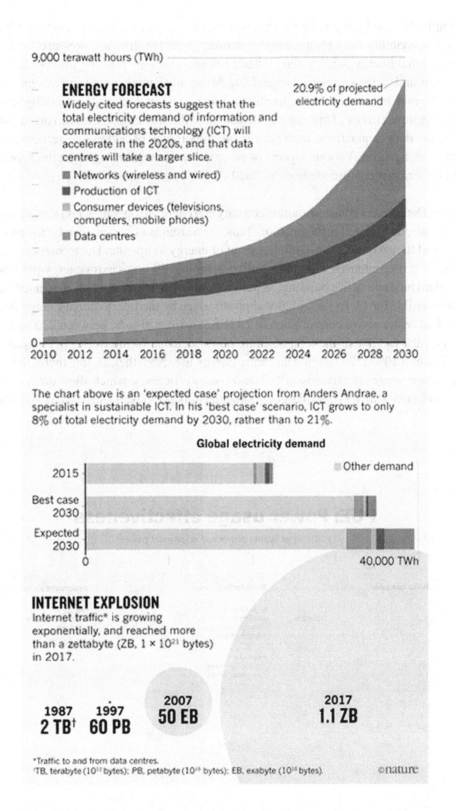

A growing body of evidence suggests that data centres can make a significant contribution to environmental sustainability through meaningful action, given the growing pressure from associations namely International Energy Agency and United Nations, which are pushing for coordinated efforts to reduce emissions and address climate change. Despite the substantial energy and resource consumption associated with powering servers and supporting infrastructure in data centers globally, the demand for their services continues to rise. This ongoing discourse within the industry, occurring at both executive levels and on the data center floor, underscores the importance of addressing resource usage and its environmental implications. Various aspects of environmental concern highlight the depth of the challenges faced by data centers (grcooling.com, 2020).

Energy Usage: Datacenters consume more electricity and energy to operate servers, storeroom, networking hardware, and related infrastructure. Tens of thousands of servers can be found in many data centres, and they require a substantial amount of energy to operate. Datacenters are in the middle of the most power-intensive structures, utilizing around 50 times extra energy per floor area than the standard business office building, as per U.S. Department of Energy. Datacenters are estimated to be responsible for 1% to 1.5% of global power usage by the International Energy Agency (IEA). Global data centre power consumption in 2021 was estimated to be between 220 and 320 Terawatt hours, or 0.9% to 1.3% of the total absolute power demand of the world. In comparison to 2015, this indicates a 10%–60% rise in data centre energy use. Nonetheless, this increase is offset by increased power usage effectiveness (PUE) and energy efficiency, which allow data centres to handle higher workloads with less negative environmental impact (Alexander and Mark Fontecchio, 2023).

Figure 3.

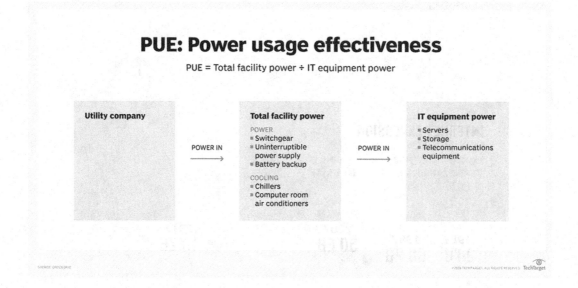

These projects, however, do not live up to the IEA's standards. Climate change and global warming are exacerbated by data centres and networks that transmit data, accounting for about 1% of energy-related greenhouse gas (GHG) emissions. A normative approach is presented by aligning with the Net Zero discharge by 2050 Scenario (NZE Scenario), itforms a roadmap for the worldwidepower sector to attain net zero CO2 discharge by 2050. This scenario is in line with imperative power oriented Sustainable Development Goals (SDGs), specifically worldwidepower usage by 2030 and considerabledevelopment in the quality of air. It also imagines superior economies to realize net zero discharge ahead of others. Additionally, it is consistent with the objective of keeping the augment in worldwidehotness to 1.5 °C, as determined by the Sixth evaluation report of the Intergovernmental Panel on Climate Change (IPCC) (iea.org, 2023).

Water Consumption

Data centres use water both directly during cooling procedures that stop servers from overheating and indirectly during the production of energy. The cooling systems, which include chillers and cooling towers, keep the temperature at its ideal level. The majority of systems use evaporative cooling, which cools incoming air and releases heat into the surrounding environment. Humidifiers are another option; they vaporize water by using electricity. Remarkably, Google datacenter utilizes almost 1703435 liters of water every day, which is the same as previously irrigating 17 acres of grass lawn. The rangecontains hyperscale datacenters that consumes ten Lakh gallons everyday and minor datacenters that utilizes 10,000 -15,000 gallons of water per day (Brien Posey, 2022).Furthermore, the location of data centres in drought-prone areas raises questions regarding the availability of water. Sometimes potable water sources are used by data centres, which create a demand for alternatives like recycled or reclaimed water, particularly in areas with limited water supplies (Jacob Roundy, 2023).

Electronic and Toxic Waste

Electrical and electronic equipment that is improperly disposed of results in electronic garbage, or "e-waste," which has an effect on resource usage, greenhouse gas emissions, and the discharge of hazardous materials. Global electronic waste generationattained 53.5 Million tons in the year 2019 and is anticipated to reach 75 Million tons by the year 2030. Roughly 8% of electronic waste is burned or ends up in land-fills. When e-waste is not properly managed, it contains dangerous materials that contribute to global warming. For example, in 2019 98 Mt of carbon dioxide (CO2) equivalents were released by discarded air conditioners. Since data centres contain a large number of IT devices, it is imperative that they set an example by recycling, refurbishing, and reusing old or broken technology (Jim O'Donnell, 2023).

Land Use

There may be environmental effects from data centre construction and location. While larger data centres can need millions of square feet and significant land clearance, smaller ones would only need 100,000 square feet. National organisations have expressed worries about the irreversible environmental implications and long-term expenses associated with the proposed Prince William Digital Gateway in Virginia, which involves rezoning 2,100 acres (Whitney Pipkin, 2022).

Greenhouse Gas Emissions

Global warming is a result of the generation conservatory gases such as CO2, methane, and nitrous oxide trapping warmth in the environment. CO2 is released by human activity, especially when fossil fuels are burned. Another major source of GHG emissions is the building of data centres, which highlights the necessity of sustainable practises (Robert McFarlane, 2021).

ENVIRONMENTAL FOOTPRINT OF DATA CENTERS

Datacenters are flattering significant parts of the digital ecosystem and this have led to a surge in their environmental effect and their prominence in talks about global sustainability. Examining the direct and indirect discharge resulting from their process is crucial to understanding their ecological footprint.

Emissions Originating Directly: Equipment and Operations in Data Centers

Direct emissions include pollutants and greenhouse gases (GHGs) that are released directly as a result of data centre activities (Katal, Dahiya, &Choudhury, 2023). The hardware and equipment that make up these centres' principal structural elements are the primary sources of these emissions. The main workhorses in datacenters are servers, which handle enormous volumes of data to make sure that cloud computing apps, e-mail, digital games, and other online services run smoothly. Their regular functioning results in a continual need for electricity, which has a direct impact on carbon discharges, particularly if the power source is derived from fossil fuels (Olujobi et al., 2023).

Chilling mechanisms make the problem of emissions much worsedirectly. Refrigerants and coolants are used in both conventional HVAC systems and certain cutting-edge cooling methods; their leakage can have a greater greenhouse effect than even CO2. The environment is disproportionately affected by these emissions, even though they may not occur as frequently as CO2 discharges due to the towering worldwide warming possible of certain refrigerants (Ewim et al., 2023). Diesel-poweredgensets, which are essential to the continuous functioning of datacenters, release pollutants into the air when they are turned on for regular testing or power outages. Diesel energy combustion consequences in a considerable discharge of greenhouse gases, even though their operating hours may be less than those of primary equipment (Nelson et al., 2022).

Emissions Indirectly Generated: Energy Sources, Construction, and Additional Contributors

As rapid insights into the ecological impact of data center process can be obtained from direct emissions, a more comprehensive view can be obtained from indirect emissions, which are frequently less obvious but equally significant.

Energy Source

The source of the electricity used to power datacenters is the largest contributor to indirect emissions. A considerable cgment of data centers globally still rely on power generated from oli products, particularly coal and usual gas, despite the fact that some contemporary buildings showcase their green energy credentials (CassadyCraighill, 2019). When these fuels are burned in power plants, a lot of greenhouse gases (GHGs), mostly CO2, are released. The carbon footprint of each digital transaction, regardless of size, is intrinsically linked to these emissions when this energy is utilised to run data centres (Mytton&Ashtine, 2022).

Construction and Infrastructure

Environmental expenditures are incurred during the whole data centre construction process. Every stage of the construction process, from raw material extraction and processing to on-site labour, has a carbon cost. Energy-intensive production methods are used to produce construction materials like metals and concrete, and transportation of these materials adds to greenhouse gas emissions (Sousa et al., 2023).

Supply Chain Emissions

Mining minerals, producing component parts, and assembling the finished product are the first steps in the lifetime of a server or several other portions of data center apparatus. According to businessnorway. com (2023), a considerable amount of the tortuous emissions connected with data center process can be attributed to the emissions produced by every phase of this chain of supply.

THE NEED FOR SUSTAINABLE DATA CENTERS

Currently, the majority of global data storage occurs in datacenters. According to the International Energy Agency (IEA), these datacenters are projected to consume around 8% of the total energy by the year 2030. Presently, data centers use approximately 1-1.5% of the global energy and contribute to 0.3% of global emissions. India is also experiencing a surge in data consumption; leading to the implementation of more localized rules and regulations within the country (iea.org, 2023).India's data industry is poised for significant growth in data center capacity. According to JLL, the sector has consistently expanded, reaching 636 MW in 2022, with expectations to reach 1318 MW by 2024. This industry is experiencing an annual growth rate of over 20%, resulting in a substantial increase in both scope 1 and scope 2 emissions. While renewable energy stands out as a viable means to reduce scope 2 emissions, certain regions face challenges due to intricate and unclear policies (jll.co.in, 2022).

In addition to scope 2 emissions, scope 1 emissions are noteworthy, given the substantial backup power required by data centers during outages for reliability and redundancy. Although major technology firms are actively procuring clean energy to power data center infrastructure through 100% renewable sources, the industry often relies on offsets (such as REC, Green tariff, Carbon Credits, etc.), which may not effectively contribute to the primary goal of grid decarbonization. The government is emphasizing increased renewable energy use through various technologies, policy relaxations, concepts like 24x7 carbon-free energy, facilitation of development in open access, and the introduction of technologies like green hydrogen. Recent policy releases, including the Green Energy Open Access Rules 2022, Electricity Amendment Rules 2022, and The Energy Conservation (Amendment) Bill 2022, aim to promote greater renewable energy adoption by easing policies, particularly for large energy consumers like data centers in India (Rashmi Singh, 2023). However, there is still room for innovation in policy and market mechanisms to facilitate clean energy transactions. An integrated approach to policy and solutions could drive the adoption of cleaner energy in the data center and Information and Communication Technology (ICT) industry, promoting decarbonization through new integrated clean technologies, business models, and incentive structures.

STRATEGIES FOR SUSTAINABLE DATA CENTER OPERATIONS WITH AN EMPHASIS ON ENERGY EFFICIENCY

The relationship between technology and the environment, especially as it relates to data centres, has taken centre stage in conversations about sustainable development. Innovative and energy-efficient solutions are becoming more and more in demand as the effects of data centres on the environment become increasingly apparent. The knowledge that, despite the energy-intensive nature of data centres, their environmental impact may be greatly impacted by the energy sources, uses, and management of this energy is the basis of this conversation. This section looks at several strategies that have been adopted or proposed to increase data centre energy efficiency and lay the groundwork for a sustainable digital future (expresscomputer.in, 2023).In the larger framework of mitigating climate change, efficiency refers to optimising resource utilisation to produce comparable or superior results with fewer inputs. Efficiency measures are designed to maintain or improve computing output in data centres while lowering energy intake and related emissions. This optimisation is essential to sustainability because it makes sure that when digitalization grows, the environmental costs do not increase proportionately (Driskell, 2022).

Cooling Technologies

Data centres have historically relied significantly on HVAC (Heating, Ventilation, and Air Conditioning) systems, which are notorious for consuming a lot of energy because they use compressors and refrigerants for mechanical cooling. The goal of creative cooling solutions is to reduce or do away with mechanical cooling. Energy-intensive methods are not necessary when employing techniques like free cooling, which make use of water or ambient air. By submerging servers in non-conductive fluids, liquid cooling reduces the need for substantial air cooling by enabling direct heat removal (Mulay et al., 2019).

Cases of Unbeaten Cooling Approaches Followed By Various Users

Google's Deep Mind AI Solution helped Google to reduce cooling-related energy consumption by 40% by optimising cooling using their DeepMind AI (DemisHassabis, Mustafa Suleyman, and Shane Legg, 2017).Facebook's Swedish Data Centre which is situated in Luleå, Sweden, uses the Arctic region to generate free cooling, which greatly reduces the requirement for automatic chilling (Harding, 2015).

Enhancement of Structural Design

The structural footprint of a datacenter plays a critical responsibility in its power demands. Plan emphasizing usual exposure to air, thermally advantageous resources, and optimal server outlines can lessen the requirement for synthetic cooling. Inherently energy-efficient architecture is enhanced by small details like thermal buffers, green roofs, and reflecting roof coatings. For more effective and energy-efficient cooling, techniques including hot/cold aisle containment and vertical server stacking are investigated (Zhang et al., 2023).

Renewable Energy Sourcing

Efficiency measures seek to lower energy usage, but it's important to take the energy's source into account. Data centres' carbon footprints can be greatly decreased by exchanging from fossil fuels to rechargeable energy basis. Large tech corporations have begun incorporating rechargeable power sources into their data centre operations as a way of acknowledging their environmental responsibilities. Since 2014, Apple has used only renewable energy from sunlight, wind, and other renewable sources to power all of its organizations worldwide, together with datacenters (Apple Inc., 2023).Amazon Web Services (AWS) uses wind farms to counter balance the power usage of its enormous system of datacenters; AWS has started comprehensive renewable power initiatives, such as wind mills (Carol Yan, 2023).

Hardware Efficiency

Server technology has been improving constantly, both in terms of recital and power efficiency. In order to reduce overall energy usage, modern servers frequently give higher computational power per watt. Simplified energy utilisation is facilitated by the switch from conventional hard drives to solid-state drives (SSDs) and the use of processors with low energy use (Ewim et al., 2023). By reducing the requirement for many physical servers, virtualization effectively reduces energy usage by enabling a single physical server to perform the functions of multiple virtual servers. By using methods like de-duplication and recognizing and accumulate redundant data just once, storage can be optimised and energy consumption can be further decreased (Rajkumar & Hanakoti, 2020). This section highlights that there is more room for creativity than debate in the discussion of data centre energy efficiency and environmental effect. A roadmap for sustainable data centre operations is created by combining strategies related to hardware efficiency, energy resources, architectural blueprint, and cooling move toward. Assimilating these tactics becomes necessary as digitalization progresses, not just desirable. For know-how and the atmosphere to coexist peacefully in the future, data centres must develop in agreement with sustainability goals (Ewim et al., 2023).

DRIVING GOVERNANCE FOR ENSURING SUSTAINABILITY IN DATA CENTERS

Beyond the domains of technologists and environmentalists, the interaction between data centres and their environmental impact now involves governance at the local, national, and international levels. Energy competence routes within the data centre industry are guided, incentivized, and occasionally mandated by policies and supremacy frameworks. Industries, particularly those like data centres that operate at the nexus of know-how and ecological concerns, can function within the framework that governance provides. By defining standards for assessing and optimising data centre operations, providing incentives for the implementation of renewable power sources, and instituting ambitious energy efficiency targets, efficient policies can stimulate modernization (Laura Sebastian, 2022).

Local policies at the city or state level deal with minute details, such as land use and encouraging the incorporation of renewable power basis locally. At the national level, nations modify their energy policy to address the benefits and problems that data centres bring. In order to encourage and facilitate energy-efficient behaviour, the US, for example, highlights its Energy Star accreditation for datacenters (U.S. Environmental Protection Agency, 2023). In a similar vein, the European Union combines datacenters with environmental responsibility as part of its Green Digital strategy. In the context of broader conversations about sustainable technological breakthroughs, international associations such as the United Nations and the IEA emphasize the importance of power efficiency in datacenters (International Energy Agency, 2023).

LEADING DATACENTERS IN ENERGY EFFICIENCY: REAL CASES

Understanding data centers' design and energy performance is crucial for future design solutions. Several studies have been conducted, including works presented by Ewim et al. (2023), which highlight innovative case studies:

Microsoft's Underwater Data Centre

The notion of an underwater datacenter made its debut at Microsoft in 2014 at ThinkWeek, an occasion where staff members exchange innovative ideas. The idea was seen as a possible means of saving energy and offering coastal communities lightning-fast cloud services. The majority of people on Earth reside 120 miles or less from the shore. Data would have a short distance to travel if datacenters were submerged close to coastal communities, enabling quick and seamless online browsing, video streaming, and gaming. The subterranean oceans' constant coolness also makes energy-efficient datacenter designs possible. They can, for instance, make use of heat-exchange piping, like that seen in submarines. In 2015, the Project Natick team from Microsoft demonstrated the viability of the underwater datacenter idea with a 105-day deployment in the Pacific Ocean. In order to demonstrate the concept's viability, Phase II of the project involved hiring experts in shipbuilding, renewable energy, and logistics related to the maritime industry. Microsoft is currently attempting to use our accomplishments rather than feeling compelled to go out and demonstrate even more. It has completed its tasks. If it makes sense, Natick is a vital component that the business can utilise.

The researchers claim that discussions on how to make datacenters utilise energy more sustainably are already influenced by the lessons learnt from Project Natick. For example, the Orkney Islands' 100% wind and solar power generation, together with experimental green energy technologies being developed at the European Marine Energy Centre, is one of the reasons the Project Natick team chose them for the Northern Isles deployment. Despite what most land-based data centres view as an unstable grid, Microsoft has been able to function quite effectively on it. It's also optimistic that after reviewing the data, it might not require as much infrastructure centred on dependability and electricity.

Cutler has previously considered situations like co-locating an offshore windfarm with an underwater datacenter. The datacenter would probably have adequate electricity even with mild winds. A powerline from the shore might be combined with the fibre optic cabling required to transfer data as a last option. The elimination of the requirement for replacement components may be one of the other sustainability-related advantages. Approximately every five years, all servers in a datacenter with no lighting would be replaced. Because of the servers' high level of dependability, the few that malfunction early are easily taken offline. Furthermore, Cutler pointed out that Project Natick has demonstrated that datacenters can be run and maintained at a cool temperature without using freshwater resources that are essential to human health, agriculture, and wildlife. Microsoft is now exploring methods to accomplish this for land datacenters (Alghamdi et al., 2023).

Google's Zero Carbon Data Centre in Finland

As an added bonus of its location and former function, Google's latest data center has located in Hamina, Finland, will only be chilled by sea water. The data centre is located near the Gulf of Finland. According to Joe Kava, Google's superior executive of datacenter building and functions, "Google's team was actually worried to make use of the chance of it being accurate near the gulf to come up with agroundbreaking and well-organized cooling mechanism (Wang et al., 2022). Google plans to use the facility's existing tunnels to bring in cool seawater, mix it with cold water to make it roughly equal in temperature to the water it is returned to after passing it from side to side heat exchangers to scatter the temperature from the servers. For the remainder of the year, Google intends to put the data centre into operation. In its efforts to maximise the efficiency of its data centres, Google is not new to leveraging the distinctive features of a website. Because the climate outside of Saint-Ghislain, Belgium, is consistently colder than that of Google's data centres, the data centre there was constructed without chillers.This makes it possible for Google to employ exterior air for cooling, and it has contributed to the facility being Google's most energy-efficient data centre (Jonathan Bardelline, 2011).

The Green Mountain Data Centre in Norway

Within the mountain, the Green Mountain Data Centre spans around 21,000 square metres of floor area. This contains nine rooms to hold data servers, a quay, administrative, and warehousing facilities. Servers generate a lot of heat as they store and process digital data, so keeping them from overheating requires a strong cooling system. While many data centres worldwide are striving for more energy-efficient solutions, it looks like the Green Mountain Data Centre has struck gold. The centre employs the surrounding rock and dirt as moderators since it is first buried beneath. Next, it cools the subterranean area effectively by using water from the nearby fjord in a closed-loop groundwater heat exchanger.

The centre is completely carbon neutral since it gets all of its electricity from a number of nearby renewable energy sources. The facility markets itself as an eco-friendly server farm thanks to its interior energy-efficient architecture. The farm also claims a 99.9997% availability due to the several close power sources that are in the area and have a high degree of redundancy. Although the majority of the data center's revenue will come from within Norway, its eco-friendly products may potentially draw customers from the US and Europe. They anticipate having the server rooms and mountain halls operational by the end of 2012. Construction is now underway on these areas.

While the standalone corporations such as Green Mountain, Google and Microsoft showcase achievements through novelty and dedication, the role of strategy in scaling up these most excellent practices across industries is crucial. As datacenters prolong to strengthen themselves in digital age, a synergy of guiding principle, process modernization, and paramount practices will be essential in routing their development towards a sustainable and green future.

DISCUSSIONS

The amount of data generated in the digital era has increased at an unparalleled rate, necessitating the development of infrastructure for data processing, management, and storage. As the hubs of our global network, data centres have experienced a rapid evolution in terms of size and intricacy. The inevitable confluence of energy demand, environmental aspirations, and technology improvements presents both possibilities and challenges for stakeholders as we look to the future. This part aims to predict the future patterns in the expansion of datacenters and the energy requirements that go along with it. It finishes with a list of suggestions to help create a sustainable digital future. The amount of data created is expected to expand exponentially due to the development of the IoT, the rise of AI, and the broad use of online services worldwide. According to estimates, the amount of data generated globally may have multiplied tenfold since the start of this decade (Daniel, 2019). Strong data centre infrastructure is required to handle this data tsunami. With the advent of new generation technologies such as driverless cars, augmented reality, and real-time AI analytics, real-time data processing becomes essential and may lead to a move towards edge and distributed computing. This suggests that there could be more localised, smaller data centres processing information nearer to the source to cut down on latency (Thangam & Chavadi, 2023). Quantum computing is unmoving in its premature stages, yet it has the potential to revolutionize current computer paradigms. Even if it is still in its formative years, quantum computing grasps the probable to surpass existing computer paradigms. Data centres must adapt as they grow, not just by replacing outdated hardware but also by addressing evolving cooling and energy requirements, especially in light of the unique operational requirements presented by quantum machines. (Karkošková, 2023). This work Contributes to the existing body of knowledge in the following manner.

With the help of digital infrastructure, we can now interact, work, and communicate on a scale that has never been possible in modern civilization. However, this expansion has come at a great cost to the environment. Because of their high water and energy consumption as well as the trash that they produce from electronic devices, data centres are becoming more widely acknowledged as major sources of pollution, including greenhouse gas emissions. The current study is significant and critical in this regard. It provides a comprehensive and integrated strategy that aims to encourage more sustainable and accountable data centre operations in addition to increasing energy efficiency. It encourages a change in our understanding of and approach to managing these infrastructures by going beyond the traditional focus on technology fixes and including more expansive environmental aspects. In order to manage and minimise their environmental effect while satisfying the increasing demand for digital services, data centres must implement environmental sustainability criteria in order to solve these problems. Establishing data centres in brownfields underused or polluted areas repurposed for new development are one creative way to further improve sustainability and can have positive effects on the environment and the economy. A number of advantages, including lower operational costs, improved company reputation, and compliance with environmental laws, might result from putting the framework into practice.

By employing sustainability metrics, data centres may get more insight into their environmental performance and identify opportunities for enhancement. By doing so, they may reduce their energy and water use as well as their garbage production, which will lower operating expenses and boost profitability. Furthermore, employing sustainability metrics may enhance a data center's reputation as a leader in the field and attract environmentally conscious customers. This can increase market share, foster customer loyalty, and create new company opportunities. Another advantage of adhering to environmental regulations is that adopting sustainability measures guarantees data centres are operating in accordance with municipal, regional, and federal environmental legislation. Data centres must continuously work to enhance their sustainability performance if they are to play a part in the data centre industry's transition to a greener and more sustainable future.

Incorporating brownfield development into data centre planning can yield other advantages, such as reducing the need for new land, limiting urban growth, and revitalising areas that are underutilised or contaminated. Functioning with business partners and stakeholders, using best performs and new technology and frequently assessing sustainability KPIs may all help achieve this. The framework's ability to encourage structural change in the industry is another factor contributing to its significance. By promoting a proactive approach to sustainability, it fosters a culture of innovation and ongoing progress. It can assist data centres in transitioning to greener operations, which will contribute to the advancement of a greener economy and the slowing down of global warming.

Consequently, the developed framework offers a comprehensive and all-encompassing method of tackling sustainability issues in data centres. The structure being offered highlights four important sustainability criterion, stresses the value of employing energy proficient machinery and green construction plan principles, and provides an overview of best practices for operations. Data centre administrators have an easy path to achieving sustainable operations with this technique. The framework also emphasises the necessity of adhering to rules, considering brownfield development as a tactical substitute for sustainable expansion, and incorporating sustainability metrics into the administrative procedures. By using this method, data centre operators may reduce their impact on the environment and make a positive contribution to a sustainable future.

RECOMMENDATIONS

From the overall observation, this study has come up with following recommendations as practical implications and the same presented in the diagram.

Adopt Holistic Design Principles: Energy efficiency, cooling strategies, and renewable energy sources should all be included into future data centres in addition to computing efficiency. Sustainability in operations and the environment may be fueled by a comprehensive strategy that takes these factors into account right on.

Invest in R&D for Cooling Technologies: Industry and academic players should make larger investments in investigating and creating cutting-edge cooling technologies, given the projected problems in cooling. The cooling industry is a prime place for innovation, whether it takes the form of using geo and thermal energy, testing with phase-change resources, or investigating novel structural concepts.

Forthrighten Collaborative Frameworks: Future obstacles are structural in nature as well as technology. Policymakers, technology companies, energy suppliers, and data centre operators must work together. By working together, we can create forward-thinking policies, standardise processes, and exchange best practises.

Give Renewable Energy combination Top Priority: By making a bold transition to renewable energy, datacenters may drastically lessen their environmental effect. Incorporating green and environment friendly power sources should be a primary concern, whether from side to side power buy concurrence with renewable power dealers or on-site renewable power generation such as solar, thermal or wind farms.

Figure 4.

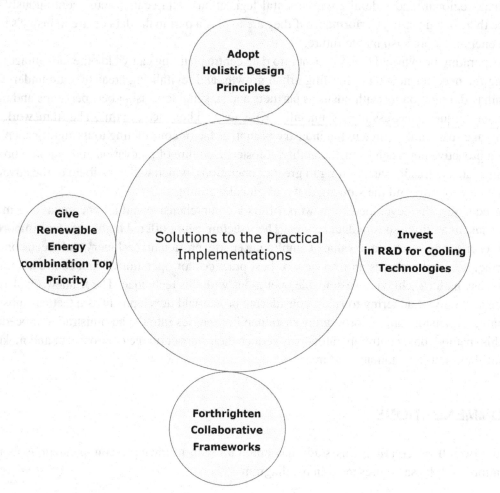

CONCLUSION

The rapid digitization of today's society has made data centres essential components of our digital existence. Nevertheless, their power footprints and the ensuing effects on environment change are closely related to this story. The many facets of data centres have been discussed in this debate, including their history, energy requirements, and environmental impact. Understanding and addressing the effects of data centre expansion is essential as they increase. Armed with their history and current knowledge, we have travelled into the future, predicting future directions and offering suggestions for balancing technology advancement with environmental awareness. One important aspect that the story emphasises is flexibility. The contemporary environmental requirements were not considered during the design of data centres in the past. Modern problems, however, have sparked inventions, ranging from state-of-the-art cooling systems to harmoniously designed buildings. These changes are evidence of human inventiveness and the desire to balance progress with the environment.

Within the dynamic realm of data centres, governance plays a pivotal role in establishing the path towards sustainability. Governance serves as a protector and a guide for everything from international entities that oversee global discussions to local laws designed for particular situations. However, obstacles and possibilities are ahead on any trip. The proposals, which are based on recent discoveries, aim to shed light on this route and make sure that the expansion of data centres is consistent with environmental responsibility. In conclusion, there is a continuous tale of evolution, introspection, creativity, and adaptation concerning data centres and our environment. Our joint effort as we traverse the intricacies of the twenty-first century is to guarantee that data centres develop not just as data storage facilities but also as emblems of sustainable development, promising a more promising and environmentally friendly future for all.

REFERENCES

Alexander & Fontecchio. (2023, October 28). *Power usage effectiveness (PUE)*. https://www.techtarget.com/searchdatacenter/definition/power-usage-effectiveness-PUE

Alghamdi, R., Dahrouj, H., Al-Naffouri, T., & Alouini, M. S. (2023). *Toward Immersive Underwater Cloud-Enabled Networks: Prospects and Challenges*. IEEE BITS the Information Theory Magazine.

Andrae, A. S., & Edler, T. (2015). On global electricity usage of communication technology: Trends to 2030. *Challenges*, *6*(1), 117–157. doi:10.3390/challe6010117

Angus Loten. (2019, August 19). *Data-Center Market Is Booming Amid Shift to Cloud*. https://www.wsj.com/articles/data-center-market-is-booming-amid-shift-to-cloud-11566252481

Apple Inc. (2023, April 5). *Apple and global suppliers expand renewable energy to 13.7 gigawatts*. https://www.apple.com/in/newsroom/2023/04/apple-and-global-suppliers-expand-renewable-energy-to-13-point-7-gigawatts/

Austen & FitzGerald. (2016, August 26). *Rackspace to Go Private in $4.3 Billion Deal*. https://www.wsj.com/articles/rackspace-to-go-private-in-4-3-billion-deal-1472218264

Belkhir, L., & Elmeligi, A. (2018). Assessing ICT global emissions footprint: Trends to 2040 & recommendations. *Journal of Cleaner Production*, *177*, 448–463. doi:10.1016/j.jclepro.2017.12.239

Birke, R., Chen, L. Y., & Smirni, E. (2012, June). Data centers in the cloud: A large scale performance study. In *2012 IEEE Fifth international conference on cloud computing* (pp. 336-343). IEEE. 10.1109/CLOUD.2012.87

Brien Posey. (2022, April 29). *Data center temperature and humidity guidelines*. https://www.techtarget.com/searchdatacenter/tip/Data-center-temperature-and-humidity-guidelines

Briscar, J. R. (2017). *Data Transmission and Energy Efficient Internet Data Centers*. Am. UL Rev., 67, 233.

businessnorway.com. (2023, October 30). *Explore Norway's green data centre industry*. https://businessnorway.com/key-industries/data-centres

Carol Yan. (2023, April 26). *AWS Collaborates with WindEurope and Accenture to Streamline Wind Permitting in Europe*. https://aws.amazon.com/blogs/industries/aws-collaborates-with-windeurope-and-accenture-to-streamline-wind-permitting-in-europe/

Cassady Craighill. (2019, February 13). *Greenpeace Finds Amazon Breaking Commitment to Power Cloud with 100% Renewable Energy*. https://www.greenpeace.org/usa/news/greenpeace-finds-amazon-breaking-commitment-to-power-cloud-with-100-renewable-energy/

Chen, X., Pan, M., Li, X., & Zhang, K. (2022). Multi-mode operation and thermo-economic analyses of combined cooling and power systems for recovering waste heat from data centers. *Energy Conversion and Management, 266*, 115820. doi:10.1016/j.enconman.2022.115820

Daniel Hamad. (2019, April 10). *Data Centres: How Will the Rising Demand for Data be Powered?* https://www.arcadis.com/en-au/knowledge-hub/blog/australia/daniel-hamad/2019/data-centres-how-will-the-rising-demand-for-data-be-powered

Demis, Suleyman, & Legg. (2017, January 3). *DeepMind's work in 2016: a round-up*. https://deepmind.google/discover/blog/deepminds-work-in-2016-a-round-up

Drew & Hufford. (2016, December 6). *Equinix to Buy Some Verizon Data Centers for $3.6 Billion*. https://www.wsj.com/articles/equinix-to-buy-some-verizon-data-centers-for-3-6-billion-1481034980?mod=article_inline

Driskell, D. (2022). *Strategies for Sustainable Data Centers*. Technology and Sustainability in Modern Society.

epa.gov. (2023, October 28). *Overview of Greenhouse Gases*. https://www.epa.gov/ghgemissions/overview-greenhouse-gases

Ewim, D. R. E., Ninduwezuor-Ehiobu, N., Orikpete, O. F., Egbokhaebho, B. A., Fawole, A. A., & Onunka, C. (2023). Impact of Data Centers on Climate Change: A Review of Energy Efficient Strategies. *The Journal of Engineering and Exact Sciences, 9*(6), 16397–01e. doi:10.18540/jcecvl9iss6pp16397-01e

expresscomputer.in. (2023, September 12). *Energy-Efficient Design Strategies for Sustainable Data Center Operations*. https://www.expresscomputer.in/data-center/energy-efficient-design-strategies-for-sustainable-data-center-operations/103415/#:~:text=Optimizing%20the%20supply%20of%20air,speed%20chillers%20and%20pumps%20and

gray.com. (2023, October 17). *The Data Center Industry Is Booming*. https://www.gray.com/insights/the-data-center-industry-is-booming/

grcooling.com. (2020). *The Effects of Data Centers on the Environment*. https://www.grcooling.com/blog/the-effects-of-data-centers-on-the-environment/

Guitart, J. (2017). Toward sustainable data centers: A comprehensive energy management strategy. *Computing, 99*(6), 597–615. doi:10.1007/s00607-016-0501-1

Guo, C., Luo, F., Cai, Z., & Dong, Z. Y. (2021). Integrated energy systems of data centers and smart grids: State-of-the-art and future opportunities. *Applied Energy, 301*, 117474. doi:10.1016/j.apenergy.2021.117474

Harding, A. C. (2015). *Improved methods for identifying, applying, and verifying industrial energy efficiency measures*. Academic Press.

Hashem, I. A. T., Yaqoob, I., Anuar, N. B., Mokhtar, S., Gani, A., & Khan, S. U. (2015). The rise of "big data" on cloud computing: Review and open research issues. *Information Systems, 47*, 98–115. doi:10.1016/j.is.2014.07.006

Hinton, K., Baliga, J., Feng, M., Ayre, R., & Tucker, R. S. (2011). Power consumption and energy efficiency in the internet. *IEEE Network, 25*(2), 6–12. doi:10.1109/MNET.2011.5730522

iea.org. (2023, October 26). *Net Zero Emissions by 2050 Scenario (NZE).* https://www.iea.org/reports/global-energy-and-climate-model/net-zero-emissions-by-2050-scenario-nze

International Energy Agency. (2023). *The World Energy Outlook 2023.* https://www.iea.org/reports/world-energy-outlook-2023

irena.org. (2019, February 15). *Innovation landscape for a renewable-powered future.* https://www.irena.org/publications/2019/Feb/Innovation-landscape-for-a-renewable-powered-future

Jacob Roundy. (2023, February 24). *A primer on hyperscale data centers.* https://www.techtarget.com/searchdatacenter/tip/A-primer-on-hyperscale-data-centers

Jim O'Donnell. (2023, March 17). *Data center, devices scrutinized for sustainability goals.* https://www.techtarget.com/sustainability/feature/Data-center-devices-scrutinized-for-sustainability-goals

Jonathan Bardelline. (2011, May 31). *Google Uses Sea Water to Cool Finland Data Center.* https://www.greenbiz.com/article/google-uses-sea-water-cool-finland-data-center#:~:text=Google's%20new%20data%20center%20in,by%20the%20Gulf%20of%20Finland

Julia Borgini. (2022, May 3). *Data center cooling systems and technologies and how they work.* https://www.techtarget.com/searchdatacenter/tip/Data-center-cooling-systems-and-technologies-and-how-they-work

Karkošková, S. (2023). Data governance model to enhance data quality in financial institutions. *Information Systems Management, 40*(1), 90–110. doi:10.1080/10580530.2022.2042628

Katal, A., Dahiya, S., & Choudhury, T. (2023). Energy efficiency in cloud computing data centers: A survey on software technologies. *Cluster Computing, 26*(3), 1845–1875. doi:10.1007/s10586-022-03713-0 PMID:36060618

Liu, L., Zhang, Q., Zhai, Z. J., Yue, C., & Ma, X. (2020). State-of-the-art on thermal energy storage technologies in data center. *Energy and Building, 226*, 110345. doi:10.1016/j.enbuild.2020.110345

Manganelli, M., Soldati, A., Martirano, L., & Ramakrishna, S. (2021). Strategies for improving the sustainability of data centers via energy mix, energy conservation, and circular energy. *Sustainability (Basel), 13*(11), 6114. doi:10.3390/su13116114

Masanet, E., Shehabi, A., Lei, N., Smith, S., & Koomey, J. (2020). Recalibrating global data center energy-use estimates. *Science, 367*(6481), 984–986. doi:10.1126/science.aba3758 PMID:32108103

Matsveichuk, N. M., &Sotskov, Y. N. (2023). *Digital Technologies, Internet of Things and Cloud Computations Used in Agriculture: Surveys and Literature in Russian.* Academic Press.

Morley, J., Widdicks, K., & Hazas, M. (2018). Digitalisation, energy and data demand: The impact of Internet traffic on overall and peak electricity consumption. *Energy Research & Social Science, 38,* 128–137. doi:10.1016/j.erss.2018.01.018

Morley, J., Widdicks, K., & Hazas, M. (2018). Digitalisation, energy and data demand: The impact of Internet traffic on overall and peak electricity consumption. *Energy Research & Social Science, 38,* 128–137. doi:10.1016/j.erss.2018.01.018

Mulay, M. R., Chauhan, A., Patel, S., Balakrishnan, V., Halder, A., & Vaish, R. (2019). Candle soot: Journey from a pollutant to a functional material. *Carbon, 144,* 684–712. doi:10.1016/j.carbon.2018.12.083

Mytton, D., & Ashtine, M. (2022). Sources of data center energy estimates: A comprehensive review. *Joule, 6*(9), 2032–2056. doi:10.1016/j.joule.2022.07.011

Nelson, B., Zytner, R. G., Dulac, Y., & Cabral, A. R. (2022). Mitigating fugitive methane emissions from closed landfills: A pilot-scale field study. *The Science of the Total Environment, 851,* 158351. doi:10.1016/j.scitotenv.2022.158351 PMID:36049680

Olujobi, O. J., Okorie, U. E., Olarinde, E. S., & Aina-Pelemo, A. D. (2023). Legal responses to energy security and sustainability in Nigeria's power sector amidst fossil fuel disruptions and low carbon energy transition. *Heliyon, 9*(7), e17912. doi:10.1016/j.heliyon.2023.e17912 PMID:37483776

Park, S., & Seo, J. (2018). Analysis of Air-side economizers in terms of cooling-energy performance in a data center considering Exhaust air recirculation. *Energies, 11*(2), 444. doi:10.3390/en11020444

Pearce, F. (2018). Energy hogs: can world's huge data centers be made more efficient. *Yale Environment, 360*(3).

Posey, B. (2022). *What Is the Akida Event Domain Neural Processor?* Academic Press.

Rajkumar, K., & Dhanakoti, V. (2020, December). Methodological Methods to Improve the Efficiency of Cloud Storage by applying De-duplication Techniques in Cloud Computing. In *2020 2nd International Conference on Advances in Computing, Communication Control and Networking (ICACCCN)* (pp. 876-884). IEEE. 10.1109/ICACCCN51052.2020.9362940

Rashmi Singh. (2023, February 28). *Green building regulations give impetus to sustainable data centers in India.* https://india.mongabay.com/2023/02/green-building-regulations-give-impetus-to-sustainable-data-centers-in-india/

Research and Markets. (2020, February 13). *Comprehensive Data Center Market Outlook and Forecast 2020-2025.* https://www.globenewswire.com/news-release/2020/02/13/1984742/0/en/Comprehensive-Data-Center-Market-Outlook-and-Forecast-2020-2025.html

Robert McFarlane. (2021, September 27). *Considerations for sustainable data center design.* https://www.techtarget.com/searchdatacenter/tip/Considerations-for-sustainable-data-center-design

Sean Ratka and Francisco Boshell. (2020, June 26). *The Nexus between Data Centres, Efficiency And Renewables: A Role Model For The Energy Transition.* https://energypost.eu/the-nexus-between-data-centres-efficiency-and-renewables-a-role-model-for-the-energy-transition/

Sebastian-Coleman, L. (2022). *The Culture Challenge: Organizational Accountability for Data. Meeting the Challenges of Data Quality Management.* Academic Press. doi:10.1016/B978-0-12-821737-5.00008-0

Škare, M., & Soriano, D. R. (2021). A dynamic panel study on digitalization and firm's agility: What drives agility in advanced economies 2009–2018. *Technological Forecasting and Social Change, 163,* 120418. doi:10.1016/j.techfore.2020.120418

The Global Renewables Outlook. (2020, April 24). *Global Renewables Outlook: Energy transformation 2050.* https://www.irena.org/publications/2020/Apr/Global-Renewables-Outlook-2020

Townend, P., Clement, S., Burdett, D., Yang, R., Shaw, J., Slater, B., & Xu, J. (2019, April). Improving data center efficiency through holistic scheduling in kubernetes. In *2019 IEEE International Conference on Service-Oriented System Engineering (SOSE)* (pp. 156-15610). IEEE. 10.1109/SOSE.2019.00030

U.S. Environmental Protection Agency. (2023, May 17). *U.S. EPA's Energy Star Program Develops Energy-Saving Guidance for Co-Location Data Centers in Collaboration with Equinix and Iron Mountain.* https://www.epa.gov/newsreleases/us-epas-energy-star-program-develops-energy-saving-guidance-co-location-data-centers

Van Heddeghem, W., Lambert, S., Lannoo, B., Colle, D., Pickavet, M., & Demeester, P. (2014). Trends in worldwide ICT electricity consumption from 2007 to 2012. *Computer Communications, 50,* 64–76. doi:10.1016/j.comcom.2014.02.008

Whitney Pipkin. (2022, October 21). *Data center decisions could have big land use impacts in Virginia's Prince William County.* https://www.bayjournal.com/news/growth_conservation/data-center-decisions-could-have-big-land-use-impacts-in-virginia-s-prince-william-county/article_51ef20a2-5166-11ed-9409-b386158a70c3.html

Wilson, D. C., Acun, F., Jana, S., Ardanaz, F., Eastep, J. M., Paschalidis, I. C., & Coskun, A. K. (2023, November). An End-to-End HPC Framework for Dynamic Power Objectives. In *Proceedings of the SC'23 Workshops of The International Conference on High Performance Computing, Network, Storage, and Analysis* (pp. 1801-1811). 10.1145/3624062.3624262

Zhang, H., Tian, Y., Tian, C., & Zhai, Z. (2023). Effect of key structure and working condition parameters on a compact flat-evaporator loop heat pipe for chip cooling of data centers. *Energy, 284,* 128658. doi:10.1016/j.energy.2023.128658

Chapter 5
Power–Aware Virtualization:
Dynamic Voltage Frequency Scaling Insights and Communication– Aware Request Stacking

Dhaarini K. N. Hathwar

(iD) https://orcid.org/0009-0006-9896-0342

REVA University, India

Srinidhi R. Bharadwaj

REVA University, India

Syed Muzamil Basha

REVA University, India

ABSTRACT

This chapter describes the central role that virtualization technologies play in promoting sustainable computing practices. The authors thoroughly explore the complexities of green data center and server operations and highlight the importance of server virtualization in collaborative integration efforts. Essential technologies such as dynamic voltage and frequency scaling (DVFS) will be examined for their potential to reduce energy consumption. Additionally, they introduce a new approach called communication-aware request stacking to optimize energy efficiency. By advocating best practices in network design, they are committed to embracing green networks and leveraging energy-efficient resources and nodes. The proposed framework integrates network virtualization and adaptive link rate, promising improved network performance and a greener operational paradigm. This chapter provides rich insights for practitioners, researchers, educators, and policy makers working to promote environmental sustainability in computing and networking.

DOI: 10.4018/979-8-3693-1552-1.ch005

INTRODUCTION

The arrival of cloud computing has completely transformed the tech scene. Now, however, it is facing a major unexpected hurdle-balancing energy use by information and communication technology (ICT). As the globalization of these systems continues, in addition to increasing electronic waste and emissions which harm public health they add pressure on energy resources around the world with resulting damage to our environment. As a result, there is an urgent need for research in the field of green computing. Including not only studying e-waste disposal itself but also how to efficiently cool servers given cloud systems' ever expanding popularity. Expert thinking has thus given rise to the idea of "green cloud computing" as a direct frontal attack on this difficult problem. An advanced ecosystem of cloud data centres. In the domain of cloud data centres, it intends to develop an ecologically pure environment that combines economic efficiency with eco-responsibility. Green cloud providers and users are pivotal performers in this vision, since they use multi-tenancy principles that create secure allocation strategies while reducing resource consumption. Coupled with the use of sustainable computing, through innovations in green ID virtualization technologies our computational activities can be made to increasingly green. We should also be able improve economic efficiency at the same time .This vital work directly confronts some very serious problems in the areas of reducing levels of energy consumption and proper management processes for electronic waste. Ultimately, through the like efforts over green cloud computing things today's high-tech industry is laying solid foundations for a greener tomorrow in which everything from technology to its application exudes sustainable life.

LITERATURE SURVEY

1. Anwar et al. (2017) literature survey emphasizes that we have to use virtualization technology in order to prevent e-waste. From the standpoint of resource management, server utilization and load balancing to energy-efficient using up e-waste serving as data devices for virtualizing hardware. These advantages notwithstanding, there are still significant challenges in the area of resource management in server and network virtualization. If this virtualization is done carefully, and e-waste can be disposed of in a responsible fashion, then computing's environmental impact might end up being lowered to the point where it will conform to an "eco" style of sustainable green technology. Nevertheless, the existing literature does not carry out an in-depth analysis of resource management problems specific to server and network virtualization. This represents a major vacuum.

2. Through their study, Nagar & Pillai (2023) highlight the extent of tech's impact on business. Information technology is now an irreversible part of doing business and increasingly dictates how corporations survive in today's permanently altered global economic environment. However, they also highlight the ecological obstacles created by technological developments: electronic wastes and increased energy needs. This research recommends that the IT sector needs to be eco-conscious, with a special emphasis on Open Source Technology. This approach seeks to reduce power consumption through creative notions such as Virtualization, Server consolidation and storage technologies. In the midst of all these research findings, when it comes to opting for a Consolidation solution in terms or server consolidation itself, this study truly gives its seal of approval to Xen's paravirtualization strategy mainly because many environmental factors are taken into account along with economic ones which involve even small-scale enterprises. By promoting this sustainable methodology in

organizations that belong to all manner of industries and businesses-however manufacturing, or service based companies may be -firms are able to reduce their carbon footprint while at the same time improving operational costs.

3. The survey indicates how urgency brought on by population growth presses upon electronic device use, and the importance of green computing solutions. According to M Ranjani (2012) the need for a green maturity model in virtualization is emphasized. Confronted with rising costs and environmental concerns, the survey recognizes that computer power consumption is on the rise. It therefore recommends power reduction strategies in response to this trend. Highlighting the importance of thoughtful design for reducing energy and emissions, such a survey pushes forward co-management of sustainability architecture in infrastructure on one hand, as well as software architecture. It finally suggests a comprehensive green approach which includes water and transport conservation for the future.

4. In creating technology, the imperative link between technological innovation and environmental conservation is discussed. It puts an emphasis on the need for green computing under conditions of ballooning carbon dioxide emissions. Intriguingly, Singh (2015) provides concepts for sustainable infrastructure in key parts of the architecture. This approach underlines the rise in carbon emissions as computer use increases. According to this reasoning, it is not until we have green IT that these become clean lights and electricity brings low prices but high satisfaction for real smiles on faces. This work plays a role as the foundation stone, and the survey helps them recognize green computing. It demands environmentally sound approach, pointing to the enormous scope in possible savings in energy. Citing relevant studies, it also points up a lacuna in thorough-going research on sustainable computing perspectives which have great potential for developments to come.

5. The study by Moedjahedy & Taroreh (2019) critically addresses the escalating energy consumption in data centres supporting Internet applications. Their innovative approach, implementing a Green Data centre at Cravat University (UNKLAB), significantly reduces energy consumption through simulation, direct observation, and the application of clustering and virtualization techniques. By incorporating VM-level metrics into the power consumption formula, a remarkable 286.66% reduction is achieved after virtualization. While the study demonstrates the efficacy of clustering and virtualization at UNKLAB, the survey lacks an explicit identification of existing gaps, necessitating a more comprehensive exploration of related work and research deficiencies in the field.

6. According to a study by B. Yamini (2010), global warming is caused by increasing carbon dioxide emissions from energy consumption, all of which underlines the need for environmentally friendly procedures and encourages carbon tax policies as well. But it has no relation with modern times. Elaborating upon this, a recent survey stresses that the design of application for "green" devices should be based on energy-saving considerations throughout its operating lifespan. Only after the fact does it mention in passing ways of building a Green IT environment, including such concepts as virtualization and energy management or remote work. However, these ideas are barely connected to what is proposed here--a cloud-based solution using novel click letter masking algorithm. The survey needs better linking of existing research. Citing recent studies, the proposed system should define its unique contribution in filling a gap not otherwise addressed by scientific study.

7. According to Wadhwa and Verma (2014), green cloud computing is not just necessary, it actually saves money. It can also be made easier to use than conventional distributed databases are today. How to improve energy efficiency and, in particular, reduce cloud computing's carbon footprint is the aim of this study. Among other things it examines various strategies such as utilization of

techniques derived from concepts surrounding virtual machines like vmuelog() scheduling and migration. The literature review offers a comprehensive survey of previous studies, delineating their major characteristics. In light of the proven advantages in virtualization, however, it is really important-a need clearly noted by this survey-to take into account not only migration costs but also one's own ability to handle system stability and performance during a migration.

8. Indeed, according to a survey by Roy & Gupta (2014) of energy consumption in cloud computing finds that because this new network is facing exponential growth it's more critical than ever necessary for green computing. While cloud services have the advantages of cost-effectiveness, concerns about carbon emissions and profitability arise in response to data centre energy use requirements. In brief, the green cloud-enabled framework proposed here attempts to achieve a balance between profitability and environmental protection through lowering carbon footprints. The study notes that future research into cloud services and green computing should highlight energy efficiency, emissions reduction. It concerns CO2 effects, presents an energy saving structure and gives objectives as well as schemes to be congruent in aspects of data processing. In the existing literature there is a glaring research gap in this regard.

9. This survey indicates a big step forward in the concept of green cloud, as represented by Goyal & Garg (2022), which introduced an ecological approach. This new approach uses cloud and green energy technology to make data storage and computing more environmentally friendly. Its a clear focus on energy, water and equipment efficiency and greenhouse gas reductions. In this field, for example, one can see the advancement in energy monitoring and virtualization as well as efficient data centres. In this regard, the literature recognizes accuracy and energy savings as beneficial effects of adoption but lacks research into what greater these may signal and how their spread would affect security. More detailed examination of this gap will lend greater depth to the survey.

10. The survey emphasizes the urgency of dealing with carbon emissions from ICTs, which account for 2% of global greenhouse gas emissions. Based on Light (2020) ' s study, it also examines the crucially important place of green networks. It describes how software-defined networking interacts with edge computing coupled with virtualizing processors. These technologies can produce substantial energy savings by simulation in Cloud Sim. Yet accurate measurement is still limited. One example is that putting edge and virtualization together has potential for improving efficiency, pointing to a type of research gap: there are more precise measurement methods needed.

11. The primary goal of the authors Devarasetty &Reddy (2018) is to address energy consumption problems for cloud computing. In other words, they propose a novel algorithm called MOACO that draws its inspiration from the biological sciences and employs Ant Colony Optimization. Their primary interest lies in optimizing placement of virtual machines onto data centres to avoid wasting resources, cut energy consumption and reduce communication costs. In order to evaluate the effectiveness of the proposed MOACO algorithm, it is compared against two existing methods: MOGA and DVFS. In addition, the study has important implications for dealing with energy efficiency issues in cloud computing systems and provides useful techniques such as bio-inspired algorithms. It also points to directions of future research.

12. Tawfeek et al. (n.d.) address cloud computing, coming up with a task arrangement policy based on an ant colony optimization (ACO) algorithm. The authors completely adjust various parameters, and the outcome is compared with ordinary FCFS (order of arrival), round-robin algorithms to minimize task make span. Cloud Sim simulations reveal ACO consistently beats traditional techniques. As future tasks, this study focuses attention on several topics. The exploration of the task

precedence relationships is highlighted as are load balancing techniques and comparisons with other meta-heuristic methods. This reflects the determination of the authors to advance optimizing cloud computing, illustrating a completely integrated way of meeting changing challenges in that area.

13. In terms of cloud computing task scheduling the authors Chintamani & Amdani (2022) introduce a proposed Ant Colony Optimization (ACO) algorithm sending tasks to virtual machines with low network cost and improved efficiency. The ACO algorithm further validates superior efficiency and execution cost. As well, the study offers an detailed survey on cloud computing use, highlighting how scheduling can optimize resource utilization within data centers. It finds multiple resource scheduling strategies and gets into the architecture of Cloud Sim, showing how well ACO can reduce make span times compared to traditional approaches.

14. In tackling the difficulties surrounding cloud data centres, Pang et al. (2017) build an integrated dynamic energy management system in their contribution to solving power consumption problems. They elaborate upon a design including DVS Management, Load Balancing and Task Scheduling modules. LET-ACO is a task oriented resource allocation method, which from the point of view of system runtime and energy consumption has been analysed through Stochastic Petri Net (SPN). Through the results of computer simulation, the efficiency of the newly proposed LET-ACO method is determined. Thus a comparison with ACO can show that saves as much at 40% in energy use compared to its competitor It stresses the importance of thinking in a comprehensive manner about energy-efficient cloud data center management. Future directions include deeper system modeling; consideration, for example, of network bandwidth and traffic; exploring possible ways to apply biocomputing toward solving problems encountered by cloudscapes.

RELATED WORK

Virtualization

In data centre environments, advanced technology including virtualization consumes resources and energy efficiently. It decomposes and abstracts hardware into components so that a single system runs as if it were several independent units. Consequently, not only can the same number of servers run more operating systems and applications at the same time, but up to 7,000 kilowatt hours (kWh) in power consumption per server per year will be cut. Where data centre efficiency is of prime concern, virtualization plays a central role in green IT strategies. Since the mainframe era, task concurrency has developed into a technology. High-end hardware such as RAM, processors and networking components are specially configured by modern cloud systems. In contrast with traditional processing in sequence, virtualization allows the sharing of resources among different virtual machines and automatically balances workloads. It also secures electronic environments. Server virtualization is perhaps the most important form of virtualization. When several systems are combined onto powerful servers, computing resources become abstracted. This not only optimizes resource usage but also lowers power and cooling consumption, contributing greatly to achieving energy efficiency. Some progress in the domain of server virtualization are Dynamic Voltage and Frequency Scaling (DVFS) which allows CPU operating frequencies and voltages to be dynamically adjusted based on workload requirements, and power monitoring. This technology adjusts the level of computing power to match current workload, which means considerable energy savings. By integrating DVFS with server virtualization, allocations and energy consumption of resources are optimized to enhance the long-term environment for your data center. Power monitoring, however, is an important part of server virtualization. In virtualized environments, patterns of power consumption are continuously monitored and analysed in real time. This is an important source of information about resource usage and energy efficiency for administrators. Energy monitoring allows green IT improvements to be repeated, with inefficiencies noted and places needing attention identified.

Dynamic Voltage Frequency Scaling (DVFS)

DVFS, or dynamic voltage and frequency scaling is one of the most widely-used power management techniques in modern computer systems (especially processors). By dynamically adjusting both voltage and clock frequency Dynamic balance between power consumption and performance. Such a adjustment makes the processor capable of responding to workload requirements in real time, increasing performance during heavy computational loads and power conservation under light ones. The varieties of DVFS approaches considered here provide a glimpse into the complexity involved in performance optimization. This study offers practical perspective on various DVFS methods and gathers wisdom about the current state of affairs in regards to such. A rigorous empirical analysis allows us to point up differentiated DVFS techniques and by providing both theoretical underpinnings as well as practical implementation, we hope our work will be of value across the board.

1) Modelling of Quality of Service (QoS) With Frequency Awareness (Mao et al., 2020)

DVFS in virtualized environments, however, requires as a major step the integration of Frequency-Aware QoS (Quality of Service) modelling framework. Taking virtual machines (VMs) as separate tasks, this frame of reference stresses real-time behaviors with deadlines and performance parameters like Worst Case Execution Time (WCET), to specify the user QoS requirements. This is an important strategy for improving performance and energy efficiency while meeting the time-critical requirements of tasks.

The QoS modelling approach takes into account the following crucial components:

1. Deadlines and Worst-Case Execution Time (WCET): This study provide a DVFS system-specific quality of service (QoS) model for real-time applications. While non-time critical jobs are based on frequency requirements, traditional tasks include time limits. The CPU frequency will always be higher than the total of the minimal specifications. Maintaining a minimal QoS frequency is sufficient for lengthy tasks without any special time constraints. This strategy prevents SLA violations while balancing performance and energy economy.

2. Frequency Ratio Model: The energy consumption of modern CMOS-based processors consists of both dynamic (E_{dyn}) and static (E_{sta}) components. The power supply voltage (V_{dd}), active gate percentage (A), load capacitance (Cl), and processor frequency (f) are used to calculate dynamic power (P_{dyn}). This frequency-voltage correlation provides a concise representation of dynamic CPU performance expressed as $P_{dyn} = (\beta. f3)$. DVFS uses frequency ratios (r = f/fmax) to accommodate discrete frequency levels.

The frequency expression of total CPU performance (P) is the combination of its static and dynamic components expressed as

$$P = P_{sta} + P_{dmax} \cdot r^3$$

3. For jobs in virtualized settings that demand frequency, a dynamic voltage and frequency scaling (DVFS) model is presented in this section, which employs the Lagrange multiplier approach to calculate energy consumption based on clock cycles, frequency, and power levels. The ideal energy usage for a single processor is derived, and the total energy usage for the entire system is calculated. By taking into account various frequency levels, the problem of determining the ideal frequency ratio is resolved.

The optimal energy consumption is calculated as follows:

$$^E opt_c = \sum_{i \in T} \Gamma_i \cdot {}^f max_c ({}^P sta_c \cdot {}^r opt_c + {}^P dmax_c \cdot ({}^r opt_c)^2)^4$$

4. Analysis of Frequency Ratios: The suggested frequency-aware DVFS model's energy consumption is computed using $\lambda = P_{sta} / (P_{dmax} + P_{sta})$, which is based on the frequency ratio (r). The flexibility of this paradigm is demonstrated by its analysis of a theoretical case. Figure demonstrates the relationship between energy consumption and frequency ratio (r). Different static power shares denote various server configurations and energy-saving measures. Higher numbers do not correspond to more energy-efficient servers; lower values do.

Mao et al (2020) creates a thorough framework for enhancing energy usage and performance in virtualized environments by integrating Frequency-Aware QoS modelling into our DVFS approach. This method enables us to meet the various demands of tasks in real-time environments while maintaining a delicate balance between power efficiency and computing power.

2) System Architecture: Energy-Aware VM Allocation and Migration (Masoudi et al., 2022)

The system architecture represents a sophisticated approach for optimizing energy consumption in cloud data centres, operating in an environment consisting of physical machines with different processing power measured in Millions Instructions per Second (MIPS). These machines serve as the foundation of your cloud infrastructure and host a set of dynamic virtual machines (VMs). This system allows a data centre administrator to interact with a user interface and request her VMs based on her specific computing needs. The core of this architecture lies in the intelligent allocation and migration of these VMs to reduce energy consumption and associated costs.

Phase 1: Particle Swarm Optimization (PSO) to Reduce Energy Consumption

Phase 1 leverages the power of particle swarm optimization (PSO) to achieve two important goals. First, the overall energy consumption within a cloud data centre must be minimized. Second, it aims to reduce the overhead costs associated with VM migration. PSO is an optimization method inspired by natural phenomena in which individuals (particles) within a swarm adjust their positions in the solution space based on their own experience and the group's collective knowledge.

In this context, Phase 1 progresses through several important steps.

1. Particle position and velocity calculation: The position and velocity of each particle is calculated based on its current state and previous experience.
2. Mapping the VM migration problem to his PSO: The VM migration problem is mapped to his PSO framework, which allows the particles to represent the specific her VM allocation configuration.
3. Define particle locations and objective functions: Each particle location corresponds to a specific VM mapping configuration. The objective function for each particle summarizes the total energy consumption and displacement.
4. Velocity calculation of particle motion: Calculate the velocity vector to determine how each particle moves in the solution space, thereby guiding the search for the optimal configuration.

The end result of these steps leads to the implementation of an energy-aware load balancing particle swarm optimization algorithm (EALBPSO) that aggregates energy-aware VMs and lays the foundation for energy-efficient allocation.

Phase 2: Load Balancing With PSO

Based on the results of Phase 1, Phase 2 will focus on fine-tuning the distribution of VMs across active physical machines. The central goal is to minimize the differences in processor and memory utilization between these machines. This is very important for balanced and efficient allocation of resources.

The objective function for this phase is carefully developed and key metrics are taken into account and it is expressed as:

$$F = \sum_{k=1}^{|PHnew|} (Uck - U\widetilde{c})^2 + (Ubk - U\breve{b})^2$$

Processor Utilization (Uck): Represents the active physical machine processor utilization.
Memory Usage (Ubk): Reflects the memory usage of these machines.

Average values across active machines (Uc˘ and Ub˘): Provides global average values to facilitate load balancing evaluation. By minimizing the deviations of these metrics, phase 2 achieves a more balanced distribution of tasks, which in turn improves the energy efficiency of the cloud data centre.

Overall, the system proposed by Masoudi et al. (2022) provides a comprehensive solution to the energy aware VM allocation problem in cloud data centres. It effectively combines heuristic algorithms, optimization techniques, and performance modelling to achieve energy-efficient resource utilization.

3) Integration of Dynamic Voltage and Frequency Scaling (DVFS) Into Genetic Algorithm (Coa et al., 2023)

E Coa et al (2023) describes a novel approach for scheduling tasks in cloud workflow applications, aiming to optimize energy consumption, reliability, make span, and memory constraints. The proposed method utilizes a Genetic Algorithm (GA) inspired by biological evolution, employing a unique three-segment integer encoding for chromosomes. These segments represent:

i)   task-to-VM mapping
ii)   VM-to-configuration mapping
iii)   task-to-CPU frequency mapping.

Initial Population Generation: This provides a description of the suggested genetic algorithm (GA)-based method for cloud workflow planning. Making an initial population of schedules first, then confirm that they adhere to storage limitations. In order to optimise the VM setup, the algorithm then goes through numerous rounds of evolution using crossover and mutation operations. In rare instances, frequency scaling operations are also carried out. This method makes use of a three-segment integer encoding that is chromosome-specific. This is a representation of the task-to-VM, task-to-configuration, and task-to-CPU frequency mappings. This coding is crucial for the optimisation of plans.

Chromosome Modification: The chromosomal modification procedure makes sure that the VM is set up in the best way possible for carrying out the task. It evaluates the compatibility of each chromosome's storage size across the population. The VM chooses the ideal memory amount if it is unable to accomplish the given task. Task execution is optimised by this routine change.

Genetic Operators (Crossover): The difficulty of maintaining task grouping information during genetic processes is addressed by the suggested three-segment crossover operation. This switches between the two chromosomes at random, selecting a task and its corresponding VM setup. Task grouping is little disrupted as a result. On paired chromosomes, the programme repeatedly performs crossovers and modifies the arrangement as necessary. optimises frequency and unused VMs. The offspring of the interbreeding are assimilated into the new population.

Genetic Operators (Mutations): Mutation in genetic engineering improves exploration by exploiting the idle rate of VMs. The higher the idle rate of VMs, the more likely they are to be selected for mutation. Each chromosome is iterated through the algorithm and may require replanning. Depending on the rate you choose, mutations will occur. Tasks are randomly distributed among VMs, resulting in varying frequency levels. When the empty VM is removed, a modified population is returned. This strategy prevents local optimization and promotes variation.

Frequency Scaling: In DVFS-enabled cloud workflows, the frequency scaling technique introduced in this method optimizes the CPU frequency during task execution. Both execution cost and energy consumption are significantly affected by this optimization. Each chromosome is evaluated repeatedly by the algorithm to ensure make span compliance. Determine potential processing margins for eligible orders at various frequencies and choose the most economical option within specified parameters. Task frequencies are updated to the population for better optimization.

The authors of this work contrasted the EES, DEWTS, and WUS DVFS-based approaches with their proposed energy- and reliability-aware task scheduling system. They expanded these techniques to take into consideration task memory constraints and workflow cost reduction. At various scales, experiments were conducted utilising various procedures (Sipht, Montage, CyberShake, and LIGO). The VM design used two of his DVFS-capable AMD processors, and checkpointing and soft error modelling were used to increase reliability. The studies employed Cloud Sim and Google Cloud custom machine types and took into account lincar and superliner pricing models. The proposed method demonstrated its efficiency in streamlining DVFS-enabled cloud operations by outperforming the baseline method in terms of overall cost, energy consumption, manufacturing time, and dependability.

In conclusion, combining DVFS with genetic algorithm frameworks offers a sophisticated approach to streamlining cloud workflow application scheduling. This strategy enables the algorithm to optimally distribute resources to jobs by dynamically adjusting processing power. The clever application of slack and empirical performance analysis strengthens the applicability and efficacy of this method in actual computer settings.

Findings and Limitations

Implementation of dynamic voltage and frequency scaling (DVFS) in cloud computing environments offers a promising avenue to optimize energy consumption and performance. Through approaches such as frequency-aware QoS modelling, energy-aware VM allocation and migration, and integration with genetic algorithm frameworks for task scheduling, we are making significant strides in enabling energy-efficient, high-performance computing in the cloud.

However, it is important to be aware that there are some limitations. First, these approaches can be resource-intensive and can cause scalability issues in large cloud environments. Furthermore, its effectiveness may depend on the specific characteristics of the workload and application being executed. Finding the right balance between energy efficiency and performance is a complex task with potential trade-offs between the two. Additionally, reliance on specific hardware or platform capabilities may limit the applicability of these approaches to specific cloud environments. Finally, the dynamic nature of workloads requires constant adjustment of parameters and strategies, making it a challenge to maintain efficiency over time.

Ant Colony Optimization (ACO) appears as the ideal choice among Particle Swarm Optimization (PSO) and Genetic Algorithm (GA) unified with Dynamic Voltage and Frequency Scaling (DVFS) for numerous reasons. ACO, unlike PSO and GA, factors in network costs, crucial for real-world cloud environments. In comparison, GA may lack the same detailed QoS modeling. ACO's detailed system architecture, inspired by ant foraging behaviour, provides a nuanced approach compared to PSO. ACO's real-time flexibility in task scheduling, vigorously adjusting CPU frequency, sets it apart from PSO and GA. Notably, ACO achieves up to 40% energy consumption savings, showcasing its dominance in minimizing energy use while meeting performance constraints.

Dynamic Voltage and Frequency Scaling (DVFS) holds promise for enhancing energy efficacy in cloud computing. Despite progressions, challenges like resource intensity and workload dependencies continue. Future efforts must prioritize refining DVFS implementation, ensuring seamless adaptation to varied workloads while conserving productivity. Addressing these challenges is vital for a strong integration of DVFS, confirming sustained efficiency in optimizing energy consumption and performance in the dynamic cloud computing landscape.

METHODOLOGY

Proposed Framework

Figure 1. Framework: Optimizing communication-aware request stacks in virtualized systems

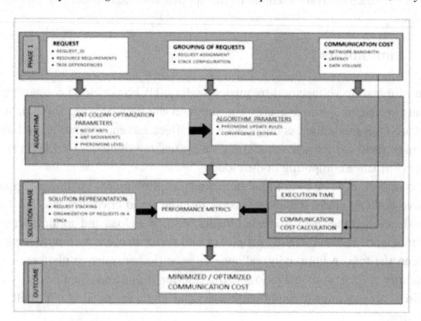

This chapter describes a new comprehensive Framework for optimizing communication-involving request stacks in Virtualized systems. The goal of our research can be thought of as being organized into four phases. Each embodies a part necessary for building an efficient communication-aware optimization framework according to the whole picture, namely from overall system optimization until finally becoming operational in practice. The phases are meant to build upon one another, offering a systematic approach for addressing the variety of demands and consequences that had to be considered in getting request processing, resource allocation and communication cost into control as part of migrating computing environments into virtualization.

Phase 1: Understanding Requests in Virtualized Systems

1. Request Attributes

Understanding how requests work in virtualized systems is the first step, focusing on the basic points: Request ID, resource requirements and task dependencies. In the end, it is simply a matter of maintaining an accurate record; allocating resources appropriately and scheduling work properly. Understanding these basic elements is very important--it's like a stable support for dealing with requests in virtualized environments. Not only does this make things run smoothly, it also increases the performance of all aspects of the system to ensure efficient operation. It's the laying of a foundation for an orderly, productive system.

2.   Request Stacking and Grouping

We thereby introduce the concept of request stacking. We systematically process requests into a queue with structure according to such characteristics as Request IDs; resource requirements (such as memory and disk space); task relationships (such that later tasks both depend on their predecessors and allow them to be run in parallel). This is an organization which optimizes the method of execution, reducing communication overhead and increasing efficiency. By so grouping requests with these common attributes, the process of handling can be more streamlined. Improved system performance is thereby realized by eliminating unnecessary communication and optimizing resource use. A strategic approach to request management forms the basis for a well-organized, smoothly operated system in virtual environments.

3.   Optimization Factors

Elements like "Request Assignment" and "Stack Configuration," play a crucial role in forming groups. We examine the tremendous effect of optimizing these configurations on improving request handling performance. The discussion also relates the case to communication costs, which are discussed thoroughly there. Network bandwidth is one of several resources humanized has in common with dynamic and batch concurrent processing which need to be accounted for so as not to add much overhead or mean that they're never fully utilized regardless how many servers you throw on it. This underlines the importance of tuning configurations, because doing so can have a direct effect on efficiency and resource economy in request processing within virtualized systems. Thus, these critical attributes are optimized as a major factor affecting the performance of request management processes.

Phase 2: Ant Colony Optimization (ACO) for Communication-Aware Request Stacks

Building on the foundational understanding from Phase 1, we introduce the use of Ant Colony Optimization (ACO) as a metaheuristic algorithm for optimizing communication-aware request stacks.

1.   ACO Parameters

Ant Colony Optimatio n (ACO) parameters are pivotal in the enhancement of communication-aware request stacks within an optimization framework. These parameters encompass the number of ants, ant movements, and pheromone levels, each playing a distinctive role in the optimization process.

- The quantity of ants determines the breadth of exploration within the solution space. While a higher number of ants enables a more extensive search for optimal solutions, it introduces the trade-off of increased computation costs.
- Ant movements, defined by specific rules, govern the exploration-exploitation balance. Well-designed movements facilitate the algorithm in efficiently discovering new solutions while exploiting promising ones, contributing to the overall optimization process's effectiveness.
- Pheromone levels, representing the intensity of communication among ants, guide the prioritization and assignment of requests. Optimizing pheromone levels effectively steers the algorithm toward solutions that minimize communication costs, enhancing the efficiency of the optimization process.

Figure 2. The request stacking and grouping process

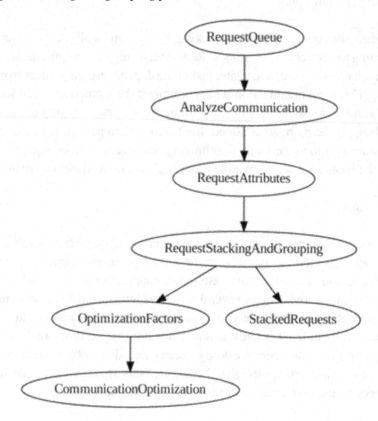

The benefits of tuning ACO parameters are multifaceted. They enable efficient prioritization of requests, focusing the algorithm on paths leading to reduced communication costs. Properly tuned parameters ensure a balanced exploration of the solution space, preventing premature convergence and allowing for the discovery of optimal solutions.

Additionally, ACO parameters influence the assignment of requests, contributing to optimal resource utilization and minimizing unnecessary communication overhead. The adaptability of these parameters to varying system conditions ensures continued effectiveness in minimizing communication costs based on the specific characteristics of the virtualized environment.

2.   Algorithm Parameters

In Ant Colony Optimization (ACO), algorithm parameters play a pivotal role in steering the decision-making process for efficient communication-aware request stacks. Two key parameters are central to this optimization framework:

Figure 3. Modelling ant behaviour for pathfinding

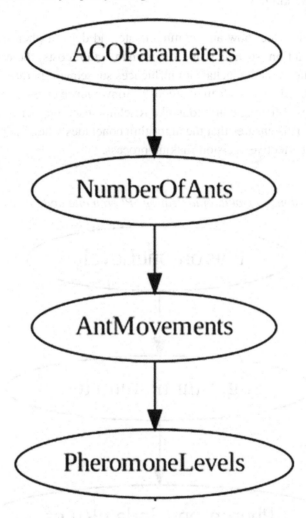

- Pheromone Update Rules:

Pheromones act as indicators of communication among ants, shaping the paths chosen during optimization. The rules governing pheromone updates dictate how these levels are adjusted. The dynamic adjustment of pheromone levels, based on the quality of explored paths, allows the algorithm to adapt to evolving conditions. This adaptability enables ants to concentrate on paths that lead to minimized communication costs, ultimately elevating the overall efficiency of the optimization process.

- Convergence Criteria:

Convergence criteria delineate the circumstances under which the optimization process is deemed complete or converged to a satisfactory solution. Establishing appropriate convergence criteria is pivotal to prevent unnecessary computations, ensuring that the algorithm halts when it reaches a satisfactory solution. This strategic approach prevents excessive processing that might lead to diminishing returns, aligning the algorithm's operation with the specified goals.

- Guiding Decision-Making:

Pheromone update rules steer how ants communicate and deposit pheromones on paths. If a path yields a more optimal solution, such as reduced communication costs, the corresponding pheromone level increases. This reinforcement mechanism influences subsequent iterations, guiding ants to prefer paths that have demonstrated success. Simultaneously, convergence criteria define when the optimization process concludes, establishing conditions like reaching a set number of iterations or achieving a specific solution quality. This ensures that the algorithm concludes when the predefined goals are met, shaping a purposeful and effective decision-making process.

Figure 4. From local decisions to global optimization: Pheromone updates in ACO

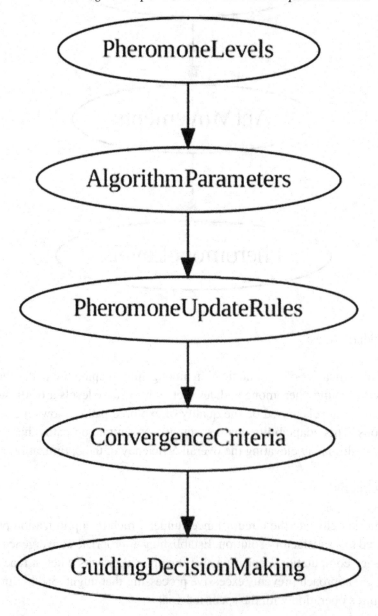

Phase 3: Solution Representation in Communication-Aware Request Stacks

In this phase, we discuss the representation of solutions within the context of communication-aware request stacks, focusing on request stacking methodology and its impact on resource allocation and efficient execution.

1.   Request Stacking Methodology

Request stacking leverages a "last-in-first-out" approach to manage resource allocation and execution. Each request is a layer on the stack, prioritized based on arrival time or performance metrics (e.g., execution speed, resource requirements). This allows efficient resource utilization, prioritizing newer requests while ensuring older ones are not forgotten. Monitoring metrics like throughput, latency, and resource utilization helps evaluate the stacking system's effectiveness and optimize resource allocation for improved performance.

2.   Performance Metrics

In the digital sphere, performance metrics serve as vital indicators, unveiling the efficiency and responsiveness of operational systems. Execution time and communication cost, key metrics influencing task speed and fluidity, play a central role in optimizing systems for seamless user experiences.

- **Execution time** acts as a stopwatch, measuring the duration for a system to process a task. A shorter execution time ensures faster responses, enhancing system efficiency. For instance, in e-commerce, swift execution during checkout correlates with user satisfaction and improved conversion rates, highlighting the importance of optimizing workflows.
- **Communication cost** represents resources used in data exchange. Minimizing this cost improves scalability and overall system performance. Optimizing communication protocols and reducing data redundancy in distributed systems diminishes communication cost, ensuring a responsive environment.

These metrics, intertwined, shape system efficiency. A balance of swift execution and minimal communication cost yields a responsive system. Vigilant monitoring enables developers to identify bottlenecks and fine-tune processes for operational excellence.

Examples of performance metrics:

- **Throughput:** Number of tasks completed per unit of time.
- **Latency:** Time between sending a request and receiving a response.
- **Memory usage:** Amount of memory consumed by a process.
- **CPU utilization:** Percentage of CPU time used for computations.
- **Network bandwidth:** Amount of data transferred per unit of time.

Figure 5. Request stacking methodology

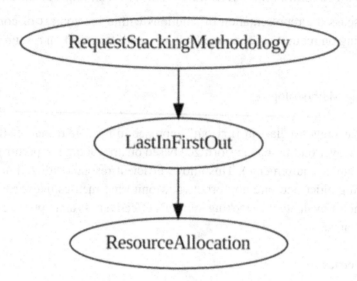

Phase 4: Minimizing Communication Cost for Enhanced System Performance

Our final phase examines the concept of minimizing communication costs within systems and how it contributes to improved overall system performance.

1.   Optimization Efforts

Optimization efforts involve scrutinizing past research endeavours dedicated to curtailing communication costs and improving system performance. This review assesses methodologies and strategies implemented in prior studies, seeking insights into successful approaches and lessons learned. By delving into the collective knowledge accumulated in the field, researchers aim to identify effective optimization techniques that contribute to streamlined communication, reduced resource expenditure, and overall enhanced operational efficiency in virtualized systems.

2.   Impact on System Performance

Minimizing communication costs profoundly impacts system performance by fostering quicker response times, mitigating delays, and optimizing resource utilization. As communication overhead diminishes, the system achieves enhanced responsiveness, significantly reducing the time required for task completion. This efficiency translates into a more streamlined and agile operation, ensuring that resources are judiciously allocated and utilized. Ultimately, the concerted effort to minimize communication costs acts as a catalyst for improved overall system performance, fostering a responsive and resource-efficient environment.

Figure 6. Understanding system efficiency: A breakdown of key performance metrics

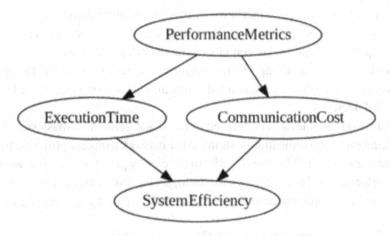

Figure 7. Impact on system performance

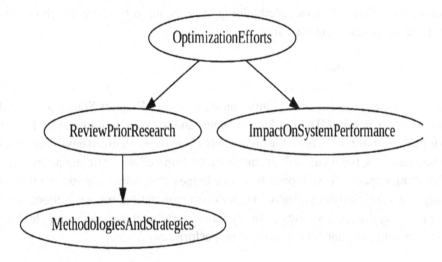

Real-World Examples

In a dynamic cloud computing platform catering to diverse user needs, the optimization framework is applied to enhance resource allocation efficiency and minimize communication costs. The real-world impact is highlighted through scenarios faced by the cloud service provider.

- Flash Sale Handling:
  Challenge: During a flash sale, the cloud platform experiences a sudden surge in VM requests. Solution: The framework efficiently manages the increased load, dynamically allocating resources to handle the spike in traffic. Communication costs are minimized, leading to a seamless shopping experience for users. For example, during a Black Friday sale, the system effectively scales resources, ensuring uninterrupted service and a positive user experience.

- Diverse User Requests:
  Challenge: Users submit varied requests with different resource demands.
  Solution: The framework's optimization factors adapt to diverse requests, ensuring that each user receives the required resources. This enhances the overall performance of the cloud service. For instance, when users with varying application requirements access the cloud, the system intelligently allocates resources based on individual needs, optimizing user satisfaction and system efficiency.
- Multi-Tenant IoT Devices:
  Challenge: Numerous IoT devices connected to the cloud generate a massive volume of data.
  Solution: Minimizing communication costs in the IoT network using the proposed framework reduces latency in data transmission. The cloud platform efficiently processes and manages IoT data, ensuring optimal performance. In a smart city scenario, where IoT devices continuously transmit data, the framework reduces communication overhead, ensuring real-time processing and responsiveness.

Superiority of Communication-Aware Request Stacks Methodology Over Dynamic Voltage Frequency Scaling

Communication-Aware Request Stacks Methodology stands out as the superior choice over DVFS for optimizing virtualized systems. Here's why:

1. Context-Aware Optimization:

   Tailored for virtualized systems, the Communication-Aware Request Stacks Methodology focuses on key attributes such as Request ID, task dependencies, and resource requirements. This context-aware optimization is purpose-built to navigate the unique intricacies of virtualized environments. By considering these crucial elements, the methodology ensures a customized and efficient approach, emphasizing precision in handling requests. This approach acknowledges the distinct context of virtualized systems, providing a targeted solution that optimally manages communication, resource allocation, and task dependencies. In essence, the methodology is finely tuned to the specific needs and challenges posed by virtualized environments, enhancing overall system performance.

2. Holistic System Understanding:

   The methodology ensures a holistic understanding of requests, addressing challenges in request processing, resource allocation, and communication costs within virtualized systems. Unlike the energy-centric focus of DVFS (Dynamic Voltage and Frequency Scaling), this approach comprehensively covers multiple facets of system optimization. By considering the intricate interplay of request dynamics, resource needs, and communication efficiency, the methodology provides a well-rounded solution. It acknowledges the interconnected nature of these elements, contributing to a more thorough and effective optimization strategy. This holistic approach ensures that the virtualized system is optimized not only for energy efficiency but also for overall performance, responsiveness, and resource utilization.

3. Efficient Communication Overhead Management:

Diverging from DVFS's exclusive emphasis on energy, the Communication-Aware Request Stacks Methodology prioritizes efficient request stacking and grouping. This targeted approach directly confronts communication challenges within virtualized systems. By organizing requests strategically and considering shared attributes, it mitigates communication overhead. The result is improved resource utilization and heightened system performance. In the dynamic landscape of virtualized systems, this methodology stands out for its ability to address communication issues comprehensively, transcending the singular focus on energy consumption seen in DVFS. It presents a more nuanced and effective strategy tailored to the intricate demands of virtualized environments.

4. Optimization Tailored to Virtualized Systems:

Crafted for virtualized environments, the Communication-Aware Request Stacks Methodology excels in tailored optimization, accounting for distinct features like request attributes and dependencies. This specificity offers a more pertinent and precise optimization strategy compared to DVFS, which may lack the nuanced adaptability needed for the intricate demands of virtualized systems. By addressing the unique characteristics inherent in virtualization, this methodology demonstrates a heightened aptitude for optimization, ensuring a more effective and contextually aware approach compared to the more generalized focus of DVFS.

5. Utilization of Metaheuristic Algorithms:

Embracing sophistication, the Communication-Aware Request Stacks Methodology integrates Ant Colony Optimization (ACO), a powerful metaheuristic algorithm renowned for navigating intricate solution spaces effectively. This intelligent approach diverges from DVFS, which predominantly hinges on voltage and frequency adjustments, potentially lacking the problem-solving finesse intrinsic to metaheuristic algorithms. ACO's adaptive nature enables it to excel in scenarios with high complexity, showcasing its prowess in optimizing communication-aware request stacks. This stands in contrast to DVFS, highlighting the Communication-Aware Request Stacks Methodology's utilization of advanced problem-solving techniques for enhanced efficiency and performance in virtualized systems.

6. Structured Implementation Approach:

The Communication-Aware Request Stacks Methodology distinguishes itself through a structured and systematic implementation approach across well-defined phases, ensuring accessibility and ease of execution. In contrast, DVFS may lack such a clear structure, potentially hindering usability and complicating successful implementation. The methodical progression of phases in the Communication-Aware Request Stacks Methodology contributes to its user-friendly nature, providing a roadmap for implementation that fosters efficiency and effectiveness in optimizing virtualized systems.

7.   Focus on Minimizing Communication Costs:

The methodology stands out for its explicit focus on minimizing communication costs, a distinct feature that aligns with the overarching goal of enhancing overall system performance. This targeted emphasis on reducing delays and optimizing resource utilization distinguishes it from DVFS, showcasing a commitment to fostering more responsive systems. The Communication-Aware Request Stacks Methodology's strategic attention to communication efficiency contributes significantly to its effectiveness in addressing key challenges within virtualized environments, providing a tailored solution for improved system responsiveness and resource utilization.

8.   Scalability:

The Communication-Aware Request Stacks Methodology excels in scalability, accommodating evolving system needs seamlessly. Its structured expansion approach optimally handles resource demands and communication intricacies as systems grow. Unique attributes consideration ensures adaptability in virtualized environments, addressing challenges in larger settings. Efficiently managing communication overhead, the methodology minimizes resource consumption, especially in extensive networks. The structured implementation approach aids in addressing communication complexities for larger systems, and the integration of Ant Colony Optimization enhances adaptability to intricate scenarios. The methodology ensures enhanced performance and responsiveness even as the system scales, dynamically adjusting to varying conditions.

The Communication-Aware Request Stacks Methodology offers distinct advantages for both small-scale and large-scale industries:

1.   Resource Optimization:
     ◦   Small Scale: Limited resources in small industries are efficiently managed by the methodology. Precise resource allocation prevents wastage, ensuring that every resource is utilized optimally, thereby enhancing overall performance.
     ◦   Large Scale: Diverse and extensive resource pools in large-scale industries are effectively handled. The methodology aids in managing and allocating resources based on demand, contributing to streamlined and efficient operations.
2.   Cost Efficiency:
     ◦   Small Scale: For smaller industries, minimizing communication costs directly impacts their financial efficiency. The methodology's focus on cost reduction ensures that available resources are used judiciously, contributing to overall cost efficiency.
     ◦   Large Scale: Large-scale industries, dealing with substantial communication overhead, benefit significantly from the methodology. By minimizing communication costs, it leads to substantial savings and improved financial efficiency on a larger scale.
3.   Scalability:
     ◦   Small Scale: Smaller industries, often facing growth challenges, find the methodology adaptable to their evolving needs. The structured implementation approach facilitates scalability, allowing them to expand operations smoothly.

    ◦   Large Scale: Large-scale industries dealing with complex operations and scalability challenges appreciate the methodology's systematic framework. It helps in addressing communication issues seamlessly as operations expand.

4. Operational Efficiency:
    ◦   Small Scale: Improved communication and resource efficiency contribute to streamlined processes, enhancing overall operational efficiency for smaller industries. This results in quicker task completion and improved customer satisfaction.
    ◦   Large Scale: The methodology's holistic approach is crucial for large-scale industries dealing with intricate operations. It ensures that communication challenges are comprehensively addressed, leading to enhanced overall operational efficiency.

5. Adaptability:
    ◦   Small Scale: Tailored optimization for virtualized systems makes the methodology adaptable to the unique needs of smaller industries. It takes into account specific attributes and dependencies, ensuring relevance.
    ◦   Large Scale: Large-scale industries benefit from the methodology's adaptability to handle extensive virtualized environments. It provides a solution that aligns with their scale and diversity, considering the complexity of their operations.

6. Competitive Edge Sustainability:
    ◦   Small Scale: Implementation of the methodology gives smaller industries a competitive edge. Improved system responsiveness and resource management contribute to a positive image and customer satisfaction, setting them apart in the market.
    ◦   Large Scale: Large-scale industries can maintain a competitive edge by efficiently managing communication costs and ensuring optimal resource utilization. This contributes to overall competitiveness in the market and positions them as leaders in their industry.

CONCLUSION

In overall, the CARS method is superior to DVFS in optimizing VMS. CARS was created specifically for challenges in the virtual worlds by taking note of key features like the request ID, dependency on mission, and resource demand. It is a comprehensive and tailored solution because of its context-aware optimization, holistic system comprehension, and focus on efficient verbal overhead control. Even though CARS borrows ideas from DVFS, it goes beyond by proposing effective request stacking and aggregation; the communication bottlenecks for better resource exploitation and machine functionality. The use of ACO allows intelligent and flexible decision-making within the framework of class. Finally, the dependent method of implementation used in CARS contributes to its ease of use, outlining a clear map for successful deployment as compared with the probably less defined scheme of DVFS. Its inherent concern reduction in talk expenses is relevant towards increasing normal instrument efficacy, therefore additional receptive systems. However, it is pertinent to mention that CARS advances on what has been achieved via DVFS though overcoming its limitations. Nevertheless, theoretical models as well as empirical researches are required for verification of practical efficiency of V-CARS in varied virtual contexts. In future, studies have to be focused on strict testing and fine-tuning of all the theoretical aspects of CARS in order to prove the functionality across different workloads and programs. Moving forward, CARS offers an avenue of refining chatty stack request for stronger, energy efficient and efficient data structures.

FUTURE WORK

The circulate closer to Communication-Aware Request Stacking (CARS) indicates a tremendous trade in how requests are controlled inside a system, bringing approximately no longer only a change but a fundamental evolution in the manner conversation needs are treated. Unlike a trifling adjustment, CARS introduces a smarter method to organizing requests inside a machine. At the coronary heart of CARS is its capability to intelligently stack requests, taking into account the complicated dynamics of communication. Instead of following a conventional request processing float, CARS introduces a extra sophisticated stacking mechanism that is inherently aware about how requests engage and rely on every different.

This shift calls for a deeper knowledge of ways requests perform and communicate inside the system. Rather than a linear processing of requests, CARS introduces a contextually aware stacking system. This means that the mechanism organizing requests is now informed by a eager consciousness of the context wherein those requests operate. This ends in greater efficient and optimized pathways for conversation in the gadget. The final purpose of this transformation is to enhance the general performance of the device by way of intelligently grouping and prioritizing requests primarily based on their verbal exchange characteristics. In different phrases, CARS aims to make the machine work smarter by means of information the nuances of how extraordinary requests communicate with each different and organizing them within the most efficient manner possible. However, this evolution goes beyond floor-degree modifications. It necessitates a complete reassessment of present methodologies. Traditional strategies may additionally fall short in completely harnessing the ability of CARS. Consequently, there may be a need to adapt cutting-edge fashions and frameworks to align with the conversation-conscious nature of request stacking. This needs a greater dynamic and responsive device structure, able to accommodating the complexities added by CARS. In essence, the transition to Communication-Aware Request Stacking is not just a minor tweak; it represents a transformative leap in the direction of a greater wise and contextually conscious approach of request management. This evolution has the capacity to redefine how systems take care of verbal exchange, paving the way for extraordinary performance and adaptableness. In summary, CARS isn't always just about converting the way requests are processed; it is about revolutionizing the entire communique shape inside a machine. It's a move toward a more wise, adaptable, and efficient manner of dealing with requests, with the ability to seriously decorate the general overall performance of the device.

REFERENCES

Bhoi, S. K., Chakraborty, S., Verbrugge, B., Helsen, S., Robyns, S., Baghdadi, M. E., & Hegazy, O. (2022). Advanced edge computing framework for grid power quality monitoring of industrial motor drive applications. *2022 International Symposium on Power Electronics, Electrical Drives, Automation and Motion (SPEEDAM)*. 10.1109/SPEEDAM53979.2022.9841966

Cao, E., Musa, S., Chen, M., Wei, T., Wei, X., Fu, X., & Qiu, M. (2023). Energy and reliability-aware task scheduling for cost optimization of DVFS-enabled cloud workflows. *IEEE Transactions on Cloud Computing*, *11*(2), 2127–2143. doi:10.1109/TCC.2022.3188672

Goyal, V., & Garg, S. (n.d.). Green cloud computing for sustainable environment: A review. *IOSR Journal of Environmental Science, Toxicology and Food Technology*.

Light, J. (2020). Green networking: A simulation of energy efficient methods. *Procedia Computer Science*, *171*, 1489–1497. doi:10.1016/j.procs.2020.04.159

Ma, L., Chen, Y., Sun, Y., & Wu, Q. (2012). Virtualization maturity reference model for green software. *2012 International Conference on Control Engineering and Communication Technology*. 10.1109/ICCECT.2012.230

Mao, J., Bhattacharya, T., Peng, X., Cao, T., & Qin, X. (2020). Modelling energy consumption of virtual machines in DVFS-enabled cloud data centres. *2020 IEEE 39th International Performance Computing and Communications Conference (IPCCC)*.

Masoudi, J., Barzegar, B., & Motameni, H. (2022). Energy-aware virtual machine allocation in DVFS-enabled cloud data centres. *IEEE Access : Practical Innovations, Open Solutions*, *10*, 3617–3630. doi:10.1109/ACCESS.2021.3136827

Moedjahedy, J. H., & Taroreh, M. (2019). Green data centre analysis and design for energy efficiency using clustered and virtualization method. *2019 1st International Conference on Cybernetics and Intelligent System (ICORIS)*.

Multi objective Ant colony Optimization Algorithm for Resource Allocation in Cloud Computing Prasad Devarasetty. (n.d.). https://www.researchgate.net/publication/277477534

Patil, A., & Patil, D. R. (2019). An analysis report on green cloud computing current trends and future research challenges. SSRN *Electronic Journal*. doi:10.2139/ssrn.3355151

Ranjani, M. (2012). Green computing - maturity model for virtualization. *International Journal of Data Mining Techniques and Applications*, *1*(2), 29–35. doi:10.20894/IJDMTA.102.001.002.002

Roy, S., & Gupta, S. (2014). The green cloud effective framework: An environment friendly approach reducing CO2 level. *2014 1st International Conference on Non Conventional Energy (ICONCE 2014)*.

Sharma, R., Bala, A., & Singh, A. (2022). Virtual machine migration for green cloud computing. *2022 IEEE International Conference on Distributed Computing and Electrical Circuits and Electronics (ICDCECE)*. 10.1109/ICDCECE53908.2022.9793067

Singh, S. (2015). Green computing strategies & challenges. *2015 International Conference on Green Computing and Internet of Things (ICGCIoT)*. 10.1109/ICGCIoT.2015.7380564

Sneha, T.V., Singh, P., & Pandey, P. (2023). Green cloud computing: Goals, techniques, architectures, and research challenges. *2023 International Conference on Advancement in Computation & Computer Technologies (InCACCT)*. 10.1109/InCACCT57535.2023.10141845

Sun, P., Wang, J., Xu, Y., & Wang, L. (2020). Research and application on power generation safety monitoring and cloud platform. *2020 IEEE 1st China International Youth Conference on Electrical Engineering (CIYCEE)*.

Tambe, A., & Shrawankar, U. (2014). Virtual batching approach for green computing. *International Conference for Convergence for Technology*-2014.

Wadhwa, B., & Verma, A. (2014). Energy saving approaches for Green Cloud Computing. *Recent Advances in Engineering and Computational Sciences*.

Yamini, B., & Vetri Selvi, D. (2010). Cloud virtualization: A potential way to reduce global warming. *Recent Advances in Space Technology Services and Climate Change 2010 (RSTS & CC-2010)*.

Chapter 6
Strategies to Achieve Carbon Neutrality and Foster Sustainability in Data Centers

K. Gopi
M.G.R. Arts and Science College, Hosur, India

Anil Sharma
iD https://orcid.org/0000-0003-4299-0340
Parul Institute of Management and Research, India

M. R. Jhansi Rani
ISBR Business School, Bangalore, India

K. Praveen Kamath
ISBR Business School, Bangalore, India

Thirupathi Manickam
iD https://orcid.org/0000-0001-7976-6073
Christ University, India

Dhanabalan Thangam
iD https://orcid.org/0000-0003-1253-3587
Presidency Business School, Presidency College, Bengaluru, India

K. Ravindran
Presidency Business School, Presidency College, Bengaluru, India

Chandan Chavadi
iD https://orcid.org/0000-0002-7214-5888
Presidency Business School, Presidency College, Bengaluru, India

Naveen Pol
iD https://orcid.org/0000-0002-6113-4665
ISBR College, India

ABSTRACT

Data centers and transmission networks are crucial in the digital age, with the market expected to grow from $50 billion in 2021 to $120 billion by 2030. However, their extensive computing infrastructure and continuous operation generate significant heat, necessitating energy-intensive cooling systems. The IEA report revealed that data center power consumption surged by over 60% between 2015 and 2021, with transmission networks experiencing a 60% usage increase. Addressing these growing energy demands poses significant challenges for the industry, with some countries considering restrictions on new data center licenses due to environmental concerns. To mitigate the climate impact, the industry must prioritize the procurement of low-carbon or carbon-free electricity to reduce Scope 2 emissions related to electricity, heating, and cooling. Tech giants like AWS, Google, and Meta/Facebook have already adopted ambitious public targets, either running on carbon-free electricity or investing in global projects for cost-effective and large-scale emissions reduction.

DOI: 10.4018/979-8-3693-1552-1.ch006

INTRODUCTION

In the age of digital transformation, the significance of data centers and communication networks has grown substantially within our globally interconnected landscape. This is evident in the anticipated market growth, set to rise from $50 billion in 2021 to $120 billion by 2030. Despite their pivotal role in supporting our digital lifestyles, there are environmental considerations associated with the essential functions they perform. Data centers, consisting of servers, storage devices, and networking equipment, consume considerable amounts of electricity due to their extensive computing infrastructure. The continuous operation of these components generates significant heat, necessitating energy-intensive cooling systems to maintain optimal temperatures. As the size and quantity of data centers increase, they have become a major contributor to the growing global demand for energy. A recent report from the IEA highlighted a substantial surge of more than 60 percent in data center power consumption between 2015 and 2021, with a similar increase of 60 percent in the usage of transmission networks.

Addressing the escalating energy requirements of data centers and their associated electricity grids poses significant confronts for the industry, with a number of countries contemplating limitations on new data center permits due to ecological distress. Simply improving energy efficiency is insufficient to efficiently alleviate the climate brunts of data centers. It is crucial for the industry to prioritize the adoption of low-carbon or carbon-free electrical energy to decrease Scope 2 emissions related to power, temperature, and breezy. This transition is already underway, as major tech companies like AWS, Google, and Meta/Facebook, which rank highest purchasers of clean energy since 2010 (Chavadi & Thangam, 2023), have set ambitious public targets surpassing common 100 percent renewable energy commitments. These companies, with extensive data center infrastructure, either operate globally on carbon-free electricity around the clock or invest in global projects to ensure cost-effective and large-scale emissions reduction.

Data centers currently contribute to nearly 2% of the global energy consumption. Despite ongoing concerns about the escalating energy use in data centers for more than a decade, large-scale cloud service providers (CSPs) have effectively addressed this issue by employing IT virtualization and optimizing power and thermal management infrastructure. Enterprises, historically less adept at data center efficiency than CSPs, are progressively transitioning their workloads and applications to the cloud (Sanyal et al, 2023). However, the COVID-19 pandemic's swift acceleration of digital transformation resulted in a heightened demand for data centers, reigniting apprehensions about their future energy consumption. Consequently, customers, regulators, and investors of data center service providers are advocating for environmentally sustainable growth within the industry. Data center sustainability serves as a framework for comprehending the overall operational costs and offers a means to decrease these costs through innovative solutions (Chavdi et al., 2023). The technological requisites related to maintaining and operating a facility significantly impact the overhead and emissions associated with providing data center services. This encompasses IT devices (such as processing and data storage systems), the foundational network grid used for transferring and managing operational data, and the thermal and electrical systems employed to operate all this equipment (device42.com, 2023).

Green or sustainable data centers stand at the vanguard of the sustainability movement, transforming the environmental impact of the tech industry. Through the application of energy-efficient methodologies, utilization of renewable energy sources, and integration of cutting-edge cooling technologies, these environmentally conscious facilities strive to diminish carbon footprints while ensuring seamless data processing. Embracing such centers is pivotal for fostering a greener and more sustainable future (Meenu, 2021). As society increasingly relies on data-driven technologies; the importance of energy-efficient and environmentally responsible solutions cannot be overstated. Sriram et al., (2023) mentioned critical aspect of achieving sustainability lies in the adoption of green data centers. These facilities prioritize the use of renewable energy sources, employ innovative cooling technologies, and optimize server utilization to minimize carbon footprints. By incorporating energy-efficient hardware and implementing responsible data management practices, green data centers provide a comprehensive approach to lessening environmental impact. This exploration delves into the realm of Green Data Centers and their role in fostering a sustainable future (Market Trends, 2020).

In our progressively digitized world, data centers' role in supporting the rapid expansion of information technologies. Through the optimization of energy consumption, utilization of renewable resources, and deployment of efficient cooling systems, these centers contribute to the reduction of carbon footprints. These fundamental measures advocate for a more sustainable and eco-friendly future for data-driven technologies and the overall health of the planet. However, concerns have arisen about their environmental impact due to energy-intensive operations and high carbon emissions. The emergence of green data centers addresses these challenges and champions sustainability (Aiswarya, 2023).

Against this backdrop, this chapter aims to explore how do the data centers impact the climate, carbon footprints produced by data centers, importance of decarbonization for data centers, Strategies For Data Centers' Decarbonization and Ensuring Sustainable Operations, importance and pertinence of embracing sustainable measures among the data centers, and use ground-breaking technologies for modern data centers, the future of data center decarbonization, and the real use cases of sustainable data centers. The study not only aims to highlight the seriousness of the problem but also to propose feasible resolutions and alleyways that can Achieve Carbon Neutrality and Foster Sustainability in Data Centers.

CURRENT LANDSCAPE OF DATA CENTERS' ENERGY CONSUMPTION

Data centers are geographically dispersed worldwide, with the United States leading at 2,701 centers, go after Germany with 487, and the UK being in the third place with 456. China closely follows with 443.5 centers (Statista, 2023). The servers and IT equipment within these data centers must operate within specific temperature and humidity ranges to prevent degradation and reduce the risk of failure. Running a data center consumes a substantial amount of electricity, generating heat in the process. The power consumption effectiveness of a data center is gauged by its Power Usage Effectiveness (PUE), which evaluates the overall caliber by calculating the ratio of power used by the entire center to that used by the network equipment alone (data4group.com, 2021).

Efficient cooling systems are crucial to prevent overheating in the confined space of a data center. Approximately 40% of the total energy used is dedicated to these cooling systems, contributing to data centers being responsible for 5% of conservatory gas emissions, similar to the aviation industry (Kevin Wee, 2023). The scale of energy consumption related to information and communication technology is significant, representing 5-9% of global electricity production. Notably, Bitcoin consumed approximately 105 TWh in 2021, a twentyfold increase from 2016, while Ethereum consumed around 17 TWh. Data centers alone account for 2.8% of the EU's electrical energy demand, with a projected 3.21% increase by 2030 (digital-strategy.ec.europa.eu, 2020).

Figure 1.

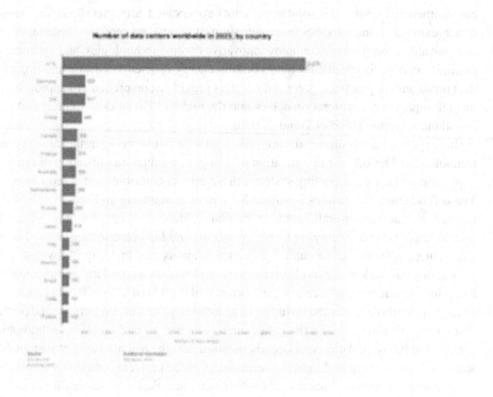

Energy consumption is rising quickly in several smaller nations that are seeing a boom in the data centre industry. In Ireland, for example, data centre electricity use tripled between 2015 and 2021, accounting for 14% of the country's overall electricity consumption. By 2025, data centre energy use in Denmark is predicted to triple, accounting for 7% of the nation's total electricity consumption. According to a report by the State Grid Energy Research Institute of China, China's data centre industry is expected to use more than 400 billion kWh of electricity by 2030, or 3.7% of the nation's overall electricity consumption. The significance comes from the source of that energy as well, since 73% of these data centres use coal, a significant supplier to greenhouse gas emissions (China Daily, 2021).

Numerous strategies are being tried to address the issue of energy sustainability in data centres. These include of implementing novel cooling techniques, investigating renewable energy sources, taking into account different sites like cold climates or submerged regions, and grouping servers into sizable hyper-scale centres. In particular, hyperscalers show energy savings by increasing utilisation while lowering the number of servers. An increasingly popular way to decrease energy usage and lower greenhouse gas emissions in data centres is to integrate renewable energy sources. This strategy reduces the environmental effect of data centres while protecting them from rising energy costs. The environmental effects of digital infrastructure are being measured, tracked, and minimised by a number of efforts (DatacenterDynamics, 2019).

With interim targets for energy efficiency and carbon-free energy usage in 2025, European data centre operators and industry associations created the Climate Neutral Data Centre Compact (CNDCP) in January 2021, pledging to achieve climate neutrality in data centres by 2030. Another effort involves major data centre operators including Google, Microsoft, and Iron Mountain and is called the 24/7 Carbon-Free

Energy Covenant. It is spearheaded by the Sustainable Energy for All Division and the United Nations (europapress.es, 2021). Not surprisingly, the operators of hyperscale data centres in particular are leading corporate initiatives to purchase renewable energy, mostly via Power Purchase Agreements (PPAs). As Rich Miller (2022) points out, the four main purchasers of corporate renewable energy PPAs are hyperscale operators: Amazon, Microsoft, Meta, and Google.

At the state level, progress is being made in fusing energy and digital technologies to create an ICT infrastructure that is greener. The largest solar-powered green data centre in Africa and the Middle East is now being built in Dubai, which serves as an example. This carbon-neutral facility will use only renewable energy sources, with a capacity more than 100 megawatts. Decarbonization efforts are also being made by digital platforms. As an example, Netflix is actively involved in the DIMPACT initiative, which intends to evaluate and reveal the carbon marks of online services (dewa.gov.ae, 2022).

THE IMPORTANCE OF DECARBONIZATION FOR DATA CENTERS

Data centers and networks for data transmission collectively contribute to approximately 1% to 1.5% of the planet's power use, as outlined by the International Energy Agency (IEA). This energy consumption, along with the resulting emissions, exerts direct and indirect impacts on the global climate. The IEA has outlined a roadmap known as the Net Zero by 2050 Scenario, aiming to achieve carbon neutrality in the global energy sector by 2050. This objective seeks to provide an opportunity to control the augment in worldwide hotness to 1.5 degree Celsius (iea.org, 2023).

The choice of 1.5 degrees Celsius as a target is based on substantial risks and concerns associated with global warming beyond this threshold, supported by observed environmental consequences and historical scientific proofs gathered by the Intergovernmental group on environmental issues. Potential risks include heightened and more frequent climate extremes, increased probabilities of extreme drought and water availability risks, expansion of areas prone to flooding, a higher likelihood of a sea ice-free Arctic Ocean during the summer, augmented ocean acidification, posing a threat to various marine species, and increased risks of local species losses and extinctions (Karumban et al, 2023).

While global commitments and actions are on the rise, they still fall short of what is necessary to restrict this temperature rise. Data center emissions have seen only modest growth since 2010, despite an escalating demand for digital services, primarily due to the adoption of more energy-efficient technologies, practices, and increased use of renewable energy. Nevertheless, to avert the severe consequences of climate change, these emissions must decrease by 50% by 2030 (Julia Borgini, 2023).

STRATEGIES FOR DATA CENTERS' DECARBONIZATION AND ENSURING SUSTAINABLE OPERATIONS

Sustainability holds paramount importance in the data center industry, with key stakeholders exploring issues such as escalating energy and water consumption, environmental impact, the objectives of eco-friendly data centers, the advantages of sustainability, the characteristics of constructing a sustainable data center, achieving carbon neutrality, and obtaining certifications that highlight sustainability initiatives. According to S&P Global Market Intelligence, as businesses scrutinize the greenhouse gas emissions associated with their operations, it becomes evident, especially for technology-centric industries that attention needs to be directed toward the data centers powering their digital infrastructure (DCS Content Team, 2023).

S&P Global's Voice of the Enterprise survey on data center sustainability reveals that more than half of the respondents, regardless of their company's size or existing sustainability goals, consider the efficiency and sustainability of their data centers to be crucial. Key findings from the survey include: 66 percent report that their company has established sustainability goals, with larger companies (over 250 employees) more likely to have set goals (70 percent) compared to smaller companies (under 250 employees) at 43 percent. 58 percent believe it is highly likely that their company will achieve its sustainability goals, while 39 percent consider it somewhat likely. 54 percent deem the efficiency and sustainability of data centers as highly important to their organization, with an additional 30 percent stating it is somewhat important (Dan Thompson, 2022).

When establishing new data centers, factors like improving the organization's carbon footprint (32 percent), extending global reach (33 percent), regional headquarters (32 percent), and supporting partner strategies/initiatives (32 percent) are equally critical drivers. The survey authors note that while the focus on sustainability has largely been on renewable energy, future considerations are expected to encompass diverse aspects such as diesel generation and water usage as companies move toward neutralizing their energy consumption with carbon-free sources (DCS Content Team, 2023; Dan Thompson, 2022).

The following are some of the strategies need to adopt to ensure data center sustainability;

Energy Efficiency

The global electricity consumption of data centers at present constitutes 1% of the total global power consumption. Therefore, emphasizing energy efficiency is a central aspect of Meta's strategy for creating and managing the most environmentally friendly data centers globally. In future power-efficient IT equipment designs and their breezy systems utilize outside air and unswerving evaporative breezy methods to conserve both power and water (Veerendra Mulay, 2018). In addition, our construction sites are carefully designed to enhance local Biological diversity, slot in indigenous vegetation and adaptive setting that replicate usual hydrology, thereby mitigating the town heat island consequence. city heat islands arise when doing arrangements, such as constructions and infrastructure, absorb and re-emit more of the sun's temperature compared to usual landscapes (Yevgeniy Sverdlik, 2020).

Renewable Energy

One other crucial component in lowering a data center's carbon footprint is the inclusion of renewable energy sources, such solar or wind power. According to Meta, every data centre in the fleet—which is spread across six nations and more than 20 US states—is fueled only by renewable energy. As our operations grow, we are dedicated to upholding this quality (Veerendra Mulay, 2018). This accomplishment is mostly attributable to our proactive strategy for obtaining renewable energy, which has led to the installation of over 70 new wind and solar power projects and provided the American grid with over 9,000 megawatts of clean electricity. The diversity of our renewable energy projects is fueled by our operating footprint, according to Amanda Yang, Meta's head of renewable energy for Americas West and the Asia-Pacific region. "Ensuring that we are adding new renewable capacity in the areas in which we operate is one of our primary values," the speaker said. Even though renewable energy powers all of their operations worldwide, we are dedicated to improving our approach to help grid decarbonization and lowering emissions per unit of electricity produced. Yang went on, "To make this happen, we are collaborating wherever possible with our utilities, developers, and regulators (Mary Riddle, 2023)."

Water Conservation

Water Conservation One of Meta's major and ambitious sustainability objectives are to achieve water positivity by 2030, and this initiative begins with efforts to minimize our water consumption (Nicole Loher, 2023). In order to minimise the amount of water needed for construction, Meta uses a number of strategies, such as using recycled water when it is practical, putting best management practises into practise, and recycling water inside our facilities to use less water overall. They use a variety of natural plant species, effective irrigation techniques, alternate water sources, and smart scheduling technologies outside of their structures. Our annual water savings come to around 130 million gallons (sustainability. fb.com, 2023).

Waste Reduction and Material Recycling

Reducing waste and reaching net-zero are related objectives. A circular approach is necessary, viewing waste as a valuable resource that may be recycled, reused, or repurposed, if a data centre is to fully exemplify sustainability. By recovering, reusing, and recycling materials, Meta actively seeks to divert waste from data centre construction and demolition projects from landfills and incineration facilities (sustainability.fb.com, 2023). In 2022, Meta diverted 91 percent of potential garbage from nearby landfills by recycling, reusing, or donating 158 thousand tonnes of materials. In the meta offices and data centers, it prioritize the use of certified Forestry Stewardship Council (FSC) products for permanently installed wood, ensuring responsible forest management, protection of indigenous people's customary rights, and the exclusion of highly hazardous chemicals (Thangam et al, 2022).

It also highlights the usage of recycled materials in data centre hardware system designs. Our main goal is to make hardware easily disassembled and reusable. It looks into ways to prolong the life of the racks and other parts that are used in our data centres in order to improve resource efficiency. It works closely with downstream partners to ensure responsible management of any residual materials by finding secondary markets for components outside our data centres (fsc.org, 2023).

Green Certifications

It also emphasises how data centre hardware system designs make use of recyclable resources. Our primary objective is to simplify hardware disassembly and reuse. In order to increase resource efficiency, it looks into ways to extend the life of the racks and other components that are employed in our data centres. By locating secondary markets for components outside of our data centres, it collaborates closely with downstream partners to guarantee responsible management of any residual materials (fsc. org, 2023).and one in Luleå, Sweden, is certified LEED Platinum. These certifications cover more than 19 million square feet in total, signifying our commitment to environmental excellence and sustainability (Nicole Loher, 2023).

Utilize Infrastructure Management for Data Centres

Reducing carbon footprints becomes easier with the help of Infrastructure Management for Data Centres (IMDC). IMDC systems provide real-time insight into data centre operations, making efficient energy use monitoring and control possible. IMDC solutions identify inefficiencies and open the door to improved data centre performance by closely examining data from IT equipment, power and cooling infrastructure, and other infrastructure. Furthermore, by automating repetitive processes like equipment provisioning and decommissioning, these systems reduce the possibility of mistakes and boost productivity. Data centres may significantly reduce their energy usage, increase operational efficiency, and lessen their carbon footprint by implementing IMDC technologies (Alexander, 2023).

Integrate Intelligent Systems

One of the most important steps towards reducing carbon footprints is the integration of the IDCM system with the Building Automation System (BAS), also known as Integrated Data Centre Management (IDCM). Building automation systems (BAS) manage and keep an eye on the HVAC (heating, ventilation, and air conditioning), lighting, and security systems. Data centre managers can see more into facility operations thanks to the IDCM and BAS systems working together, which makes it easier to optimise energy use and save carbon emissions. The ability to precisely monitor cooling system performance thanks to this integration gives managers the ability to adjust cooling capacity in accordance with actual data centre requirements. Data centres may minimise energy usage and carbon emissions while maintaining ideal working conditions for IT equipment by optimising cooling capacity. In summary, a potent tactic for lowering a data center's carbon footprint and increasing overall energy efficiency is the successful integration of the DCIM system with the BAS system (Modius, 2023).

IMPORTANCE AND PERTINENCE OF EMBRACING SUSTAINABLE MEASURES IN DATA CENTERS

The increasing usage of digital technology has led to a significant increase in energy consumption and carbon emissions, making sustainability in data centres even more important. As a result, steps have been made to improve data centre effectiveness, reduce energy consumption, and switch to renewable energy sources (Srikanth, 2023).

The tremendous growth of data has made the need for sustainable data centres imperative. Every year, society requires the computation of an ever-increasing amount of data. Data centre expansion is thus a result of this increase. Recent studies and data, including Arizton's global data centre market prediction, indicate that this trend is not abating. Conversely, it is more likely to keep growing exponentially going forward (Aiswarya, 2023). A data center's sustainability encompasses a variety of elements, such as monetary aspects, ecological aspects, and technological aspects.

Numerous studies emphasise that taking a holistic approach is the best course of action. Resources are closely related to one another and cannot be viewed as autonomous entities. For instance, increasing productivity necessitates using less hardware, energy, heat dissipation, and space. All of these factors have an effect on the environment.

The first stage is to determine the data center's sustainability level, which calls for a set of measurements. As the Uptime survey shows, the majority of data centre operators gather power data for energy efficiency goals, lowering operating costs; yet, only a small percentage of data centre operators now gather data that can be utilised to compute carbon emissions.

Water usage is another significant issue that many governments support, even if the data centre is not situated in a drought-prone area. Since there are currently 15 MW data centres with water consumption equivalent to three medium-sized hospitals, cooling systems will need to become more effective due to the trend towards ever increasing power density hardware. Water-cooling systems provide superior heat conveyance and higher-quality heat for heat reuse, enhancing the data center's sustainability while air-cooling systems have reached their limit.

Since most data centre components age quickly and need to be replaced in order for the facility to continue operating effectively, waste generation is another crucial issue to keep an eye on. According to Microsoft, the typical lifespan of parts used in data centres was five years in 2020. Two strategies could be used to address this problem: creating durable systems and putting in place a circular economy that emphasises material recycling. Because the pollution produced during data centre operations may occasionally be insignificant in comparison to the pollution produced by suppliers during component production, supply chain analysis should be included in sustainability studies. Consequently, it becomes evident that if data centres are to become more sustainable, supplier selection is crucial.

INNOVATIVE ENERGY-EFFICIENT TECHNOLOGIES FOR MODERN DATA CENTERS

Addressing energy consumption challenges in data centers necessitates strategic exploration of various avenues to achieve efficiency (Srikanth, 2023). Key recommendations for fostering energy efficiency in data centers include:

Cooling Systems Upgrade

Upgrading cooling systems comes out as a critical tactic for improving data centre energy efficiency. Adopting strategies such as hot and cold aisle containment optimises energy use by managing ambient temperatures with little overhead. Sustainable operations in smaller server rooms need the purchase of gear that is energy-efficient, with affordability and functionality being secondary considerations. Another workable option is to investigate AI-assisted automatic cooling control, which makes use of real-time sensor data to forecast consumption trends and optimise energy use for cooling.

Renewable Energy Integration

Purchasing data centres' electricity from renewable energy sources is a big step towards energy efficiency. This commitment significantly supports environmental efforts and is in line with carbon neutrality aspirations. All of these steps taken together provide a holistic strategy for promoting creativity and attaining energy efficiency in contemporary data centres, which benefits the environment and operational cost control.

Periphery Computing

Edge data centres provide an effective way to satisfy the demands of Internet of Things (IoT) components, AR/VR technology, and applications including artificial intelligence and machine learning, as market demand and innovation soar. Smaller edge data centres benefit from lower cooling energy use relative to processing output, however sustainability is still an issue. Large data centres, on the other hand, are typically inefficient in terms of electricity, carbon emissions, and electronics waste.

Ensuring Modular Design

Choosing modular data centres over their typical purpose-built equivalents could result in increased energy efficiency. Furthermore, a scalable and modular design makes it possible to include advanced and long-lasting features right away. Modular and scalable data centre construction improves resource distribution efficiency by allowing components to be added or removed as needed, reducing energy waste.

THE FUTURE OF DATA CENTER DECARBONIZATION

According to one estimate, every person on Earth produces 1.7 gigabytes of fresh data per second. As a result, data centres rank among the major and fastest-growing sectors of the world economy, and there are no signs that they will slow down. Sam Mackilligin (2023) projects that the global data centre market would expand at a rate of more than 18% per year, with a predicted value of $250 billion by 2025. The increased demand, especially from AI, automation, and cryptocurrencies, is the main driver of this rapid rise. The International Energy Agency estimates that data centres globally currently use more than 250 terawatt-hours (TWh) of electricity, which equates to 0.3% of carbon emissions and 1% of the world's electrical demand. By 2030, predictions indicate that this percentage may increase to 8% (prnewswire.com, 2021). Data centres must manage their rapid growth while also implementing more environmentally friendly energy practises to lessen their growing carbon footprint. In order to tackle this challenge, stakeholders need to start setting priorities and getting ready for an energy-efficient future.

Intensifying Demands for Heightened Efficiency

The unpredictability of the energy market is driving data centres' increasing need to implement sustainable practises. Between 2021 and 2022, there was a roughly 50% increase in petrol prices in the United States, which led to a 34.6 percent increase in energy costs. Data centres, being large energy consumers, are facing the financial fallout from these price increases and an urgent need to take corrective action. Stricter government rules are adding to the pressure to adopt more environmentally friendly practises. The U.S. Department of Energy (DOE) imposed stricter national criteria for the HVAC equipment necessary for data centre operations on January 1, 2023. Furthermore, it is expected that the SEC will make a decision in 2024 about data centres' mandatory disclosure of greenhouse gas emissions and other hazards related to climate change. It is anticipated that government regulations will force data centres to assess their carbon footprint and develop plans for lowering it (Sam Mackilligin, 2023).

Additionally, the public and their clientele are putting pressure on data centre operators. In a consumer poll conducted in the United States, 87 percent of participants said they preferred businesses that provided social and environmental benefits, and 92 percent said they were more likely to trust businesses that addressed environmental issues. The transition to sustainability is beneficial for businesses as well as the environment. This change in consumer perception should be recognised by data centres (mckinsey.com, 2022).

Navigating Challenges in Energy Efficiency

Overcoming obstacles to improve energy efficiency Data centres are energy-hungry machines that need a lot of energy to run at optimal efficiency all the time. They can use up to 50 times as much energy per square foot of floor area as a commercial building. IT equipment accounts for about half of the average energy usage in data centres, with HVAC equipment accounting for the remaining third. Older data centre operators frequently use backup generators that run on diesel, which greatly increases carbon emissions (uptimeinstitute.com, 2021). The need for data centres to manage their energy more effectively is obvious, but it's still unclear how to cut energy use while keeping operations running smoothly and staying within budget. It makes sense that many stakeholders are reluctant to give sustainability top priority given this complex issue. Just one-third of data centre managers track their carbon footprint, despite the fact that 70% track power usage effectiveness (PUE) and 82% track electricity usage, according to the Uptime 2021 Global Data Centre Survey (Peter Judge, 2022).

The expense of making the required technological expenditures seems to be one of the main barriers to lowering data centre energy use. Stakeholders placed the adoption of new technologies as their third most urgent concern in a 2023 Data Centre & Infrastructure poll, behind security risks and a shrinking workforce. As a result, data centre managers are responsible for extending the life of current systems and equipment rather than swapping them out for more recent, energy-efficient models. According to the poll, the top sustainability goal was to extend the life of equipment (39 percent), with only 16 percent of respondents ranking the decrease of energy usage as the most important sustainability goal (conecomm.com, 2017).

Considering the reluctance and apprehension surrounding the use of novel technology aimed at enhancing energy efficiency, what substitute options are accessible to stakeholders in data centres? Improved building energy management is the answer. Despite their complexity, data centres offer many chances to find and extract energy efficiencies across their whole facility. Technological advances are being aggressively explored by forward-thinking stakeholders to satisfy public expectations, regulatory needs, and budgetary constraints. Numerous of these improvements entail retrofits, upgrades, and finding ways to lower energy use in the facility's current systems. For this endeavour, here is where to begin (Peter Judge, 2022).

Enhanced Energy-Efficient Cooling and Storage Practices

A substantial portion, approximately 35 percent, of the energy consumption in large data centers is dedicated to cooling server infrastructure. To curtail this energy expenditure, stakeholders in data centers are implementing diverse cooling innovations (Paul Evans, 2018). These include tightly coupled cooling and heat removal, cooling towers, high-efficiency chillers, underfloor air distribution systems, and better coordination of currently installed air conditioners. According to a Gartner analysis, data centres that implement specialised cooling systems such as these may experience a significant decrease in operating costs by 20–40 percent by 2025 (Meghan Rimol, 2021). Among the choices are:

- Immersion cooling is a highly effective, dependable, and expandable technique that involves submerging servers into a nonconductive liquid coolant.
- Variable speed fans, which can reduce fan energy requirements by 20–25 percent.
- Control loops that only deliver as much air as the area requires, controlling temperatures at the rack's most important location. Raising the temperature set points can also lower the energy consumption for racks and the size of HVAC units required for cooling.
- Centralized air handlers that are incorporated into a central system to reduce maintenance requirements and increase efficiency.
- Artificial intelligence (AI)-driven machine learning and cooling algorithms that determine and apply the ideal amount of cooling needed at any given time, possibly saving up to 40% on cooling expenses.

Diesel generators enabled backup power delivering systems are a significant contributor to energy consumption and emissions, especially in data centres that are older. Data centres with a forward-thinking mindset are investigating ways to switch from these harmful generators to Battery Energy Storage Systems (BESS). Lithium-ion batteries are used in this cutting-edge, emission-free technology, which offers less startup and running costs. Flow batteries are an effective alternative for data centres, while not being as scalable as lithium-ion batteries. They keep their electrolyte tanks charged indefinitely by storing energy there (Nabil Taha, 2023).

SUSTAINABLE DATA CENTERS: THE REAL USE CASES

The sustainability practices implemented by cloud service providers in the United States and China are exemplified through various use cases. Despite the significant data center presence and influence of U.S.-based providers, the Chinese cloud data center market is still in its early stages, presenting ample opportunities for long-term growth. Google, a major player operating data centers in nearly 40 global regions to support its advertising and emerging public cloud endeavors, serves as an illustrative example.

Google

As Google expands its infrastructure, it prioritizes efficiency gains, compliance with environmental regulations, and fulfillment of corporate social responsibility (CSR) targets. A key aspect of Google's sustainability efforts involves adopting circular economy principles and effective waste management strategies. Notably, a substantial portion of server upgrade components up to 25% is sourced from refurbished inventory, with millions of components resold in secondary markets. These initiatives have led to a remarkable 78% diversion of waste generated by Google's data center operations from landfills. Additionally, Google has undertaken comprehensive reliability studies and optimized server designs, extending the useful life of servers by up to five years. These measures have resulted in reduced server purchases, yielding significant capital expenditure savings and contributing to a gradual reduction in carbon emissions over time (Baron Fung & Lucas Beran, 2022).

Meta

Meta, formerly known as Facebook, has been rapidly expanding its global network of data centers, which house millions of servers. This expansion aligns with the company's increased focus on AI infrastructure to improve user engagement on its social media platforms and prepare for the emerging metaverse. The energy-intensive nature of AI infrastructure poses sustainability challenges, prompting Meta to prioritize eco-friendly solutions. Currently, Meta's worldwide data center footprint, fully powered by renewable energy, includes six recently constructed buildings covering a total area of 3 million square feet. These new data centers have received the Leadership in Energy and Environmental Design (LEED) Gold certification, boasting an impressive Power Usage Effectiveness (PUE) of 1.09 and a Water Usage Effectiveness (WUE) of 0.26. To conserve water, Meta employs direct evaporative cooling, utilizing ambient air to cool data center equipment. In environments unsuitable for ambient air usage due to factors like dust, humidity, or salinity, Meta employs indirect evaporative cooling. The specific solution Meta adopts for indirect evaporative cooling utilizes less water than typical systems, as it employs air to cool water instead of the reverse. Additionally, optimizing relative humidity, combined with direct evaporative cooling, has led to a significant reduction in water usage across Meta's data centers (Baron Fung & Lucas Beran, 2022).

Huawei

In 2020, Huawei and China Telecom joined forces on a project aimed at a facility with 1,570 rack cabinets, spanning 7,000 square meters, designated for a carrier end user. Huawei introduced an innovative approach to decrease the Power Usage Effectiveness (PUE) of the data center through AI-based optimization. The AI optimization successfully adjusted the data center's operating parameters to accommodate diverse ambient conditions, resulting in a reduction of PUE from 1.57 to 1.42. This enhanced PUE translated into substantial electricity savings of 2.89 million kWh, equivalent to 1.16 million RMB per year for the customer. Huawei has actively advocated for the industry's adoption of prefabricated modular construction and environmentally conscious, intelligent energy-saving practices. A notable instance is the Fujian data center for a telecom operator, encompassing approximately 110,000 square meters of floor space upon completion, with an initial implementation featuring 2,910 IT cabinets. Leveraging prefabricated data center construction, digital delivery, and an AI-optimized design incorporating high-efficiency cooling, modular power distribution, and intelligent energy-saving optimization technologies, the data center achieved a reduction in annual PUE from 1.6 to below 1.31. The prefabrication and digital delivery approach also slashed the facility's Time to Market (TTM) by six months.

These highly scalable use cases complement a range of best practices for data center sustainability, addressing stringent requirements related to PUE, Water Usage Effectiveness (WUE), and carbon footprint.

RECOMMENDATIONS

In spite of reservations, data center stakeholders have compelling reasons to proactively enhance their energy management and decrease energy consumption. Experts in the industry unanimously assert that those data center operators who swiftly respond to enhance efficiency are poised to reap the most benefits, especially as sustainability has emerged as a paramount concern for numerous leading companies (timesofindia.com, 2023).

For example, Microsoft has stated that by 2030, all greenhouse gas emissions from its supply chain and facilities will be eliminated. Google and Amazon Web Services have comparable objectives: the latter wants to be carbon-free by 2030, while the former wants to run all of its operations on renewable energy by 2025. Many other operators of data centres are copying their low-carbon efforts (Maryam Arbabzadeh, 2022).

Stakeholders in data centres that are slow to move towards sustainability run the danger of falling behind. Data centres are strengthening their operations with an architecture that can survive the combined pressures of climate change, regulatory obligations, and public perception by acting quickly to improve building energy management.

There are doable, progressive steps existing to ease the transition, minimise disruptions, and lower capital expenditures for data centre stakeholders who are still reluctant to prioritise energy efficiency because of its complexity and cost. To further lessen the financial strain, a variety of funding options are available, such as power purchase agreements, energy procurement contracts, federal and local incentive programmes, green loans, decarbonization projects, equipment rebates, and energy procurement contracts (Maryam Arbabzadeh, 2022).

CONCLUSION

The venture toward achieving sustainability in data centers is in its early stages. While the sustainability challenges may seem substantial, they also bring forth various opportunities for innovation as data center managers strive to meet their objectives. The data center industry has the potential to take the lead, aiming to not only be among the first but also the foremost sector in setting ambitious sustainability targets. These objectives encompass achieving carbon neutrality, utilizing 100% renewable energy, and minimizing waste. By successfully reaching these milestones, the data center industry can establish a comprehensive sustainability framework that serves as a model for other sectors.

The journey toward data center sustainability is in its early stages, presenting both challenges and numerous opportunities for innovation in reaching specific objectives. The data center industry has the potential to lead the way, not only by setting ambitious sustainability goals such as carbon neutrality, reliance on 100% renewable energy, and waste elimination but also by successfully realizing these objectives. By doing so, the data center industry can establish a model for sustainability, providing a framework for other industries to emulate. Collaboration is a crucial element in achieving these goals. It is imperative for the data center industry to take the initiative in defining, establishing, and standardizing sustainability standards and metrics for reporting. Sustainability should not be treated as a mere marketing tactic but as a commitment to accountability and decisive action. Without this commitment, the industry's growth could face challenges from regulatory measures. Given the interconnected nature of sustainability, no single solution, technology, or company can achieve it in isolation. Therefore, it is recommended that data centers engage in meaningful conversations with their partners and customers to exchange ideas and insights, fostering sustainable growth in the data center industry.

REFERENCES

Aiswarya. (2023, August 2). *Green Data Centers: A Key to Sustainability.* https://www.analyticsinsight.net/green-data-centers-a-key-to-sustainability/

Alexander. (2023). *Data center infrastructure management (DCIM).* https://www.techtarget.com/search-datacenter/definition/data-center-infrastructure-management-DCIM

Baron & Beran. (2022). *Achieving sustainable data Center growth.* https://carrier.huawei.com/~/media/CNBGV2/download/products/servies/Achieving-Sustainable-Data-Center-Growth.pdf

Chavadi, C., Manoj, G., Ganesan, S. K., Manoharan, S., & Thangam, D. (2023). Global Perspectives on Social Media Usage Within Governments. In Global Perspectives on Social Media Usage Within Governments (pp. 1-19). IGI Global.

Chavadi & Thangam. (2023). Global perspectives on social media usage within governments. *Global Perspectives on Social Media Usage Within Governments, 2023,* 1–353.

China Daily. (2021, December 9). *Green data centers in focus.* http://english.www.gov.cn/statecouncil/ministries/202112/09/content_WS61b13edac6d09c94e48a1f81.html

conecomm.com. (2017). *2017 Cone Communications CSR Study.* https://conecomm.com/2017-csr-study/#download-the-research

Dan Thompson. (2022, August 2). *The rising importance of data center sustainability.* https://www.spglobal.com/marketintelligence/en/news-insights/research/the-rising-importance-of-data-center-sustainability

data4group.com. (2021, May 6). *What Is PUE?* https://www.data4group.com/es/diccionario-del-centro-de-datos/que-es-pue/

Datacenter Dynamics. (2019, September 13). *Greenpeace: China's data centers run mainly on coal, emit 99 million tons of CO2.* https://www.datacenterdynamics.com/es/noticias/greenpeace-los-centros-de-datos-de-china-funcionan-principalmente-con-carb%C3%B3n-emiten-99-millones-de-toneladas-de-co2/

DCS Content Team. (2023, July 7). *The Importance of Sustainable Data Centers: Why It Matters.* https://blog.datacentersystems.com/the-importance-of-sustainable-data-centers-why-it-matters-for-the-environment-and-your-organization

device42.com. (2023, November 10). *Best Practices for Data Center Sustainability.* https://www.device42.com/data-center-infrastructure-management-guide/data-center-sustainability/

europapress.es. (2021, September 28). *Energy sector leaders launch the 24/7 Carbonless Energy Pact (1).* https://www.europapress.es/comunicados/internacional-00907/noticia-comunicado-lideres-sector-energetico-lanzan-pacto-energia-carbono-24-20210928160229.html

fsc.org. (2023, November 17). *The future of forests is in our hands.* https://fsc.org/en/how-to-be-a-forest-steward

iea.org. (2023, October 22). *Data Centres and Data Transmission Networks*. https://www.iea.org/energy-system/buildings/data-centres-and-data-transmission-networks

Julia Borgini. (2023, May 22). *Navigate Energy Star data center standard and certification*. https://www.techtarget.com/searchdatacenter/tip/Navigate-Energy-Star-data-center-standard-and-certification

Karumban, S., Sanyal, S., Laddunuri, M. M., Sivalinga, V. D., Shanmugam, V., Bose, V., . . . Murugan, S. P. (2023). Industrial Automation and Its Impact on Manufacturing Industries. In Revolutionizing Industrial Automation Through the Convergence of Artificial Intelligence and the Internet of Things (pp. 24-40). IGI Global.

Kevin Wee. (2023, November 3). *The challenges in keeping data centers sustainable*. https://www.johnsoncontrols.com/en_id/insights/2021/in-the-news/the-challenges-in-keeping-data-centres-sustainable

Market Trends. (2020, July 8). *Delivering World-Class Insights with Data-Driven Technologies*. https://www.analyticsinsight.net/delivering-world-class-insights-with-data-driven-technologies/

Mary Riddle. (2023, February 7). *How Meta Sources 100 Percent Renewable Energy Around the World*. https://www.triplepundit.com/story/2023/meta-renewable-energy/765686

Maryam Arbabzadeh. (2022). *Clouding the issue: Are Amazon, Google, and Microsoft really helping companies go green?* https://www.climatiq.io/blog/cloud-computing-amazon-google-microsoft-helping-companies-go-green

mckinsey.com. (2022, June 26). *How to navigate rising energy costs and inflation*. https://www.mckinsey.com/featured-insights/themes/how-to-navigate-rising-energy-costs-and-inflation

Meenu. (2021, February 9). *Green Data Centers: Enhancing Digital Transformation by Conserving Environment*. https://www.analyticsinsight.net/green-data-centers-enhancing-digital-transformation-by-conserving-environment/

Meghan Rimol. (2021, May 3). *Your Data Center is Old. Now What?* https://www.gartner.com/smarterwithgartner/your-data-center-is-old-now-what

Modius. (2023, March 10). *The Role of Data Center Infrastructure Management (DCIM) Software in Running a More Profitable Data Center*. https://www.linkedin.com/pulse/role-data-center-infrastructure-management-dcim-software-running/

Nabil Taha. (2023). *The path to data center decarbonization starts now*. https://www.datacenterdynamics.com/en/opinions/the-path-to-data-center-decarbonization-starts-now/

Nicole Loher. (2023, March 15). *What Does it Mean to Be 'Water Positive'?* https://sustainability.fb.com/blog/2023/03/15/what-does-it-mean-to-be-water-positive/

Paul Evans. (2018, April 19). *Data center HVAC cooling systems*. https://theengineeringmindset.com/data-center-hvac-cooling-systems/

Peter Judge. (2022, March 22). *SEC proposals could force disclosure of Scope 1, 2, and 3 emissions.* https://www.datacenterdynamics.com/en/news/sec-proposals-could-force-disclosure-of-scope-1-2-and-3-emissions/

prnewswire.com. (2021, March 23). *Data Center Market Size to Cross over $ 519 Billion by 2025.* https://www.prnewswire.com/news-releases/data-center-market-size-to-cross-over--519-billion-by-2025---technavio-301253991.html

Rich Miller. (2022, February 12). *Cloud Titans Were the Largest Buyers of Renewable Energy in 2021.* https://www.datacenterfrontier.com/featured/article/11427604/cloud-titans-were-the-largest-buyers-of-renewable-energy-in-2021

Sam Mackilligin. (2023, November 16). *Decarbonising Data Centers.* https://infrastructure.aecom.com/2020/decarbonising-data-centres

Sanyal, S., Kalimuthu, M., Arumugam, T., Aruna, R., Balaji, J., Savarimuthu, A., & Patil, S. (2023). Internet of Things and Its Relevance to Digital Marketing. In *Opportunities and Challenges of Industrial IoT in 5G and 6G Networks* (pp. 138–154). IGI Global. doi:10.4018/978-1-7998-9266-3.ch007

Srikanth. (2023, September 12). *The concept of data center sustainability is revolutionizing the IT industry: Mohammed Atif, Director, Park Place Technologies.* https://www.expresscomputer.in/features/the-concept-of-data-center-sustainability-is-revolutionizing-the-it-industry-mohammed-atif-director-park-place-technologies/103376/

Sriram, V. P., Sanyal, S., Laddunuri, M. M., Subramanian, M., Bose, V., Booshan, B., . . . Thangam, D. (2023). Enhancing Cybersecurity Through Blockchain Technology. In Handbook of Research on Cybersecurity Issues and Challenges for Business and FinTech Applications (pp. 208-224). IGI Global.

Statista. (2023, October 23). *Number of data centers worldwide in 2023, by country.* https://www.statista.com/statistics/1228433/data-centers-worldwide-by-country/

Thangam, D., Malali, A. B., Subramanian, G., Mohan, S., & Park, J. Y. (2022). Internet of things: a smart technology for healthcare industries. In *Healthcare Systems and Health Informatics* (pp. 3–15). CRC Press.

Thangam, D., Malali, A. B., Subramaniyan, S. G., Mariappan, S., Mohan, S., & Park, J. Y. (2021). Blockchain Technology and Its Brunt on Digital Marketing. In Blockchain Technology and Applications for Digital Marketing (pp. 1-15). IGI Global. doi:10.4018/978-1-7998-8081-3.ch001

timesofindia.com. (2023, September 25). *Sustainable Data Centers: Paving the Path for a Greener Tomorrow.* https://timesofindia.indiatimes.com/gadgets-news/sustainable-data-centers-paving-the-path-for-a-greener-tomorrow/articleshow/103925381.cms

uptimeinstitute.com. (2021, October 22). *2021 Data Center Industry Survey Results.* https://uptimeinstitute.com/2021-data-center-industry-survey-results

usgbc.org. (2023, November 14). *LEED-certified green buildings are better buildings.* https://www.usgbc.org/leed

Veerendra Mulay. (2018, June 5). *StatePoint Liquid Cooling system: A new, more efficient way to cool a data center.* https://engineering.fb.com/2018/06/05/data-center-engineering/statepoint-liquid-cooling/

Yevgeniy Sverdlik. (2020, February 27). *Study: Data Centers Responsible for 1 Percent of All Electricity Consumed Worldwide.* https://www.datacenterknowledge.com/energy/study-data-centers-responsible-1-percent-all-electricity-consumed-worldwide#close-modal

Chapter 7
Computational Intelligence for Green Cloud Computing and Digital Waste Management:
Intelligent Computing Resource Management in Cloud/Fog/ Edge Distributed Computing

Sana Dahmani

Independent Researcher, Germany

ABSTRACT

Intelligent resource management across fog, edge, and cloud computing entails dynamic allocation, optimal utilization, and robust security measures. Cloud computing adopts a centralized approach to provision resources, whereas edge and fog computing allocate resources at the periphery, strategically minimizing latency. The incorporation of AI/ML algorithms has a pivotal role, enabling resource prediction for anticipating demand, detecting anomalies, and optimizing allocation efficiently. The self-organizing management aspect facilitates autonomous adaptation. Resource virtualization abstracts physical resources into flexible virtual counterparts, complemented by meticulous accounting that tracks consumption and costs. The inclusion of security-aware measures ensures protection against unauthorized access. This comprehensive approach not only enhances performance, scalability, and security but also promotes adaptive scaling and proactive decision-making. Additionally, the implementation of green IT practices optimizes resource utilization, effectively reducing environmental impact.

DOI: 10.4018/979-8-3693-1552-1.ch007

I. INTRODUCTION

As the Internet of Things (IoT) paradigm continues to expand, there has been a notable transformation in the operation of distributed devices towards a common goal. It remains imperative for these devices to function collaboratively to achieve collective objectives. Cloud computing has emerged as the predominant approach for managing extensive computing tasks on a large scale, providing a variety of services over the internet through distributed infrastructure. This infrastructure comprises virtualized computing nodes that can dynamically allocate resources based on negotiated service-level agreements (SLAs) between providers and consumers. The evolution of cloud computing from earlier paradigms such as distributed computing, cluster computing, grid computing; to its current state involves utility computing services that cover essential requirements such as computing power, data storage, memory, licensed software, and modern software development platforms. Cloud computing services are generally classified into three main types: Infrastructure as a Service (IaaS), Platform as a Service (PaaS), and Software as a Service (SaaS) and more services such as data as a services or development as a service may be defined. However, the proliferation of Internet of Things (IoT) devices has generated a demand for reduced response times and real-time data processing, which is a challenge traditional cloud computing struggles to tackle. This has prompted the emergence of fog and edge computing, aiming to complement the centralized cloud model by bringing computing and storage closer to IoT devices (Elhoseny et al, 2018; Choi et al, 2019).

Traditionally, data originating from user devices such as smartphones, wearables, or sensors in smart city environments and factories is transmitted to distant cloud locations for processing and storage. However, this prevailing computing paradigm may prove impractical for the future as the connection of billions of devices to the internet is likely to escalate communication latencies. Such an increase in latencies could have adverse effects on applications, leading to a degradation of overall Quality-of-Service (QoS) and Quality-of-Experience (QoE). This highlights the need for alternative models which address potential latency issues and enhance the performance of internet-connected devices. Proposing a remedial computing model entails the proximate allocation of computing resources to user devices and sensors, utilizing them for data processing, albeit in a partial capacity. This strategy is designed to curtail the volume of data transmitted to the cloud, thereby mitigating communication latencies. The underlying principle involves the decentralization of a subset of computing resources from extensive data centers, disseminating them towards the periphery of the network, closer to end-users and sensors. These resources can manifest as dedicated 'micro' data centers strategically positioned within public or private infrastructure, or as Internet nodes—such as routers, gateways, and switches—enhanced with computing capabilities. Edge computing demonstrates a computing model reliant on resources situated at the network's periphery. Moreover, a model that amalgamates both edge resources and cloud resources is termed Fog computing. The Fog and Edge conceptual frameworks seek to optimize data processing by strategically distributing computational resources across the network hierarchy (Hong & Varghese, 2018 ; Varghese & Buyya, 2018).

Effective resource management is crucial for Fog Computing due to the resource constraints and heterogeneity of fog devices alternatively referred to as fog nodes or foglets, which are tangible entities, responsible for executing computational functions, storing data, and facilitating communication services at the periphery of the network, in proximity to the data origins. These devices serve as intermediaries, connecting the cloud and end-user devices, thereby minimizing the distance between the two. Resource management strategies for Fog Computing include resource allocation, provisioning, scheduling, and task offloading (Rezazadeh et al, 2023; Fahimullah et al, 2022; Tadakamalla & Menascé, 2022).

Edge computing is a distributed computing paradigm that brings computing and storage resources to the edge of the network, where data is generated and consumed. This can help to reduce latency and improve performance for edge applications. Resource management techniques for edge computing include load balancing, dynamic resource allocation, edge caching, QoS policies, edge analytics, network optimization, predictive maintenance, energy-efficient measures, security protocols, edge orchestration, collaboration between edge and cloud resources, and the integration of AI and machine learning models at the edge. Continuous resource monitoring and auto-scaling policies are used to adapt to evolving conditions and ensure efficient and reliable operation in edge computing environments (Chen el al, 2018).

Within the domain of cloud computing, resource provisioning encompasses the allocation and oversight of computing resources, including virtual machines, storage, and network resources. This occurs within a centralized and scalable cloud infrastructure. In contrast, edge resource provisioning involves the allocation of computing resources at the periphery of the network, in close proximity to end-users or IoT devices. This strategic approach aims to enhance the efficiency of low-latency processing and reduce reliance on centralized cloud resources. Furthermore, fog resource provisioning extends the management of resources to the intermediate fog layer, strategically positioned between the cloud and edge devices. The primary objective is to optimize efficiency and responsiveness in this intermediate layer, facilitating distributed and hierarchical resource allocation. Static provisioning involves the upfront allocation of resources to customers, making it well-suited for applications with predictable workloads. In contrast, dynamic provisioning scales resources based on demand, adapting to applications with fluctuating requirements. Self-service provisioning empowers customers to independently manage their resources through a web interface, streamlining the provisioning process.

Efforts are currently underway to standardize Infrastructure as a Service (IaaS), with initiatives such as the Open Cloud Computing Interface (OCCI) and the Cloud Infrastructure Management Interface (CIMI). These endeavors aim to enhance comparability among cloud offerings and establish interoperability standards. In the realm of Agent-Based Cloud Computing (ABCC), a novel concept known as the "intelligent intercloud" is emerging. Cloud intelligence refers to the defining attributes of cloud agents, encompassing their capacity to exhibit intelligent behaviors and functionalities. This involves making independent decisions, autonomously executing actions, and engaging in social interactions. In today's increasingly digital landscape, characterized by elements like smart homes, remote work, and streaming services, there is a growing demand for computing power and data storage. Yet, the environmental impact of this escalating data consumption is often overlooked. Data centers, tasked with housing and operating digital infrastructure, commonly exhibit inefficiencies in resource utilization. Achieving cost reductions, especially through on-premises solutions for data center construction and cooling technologies, poses a formidable challenge in this context. Cloud, fog, and edge computing represent distinct distributed computing paradigms capable of enhancing the performance and efficiency of applications. However, each paradigm has its unique characteristics and specific requirements (Sim, 2011).

Intelligent computing resource management plays a pivotal role in optimizing performance and efficiency across cloud, edge, and fog computing environments. This strategic approach involves the dynamic allocation and optimization of computing resources using advanced algorithms tailored to real-time conditions and application demands. In cloud computing, this entails autoscaling and predictive analytics within a centralized infrastructure. At the edge, intelligent resource management focuses on low-latency processing and reduced dependence on centralized resources through edge intelligence and dynamic adaptation. In fog computing, the intermediate fog layer optimizes resources in a hierarchical manner, ensuring efficient processing closer to data sources, and fog nodes dynamically scale resources for responsive provisioning. This comprehensive strategy fosters adaptive scaling, proactive resource allocation, and autonomous decision-making to meet the evolving demands of diverse applications in dynamic computing landscapes.

Cloud computing's popularity arises from its efficacy in addressing substantial computational challenges via resource pooling and offering a diverse array of online services. The drive to enhance cloud performance emerges from the need for swift and responsive solutions in real-time applications, driven in part by the practice of computational offloading. Moreover, the exponential surge in data and requests has led to the proliferation of numerous global data centers. These data centers, however, consume substantial energy, resulting in heightened economic and environmental costs. This energy consumption exacerbates the emission of CO_2 and other greenhouse gases, widening the carbon footprint and contributing to the greenhouse effect. These factors underscore the imperative for the development of green technology, demanding energy-efficient solutions that optimize resource utilization. Resource management is a crucial aspect of all three paradigms, but it is especially important for fog and edge computing, due to the resource constraints of these environments. Cloud/Fog and Edge Computing technologies have the potential to address the challenges of green cloud computing by optimizing the utilization of allocated resources through the implementation of intelligent methods. Numerous studies and cloud service providers have demonstrated that the effective organization of allocated computing resources, facilitated by intelligent techniques, can not only enhance resource utilization but also significantly reduce carbon dioxide emissions (Anand et al., 2019).

II. STATE-OF-THE-ART: CLOUD KEY CONCEPTS

1. Cloud Computing (CC)

According to the National Institute of Standards and Technology (NIST), cloud computing is characterized as a model that facilitates accessible, on-demand network entry to a collective reservoir of adaptable computing resources such as networks, servers, storage, applications, and services. These resources can be swiftly allocated and relinquished with minimal managerial involvement or interaction with service providers. The framework of cloud computing leverages distributed resources, amalgamating them to address intricate, large-scale computation challenges and attain enhanced throughput (Mell & Grance, 2011).

Cloud computing (CC) has emerged as the most effective method for overseeing extensive computing paradigms on a large scale. It has gained widespread popularity as it effectively addresses significant computational challenges by aggregating resources and offers a diverse range of supplementary services accessible via the internet. A cloud system can be delineated as a distributed infrastructure, comprised of a network of interconnected and virtualized computing nodes. These nodes are subject to dynamic provisioning, enabling them to function collectively as one or more cohesive computing resources. This provisioning process is orchestrated through the negotiation and establishment of service-level agreements (SLAs) between the cloud service providers and their consumers (Buyya et al., 2009). Since the advent of the Internet, the progression of computing technologies has undergone significant advancement. This evolution has encompassed the interconnection of computers, giving rise to distributed computing, which subsequently evolved into cluster computing, grid computing, and ultimately, cloud computing. The cloud environment offers utility computing services to customers, encompassing essential requisites such as computing power, data storage, memory, licensed software, and contemporary software development platforms. This cloud infrastructure encompasses various deployment and service models (Rani et al., 2015).

2. Cloud Deployment Models

They delineate the manner in which cloud resources are furnished:

- **Private cloud**: The deployment paradigm stemming from the on-premises service model is termed the private cloud. It constitutes an environment or infrastructure that is both constructed and managed by a single organization, exclusively intended for internal utilization.
- **Public cloud**: The deployment model rooted in the SaaS, IaaS and PaaS service models is denoted as the public cloud. A public cloud is a service offering provided by a service provider, such as Microsoft Azure, accessible to a broad audience encompassing individuals and enterprises alike.
- **Hybrid cloud**: A third deployment model, known as the hybrid cloud, remains an option. The hybrid cloud integrates elements from both private and public cloud architectures, constituting a configuration wherein a private cloud environment resides at the consumer's location while concurrently utilizing the public cloud infrastructure.
- **Community cloud**: In a community cloud, multiple organizations with common interests like security and compliance share the cloud infrastructure. It's typically managed by these organizations or a third party and can be located either on-site or off-site1. Community clouds offer increased security compared to public clouds and allow resource sharing among multiple organizations. However, they are less secure than private clouds and require specific governance policies for management.

3. Types of Cloud Computing

- Cloud computing has emerged as an innovative computing paradigm with two main entities: cloud service providers and cloud end-users. The fundamental goal of cloud computing is to provide consumers or end-users with a computing environment marked by adaptable Quality of Service (QOS), leading to advantages for the cloud service providers. This involves defining how resource management is managed, whether autonomously by the user or delegated to cloud service providers. These services are typically categorized into the following service models:
- **On-Premises:** This model pertains to a scenario in which the user assumes full responsibility for the management of all resources.
- **Infrastructure as a Service (IaaS):** IaaS outlines a model where the cloud provider enables the consumer to create and configure resources from the computing layer and beyond. This comprehensive provision involves virtual machines, containers, networks, appliances, and other infrastructure-related assets. The critical task of resource provisioning involves ensuring that a cloud service has sufficient resources to effectively meet increasing customer demands. To fulfill Service Level Agreements (SLAs), a Cloud Service Provider (CSP) must implement specific strategies. For example, the CSP might establish protocols for increasing server capacity or storage as demand grows, ensuring smooth service delivery (Rountree and Castrillo, 2014).

- **Platform as a Service (PaaS):** PaaS provides the user with an environment that spans from the operating system layer and beyond. As a result, the user is freed from the responsibility of managing the underlying infrastructure.
- **Software as a Service (SaaS):** SaaS represents a model with the least amount of control and management required. In this model, a SaaS application is accessible from various clients, and the user has minimal control over the backend infrastructure, primarily focusing on tasks related to application management.

4. Characteristics of Cloud Computing

The National Institute of Standards and Technology (NIST) delineates five distinct characteristics of cloud computing (Rani et al., 2015):

- **On-Demand Self-Service:** is a functionality that involves providing computing capabilities to users based on their specific needs. This includes allocating resources such as storage and server time without requiring direct human intervention.
- **Broad Network access:** access is a feature that allows access to cloud resources through standard mechanisms via network connections. This supports various client devices, including tablets, laptops, mobile phones, and workstations.
- **Resource pooling:** within the multitenant model, involves aggregating and sharing computing resources to meet the requirements of multiple consumers. These resources may be distributed across different geographical locations, with the specific details undisclosed to end users.
- **Rapid Elasticity:** is a characteristic of cloud environments that demonstrates the capability for automated scaling and adjustment of resources to accommodate changing user demands. This ensures effective resource allocation.
- **Measured Service:** is a feature where cloud services undergo systematic metering by the cloud system. Usage data is reported to both the user and the service provider, and this information is utilized by the cloud system to optimize and regulate resource utilization according to the specific service type.
- **Self-Service and Automation:** characterize the capability for individuals to assign and supervise resources through a web-based interface or API. This promotes autonomy and the implementation of automated processes.
- **Elasticity and Scalability:** Cloud solutions exhibit the ability to efficiently adjust to fluctuating workloads, allowing for optimal utilization of resources by easily expanding or contracting as needed.
- **Redundancy and Reliability:** represent the capacity of cloud solutions to effectively adapt to varying workloads, enabling optimal resource utilization through easy expansion or contraction as required.
- **Network Connectivity:** involves accessing cloud services through the internet, requiring a reliable and robust network infrastructure to guarantee uninterrupted connectivity.

5. Cloud Computing Types (Inter-Clouds, Multicloud, Agent-Based Cloud Computing)

Inter-Cloud Computing establishes an interconnected network linking diverse cloud infrastructures from different service providers, enabling seamless data and service exchange across varied cloud environments. This methodology prioritizes enhancing compatibility and communication among different platforms (Interoperability), providing users with the flexibility to migrate workloads and data between distinct clouds (Resource Mobility), and introducing redundancy to bolster reliability and fault tolerance in interconnected cloud setups. In contrast, Multicloud strategy involves leveraging services and resources from multiple cloud providers, emphasizing vendor diversity to avoid lock-in and allowing organizations the flexibility to select the most suitable services from diverse providers. This approach mitigates risks associated with service disruptions by distributing workloads across multiple clouds, offering users the opportunity to optimize costs and performance based on their specific needs. Meanwhile, Agent-based Cloud Computing employs autonomous software agents to execute tasks within a cloud environment, emphasizing agent autonomy and adaptive behavior. These agents operate independently, making decisions without human intervention and dynamically adjusting to changing conditions to optimize performance. The decentralized decision-making process, often involving multiple agents, facilitates parallel and distributed processing within the cloud system, introducing a layer of intelligence and adaptability that significantly contributes to optimizing cloud resources and overall system performance.

5.1 Inter-Clouds Short for "Interoperable Cloud"

Refers to a cloud computing concept where multiple cloud service providers or clouds are interconnected, allowing for seamless data and resource sharing between them. In an Inter-Cloud environment, various cloud platforms, whether public, private, or hybrid, can work together to provide enhanced flexibility, scalability, and redundancy. This interconnectedness enables organizations to distribute workloads, share data, and access resources across different cloud providers or environments, promoting efficiency and resilience in cloud computing operations. Inter-Cloud architectures aim to address issues related to vendor lock-in, data portability, and the ability to select the most suitable cloud services for specific tasks or requirements. An intercloud is described as "cloud of clouds" inhabited by a community of autonomous agents, responsible for automating intercloud resource management tasks and operations. It facilitates collaboration among cloud environments.

Analogous to the Internet's paradigm of being a "network of networks," the intercloud concept engenders a mechanism wherein individual cloud infrastructures gain access to supplementary resources from disparate cloud entities, ameliorating resource scarcity scenarios precipitated by burgeoning consumer demands. Nonetheless, resource governance within an intercloud milieu entails a multifaceted spectrum of endeavors, encompassing: 1) the discernment and curation of requisite resources, 2) the collaborative amalgamation of resources across heterogeneous cloud service providers, 3) the harmonization, orchestration, and synchronization of shared resources to enable concurrent task execution, and 4) the formulation and validation of service-level agreements (SLAs) through intricate negotiation protocols (Sim, 2020 ; Bernstein, et al.,2009 ; Sim, 2015).

Intercloud resource management poses a significant challenge due to the inherent complexity of the intercloud system. This complexity arises from the extensive distributed and interconnected nature of the intercloud, as well as the intricate interactions among its constituent components, including clouds, and its various stakeholders, such as consumers, providers, and brokers. The amalgamation of diverse intercloud resources from various cloud providers to cater to multiple consumers necessitates the management of interoperations, such as service composition and the coordination of concurrent workflows, across distributed computing resources under the jurisdiction of distinct clouds. This endeavor presents a formidable challenge, underscoring the imperative for the deployment of intelligent resource management approaches that facilitate 1) autonomous operations and 2) and effective interactions (Sim, 2020).

5.2 Multicloud

Multicloud entails the utilization of cloud services from diverse providers, ranging from simple software-as-a-service (SaaS) applications like Salesforce and Workday to more complex enterprise deployments involving platform-as-a-service (PaaS) or infrastructure-as-a-service (IaaS) solutions from various cloud service providers such as Amazon Web Services (AWS), Google Cloud Platform, IBM Cloud, and Microsoft Azure. A multicloud solution, in turn, represents cloud computing portability across multiple cloud provider infrastructures and is typically constructed upon open-source, cloud-native technologies like Kubernetes, universally supported by public cloud providers. These solutions often encompass centralized workload management capabilities, accessible through a single console, and are available across various domains, including compute infrastructure, development, data warehousing, cloud storage, artificial intelligence (AI), machine learning (ML), and disaster recovery/business continuity, offered not only by leading cloud providers but also by cloud solution providers such as VMware.

Hybrid Cloud integrates public and private cloud environments to establish a unified and optimized IT infrastructure, allowing organizations to harness the advantages of both realms. Offering seamless management, orchestration, and portability, it simplifies IT operations and reduces costs through a single pane of glass for resource management. Hybrid cloud enhances developer productivity by supporting Agile and DevOps practices, facilitating quick application development and deployment. Leveraging cloud-native technologies like microservices and containers improves application performance, scalability, and resilience. Furthermore, it addresses security and compliance requirements, ensuring secure deployment and scaling of workloads across cloud environments. With precise workload control, hybrid cloud optimizes resource utilization, leading to improved performance and cost-effectiveness. It enables rapid application modernization and ensures seamless data connectivity across cloud and on-premises infrastructure. As a strategic choice for enterprises, hybrid cloud offers flexibility, scalability, and cost-effectiveness, aiding in cost reduction, agility improvement, innovation acceleration, and enhanced security. While considering a hybrid cloud strategy, organizations should evaluate their needs, budget, and expertise, but the clear benefits make it a compelling solution for any enterprise (IBM, 2023).

5.3 Agent-Based Cloud Computing

In 2009, Sim conducted pioneering research focused on advancing intelligent agents for improved resource discovery and negotiation in the realm of cloud computing. Sim introduced the innovative concept of "agent-based cloud computing" (ABCC), which has since evolved into a thriving domain marked by the creation and deployment of intelligent agents. These agents demonstrated a range of functionalities, such as cooperation protocols, automated negotiation mechanisms, coordination mechanisms, reasoning techniques, and heuristics, all designed to automate tasks related to resource discovery, selection, composition, negotiation, scheduling, workflow management, and resource monitoring in both intra-cloud and inter-cloud environments. The context of ABCC is intricately connected with the notion of "intelligent intercloud." Cloud intelligence encompasses the defining attributes of cloud agents, highlighting their ability to exhibit intelligent behaviors and functionalities, enabling independent decision-making, autonomous execution of actions, and participation in social interactions (Sim, 2009 ; Sim, 2012 ; Sim, 2019). Table 1 describes the differences between inter-clouds, multicloud and agent-based cloud computing.

A comparaison between those three cloud concepts is shown in Table 2. Inter-Clouds entail integrating numerous cloud computing infrastructures for enhanced interoperability and the flexibility to move resources. In contrast, Multicloud involves utilizing services from different cloud providers concurrently to harness a range of capabilities, while Agent-based cloud computing utilizes autonomous software agents to improve automation and decision-making in the cloud environment.

Table 1. Description of inter-clouds, multicloud, and agent-based cloud computing

Feature	Inter-Clouds	Multicloud	Agent-Based Cloud Computing
Definition	A collaboration of various cloud providers enabling users to access resources from diverse cloud platforms.	Implementing a tactic that involves leveraging various cloud service providers to distribute workloads and applications.	A distributed computing framework employing software agents for the management and automation of cloud resources.
Benefits	Increased scalability, flexibility, and fault tolerance.	Reduced vendor lock-in, improved performance, and cost savings.	Improved resource utilization, workload optimization, and self-management capabilities.
Challenges	The challenges associated with overseeing multiple cloud platforms, encompassing issues related to security, data sovereignty, and the overall complexity of management.	Overseeing diverse cloud providers, orchestrating data governance, and guaranteeing uniform performance.	Creating and upkeeping agent software, addressing security vulnerabilities, and seamlessly integrating with established cloud infrastructure.
Use cases	Major corporations with intricate IT frameworks and worldwide business activities.	Companies in need of robust availability, superior performance, and scalable solutions.	Apps demanding dynamic resource provisioning and efficient workload optimization.
Examples	OpenStack, Apache CloudStack, and Eucalyptus.	Amazon Web Services (AWS), Microsoft Azure, and Google Cloud Platform (GCP).	IBM Watson AIOps, HPE OneSphere, and Cisco Intercloud Orchestrator.
In-depth explanations	Inter-clouds represent a network of interconnected clouds designed to provide users with smooth access to resources across various cloud providers. This strategy presents numerous benefits, such as heightened scalability, flexibility, and fault tolerance. The federated nature of multiple clouds allows organizations to access a more extensive pool of resources, mitigating the risk of vendor lock-in. Nonetheless, effectively managing inter-clouds can be intricate, requiring careful consideration of security and data sovereignty concerns.	Multicloud involves strategically utilizing multiple cloud providers to distribute workloads and applications. This strategy yields several advantages, such as mitigated vendor lock-in, enhanced performance, and cost savings. Diversifying across multiple clouds allows organizations to capitalize on the unique strengths of each provider and negotiate more favorable pricing. Nevertheless, effectively managing multiple cloud providers poses challenges, and it necessitates careful planning and implementation to coordinate data governance and ensure consistent performance across different clouds.	Agent-based cloud computing is an architecture for distributed computing that utilizes software agents to automate the management of cloud resources. These agents operate autonomously, monitoring resource utilization, optimizing allocation, and executing tasks like workload migration and fault tolerance management. The approach provides benefits such as enhanced resource utilization, workload optimization, and self-management capabilities. Despite these advantages, the complexity of developing and maintaining agent software requires careful attention, and security vulnerabilities must be addressed diligently.

Table 2. Key differences between inter-clouds, multicloud, and agent-based cloud computing

Feature	Inter-Clouds	Multicloud	Agent-Based Cloud Computing
Focus	Federating multiple cloud providers	Using multiple cloud providers strategically	Employing software agents for cloud resource management
Benefits	Scalability, flexibility, fault tolerance	Reduced vendor lock-in, improved performance, cost savings	Improved resource utilization, workload optimization, self-management
Challenges	Complexity, security, data sovereignty	Managing multiple providers, data governance, consistent performance	Agent development, security vulnerabilities, integration with existing cloud infrastructure
Use cases	Large enterprises with complex IT infrastructures	Businesses with high availability, performance, scalability requirements	Applications that require dynamic resource provisioning and workload optimization

6. Cloud Architecture

Within a cloud system, there are two distinct elements: (1) the front end, which is the user-facing interface visible to end users, and (2) the back end, which constitutes the cloud infrastructure itself. These components are interconnected through a network, typically the Internet. The cloud-based delivery model encompasses the method of providing cloud services to users, whether through the internet, private networks, or a combination of both. This model ensures users can access their cloud resources globally. The essential middleware serves a crucial role by providing a runtime environment for applications, with its primary aim being the efficient allocation and utilization of resources (Rani et al., 2015 ; Strickland, 2008 ; Buyya et al., 2013). Table 3 describes more details of the cloud components:

Table 3. Cloud components

Cloud Component	Front End	Back End
Responsibility	Provides a user-friendly interface and presents information to users	Handles the processing and storage of data
Examples	Web applications, mobile apps, desktop applications	Servers, storage, networking equipment, databases
Communication	Communicates with the backend through a network	Receives requests from the frontend and sends responses back
Key Components	User Interface	Servers, Databases, Networking, Security Systems
Detailed Description	The User Interface serves as the visual and interactive component enabling user interaction with the cloud system. This interface may take the form of a web browser, mobile application, or other interfaces that provide users with access to and control over their cloud resources.	The Back End encompasses the tangible infrastructure supporting the cloud system, comprising servers, storage systems, and hardware components responsible for processing, storing, and managing data and applications. Databases play a crucial role, offering storage and retrieval functions. Networking components facilitate smooth communication between the front end and back end, ensuring efficient data transfer. Integrated security systems are vital in the back end to safeguard data, preventing unauthorized access and upholding confidentiality, integrity, and availability.

7. Fog and Edge Computing Environment

The need of IoT devices for reduced response time triggered technological advancements towards fog and edge computing. While cloud computing plays a pivotal role in facilitating applications through its provision of on-demand computing, storage, and networking services, it falls short in meeting the exigencies for immediate responses and actions necessitated by the multitude of real-time Internet of Things (IoT) devices. The pressing requirement of IoT devices to engage in distributed collaborations and perform data processing tasks with minimal latency has spurred technological progress in the realms of fog computing and edge computing which aim at complementing the centralized cloud computing model (Iorga et al., 2018 ; Cisco, 2015). Fog and edge, even though likely to be confused, are different.

7.1 Fog Computing

Fog computing is characterized as a decentralized infrastructure model, strategically positioning storage and processing elements at the periphery of the cloud edge, precisely where data sources like application users and sensors are situated. It represents a distributed computing layer positioned between the cloud and the edge, characterized by devices with greater computational power than edge devices. These fog devices offer additional services like data filtering, aggregation, and analysis. Fog nodes, which constitute autonomous entities responsible for collecting generated data. They can be categorized into three distinct classes: fog devices, fog servers, and gateways. Within this classification, fog devices primarily serve as data repositories, while fog servers assume the additional function of processing the stored data to formulate decision-making processes. Fog computing finds utility in applications demanding real-time data processing and response, such as smart cities, industrial automation, and connected vehicles. Table 4 gives more details on the Fog's feature.

Table 4. Fog chacateristics

Characteristic	Fog Device	Fog Server	Fog Gateway
Form factor	Typically small, lightweight, and low-power	More powerful than fog devices	More powerful than fog devices, but less powerful than cloud servers
Location	Deployed close to the source of data	Deployed in central locations within a fog computing environment	Connects fog devices to fog servers and the cloud
Services provided	Collect, process, and store data locally	Data filtering, aggregation, and analysis	Aggregate data from fog devices and forward it to fog servers. Provide security and authentication services.
Typical use cases	Smart sensors, cameras, and other IoT devices	Industrial automation, video surveillance, and other applications that require real-time data processing	Smart cities, industrial IoT, and connected vehicles

7.2 Edge Computing

Conversely, edge computing constitutes a subset of the broader fog computing paradigm, focusing on the immediate processing of data at its point of origin. Edge devices encompass a diverse range of hardware, including routers, cameras, switches, embedded servers, sensors, and controllers. In the context of edge computing, the data generated by these devices is both stored and processed locally on the respective device, without necessitating data sharing with the cloud infrastructure.It is a computing paradigm that extends computation and data storage to the network's periphery, where data originates and is consumed. Edge devices, typically compact, lightweight, and energy-efficient, are deployed in proximity to data sources. Edge computing serves real-time data processing needs, seen in applications like self-driving vehicles, industrial IoT, and augmented reality. A gist comparaison of Fog and Edge computing is shown in Table 5.

Table 5. Fog and edge computing comparison

Characteristic	Fog Computing	Edge Computing
Location	Between the cloud and the edge	At the edge of the network
Device capabilities	More powerful	Less powerful
Services provided	Data filtering, aggregation, analysis	Real-time data processing and response
Typical use cases	Smart cities, industrial automation, connected vehicles	Self-driving cars, industrial IoT, augmented reality

Fog and Edge relate to each other and to the cloud. Cloud computing has proven invaluable in enabling robust data processing and warehousing capabilities. Among the newer entrants into the cloud computing landscape, fog computing and edge computing stand out, as they inherit certain foundational principles from traditional cloud computing. These innovations are designed to streamline some of the intricacies associated with conventional cloud computing by harnessing the computational potential residing within the local network. This approach facilitates localized computation, rather than relying solely on distant cloud resources. Such an alignment is particularly well-suited to the characteristics of IoT systems. However, the adoption of these technologies brings forth a set of novel security and privacy challenges, which could pose significant hurdles to their widespread implementation (Alwakeel, 2021).

8. Data Flow

Data traverses different architectures in cloud computing, edge computing, and fog computing. In cloud computing, data is centrally stored in data centers owned by service providers, and processing occurs on robust servers within these centers. Edge computing shifts processing closer to the data source, reducing latency and enabling local storage on edge devices or nearby servers—especially beneficial for real-time processing applications like IoT devices. Fog computing introduces an intermediate layer, the fog layer, between the edge and the cloud. Fog nodes or servers near edge devices perform localized processing, optimizing bandwidth and reducing latency. Data transfer in fog computing involves both local and centralized data centers, with fog nodes facilitating efficient communication between edge devices and the cloud. These computing models offer diverse solutions considering latency, bandwidth, and processing capabilities, addressing specific requirements across various applications. Table 6 compares Edge, Fog and Cloud computing data flows.

Table 6. Comparaison between edge, fog, and cloud computing data flows

Data Flow Stage	Edge Computing	Fog Computing	Cloud Computing
Data Generation	Edge devices generate data	Edge devices and fog nodes receive data from edge devices	Cloud systems receive data from fog nodes and other cloud instances
Data Preprocessing	Edge nodes perform initial data filtering, aggregation, and preprocessing	Fog nodes perform more advanced data analytics, filtering, and aggregation than edge nodes	Cloud platforms perform large-scale data processing, machine learning, and pattern recognition
Data Storage	Edge nodes store a limited amount of data for local processing or caching	Fog nodes provide more storage capacity than edge nodes, allowing for intermediate data retention	Cloud systems provide vast storage capacity for long-term data retention and archiving
Data Communication	Relevant and preprocessed data is transmitted to fog or cloud layers for more complex analysis or storage	Fog nodes communicate with cloud layers for centralized data storage, collaboration, and sharing	Cloud-based applications and services access and utilize data stored in the cloud
Data Flow Interactions	Device-to-Edge	Edge-to-Fog, Fog-to-Cloud, Cloud-to-Fog/Edge	Cloud-to-Cloud, Edge-to-Edge, Fog-to-Fog

III. RESOURCE MANAGEMENT TECHNIQUES FOR CLOUD/FOG AND EDGE COMPUTING

Resource management techniques for Cloud, Fog, and Edge Computing encompass the strategies and mechanisms utilized to efficiently allocate, monitor, and optimize computing resources within these distributed computing paradigms. The objective is to augment overall performance, reliability, and resource utilization in dynamic and diverse environments. In Cloud Computing, virtualization technologies abstract physical resources into virtual instances, facilitating flexible resource allocation based on demand, while elasticity dynamically scales resources up or down to meet varying workloads. Edge Computing prioritizes proximity-based allocation, focusing on allocating computing resources near end-users or data-generating devices to reduce latency, particularly crucial for real-time processing applications. Fog Computing introduces an intermediate layer between edge devices and the cloud, optimizing resource allocation within this fog layer and adapting to changing conditions, crucial for handling variable data processing needs across the distributed architecture. These resource management techniques collectively address challenges such as load balancing, high availability, and dynamic demand responsiveness, contributing to enhanced Quality of Service (QoS), diminished latency, and overall improved system efficiency in Cloud, Fog, and Edge Computing environments.

To address the looming latency issue and ensure the continued effectiveness of internet-based applications, alternative computing models are urgently needed. These models should prioritize processing and analyzing data closer to the source, thereby reducing the need for long-distance data transfers and minimizing latency. This shift towards edge computing and fog computing will not only enhance the performance of internet-connected devices but also lay the foundation for a more scalable and efficient computing infrastructure.

In the domain of cloud computing, managing resources involves overseeing the utilization of various elements within a cloud environment. These elements include computing components such as CPU and memory, storage resources like disk space, and network resources, including bandwidth. The process of allocating resources and services from a cloud provider to a customer is termed resource provisioning in cloud computing. This encompasses the optimal deployment of software and hardware to meet application performance requirements. The provisioning process includes static and dynamic provisioning, along with resource allocation based on application needs, aiming to avoid both over and under-provisioning while considering power usage efficiency. Cloud users seek to minimize costs, while cloud service providers strive to maximize profit through efficient resource distribution (Geeks for Geeks, 2023).

Resource management algorithms are essential in fog, edge, and cloud computing, as they optimize the utilization of computing resources throughout the distributed network. These algorithms have diverse goals, including load balancing, offloading, and ensuring Quality of Service (QoS) requirements. According to (Mijuskovic et al., 2021), an evaluation framework and classification for the managing resources in cloud/fog and edge computing involves achieving resource coordination, emphasizing supervisory actions conducted by both service providers and users was created. The resource management algorithms can be succinctly outlined as follows on Table 7:

- **Resource Allocation Optimization:** Resource allocation is a methodology employed to enhance resource utilization and minimize processing costs. Task completion time is a critical consideration, impacting the overall resource allocation process. Various algorithms, including RR, ESCE, SJF, GPRFCA, ERA, Priority-based Resource Allocation (PBSA), and Feedback-Based Optimized Fuzzy Scheduling (FOFSA), implement resource allocation techniques, as detailed in Table 2.
- **Effective Workload Balancing:** Workload balancing plays a pivotal role in managing energy efficiency and mitigating issues like congestion, low-load resource management, and overload. This is particularly challenging for processing nodes within the fog environment. For example, a workload balancing algorithm designed for fog computing aims to minimize data flow latency during transmission procedures by connecting IoT devices to appropriate base stations (BSs). Several workload balancing algorithms from the literature, such as RR, SJF, ESCE, GPRFCA, DRAM, ERA, PBSA, FOFSA, Hill Climbing (HCLB), Efficient Load Balancing (ELBA min-min), and Tabu Search, are discussed in the article.
- **Efficient Resource Provisioning:** Resource provisioning is an approach that addresses the management of task and data requests among fog nodes, representing a subsequent step in the resource allocation process. While resource allocation focuses on assigning a set of resources to a task, resource provisioning involves activating the allocated resources. Algorithms like Remote Sync Differential (RSYNC), Fog Sync Differential (FSYNC), Reed–Solomon Fog Sync (RS-FSYNC), ERA, and Energy-aware Cloud Offloading (ECFO) are specifically designed to handle resource provisioning.
- **Task Scheduling for Efficient Operations:** Task scheduling algorithms are proposed to manage a large set of interdependent tasks, preventing issues such as deadlocks. Table 2 highlights various algorithms managing resources based on task scheduling, including RR, SJF, ESCE, GPRFCA, DRAM, PBSA, FOFSA, ELBA, Tabu, and ECFO. These algorithms play a crucial role in defining schedules to efficiently service tasks dependent on specific sets of resources.

Table 7. Resource management algorithms based on the work of Mijuskovic et al. (2021)

Algorithm	Deployment	Classification	Environment
Round-robin (RR)	Simulation (Cloud Analyst)	Discovery	Cloud–Fog
Equally Spread Current Execution (ESCE)	Simulation (Cloud Analyst)	Load balancing	Cloud–Fog
Shortest Job Next (SJF)	Simulation (Cloud Analyst)	Discovery	Cloud–Fog
Gaussian Process Regression for Fog–Cloud Allocation (GPRFCA)	Simulation iFogSim (a Java-based open-source simulation toolkit for simulating fog computing scenarios)	Discovery & Load-balancing	Cloud–Fog
RSYNC	Experiments in different conditions, two situations of synchronization	Discovery & Off-loading	Fog
FSYNC	Experiments in different conditions, two situations of synchronization	Off-loading	Fog
RS - FSYNC	Experiments in different conditions, two situations of synchronization	Off-loading	Fog
Dynamic Resource Allocation and Matching (DRAM)	Evaluation done with three different types of computing nodes	Load-balancing	Fog
Efficient Resource Allocation (ERA)	Simulation (Cloud Analyst)	Load-balancing	Cloud–Fog
Priority-Based Scheduling Algorithm (PBSA)	Simulation (CloudSim 3.0.3)	Load-balancing	Cloud
Iterative Algorithm	Evaluation done in three different topologies with different workloads	Placement	Cloud–Fog
Feedback-based Optimized Fuzzy Scheduling Algorithm (FOFSA)	Simulation iFogSim	Load-balancing	Fog
Hill Climbing Load Balancing (HCLB)	Simulation CloudAnalyst tool	Load-balancing	Cloud–Fog
Efficient Load Balancing Algorithm (ELBA)	Simulation CloudAnalyst tool	Load-balancing	Cloud–Fog
Tabu Search	Simulation Cloudlet Tool	Load-balancing	Cloud–Fog
Efficient Cloud-fog Offloading (ECFO)	Cloud server and three Raspberry Pi3 devices	Off-loading	Fog–Edge

1. Resource Management Techniques for Cloud

Cloud services are commonly categorized as Infrastructure as a Service (IaaS), Platform as a Service (PaaS), or Software as a Service (SaaS), but the interpretations of these classifications vary, creating challenges in comparing cloud offerings and achieving interoperability. Recent standardization endeavors for IaaS, exemplified by the Open Cloud Computing Interface (OCCI) from the Open Grid Forum (OGF) and the Cloud Infrastructure Management Interface (CIMI) from the Distributed Management Task Force (DMTF), alongside established standards like Amazon Web Services (AWS), are establishing a more robust technical basis for terms like PaaS and IaaS. This shift toward standardized interfaces holds the potential for a clearer separation of responsibilities within cloud environments, encompassing both public and private domains. Key roles within this framework include the Cloud Provider, responsible for furnishing cloud infrastructure and services; the Cloud User, a customer utilizing these resources to deploy and manage applications and data; and the End User, who interacts with the applications and services deployed on the cloud by the Cloud User. This separation of roles is crucial for addressing security, compliance, and manageability concerns, with the Cloud Provider tasked with securing the infrastructure while the Cloud User assumes responsibility for safeguarding their applications and data. Standardized interfaces facilitate this separation by facilitating secure and controlled interactions between the Cloud Provider and Cloud User (Jennings and Stadler, 2014).

2. Resource Management Techniques for Fog/Edge

2.1 Resource Management Techniques for Fog

Fog computing holds promise for addressing the data bandwidth and latency requirements of IoT applications, but effective resource management is crucial. This paper offers a comprehensive review of resource management strategies in fog computing, taking into account resource constraints, heterogeneity, and variable traffic patterns. It discusses recent advancements in fog computing resource management, encompassing aspects like resource allocation, provisioning, scheduling, and task offloading. Additionally, the authors conduct a comparative analysis of related studies and outline potential avenues for future research in this domain (Alsadie, 2022).

Resource management is a vital aspect of research, focusing on the efficient utilization of existing resources. It encompasses a series of steps aimed at achieving a balance between computational resource allocation and ease of implementation. In the context of fog computing, resource management gains significant prominence due to the prevalent constraints on computational resources, necessitating their judicious allocation and utilization. This stands in contrast to cloud computing, which may create the illusion of limitless resources from the perspective of an individual user (Bachiega et al., 2023).

We conduct a thorough examination of security concerns prevalent in both centralized and decentralized computing paradigms, with a particular focus on elucidating how Fog computing can effectively mitigate these concerns, especially in the context of data-intensive services and applications. Fog computing is positioned as a robust platform capable of efficiently managing long-term analytical data, thus significantly reducing the demands on cloud storage and transmission resources. Its successful implementation has been observed across diverse domains, including but not limited to healthcare, smart cities, and geospatial information systems. Our forthcoming research endeavors will center on further enhancing data storage efficiency and security through the implementation of resource management, utilization, and cost optimization strategies. It is noteworthy that Fog computing is rapidly gaining ground in various sectors, encompassing smart traffic lighting systems, connected vehicles, Smart Grid, IoT, Cyber-physical systems, Wireless Sensor and Actuator Networks, as well as Software Defined Networks (SDN). Nevertheless, it is imperative to underscore that Fog Computing functions most optimally when in synergy with cloud computing, a collaboration that proves indispensable due to the intricacies of the tasks involved and the requisite optimization of resource management services (Rashid, 2019).

2.2 Resource Management Techniques for Edge

Effective resource management is a crucial component within the domain of edge computing, where optimizing the use of limited resources at the edge is essential to cater to the diverse demands of various applications and services. The implementation of efficient resource management techniques plays a pivotal role in improving performance, scalability, and cost-effectiveness in edge computing environments. Notable challenges in edge resource management include addressing the heterogeneity of edge devices, ranging from resource-constrained IoT devices to powerful edge nodes, requiring adaptable resource management strategies. Additionally, the resource constraints on edge devices, characterized by limited processing power, memory, and storage compared to cloud servers, demand careful allocation to meet application requirements and minimize contention. The dynamic nature of workloads in edge environments, influenced by user interactions, device activation, and network conditions, necessitates responsive resource management techniques capable of adjusting allocations in real-time. Moreover, considering the sensitivity to latency in edge applications, resource management must prioritize low-latency demands, ensuring timely task execution and minimizing delays. Energy efficiency is crucial, given that edge devices often operate on batteries or have limited power budgets; therefore, resource management techniques must consider energy consumption and optimize allocation to extend device lifetimes. Common resource management techniques for edge computing include resource provisioning, determining optimal resource configurations based on application requirements; resource allocation, assigning specific resources to applications; workload balancing, distributing workloads to prevent bottlenecks; load prediction, anticipating future demands; resource monitoring, continuously tracking usage patterns; offloading, transferring tasks to more powerful cloud or fog resources; edge caching, storing frequently accessed data closer to edge devices; and edge intelligence, enabling intelligent decision-making at the edge. Employing these techniques, tailored to the unique challenges of edge computing, is essential to ensure that edge systems effectively support a variety of applications and services, meeting the evolving demands of our interconnected world.

Amidst the rapid growth of the Internet of Things (IoT), distributed devices are collaborating to accomplish common goals. However, ensuring the seamless operation of these devices for collective action remains critical. This emphasizes the importance of resource provisioning in the IoT, especially considering the limited computational and communication capabilities of end-user devices. Inefficient resource utilization in such scenarios strains the entire system, compromising service quality. Cloud computing techniques have proven effective in addressing these challenges, and newer cloud-based technologies like edge and fog computing have further enhanced resource management effectiveness in the IoT. These technologies bring computation and communication capabilities closer to IoT devices, enabling the offloading of certain services to edge devices. Despite facing unique challenges, incorporating IoT edge devices into traditional IoT paradigms results in optimized resource utilization and improved overall service quality. This special issue delves into various aspects of resource management in IoT edge devices, addressing challenges and proposing techniques and solutions through rigorous reviews and high-quality publications (Kumar et al., 2021).

AI-based fog and edge computing is an aspect of intelligence. Effectively managing resources in fog and edge computing proves intricate due to constraints such as limited resources, heterogeneity, dynamic workloads, and unpredictable environments. Addressing this complexity, AI/ML-based techniques, with a focus on reinforcement learning, demonstrate promising solutions for navigating sequential decision-making in resource management as shown by Figure 1. Table 8 describes a range of algorithms and the way they contribute to resource management. Despite their efficacy, these algorithms encounter challenges, including high variance, explainability issues, and the complexities of online training. Our exploration encompasses diverse machine learning, deep learning, and reinforcement learning approaches for edge AI management, culminating in the presentation of a taxonomy categorizing AI/ML-based resource management techniques. Through a comparative analysis of existing methods, we discern prevalent challenges and highlight potential avenues for future research in AI/ML-based fog and edge computing, aiming to contribute to the ongoing advancement of these technologies (Iftikhar et al., 2023).

Figure 1. AI/ML-based fog and edge computing categories (Iftikhar et al., 2023)

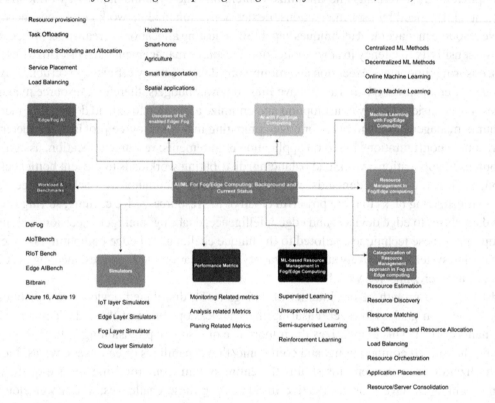

3. Resource Management Techniques for Intercloud

Intelligent Resource Management in Intercloud refers to the automated coordination and oversight of resources within a sophisticated "cloud of clouds," wherein a network of agents collaborates to streamline intercloud resource management processes. This involves efficiently utilizing resources to optimize costs and maximize usage, dynamically provisioning and releasing resources to adapt to varying workloads and demands, ensuring reliable service availability and performance even in the event of failures or disruptions, and prioritizing security measures to safeguard resources and data from unauthorized access and potential attacks. The overarching goal is to create a seamlessly functioning intercloud ecosystem that achieves efficiency, scalability, reliability, and security in the delivery of services to users (Sim, 2020).

Table 8. Algorithms for resource management

Algorithm	Description	Advantages	Disadvantages
First-come, first-served (FCFS)	Allocates resources to applications in the order that they arrive.	Simple and efficient	Can lead to situations where some applications are starved for resources.
Longest job first (LJF)	Allocates resources to applications with the longest expected runtime first.	Can improve performance for long-running applications	Can lead to situations where short-running applications are delayed.
Priority	Allocates resources to applications based on their priority.	Can ensure that critical applications receive the resources they need	Can lead to situations where non-critical applications are starved for resources.
Round robin (RR)	Allocates resources to applications in a round-robin fashion.	Can ensure that all applications receive a fair share of the resources	Can lead to situations where applications are not able to complete their tasks in the time allotted to them.
Lyapunov optimization	Uses Lyapunov functions to control the resource allocation process.	Can ensure that the resource allocation process is stable and that applications receive the resources they need	Can be complex to implement and can require a lot of computational resources.
Game theory	Uses game theory to model the interactions between applications and the resource allocator.	Can find Nash equilibria, which are points where no application can improve its own performance by unilaterally changing its strategy.	Can be complex to implement and can require a lot of computational resources.
Reinforcement learning	Uses reinforcement learning to learn how to allocate resources.	Can find resource allocation strategies that optimize performance, cost, and energy efficiency	Can be slow to learn and can require a lot of data.
Deadline-based scheduling	Allocates resources to applications based on their deadlines.	Can ensure that applications meet their deadlines.	Can be complex to implement and can require a lot of information about the applications.
Backfilling	Allocates resources to applications that are ready to run, even if they do not have the highest priority.	Can improve resource utilization and can reduce the time that applications spend waiting for resources.	Can lead to situations where applications are not able to complete their tasks in the time allotted to them.
Hierarchical scheduling	Allocates resources to applications at multiple levels of abstraction.	Can be used to manage complex applications that are composed of multiple components.	Can be complex to implement and can require a lot of information about the applications.
Market-based scheduling	Allocates resources to applications based on a market-based mechanism.	Can be used to achieve efficient resource allocation and can provide incentives for applications to cooperate.	Can be complex to implement and can require a lot of information about the applications.
Self-organizing scheduling	Allocates resources to applications in a decentralized manner.	Can be used to manage large-scale systems and can be robust to failures.	Can be difficult to predict and control.
Greedy algorithm: Allocates resources to the application that will benefit the most from them at the current time.	Simple and efficient	Can lead to suboptimal solutions in some cases.	
Heuristic algorithm: Uses heuristics to guide the search for a good resource allocation solution.	Can find good solutions quickly	May not find the optimal solution.	

continues on following page

Table 8. Continued

Algorithm	Description	Advantages	Disadvantages
Metaheuristic algorithm: Uses metaheuristics to search for a good resource allocation solution.	Can find good solutions even for complex problems	May be more computationally expensive than greedy or heuristic algorithms.	
Ant colony optimization (ACO): A metaheuristic algorithm inspired by the foraging behavior of ants.	Can find good solutions for complex optimization problems	May be slow to converge.	
Genetic algorithm (GA): A metaheuristic algorithm inspired by natural selection.	Can find good solutions for complex optimization problems	May be slow to converge.	
Simulated annealing (SA): A metaheuristic algorithm inspired by the annealing process in metallurgy.	Can find good solutions for complex optimization problems	May be slow to converge.	

3.1 Cloud-Level Intercloud

Operating at the highest abstraction layer, the cloud-level intercloud seamlessly links public clouds, private clouds, and hybrid clouds, offering a consolidated perspective on resources spanning various cloud providers. This framework streamlines resource pooling, facilitates smooth workload migration, and enhances disaster recovery capabilities. Despite these advantages, it is important to note that interactions within the cloud-level intercloud may experience increased latency owing to the geographical separation of cloud data centers.

Cloud-level intercloud resource management concentrates on the harmonious coordination and distribution of resources among diverse cloud providers, encompassing critical tasks such as resource discovery and monitoring. This involves the identification and continuous tracking of available resources, including CPU, memory, storage, and network bandwidth, across various cloud providers. Additionally, the process incorporates resource negotiation and brokering, where agreements are established to procure resources in a manner that is both cost-effective and efficient, factoring in considerations like price, performance, and reliability. Another integral aspect is dynamic workload migration, which entails the strategic allocation and movement of workloads across cloud providers to optimize resource utilization, balance load, and effectively handle peak demands. Furthermore, cloud-level intercloud resource management encompasses SLA (Service Level Agreement) management, ensuring adherence to predefined SLAs by vigilant monitoring of resource usage, prompt identification of performance bottlenecks, and the implementation of corrective actions as needed.

3.2 Fog-Level Intercloud

Centered on linking distributed computing resources known as fog nodes, the fog-level intercloud prioritizes connectivity closer to edge devices. This strategy empowers fog nodes to efficiently share resources and engage in collaborative processing tasks, ultimately lowering bandwidth consumption and enhancing the responsiveness of applications. Nevertheless, the intricacies of overseeing and coordinating a vast array of geographically dispersed fog nodes pose a notable challenge within this framework.

Fog-level intercloud resource management expands the intercloud paradigm into the realm of fog computing, emphasizing the collaborative sharing of resources among interconnected fog nodes. Fundamental tasks within this framework include fog node discovery and registration, involving the identification and registration of fog nodes within the intercloud network to facilitate the exchange of information on resource availability and utilization. Resource pooling and sharing are integral processes, consolidating resources from multiple fog nodes into a shared pool for the efficient allocation and utilization of fog resources. Additionally, collaborative task processing is emphasized, with tasks being distributed among fog nodes based on their capabilities and resource availability, thereby reducing latency and enhancing application responsiveness. A key aspect is the establishment of seamless communication and resource exchange mechanisms between fog nodes and cloud providers, facilitating the efficient offloading of resource-intensive tasks to the cloud.

3.3 Edge-Level Intercloud

At the forefront of data generation and consumption, the edge-level intercloud establishes connections among edge devices. This methodology empowers edge devices to directly exchange data and collaborate on processing tasks, thereby diminishing latency and decreasing dependence on cloud or fog resources. Nevertheless, the effectiveness of edge-level intercloud interactions is susceptible to challenges arising from the constrained processing power and storage capacity inherent in edge devices.

Edge-level intercloud resource management centers around enhancing resource utilization and fostering collaboration among edge devices while taking into account their inherent limitations in processing power and storage capacity. The core tasks encompass the identification and oversight of edge devices within the intercloud network, ensuring their security, managing software updates, and monitoring resource usage. This approach involves context-aware resource management, dynamically adjusting resource allocation and utilization strategies based on factors such as the location of edge devices, their network connectivity, and the resources available to them. Moreover, edge-to-fog collaboration is emphasized, establishing effective communication and resource exchange mechanisms between edge devices and fog nodes to enable the offloading of tasks requiring greater processing power or storage to the fog layer. Additionally, the strategy involves resource caching and replication, aiming to store and duplicate frequently accessed data at edge devices for improved efficiency.

4. Resource Management Techniques for Cloud/Fog and Edge Computing for Green IT Solutions/Green Cloud Computing

Cloud/Fog and Edge Computing technologies have the potential to address the challenges of green cloud computing by optimizing the utilization of allocated resources through the implementation of intelligent methods. Numerous studies and cloud service providers have demonstrated that the effective organization of allocated computing resources, facilitated by intelligent techniques, can not only enhance resource utilization but also significantly reduce carbon dioxide emissions.

In an increasingly digital world encompassing facets like smart homes, remote work, and streaming services, the demand for computing power and data storage is steadily on the rise. However, what often goes unnoticed is the environmental impact of this escalating data consumption. Data centers, responsible for housing and operating digital infrastructure, typically exhibit inefficiencies in resource utilization.

Achieving cost reductions, especially through on-premises solutions for data center construction and cooling technologies, is a challenging endeavor in this context.

Cloud computing's popularity arises from its efficacy in addressing substantial computational challenges via resource pooling and offering a diverse array of online services. The drive to enhance cloud performance emerges from the need for swift and responsive solutions in real-time applications, driven in part by the practice of computational offloading. Additionally, the exponential surge in data and requests has led to the proliferation of numerous global data centers. These data centers, however, consume substantial energy, resulting in heightened economic and environmental costs. This energy consumption exacerbates the emission of CO_2 and other greenhouse gases, widening the carbon footprint and contributing to the greenhouse effect. These factors underscore the imperative for the development of green technology, demanding energy-efficient solutions that optimize resource utilization (Kumbhare et al., 2022).

IV. CONCLUSION

Intelligent resource management in fog, edge, and cloud computing involves dynamic allocation, optimal utilization, and security measures. In cloud computing, resources are centrally provisioned, while edge and fog computing allocate resources at the periphery, minimizing latency. Intelligent algorithms tailor resource allocation, ensuring efficiency. This holistic approach enhances performance, scalability, and security across diverse computing layers, fostering adaptive scaling and proactive decision-making. Green IT solutions further optimize resource utilization and reduce environmental impact, addressing the challenges of modern computing paradigms.

Machine learning algorithms are leveraged for resource prediction, enabling the anticipation of resource demand, anomaly detection, and optimization of resource allocation. Self-organizing resource management is implemented, allowing systems to adapt autonomously to changing conditions, particularly valuable in dynamic intercloud settings. Resource virtualization involves abstracting physical resources into virtual counterparts, facilitating flexible allocation and utilization across different layers of the intercloud. Additionally, resource accounting and monitoring are crucial, involving the tracking of resource consumption and costs to enable optimal cost management and billing models. Security-aware resource management is emphasized, incorporating policies and mechanisms to safeguard resources and data from unauthorized access and potential attacks. As an evolving field, resource management in intercloud environments continually adjusts to new technologies, applications, and usage patterns. By efficiently managing resources across cloud, fog, and edge layers, interclouds can deliver seamless, scalable, and cost-effective services to meet a variety of user demands.

In the domain of intercloud resource management, Intelligent Resource Management in Intercloud orchestrates automated coordination within a network of agents, simplifying procedures for maximizing resource efficiency, implementing dynamic provisioning, and giving priority to security measures. Extending across cloud, fog, and edge layers, this methodology strives to establish a seamlessly operating ecosystem characterized by efficiency, scalability, reliability, and security in service delivery. The cloud-level intercloud interconnects diverse clouds, streamlining resource pooling and workload migration. The fog-level intercloud emphasizes collaborative resource sharing among dispersed fog nodes, enhancing application responsiveness. On the edge level, direct connections among edge devices aim to minimize latency, with resource management prioritizing collaboration, context-aware strategies, and efficient offloading. Green IT solutions in Cloud/Fog and Edge Computing leverage intelligent resource management to optimize resource utilization, mitigate carbon emissions, and address environmental concerns stemming from escalating data consumption and inefficient data centers.

Distributed Computing where data processed and store dis distributed over several systems and physical devices can be hosted in Cloud, Fog and Edge : Cloud being centrally situated and processing, with distributed resources allowing horizontal scaling ; Fog aiming at mitigating latencies through the local data processing before its pull-up to the cloud for further analytics and Edge described as highly-distributed by data processing taking place at the hardware which synthetizes data. This decentralized architecture are composed of devices or systems, the network through which data is exchanges and the resource management system coordinating the computing resources such as power and storage. This layout allows to each system or device to operate independently and gets empowered on the optimization and energy-efficiency side by the integration of cutting-edge technologies including AI and ML as part of it.

The Cisco Global Cloud Index (GCI) and the Global Cloud Ecosystem Index (GCEI) serve as essential tools for Intelligence Computing Resource Management (ICRM). GCI, an annual study, tracks global cloud traffic growth, aiding in planning for future resource needs, optimizing utilization, and informing decisions about cloud investments. On the other hand, GCEI, a biannual ranking, assesses 76 countries based on cloud ecosystem maturity, assisting in identifying partners, evaluating regulatory environments, and developing tailored cloud strategies. Together, these indices contribute to ICRM by providing data-driven insights, facilitating informed decision-making, and enabling proactive resource management, ultimately assisting organizations in optimizing cloud resource utilization, making informed investment decisions, and ensuring compliance with regulations (MIT Technology Review Insights, 2022).

REFERENCES

Alwakeel, A. (2021). An Overview of Fog Computing and Edge Computing Security and Privacy Issues. *Sensors (Basel), 21*(24), 8226. doi:10.3390/s21248226 PMID:34960320

Anand, M., Kinranangia, M., & Khanna, D. (2019). The Role of Cloud and Fog Computing in IoT. *International Journal of Scientific & Technology Research, 7.*

Bachiega, J. Jr, Costa, B., Carvalho, L. R., Rosa, M. J. F., & Araujo, A. (2022). Computational Resource Allocation in Fog Computing: A Comprehensive Survey. *ACM Computing Surveys, 55*(14s), 1–31. doi:10.1145/3586181

Bernstein, D. (2009). Blueprint for the Intercloud—protocols and formats for cloud computing interoperability. *Proc. 4th Int. Conf. Internet Web Appl. Services*, 328–336. 10.1109/ICIW.2009.55

Buyya, R., Yeo, C. S., Venugopal, S., Broberg, J., & Brandic, I. (2009). Cloud Computing and Emerging IT Platforms: Vision, Hype, and Reality for Delivering Computing as the 5th Utility. *Future Generation Computer Systems, 25*(6), 599–616. doi:10.1016/j.future.2008.12.001

Choi, Y.-J., Kang, H.-J., & Lee, I.-G. (2019). Scalable and Secure Internet of Things Connectivity. *Electronics (Basel), 8*(7), 752. doi:10.3390/electronics8070752

Cisco. (2015). *Fog Computing and the Internet of Things: Extend the Cloud to where the Things are.* Cisco Whitepaper. Retrieved from https://www.cisco.com/c/dam/en_us/solutions/trends/iot/docs/computing-overview.pdf

Dar, A. R., & Ravindran, D. (2019). Fog Computing Resource Optimization: A Review on Current Scenarios and Resource Management. *Baghdad Science Journal, 16*(2).

Elhoseny, M., Abdelaziz, A., Salama, A. S., Riad, A. M., Muhammad, K., & Sangaiah, A. K. (2018). A hybrid model of Internet of Things and cloud computing to manage big data in health services applications. *Future Generation Computer Systems, 86*, 1383–1394. doi:10.1016/j.future.2018.03.005

FahimullahM.AhvarS.TrocanM. (2022). A Review of Resource Management in Fog Computing: Machine Learning Perspective. *arXiv:2209.03066*

Geeks for Geeks. (2023). *Resource Allocation Methods in Cloud Computing.* Retrieved from https://www.geeksforgeeks.org/resource-allocation-methods-in-cloud-computing/

IBM. (2023). *Multicloud.* Retrieved from https://www.ibm.com/topics/multicloud

Iftikhar, S., Gill, S. S., Song, C., Xu, M., Aslanpour, M. S., Toosi, A. N., Du, J., Wu, H., Ghosh, S., Chowdhury, D., Golec, M., Kumar, M., Abdelmoniem, A. M., Cuadrado, F., Varghese, B., Rana, O., Dustdar, S., & Uhlig, S. (2023). AI-based fog and edge computing: A systematic review, taxonomy and future directions. *Internet of Things : Engineering Cyber Physical Human Systems, 21*, 100674. doi:10.1016/j.iot.2022.100674

Iorga, M. (2018). Fog computing conceptual model - Recommendations of the national institute of standards and technology. *NIST Special Publication 500-325.* doi:10.6028/NIST.SP.500-325

Jennings, B., & Stadler, R. (2014). Resource Management in Clouds: Survey and Research Challenges. *Journal of Network and Systems Management, 23*(3), 567–619. doi:10.1007/s10922-014-9307-7

Kumar, N., Jindal, A., Villari, M., & Srirama, S. N. (2021). Resource management of IoT edge devices: Challenges, techniques, and solutions. *Software, Practice & Experience, 51*(12), 2357–2359. Advance online publication. doi:10.1002/spe.3006

Mell, P., & Grance, T. (2011). *The NIST Definition of Cloud Computing. National Institute of Standards and Technology, Special Publication.* doi:10.6028/NIST.SP.800-145

Mijuskovic, A., Chiumento, A., Bemthuis, R., Aldea, A., & Havinga, P. (2021). Resource Management Techniques for Cloud/Fog and Edge Computing: An Evaluation Framework and Classification. *Sensors (Basel), 21*(5), 1832. doi:10.3390/s21051832 PMID:33808037

Rani, B. K., Rani, B. P., & Babu, A. V. (2015). Cloud Computing and Inter-Clouds – Types, Topologies and Research Issues. *Procedia Computer Science, 50*, 24–29. doi:10.1016/j.procs.2015.04.006

RezazadehB.AbyanehZ.G.AsghariP. (2023). Resource Management in Fog Computing: A Systematic Review of Techniques, Open Issues, and Future Trends. Authorea. doi:10.22541/au.169816176.69149377/v1

Rountree, D., & Castrillo, I. (2014). Evaluating Cloud Security: An Information Security Framework. In The Basics of Cloud Computing. Syngress. doi:10.1016/B978-0-12-405932-0.00006-2

Sim, K. M. (2009). Agent-based cloud commerce. *Proc. IEEE Int. Conf. Ind. Eng. Eng. Manage.*, 717–721.

Sim, K. M. (2015). Agent-based interactions and economic encounters in an intelligent Intercloud. *IEEE Transactions on Cloud Computing*, 3(3), 358–371. doi:10.1109/TCC.2015.2389839

Sim, K. M. (2019). Agent-based approaches for intelligent intercloud resource allocation. *IEEE Transactions on Cloud Computing*, 7(2), 442–455. doi:10.1109/TCC.2016.2628375

Sim, K. M. (2020). Intelligent Resource Management in Intercloud, Fog, and Edge: Tutorial and New Directions. *IEEE Transactions on Services Computing*. Advance online publication. doi:10.1109/TSC.2020.2975168

Sim, K. M. (2011). Agent-Based Cloud Computing. *IEEE Transactions on Services Computing*, 5(4), 1–1. doi:10.1109/TSC.2011.52

Sim, K. M. (2022). Agent-based Cloud Computing. *IEEE Transactions on Services Computing*, 5(4), 564–577. doi:10.1109/TSC.2011.52

Technology Review Insights, M. I. T. (2022). *Global Cloud Ecosystem Index 2022*. Retrieved from https://www.technologyreview.com/2022/04/25/1051115/global-cloud-ecosystem-index-2022/

Chapter 8
Modern Technological Innovation in Digital Wase Management

V. Vijayalakshmi

Department of DSBS, SRM Institute of Science and Technology, Kattankulathur, India

R. Radha

Department of DSBS, SRM Institute of Science and Technology, Kattankulathur, India

S. Sharanya

Department of DSBS, SRM Institute of Science and Technology, Kattankulathur, India

ABSTRACT

As the rapid pace of technological advancement continues to propel society into the digital age, the surge in electronic waste (e-waste) poses significant challenges to environmental sustainability. This research explores modern technological innovations in digital waste management that contribute to the reduction, recycling, and responsible disposal of electronic devices. Special attention is given to advancements in recycling methods, the application of artificial intelligence (AI), machine learning (ML), deep learning (DL), robotics, and IoT in automated e-waste processing. The research investigates the utilization of exploring how materials can facilitate easier recycling and reduce the environmental impact of electronic devices. The research also explores the role of extended producer responsibility (EPR) in adapting to sustainable practices in product design, disposal, and recycling. This research contributes by offering an understanding of the tools, strategies, and policies that can contribute to a more sustainable volume of e-waste in our increasingly digitized world.

DOI: 10.4018/979-8-3693-1552-1.ch008

INTRODUCTION

E-waste poses a significant danger to human beings, animals, and even the environment due to its composition. Typically, e-waste contains materials such as metals, plastics, Cathode Ray Tubes (CRTs), circuit boards and printed cables. The precious metals such as platinum, gold, silver, and copper can be reclaimed through scientific processing of E-waste. However, the occurrence of hazardous substances including but not limited toarsenic, lead, copper, chrome, cadmium, barium, cobalt, brominated flame retardants, mercury, polychlorinated biphenyls (PCBs), nickel, selenium, lithium, and liquid crystal renders electronic waste (e-waste) highly perilous, particularly when subjected to crude dismantling methods and rudimentary processing techniques (Hindrise, 2023).

When electronic devices like Air Conditioners, Printers, Washing Machines, IPods, Medical apparatus, Televisions, Copiers, Servers, Cellular Phones, Transceivers, Fax Machines, Refrigerators, Battery Cells, Calculators, Compact Discs (CDs), Scanners, Monitors, Mainframes, and Computers become unfit for use, they contribute to the category of e-waste. Even in minute quantities, the occurrence of extremely toxic substances and heavyweight metals like lead, mercury, cadmium, and beryllium pose an important threat to the environment. Figure 1 clearly depicts the electronic waste materials.

Figure 1. E-waste materials

E-waste, or electronic waste, emerges when electronic or electrical devices become obsolete or surpass their expiration dates. The swift pace of technological progress and the constant introduction of new electronic devices contribute to the frequent replacement of older models, resulting in a substantial surge in e-waste generation in India. The inclination of individuals to embrace newer models and trending technologies, coupled with the natural decline in product lifespan, exacerbates the challenge of managing E-waste in the nation. The effective methods of E-waste in country like India hinges on the active participation of consumers. Various initiatives, like Extended Producer Responsibility (EPR), Design for environment, and the technology platform promoting the 3Rs (Reduce, Reuse, Recycle), aim to create a circular economy by encouraging customers to responsibly discard of E-Waste, enhance rates of repurpose and recycling, and foster sustainable customer practices. While established nations accord high priority to the E-Waste management, the situation in emerging nations, including India, is complicated by attempts to directly adopt or replicate the systems of developed counterparts. This approach exacerbates challenges due to insufficient investment, a shortage of technically skilled human resources, inadequate framework, and the nonattendance of specific legislature addressing E-Waste. Furthermore, there is a lack of clarity regarding the functions and obligations of individuals and entities engaged in the management of electronic waste (Hindrise, 2023).

IMPACTS OF E-WASTE ON ENVIRONMENT AND HUMAN HEALTH

Handling E-Waste is an extremely intricate task due to its complex structure, comprising multiple components, some of which harbor toxic substances. If not managed appropriately, improper recycling and disposal methods can have adverse effects on human health and the environment. Therefore, there is a very essential for suitable technology to handle and dispose of these types of chemicals. According to the Basel Convention, the E-Waste is classified as dangerous when it contains or is contaminated with substances such as mercury, lead, cadmium, polychlorinated biphenyls, and more. Wastes that include metal or insulation cables covered with plastics polluted by or comprising coal, cadmium, lead, Polychlorinated biphenyls (PCBs), and tar are also designated as hazardous. Additionally, very precious metal ash collected from boards from printed circuit, glass waste from cathode-ray tubes, Liquid-Crystal Display (LCD) screens, and also some other activated glasses fall into the category of hazardous wastes(Ram Krishna, 2015). The following table 1 outlines the consequences of few of the primary dangerous components in e-waste:

GENERATION OR ESTIMATION OF E-WASTE

The initial way for planning the effective E-waste management system is to estimate or calculate the amount of E-waste generated (Alavi et al., 2015). This estimation not only facilitates management planning but also contributes to understanding the flow of materials in e-waste. Various works have concentrated on methods for calculating or estimating generation of e-waste, including works by (Araujo et al., 2012; Crowe et al., 2003; EEA 2003; Lau et al., 2013; Matthews et al., 1997; Schluep et al., 2012; Wang et al., 2013; Widmer et al., 2005). E-waste generation or estimation is prominent topics in the literature survey. (Perez-Belis et al., 2015) highlighted, through a wide spread literature work, that many countries less understanding the consistent methods or techniques for e-waste generation or estimation. In another study, (Ikhlayel, M. 2016) assessed the merits and demerits of five E-waste generation or estimation methods, focusing on their capability to evolving countries. The author proposed an extended edition of the Consumption & Use (C&U) technique for approximating generation of E-waste. Applying this modified method to six appliances in Jordan, (Ikhlayel, M. 2016) found that while most methods provided alike estimates of whole e-waste generation, they yielded multiple outcomes for calculating the quantities of E-waste collected from each Electrical and Electronic Equipment (EEE). The study emphasized the need for cautious attention of marketplace situations for each EEE when employing these estimation methods.

Table 1. The effects of dangerous components

S. No.	Dangerous Components	Consequence of Dangerous Components of E-Waste
1	Arsenic	The presence of arsenic in e-waste lead to cancer, neurological issues, and respiratory problems in humans. Arsenic contaminates soil and water, affecting ecosystems and wildlife. Airborne exposure during e-waste processing harms respiratory health.
2	Barium	Barium in e-waste poses health risks, including respiratory and gastrointestinal issues in humans.
3	Beryllium	Beryllium in e-waste poses serious health risks. Exposure can lead to lung and skin diseases, and chronic exposure may result in lung cancer.
4	BFR (Brominates flame retardants)	BFR has the potential to adversely affect the reproductive and immune systems and may lead to hormonal disorders.
5	Chromium	Chromium can potentially harm the kidneys and liver, and may lead to conditions like lung cancer and asthmatic bronchitis.
6	Cadmium	Cadmium can induce significant hurt in the spine and joints, impacting the kidneys and causing bone softening.
7	Chlorofluorocarbon (CFC)	CFC has the potential to impact the ozone layer and may result in skin cancer in humans, as well as genetic damage in organisms.
8	Dioxin	These substances pose a high toxicity risk to animals and can result in fetal malfunctions, reduced reproduction and growth rates, and adverse effects on the immune system.
9	Lead	Lead has the potential to impact reproductive systems, kidney, and nervous systems. It may lead to blood & brain disorders, and in some cases, it can be fatal.
10	Mercury	Mercury makes impact on the kidneys, immune system, and central nervous system, leading to impaired fetal growth. It has the potential to cause brain or liver damage.
11	Polychlorinated Biphenyl (PCB)	PCB may induce dangerous disease like cancer in birds and animals and has the capacity to disrupt the reproductive, immune, nervous, and endocrine system. PCBs insistently pollute the surrounding, resulting in profound and enduring damage.
12	Polyvinyl Chloride (PVC)	PVC containing up to 56 percent of chlorine, PVC emits Hydrogen chloride gas when burned, ultimately forming hazardous hydrochloric acid that can adversely affect the respiratory system.

The rapid increase in internet handlers from five hundred and one million in 2006 to above 1.3 billion in 2011 especially in evolving nations reflects a substantial surge in the sale of computers and related devices. There are 44% of internet handlers were in emerging countries in 2006, and by 2011, this figure had risen to 62%. The sales of personal computers witnessed a significant upswing from one hundred and seventy million units universally in 2000 to approximately three hundred and seventy million units in 2010. Projections indicate that sales value in 2014 is expected to the range around four hundred and seventy million units, extra than doubling in the past decade. In India, an estimated 1.42 million PCs become obsolete each year.

In June 2012, data from the International Telecommunication Union (ITU) revealed that the cumulative count of subscriptions of mobile cellular attained nearly six billion at the end of 2011, with about 80 percent of the 660 million extra subscriptions in developing countries. A report titled "Recycling - from E-Waste to Resources" presented at a Basel Convention meeting predicted a substantial increase in e-waste from computer systems by 2020 in South Africa and China, ranging between 200% and 400% over 2007 figures and a 500% increase in India. The report projected that India's E-waste collected from mobile phones would surge to eighteen times the 2007 levels by 2020, with China experiencing a sevenfold increase. The Basel Action Network highlighted that a significant portion, ranging from 50 percent to 80 percent of the e-waste produced in the USA is sell to other countries such as China, India, Taiwan, Pakistan, and various nations in Africa.

According to the United Nations Environment Programme (UNEP), E-waste is increasing at a rate of 40 percent annually worldwide, making it the fastest-growing type of waste. On a global scale, the annual generation of e-waste ranges between 20 and 50 million metric tons, accounting for over 5 percent of urban solid waste. This is especially notable in the developing nations, where the value is projected to increase by up to five hundred percent in the next ten years. In the case of country like India, the estimated volume or amount of E-waste was 0.8 million tons in the year 2012. The report provided by CAG (Controller and Auditor General's) highlighted that India generates every year dangerous waste from manufacturing company, electronic equipments waste, plastic waste, hospital or medical waste, municipality waste by 400,000 tonnes, 1.5 million metric tons, 1.7 million metric tons, 48 million metric tons respectively.

E-WASTE MANAGEMENT

E-waste management systems and practices vary across different countries, reflecting diverse approaches and strategies to address the challenges posed by electronic waste. These variations encompass regulatory frameworks, recycling initiatives, and public awareness campaigns. Below are brief summaries of E-waste management practices in numerous countries in table 2:

The worldwide production of electronic and electrical equipment is experiencing fast development, driven by advancements in the sectors like electronics and (IT) information technology, evolving consumer habits, brief product lifespan resulting from technological advancements and economic progress(Hossain M., 2015; Needhidasan, 2014; NSWMA, 2013; Terazono et al., 2006). This surge in EEE production poses significant challenges for many developing countries, leading to issues in the managing of e-waste produced domestically or illegitimately imported as utilized products (Nnorom, I.C. & Osibanjo, O., 2008). In numerous little and middle revenue developing nations, a considerable portion of E-waste apparatus is discarded in unhygienic land places, while unethical recycling methods, such as burning wires and acid extraction for precious metals, are prevalent. This practice is notably observed in nations like India, China, Philippines, Pakistan, Vietnam, Ghana and the Nigeria, where individuals lacking proper facilities dismantle of e-waste using rudimentary methods, posing risks to the atmosphere and public healthiness (SEPA 2011; Leung et al., 2006). Examples of inappropriate e-waste management procedures can be seen in India, as illustrated in Figure 2 (The e-waste guide, 2017). Analysis conducted across ten countries in MENA (Middle East and North Africa), revealing a rising concern regarding e-waste (Seitz, J., 2014). The hazards to healthiness of human beings and the surroundings from inadequate e-waste system in these countries are not yet very well understood, and consciousness are remains limited. The author (Heeks et al., 2015), observed that there is a distinctions between already well developed and currently developing nations considering e-waste. In developing nations, the primary challenges include: (i) higher threats due to inadequate handling, (ii) a lack of formal or old recycling methods in various cases, and (iii) weak or absent recycling legislations. Additionally, (Osibanjo, O. & Nnorom, I.C. 2007) recognized five critical challenges related to e-waste in just beginning nations: (i) the swift expansion of (ICT) Information & Communication Technology influencing E-waste volumes, (ii) the growing pace of generation of e-waste, (iii) the constituents establish in E-waste, (iv) organization hurdles, and (5) the pollution arising from current management practices.

Table 2. Summaries of e-waste management practices in several countries

S. No.	Country	Reference	Practices
1	Malaysia	(Suja et al., 2014; Ismail, H., Hanafiah, M.M, 2019)	In Malaysia, e-waste management involves regulatory frameworks, recycling initiatives, and public awareness programs. The government has established regulations, encourages responsible disposal, and collaborates with industry stakeholders. Efforts include improving recycling infrastructure and addressing challenges posed by informal recycling practices. Extended Producer Responsibility (EPR) principles are considered, and public awareness campaigns aim to promote proper e-waste management.
2	Brazil	(de Souza et al., 2016)	Brazil has established E-waste legislation, focusing on collection, recycling, and disposal. Extended Producer Responsibility is a key principle. The country faces challenges in ensuring effective enforcement and infrastructure.
3	China	(Zeng et al., 2017)	China is a significant producer of e-waste but faces challenges in managing it effectively. The country has implemented regulations, but informal recycling practices persist in certain regions. Government initiatives aim to improve formal recycling infrastructure.
4	France	(Bahers et al, 2018)	E-waste management in France adheres to regulations outlined in the Waste Electrical and Electronic Equipment (WEEE) directive. The nation adopts Extended Producer Responsibility (EPR) principles, holding manufacturers accountable for the recycling of their products. Collection and recycling systems are well-established, and public awareness campaigns promote proper e-waste disposal.
5	Costa Rica	(Abarca-Guerrero et al., 2018)	Costa Rica has been working on improving e-waste management. Efforts include regulatory measures and public awareness campaigns.
6	USA	(Schumacher,K.A., Agbemabiese, L., 2019)	E-waste management practices vary across states. Some states have implemented e-waste recycling programs, while others rely on private initiatives.EPR laws are not federally mandated, leading to differences in approaches.
7	Bangladesh	(Masud et al., 2019)	Bangladesh has been addressing e-waste challenges through regulatory measures and public awareness campaigns.

Figure 2. PCBs open burning in India

E-WASTE RECYCLING METHODS

During an age marked by swift technological progress, a substantial volume of e-waste is generated due to increased production of EEE, which becomes obsolete at the completion of its essential life. The Effective or best strategies play a crucial role in ensuring environmentally responsible e-waste management, encompassing both the production phase and post-use considerations. While recycling serves as a remarkable tool for tackle the mounting volumes of waste materials, suboptimal recycling systems contribute to heightened stage of toxic or dangerous pollutants in the environment. The prevailing worldwide pattern in e-waste managing system, where informal procedure constituted 82.6 percent of total world e-waste in the year 2019, is not sustainable, highlighting the need for a multifaceted approach to achieve a comprehensive solution. A nuanced perceptive of e-waste gathering or collection and also recycling processes becomes imperative in addressing this challenge. This portion gives small information of the formal methods employed for e-waste collection and the procedures involved in recovering important metals from the E-waste.

Recycling methods of E-waste involves three main steps: (a) gathering of e-waste components, (b) handling or utilizing, and (iii) retrieval of expensive materials (Meskers et al., 2009). The consolidation and collection of perfect EEE are crucial steps requiring public awareness and regulatory measures, efficiently gathering household or institutional e-waste requires well-defined policies by government and available gathering services. In formal collection or gathering processes, organizations authorized by government, providers or producers utilizing community collection spots, take-up work, and the sellers play a pivotal role. Various nations have implemented the Extended Producer Responsibility (EPR) norm, requiring earliest apparatus producers to set aside a specific percentage of their sales as payment.

The subsequent phase in e-waste recycling involves the enhancement of valued metals by removing unimportant and non-recyclable portions. This process includes dismantling Electronic and Electrical Equipment (EEE) to gather separate components. Based on the type, producer, and stage of the apparatus, e-waste comprises different amounts of components such as ferrous (iron or steel), non-ferrous metals (copper, silver, aluminium, gold, platinum), wood, plastics, rubber, glass, and other materials. These components undergo sorting into distinct batches, including paper, plastics, metals, ceramics items, wood products, and specific types such as capacitors, PCBs, batteries, and LCDs. Mechanical processes, like large-scale granulation machines, scrap them into minor pieces ranging from 0.177 millimetres to 5 centimetres (Cui, J., Forssberg, E., 2003). To enhance recovery, the processed e-waste is then separated into the non metallic & metallic portions that utilizing techniques like current based, magnetic, or separation of density (Li, J., 2007; Zhang, S., Forssberg, E., 1997). The volume of plastic related e-waste is primarily reclaimed for power or transformed into bottles.

The materials recovered from the extracted E-waste undergo pyrometallurgical and hydrometallurgical methods. These hydrometallurgical techniques employ processes such as leaching, solvent extraction, ion exchange and adsorption to selectively dissolve valuable metals. Subsequently, recovery is achieved through chemical reduction or electro-refining processes (Ritcey, G.M., 2006; Sadegh Safarzadeh, 2007). The revival of precious and clean based metals from electronic waste involves the use of numerous mineral acids, oxidants (e.g., $HClO_4$, H_2SO_4, HNO_3/H_2O_2, HCl, NaClO, aqua-regia), and lixiviants (e.g., thiourea,cyanide, halide, thiosulfate) (Ding, Y., 2019; Islam, 2020). Though, it's important to note that hydrometallurgical processes often utilize toxic, very highly alkaline/acidic, or inflammable reagents, leading to the generation of substantial solid/garbage wastes and effluents/sewage wastes.

The Off-gas action is required in hydrometallurgy due to the construction of furans, toxic dioxin, and volatile metals due to the presence of halogenated flame retardants utilized in printed circuit boards (Hageluken, C., 2006). It is difficult to achieve great extraction effectiveness and recyclability with a single process, let alone cost-effective extraction with low emissions of hazardous components into the environment. To reduce the amount of hazardous trash and wastewater that is disposed of, techniques including and incineration are used.

Informal recycling poses a significant challenge in developing countries, often overlooked within a municipality's dense waste and resources management system. (Velis et al., 2012; Fei et al., 2016) investigated the stream and destiny of resources collected through unsanctioned recycling, where individuals engage to yield a modest revenue. The studies suggested that administrations should officially integrate unfamiliar recycling into community waste management by providing exercise to participants and enhancing recycling facilities. Optimization of existing recycling schemes was also recommended. Another study by (Parajuly, 2017) expected to examine the opportunities and challenges of launching a renewable Electronic-Waste management system, proposing a waste management systems using resource oriented model. This work found that the lack of resource recovery policies in conventional waste management systems has an impact on the contribution of unofficial labourers in garbage recycling/reproducing.

OPPORTUNITIES AND CHALLENGES IN MANAGING E-WASTE

The glass, plastics products, metal items, and uncommon ground elements found in E-waste represent worthy resources secondarily. The abundance of valuable elements in waste becomes apparent when considering estimates, such as a classic Cathode-Ray tube television containing roughly 450 gram copper, 227 gram Aluminium, and 5.6 gram gold (Zeng, 2018). likewise, PC (Personal Computers) waste of metric ton volume has the potential to yield more gold than the amount collected from seventeen metric tons of Au/gold ore (Bleiwas, D.I., Kelly, T., 2001). However, the recovery of easily extractable components, such as metals with moderately very high content and available forms, has been the main emphasis of standard e-waste recycling initiatives (Baccini, P., Brunner, P.H., 2012).

Challenges

Precious and very uncommon earth metals/materials have historically been extracted from electronic components, which include printed graphic and memory cards, circuit boards, cables,hard drives, and connections frequently seen in gadgets like smartphones and personal computers. Nevertheless, throughout its processing and recycling, e-waste releases a number of hazardous compounds in addition to being a source of valuable commodities. Significant risks to the surroundings and public people health issues are posed by hazardous composites found in electronic waste, primary emissions namely Hg, Pb, Cd, As, and PCBs, as well as by-products created during processing and recycling, secondary emissions such as furans, dioxins, and PAH (Polycyclic Aromatic Hydrocarbon) (Orlins, S., Guan, D., 2016). Achieving sustainable e-waste management requires a comprehensive procedure that involves collecting all the classes of e-waste at the complete usage of their lifetime and reprocessing them completely to minimize material wastage (Reuter, 2013; Song, Q., Li, J., 2015). The prime difficulty in managing e-waste lies in its massive capacity and complication. E-waste production is constantly rising at a predictable value of 2 million tons annually, with India, the USA, China, Brazil, and Japan collectively producing fifty percent of the total electronic waste in the year 2019 (Forti, 2020). The escalating volume of e-waste is particularly alarming in nations like India and China; these are both leading e-waste originators and carriers from around the world (Wang, 2016). Compounding this challenge is the very difficult in segregating e-waste materials due to different attachment methods, such as screwing, bolting, snapping, soldering, or gluing. Additionally, the tendency of households to discard of all varieties of waste collected turned it very difficult to divide components for effective product-specific recycling (Chi, 2014).

Households are either selling e-waste to dealers or disposing of these as solid or raw waste due to a lack of ecological alertness, inadequate efforts by local governments to promote collection strategies and e-waste recycling policies, and limited knowledge about e-waste disposal options (Wang, 2017). In addition, the lack of efficient networks and collecting systems has led to the emergence of unofficial setups that are limited in their ability to recycle specific types of e-waste because they lack architecture and the technology related processes. Resultantly, hazardous compounds from first, second, and third emissions are discharged into the atmosphere, or a sizable quantity of worthy secondary elements are vanished. Because extremely programmed and high technically competent proper zone struggle with huge running expenses for organized e-waste gathering and management, rent of location, and the qualified staff, a large portion of e-waste is managed by the improper zone. Because of this, unofficial setups that lack the fundamental technology and infrastructure needed to improve all valued substances are incapable to put preventative measures in place to stop the release of contaminants into the environment. Effective e-waste management is seriously threatened by the informal sector's crude recycling practices, which include extensive manual dismantling with chisels, hammers, and screwdrivers used only by the hands, as well as methods for recovering metals like flaming in the open or stripping in open-pit (Wong, 2007; Chi, 2011; Awasthi, A.K., Li, J., 2017).

Opportunities

Despite numerous challenges, the e-waste recovering presents many more opportunities or chances due to their vast resource prosperity and its responsibility in slowing down the step of generation of e-waste. A significant advantage of methodically handled e-waste recycling is its provision of a guarded processing system for the e-waste that lead to contribute to the environmental contamination in landfills or incinerators (Widmer, 2005). This approach disrupts the conventional linear resource extraction model, helping to extend the lifespan of materials. Consequently, it diminishes waste production and encourages the utilization of secondary raw resources obtained by recycling of materials (Klinglmair, M., Fellner, J., 2010; Cossu et al., 2012). Numerous social and environmental profits result from keeping dangerous waste out of landfills, reducing greenhouse gas secretion, and decreasing the carbon footmark by lowering the mandate for extra metal or materials (Debnath, et al., 2015). Furthermore, by removing unprocessed materials from the e-waste, less waste is disposed of, and unrecyclable parts may be utilized as fuel in industrialized manufacture to save a lot of energy. E-waste recycling has substantial economic implications. For example, the estimated price of untreated materials available in the 53.6 million tons of e-waste generated in the year 2019 is around 57 billion USD. By enhancing the developing economies, recycling and collection of e-waste, heavily reliant on imports for energetic minerals like nickel, cobalt, and copper can curb more-extraction, more-dependence on imports, and withstand challenges such as value increases and strategic behavior by exporters. Another economic aspect is the potential creation of additional jobs in the e-waste recycling factories. Moreover, one of the most significant benefits of e-waste reprocessing is the concentrated requirement for virgin extraction to mine metals. With iron and copper resources in geological stores nearly depleted, an instant shift to big-scale resource regaining is crucial to reserve enduring natural resources (Krook, J., Baas, L., 2013).

Additionally, due to declining ore grades and increasing global demand, the mining of small-piece and additional compound ore covering is necessary, a demand that may be partially met through recycling e-waste, given that the average grade and concentration of materials in electronic waste are generally more than those in ores extracted from mines. Furthermore, automated e-waste recycling can restrict the environmental release of toxic components during natural decomposition or through improper disposal. This approach reduces various kinds of pollution related with e-waste and also prevents the exposure of vulnerable populations to hazardous e-waste pollutants.

E-WASTE MANAGEMENT USING ARTIFICIAL INTELLIGENCE, MACHINE LEARNING, DEEP LEARNING, IOT, AND ROBOTICS

Managing electronic waste (E-waste) has become a critical challenge in our technologically advanced society. To address this issue effectively, a holistic approach incorporating cutting-edge technologies is imperative. This paper explores the integration of Artificial Intelligence (AI), Machine Learning (ML), Internet of Things (IoT), Deep Learning (DL), and Robotics in E-Waste Management. Figure 3 shows the technologies used for E-Waste Management.

1. Artificial Intelligence (AI): AI algorithms play a pivotal role in automating processes related to E-Waste Management. These algorithms can be employed for data analysis, decision-making, and optimization of waste sorting and recycling procedures.
2. Machine Learning (ML): Machine Learning algorithms permit systems to learn and improve from experience. In the context of E-Waste Management, ML can be utilized for predictive modeling, identifying patterns in waste generation, and optimizing collection routes.
3. Deep Learning (DL): Deep Learning, a subset of ML, is particularly effective in image and pattern recognition. In E-Waste Management, Deep Learning models can be applied for the visual identification of electronic components, facilitating efficient sorting and recycling.
4. Internet of Things (IoT): IoT devices equipped with sensors can be embedded in electronic devices to monitor their usage, lifespan, and potential for recycling. These devices provide real-time data, enabling proactive and informed decision-making in waste management.
5. Robotics: Robotics offers a hands-on solution for the physical aspects of E-Waste Management. Robots can be deployed for the safe dismantling, sorting, and handling of electronic components. Their precision and efficiency contribute to a safer and more streamlined process.

The synergy of these technologies creates a complete E-Waste processing system. AI and ML algorithms process vast amounts of data, while IoT devices gather real-time information. Deep Learning models enhance visual recognition, and robotics provide the physical capabilities needed for effective waste handling. Implementing this integrated approach offers several advantages. It enhances the accuracy and efficiency of waste sorting, reduces reliance on manual labor, and promotes the extraction of valuable materials from electronic devices. Moreover, the continuous learning capabilities of AI and ML contribute to the evolution and improvement of E-Waste Management strategies over time. The convergence of AI, ML, Deep Learning, IoT, and Robotics presents a transformative solution to the challenges posed by E-Waste. This interdisciplinary approach not only addresses the environmental worries connected with electronic waste but also establishes a framework for sustainable and technologically advanced waste management practices.

Figure 3. Technologies for e-waste management

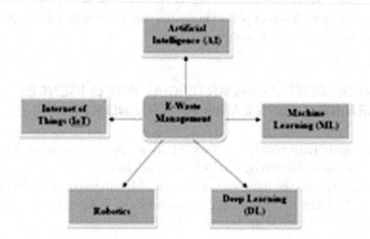

In the empire of technological advancements, as people adapt to new innovations, progress seamlessly integrates into their lives. Each technological shift brings improvements, leading to the replacement and abandonment of old devices. Electronic and Electrical Equipments rejected by users are categorized as e-waste. EEEs consist of numerous components; few comprising contaminated materials that may harmfully affect environment and the health of human if not handled appropriately. The global production of EEEs has prompted governments worldwide to enact stringent strategies to confirm the effective discarding of generated e-waste. In order to create a predictive model; this paper investigated the use of machine learning in the field of E-waste management system. This works compares the performance of Neural Network, ML algorithms and Gradient Boosting Regression Tree (GBRT) to evaluate weekly e-wastage for each urban sub-segment. Various data-driven techniques, including the existing engineered model and its modification, along with common machine learning algorithms, are examined. The utilization of Machine Learning algorithms enhances prediction accuracy, reaching 99.1% with the best-performing algorithm. Beyond significant quantitative improvements, the proposed design can contribute to optimizing long-term e-waste management in a smart city environment using historical data (Li et al., 2021).

The global volume of E-waste is rapidly increasing and is anticipated to achieve 74.7 million tons by the year 2030. However, a 2019 United Nations study revealed that the recycling and collection price of e-waste is merely 17.4 percent. A significant hurdle in significant e-waste reprocessing is the labour concentrated dismantling, arranging, and harmfulness eliminated processes that has led to the illicit transfer of substantial e-waste quantities from the developed nations to African and Asian nations. An automated approach to e-waste sorting and separation is desperately needed in order to protect the environment and worker safety. The image detection algorithm Convolutional Neural Network (CNN) was created in this study to effectively categorize e-waste into various divisions. A demonstration setup and an image input database with four distinct classifications of waste were developed. The correctness of the CNN model, the choice of hyperparameters, and the pre-processing of image data were all covered in detail. A testing accuracy is 93.9% and a training accuracy is 96.9% were shown by the created CNN model. To improve the performance of the model, a number of tests were carried out, including the utilization of various image sizes, the implementation of augmentation of data over rotation, and the elimination of backdrops (Zhou, E. P., 2022).

Improper discard of e-waste is now evolving into a public health and environmental concern, constituting the fastest-growing segment of municipal waste globally. Despite its complex nature, this escalating waste stream contains valuable metals like Indium, Neodymium, Tantalum, Palladium, Gold, Platinum, Silver, Copper and Aluminium. These precious materials can be reclaimed from the waste and reintegrated into the construction cycle for everyday use. Consequently, there is a pressing requirement for effective e-waste separation and management to improve these valuable resources. This work introduced a Deep Learning model applied using NVIDIA's Jetson Nano development kit, aiming to categorize waste materials into two groups depends on the presence of Non-Precious or Precious metals. The prototype model achieves accurate waste segregation with minimal time consumption (Elangovan et al., 2021).

Electronic equipment, the production of Electronic-Waste (E-waste) has escalated significantly. This surge poses environmental and health risks. This paper introduces a robust E-waste classification approach, the Fractional Horse Herd Gas Optimization-based Shepherd Convolutional Neural Network (FrHHGO-based ShCNN), designed for categorizing E-wastes in the domain of Internet of Things (IoT) and cloud computing platform. In this system, IoT nodes collect images of E-waste, then store them in cloud computing, and employ the new FrHHGO algorithm for routing. Efficient features are taken and improved for E-waste classification. FrHHGO combines both the algorithms such as FHGO (Fractional Henry Gas Optimization) and HOA (Horse Herd Optimization Algorithm). The established procedure surpasses previous techniques with minimal delay and energy of 0.666 sand 0.301 J, along with highsensitivity, specificity, and accuracy of 0.934, 0.967, and 0.950, respectively. Thus, the FrHHGO-based ShCNN enhances environmental, social, and economic sustainability in emerging economies through effective E-waste classification (Ramya, P., & Ramya, V., 2023).

The significant motive of this research is to generate and deploy an innovative pre-programmed E-Waste monitoring System leveraging machine learning algorithms. Traditional Waste Management Systems (WMS) are predominantly manual process and have been augmented with an Internet of Things (IoT) sensor-based alert system to notify relevant personnel about waste overflow for prompt action. However, this system lacks efficiency in terms of time and cost and neglects the recycling or disposal of E-waste. As the utilization of electronic devices has surged, there's a corresponding increase in the production of E-waste, including laptops, monitors, mobiles, headphones, and tablets. Despite posing an environmental threat, E-waste offers the potential for extracting valuable minerals by recycling technique.

Several systems have been developed to optimize the handling of electronic waste. Ongoing research explores the utilization of E-waste through image processing techniques, particularly with many deep-learning algorithms that enhance prediction accuracy and classification. This work introduced a model based on deep learning to enhance the precision of waste management and the prediction systems. The projected technique incorporates a Convolutional Neural Network (CNN) algorithm for feature extraction, and these gathered features are further processed utilizing a RBN (Restricted Boltzmann Machine) model to improve accuracy. The validation of the proposed system employs an open-source dataset, and the outcomes are related with different existing methods. The comparative analysis reveals that the proposed algorithm achieves superior prediction accuracy, reaching 96% for E-waste (Khatiwada et al., 2023).

The creation of E-waste has increased significantly each year due to the increase in the manufacturing of electronic products and the growing demand for the newest technologies. India alone created 3.2 million tonnes of e-waste in the year 2020; the nation's largest cities were Bangalore, Mumbai, and Delhi. Modern urban ecosystems cannot be sustained without effective E-waste recycling and management. There are very few options available for collecting waste from private homes, despite the focus being on commercial and industrial E-waste collections. This article proposes the use of transfer learning to identify frequent electrical wastes by deploying a mobile robot. The robot, which is meant to be an add-on for already-existing municipal garbage trucks, goes around, detects electronic waste, and uses a storage mechanism and arm based lift to separate the stuff it finds. The system uses a Convolutional Neural Network (CNN) to classify E-waste, and it does it with an astounding 96% accuracy. Specifically for India, this innovative programme seeks to gather and separate E-waste from residences and individuals, freeing up unskilled labour from dangerous procedures and providing a 20% cost savings over a five-year period. The goal is to offer a workable mobile solution that requires the least amount of human participation for collecting e-waste from homes (Shreyas Madhav et al., 2022).

This system uses Internet of Things devices and also the sensors to manage and monitor the process of collecting, arranging, and discarding of e-waste. These Internet of Things devices have embedded sensors that can measure the amount of e-waste in a specific region and provide real-time data to improve the efficiency of collection and disposal. This technology uses machine learning to discern between metallic and plastic components, seeing e-waste as a valuable resource. Plastic waste can be converted into biofuel by using pyrolysis, and the metallic parts can be used to produce solar batteries. This method offers a viable plan for resource recovery and sustainable waste management since it maximises resource utilisation while minimising environmental effect. The system also makes use of cloud-based platforms for trend and pattern analysis of data. By using the cloud-based statistical technique Autoregressive Integrated Moving Average, waste collection schedules may be optimised and overall processes can be improved by gaining insights into future garbage levels (Farjana et al., 2023).

As a result of the dangerous materials found in electronic equipment, electronic trash is currently the speediest increasing waste flow in the world. It puts the environment and human health at danger if improperly managed. Thus, there is a pressing demand for e-waste management technologies that are safer, more inventive, and ecologically benign. This research presents an intelligent e-waste management system that combines object detection based on DL with the IoT. For e-waste object detection, three high-tech models for detecting objects have been used: YOLO version 5s, YOLO version 7-tiny, and YOLO version 8s. YOLO version 8s reaches the maximum mAP@50 of 72% and mAP@50-95 of 52%, according to the data. This innovative approach could improve the effectiveness of e-waste management, supporting sustainability and being in line with green city policies. This technique tackles contamination issues by implementing an intelligent green city vision, which helps both the environment and humankind (Voskergian, Daniel and Ishaq, Isam, 2023).

The main tactics used in modern E-waste management procedures are reprocessing, reutilizing, and minimising electronic trash or waste. Nevertheless, there isn't a perfect solution for managing e-waste just yet, and the approaches that are now in use frequently need a large amount of labour and resources. This article presents a new method for classifying e-waste utilizing a Deep Convolutional Neural Network (DCN) that is depends on Fractional Henry Gas Optimisation-based deep CNN. The energy-aware FHGO routing algorithm is applied to the E-waste photos in order to determine the best route. Local Gabor Binary Pattern (LGBP), Grey Level Co-occurrence Matrix (GLCM) feature, and Histogram of Oriented Gradient (HOG) are some of the procedures used in feature extraction. The pre-processing stage is finished with an average filter. The inferred attribute size is improved through the use of data augmentation process. Furthermore, deep CNN trained by the FHGO algorithm—a combination of fractional calculus (FC) and Henry gas solubility optimisation (HGSO)—is used to classify e-waste. The suggested method shows better accuracy compared to current approaches including TensorFlow deep learning, machine learning, deep learning, Cuckoo search neural network, with improvements of 18.05%, 19.49%, 7.89%, and 12.77% respectively (Puppala Ramya et al., 2023).

The waste management involves the discard of waste through recycling and landfilling processes. The integration of two different domains such as the Internet of Things and deep learning provides a dynamic resolution for waste real-time data monitoring and classification respectively. This study presents a robust architecture for a waste management system, leveraging deep learning and IoT technologies. The projected model introduced an intelligent approach to categorize indigestible and digestible waste, employing a CNN model, a broadly used framework based on deep learning. Additionally, the system includes a smart trash bin design featuring a microcontroller equipped with many more sensors. Bluetooth connectivity and IoT are utilized for data monitoring, allowing real-time control from any location and short-range data monitoring with the help of an Android application. To evaluate the effectiveness of the generated model, metrics such as sensor data estimation, waste label classification accuracy, and System Usability Scale (SUS) are measured and analyzed. The classification accuracy achieved by the CNN-based architecture is 95.3125%, and the SUS score is 86%. This adaptive smart system is designed for seamless integration into household activities, providing real-time monitoring of waste disposal (Rahman et al., 2022).

A viable compromise is to involve both a robot and human operator in the process. This research aims to optimize the recycling of electronic equipment, considering both technical and economic factors, and incorporating advancements in collaborative robot technology (Alvarez-de-los-Mozos, E., & Renteria, A., 2017).E-waste pertains to discarded electrical or electronic devices. Robotics encompasses the design, construction, operation, and control of robots, along with computer systems for feedback and information processing. Integrating these fields, employing robotics for E-waste elimination contributes to a healthier world by reducing harmful E-wastes (P. SHINY ESTHER et al., 2018).

CONCLUSION

The surge in Electronic Waste (E-Waste) propelled by rapid technological advancements presents a critical challenge to environmental sustainability. This research has thoroughly examined modern technological innovations in digital waste management, with an attention on maintainable and efficient solutions to discourse the escalating concerns surrounding e-waste. The study underscores the adversative properties of e-waste on the environment and human health, emphasizing the need for suitable technology to handle and dispose of hazardous substances present in e-waste. It also highlights the importance of accurate estimation of e-waste generation for efficient management planning. By exploring e-waste management practices in different countries, the research provides valuable insights into the varied approaches, regulatory frameworks, and initiatives adopted globally. The summaries of practices in Malaysia, Brazil, China, France, Costa Rica, the USA, and Bangladesh reveal the multifaceted nature of e-waste management, reflecting a combination of regulatory measures, recycling initiatives, and public awareness campaigns.

The study concludes that a holistic approach is imperative, integrating sustainable practices, technological innovations such as Machine Learning, Artificial Intelligence, Deep Learning, Robotics, and the

Internet of Things, and effective policies. Design principles fostering a circular economy, emphasizing product longevity, reparability, and recyclability, play a pivotal role in minimizing e-waste generation. As the global production of electrical and electronic equipment continues to rise, collaborative efforts from policymakers, industry stakeholders, and environmental advocates are essential. By implementing the recommendations derived from this research, we can collectively move towards a more sustainable and responsible approach to managing the challenges posed by the escalating volume of e-waste in our increasingly digitized world. This research contributes valuable insights to the ongoing discourse on digital waste management, offering a roadmap for a healthier, more sustainable future.

REFERENCES

Abarca-Guerrero, L., Roa-Gutiérrez, F., & Rudín-Vega, V. (2018). WEEE resource management system in Costa Rica. *Resources*, *7*(1), 1–14. doi:10.3390/resources7010002

Alavi, N., Shirmardi, M., Babaei, A., Takdastan, A., & Bagheri, N. (2015). Waste electrical andelectronic equipment (WEEE) estimation: A case study of Ahvaz City, Iran. *Journal of the Air & Waste Management Association*, *65*(3), 298–305. doi:10.1080/10962247.2014.976297 PMID:25947126

Alvarez-de-los-Mozos, E., & Renteria, A. (2017). Collaborative robots in e-waste management. *Procedia Manufacturing*, *11*, 55–62. doi:10.1016/j.promfg.2017.07.133

Araujo, M. G., Magrini, A., Mahler, C. F., & Bilitewski, B. (2012). A model for estimation of potential generation of waste electrical and electronic equipment in Brazil. *Waste Management (New York, N.Y.)*, *32*(2), 335–342. doi:10.1016/j.wasman.2011.09.020 PMID:22014584

Awasthi, A. K., & Li, J. (2017). Management of electrical and electronic waste: A comparative evaluation of China and India. *Renewable & Sustainable Energy Reviews*, *76*, 434–447. doi:10.1016/j.rser.2017.02.067

Baccini, P., & Brunner, P. H. (2012). *Metabolism of the Anthroposphere: Analysis, Evaluation, Design*. MIT Press.

Bahers, J. B., & Kim, J. (2018). Regional approach of waste electrical and electronic equipment (WEEE) management in France. *Resources, Conservation and Recycling*, *129*, 45–55. doi:10.1016/j.resconrec.2017.10.016

Bleiwas, D. I., & Kelly, T. (2001). Obsolete Computers, "Gold Mine", or High-tech Trash? Resource Recovery From Recycling. Fact Sheet.

Chi, X., Streicher-Porte, M., Wang, M. Y. L., & Reuter, M. A. (2011). Informal electronic waste recycling: A sector review with special focus on China. *Waste Management (New York, N.Y.)*, *31*(4), 731–742. doi:10.1016/j.wasman.2010.11.006 PMID:21147524

Chi, X., Wang, M. Y. L., & Reuter, M. A. (2014). E-waste collection channels and household recycling behaviors in Taizhou of China. *Journal of Cleaner Production*, *80*, 87–95. doi:10.1016/j.jclepro.2014.05.056

Cossu, R., Salieri, V., & Bisinella, V. (2012). *Urban Mining: a Global Cycle Approach to Resource Recovery From Solid Waste*. CISA Publ.

Crowe, M., Elser, A., Gopfert, B., Mertins, L., Meyer, T., Schmid, J., Spillner, A. & Strobel, R. (2003). *Waste from Electrical and Electronic Equipment (WEEE) – Quantities, Dangerous Substances and Treatment Methods*. Academic Press.

Cui, J., & Forssberg, E. (2003). Mechanical recycling of waste electric and electronic equipment: A review. *Journal of Hazardous Materials*, *99*(3), 243–263. doi:10.1016/S0304-3894(03)00061-X PMID:12758010

de Souza, R. G., Clímaco, J. C. N., Sant'Anna, A. P., Rocha, T. B., do Valle, R. de A.B., & Quelhas, O. L. G. (2016). Sustainability assessment and prioritisation of e-waste management options in Brazil. *Waste Management (New York, N.Y.)*, *57*, 46–56. doi:10.1016/j.wasman.2016.01.034 PMID:26852754

Debnath, B., Baidya, R., Biswas, N. T., Kundu, R., & Ghosh, S. K. (2015). E-waste recycling as criteria for green computing approach: analysis by QFD tool. In K. Maharatna, G. K. Dalapati, P. K. Banerjee, A. K. Mallick, & M. Mukherjee (Eds.), *Computational Advancement in Communication Circuits and Systems* (pp. 139–144). Springer. doi:10.1007/978-81-322-2274-3_17

Ding, Y., Zhang, S., Liu, B., Zheng, H., Chang, C., & Ekberg, C. (2019). Recovery of precious metals from electronic waste and spent catalysts: A review. *Resources, Conservation and Recycling*, *141*, 284–298. doi:10.1016/j.resconrec.2018.10.041

EEA. (2003). Waste from electrical and electronics equipment (WEEE)-quantities, dangerous substances and treatment methods. EEA.

Elangovan, S., Sasikala, S., Kumar, S. A., Bharathi, M., Sangath, E. N., & Subashini, T. (2021, August). A deep learning based multiclass segregation of e-waste using hardware software co-simulation. *Journal of Physics: Conference Series*, *1997*(1), 012039. doi:10.1088/1742-6596/1997/1/012039

Farjana, M., Fahad, A. B., Alam, S. E., & Islam, M. M. (2023). An IoT- and Cloud-Based E-Waste Management System for Resource Reclamation with a Data-Driven Decision-Making Process. *IoT.*, *4*(3), 202–220. doi:10.3390/iot4030011

Fei, F., Qu, L., Wen, Z., Xue, Y., & Zhang, H. (2016). How to integrate the informal recycling system into municipal solid waste management in developing countries: Based on a China's case in Suzhou urban area. *Resources, Conservation and Recycling*, *110*, 74–86. doi:10.1016/j.resconrec.2016.03.019

Forti, V., Balde, C. P., Kuehr, R., & Bel, G. (2020). The Global E-waste Monitor 2020: Quantities, Flows and the Circular Economy Potential. United Nations University.

Hageluken, C. (2006). Improving metal returns and eco-efficiency in electronics recycling - a holistic approach for interface optimisation between pre-processing and integrated metals smelting and refining. *Proceedings of the 2006 IEEE International Symposium on Electronics and the Environment*, 218–223. 10.1109/ISEE.2006.1650064

Heeks, R., Subramanian, L., & Jones, C. (2015). Understanding e-Waste Management in Developing Countries: Strategies, Determinants, and Policy Implications in the Indian ICT Sector. *Information Technology for Development*, *21*(4), 653–667. doi:10.1080/02681102.2014.886547

Hindrise. (2023). *E-Waste Management in India*. https://hindrise.org/resources/e-waste-management-in-india/

Hossain, M., Al-Hamadani, S., & Rahman, R. (2015). E-waste: A Challenge for Sustainable Development. *Journal of Health & Pollution*, 5. PMID:30524771

Ikhlayel, M. (2016). Differences of methods to estimate generation of waste electrical and electronic equipment for developing countries: Jordan as a case study. *Resources, Conservation and Recycling*, *108*, 134–139. doi:10.1016/j.resconrec.2016.01.015

Islam, A., Ahmed, T., Awual, M. R., Rahman, A., Sultana, M., Aziz, A. A., Monir, M. U., Teo, S. H., & Hasan, M. (2020). Advances in sustainable approaches to recover metals from e-waste-A review. *Journal of Cleaner Production*, *244*, 118815. doi:10.1016/j.jclepro.2019.118815

Ismail, H., & Hanafiah, M. M. (2019). Discovering opportunities to meet the challenges of an effective waste electrical and electronic equipment recycling system in Malaysia. *Journal of Cleaner Production*, *238*, 117927. doi:10.1016/j.jclepro.2019.117927

Khatiwada, B., Jariyaboon, R., & Techato, K. (2023). E-waste management in Nepal: A case study overcoming challenges and opportunities. e-Prime-Advances in Electrical Engineering. *Electronics and Energy*, *4*, 100155.

Klinglmair, M., & Fellner, J. (2010). Urban mining in times of raw material shortage. *Journal of Industrial Ecology*, *14*(4), 666–679. doi:10.1111/j.1530-9290.2010.00257.x

Krishna, R. (2015). *Study Paper On e-waste management, DDG(FA)*. TEC.

Krook, J., & Baas, L. (2013). Getting serious about mining the technosphere: A review of recent landfill mining and urban mining research. *Journal of Cleaner Production*, *55*, 1–9. doi:10.1016/j.jclepro.2013.04.043

Lau, W. K., Chung, S. S., & Zhang, C. (2013). A material flow analysis on current electrical and electronic waste disposal from Hong Kong households. *Waste Management (New York, N.Y.)*, *33*(3), 714–721. doi:10.1016/j.wasman.2012.09.007 PMID:23046876

Leung, A., Cai, Z. W., & Wong, M. H. (2006). Environmental contamination from electronic waste recycling at Guiyu, southeast China. *Journal of Material Cycles and Waste Management*, *8*(1), 21–33. doi:10.1007/s10163-005-0141-6

Li, H., Jin, Z., & Krishnamoorthy, S. (2021). E-waste management using machine learning. In *Proceedings of the 6th International Conference on Big Data and Computing* (pp. 30-35). 10.1145/3469968.3469973

Li, J., Lu, H., Guo, J., Xu, Z., & Zhou, Y. (2007). Recycle technology for recovering resourcesand products from waste printed circuit boards. *Environmental Science & Technology*, *41*(6), 1995–2000. doi:10.1021/es0618245 PMID:17410796

Masud, M. H., Akram, W., Ahmed, A., Ananno, A. A., Mourshed, M., Hasan, M., & Joardder, M. U. H. (2019). Towards the effective E-waste management in Bangladesh: A review. *Environmental Science and Pollution Research International*, *26*(2), 1250–1276. doi:10.1007/s11356-018-3626-2 PMID:30456610

Matthews, S., Francis, C., McMichael, C., Hendrickson, T., & Hart, D.J. (1997). Disposition and End-of-Life Options for Personal Computers: Green Design Initiative Technical Report. Carnegie Mellon University.

Meskers, C., Hagelüken, C., Salhofer, S., & Spitzbart, M. (2009). *Impact of Pre-processing Routes on Precious Metal Recovery From PCs*. Academic Press.

Needhidasan, S., Samuel, M., & Chidambaram, R. (2014). Electronic waste – an emerging threat to the environment of urban India. *Journal of Environmental Health Science & Engineering*, *12*(1), 12. doi:10.1186/2052-336X-12-36 PMID:24444377

Nnorom, I. C., & Osibanjo, O. (2008). Overview of electronic waste (e-waste) management practices and legislations, and their poor applications in the developing countries. *Resources, Conservation and Recycling*, *52*(6), 843–858. doi:10.1016/j.resconrec.2008.01.004

NSWMA. (2013). *Waste characterization and per capita generation rate report*. NSWMA.

Orlins, S., & Guan, D. (2016). China's toxic informal e-waste recycling: Local approaches to a global environmental problem. *Journal of Cleaner Production*, *114*, 71–80. doi:10.1016/j.jclepro.2015.05.090

Osibanjo, O., & Nnorom, I. C. (2007). The challenge of electronic waste (e-waste) management in developing countries. *Waste Management & Research*, *25*(6), 489–501. doi:10.1177/0734242X07082028 PMID:18229743

Parajuly, K., Thapa, K. B., Cimpan, C., & Wenzel, H. (2017). Electronic waste and informal recycling in Kathmandu, Nepal: Challenges and opportunities. *Journal of Material Cycles and Waste Management*.

Perez-Belis, V., Bovea, M. D., & Ibanez-Fores, V. (2015). An in-depth literature review of the waste electrical and electronic equipment context: Trends and evolution. *Waste Management & Research*, *33*(1), 3–29. doi:10.1177/0734242X14557382 PMID:25406121

Puppala Ramya, V. (2023, February 10). Optimized Deep Learning-Based E-Waste Management in IoT Application via Energy-Aware Routing. *Cybernetics and Systems*, 1–30. Advance online publication. doi:10.1080/01969722.2023.2175119

Rahman, M. W., Islam, R., Hasan, A., Bithi, N. I., Hasan, M. M., & Rahman, M. M. (2022). Intelligent waste management system using deep learning with IoT. *Journal of King Saud University. Computer and Information Sciences*, *34*(5), 2072–2087. doi:10.1016/j.jksuci.2020.08.016

Ramya, P., & Ramya, V. (2023). E-waste management using hybrid optimization-enabled deep learning in IoT-cloud platform. *Advances in Engineering Software*, *176*, 103353. doi:10.1016/j.advengsoft.2022.103353

Reuter, M., Hudson, C., Hagelüken, C., Heiskanen, K., Meskers, C., & Schaik, A. (2013). *Metal Recycling - Opportunities, Limits, Infrastructure*. United Nations Environment Programme.

Ritcey, G. M. (2006). Solvent extraction in hydrometallurgy: Present and future. *Tsinghua Science and Technology*, *11*(2), 137–152. doi:10.1016/S1007-0214(06)70168-7

Sadegh Safarzadeh, M., Bafghi, M. S., Moradkhani, D., & Ojaghi Ilkhchi, M. (2007). A review on hydrometallurgical extraction and recovery of cadmium from various resources. *Minerals Engineering*, *20*(3), 211–220. doi:10.1016/j.mineng.2006.07.001

Schluep, M., Müller, E., & Rochat, D. (2012). *E-Waste Assessment Methodology Training & Reference Manual*. Empa - Swiss Federal Laboratories for Materials Science and Technology.

Schumacher, K. A., & Agbemabiese, L. (2019). Towards comprehensive e-waste legislation inthe United States: Design considerations based on quantitative and qualitative assessments. *Resources, Conservation and Recycling, 149*, 605–621. doi:10.1016/j.resconrec.2019.06.033

Scitz, J. (2014). *Analysis of Existing E-Waste Practices in MENA Countries. The Regional Solid Waste Exchange of Information and Expertise Network in Mashreq and Maghreb Countries*. SWEEP-Net.

SEPA. (2011). Recycling and disposal of electronic waste: Health hazards and environmental impacts. SEPA.

Shreyas Madhav, A. V., Rajaraman, R., Harini, S., & Kiliroor, C. C. (2022). Application of artificial intelligence to enhance collection of E-waste: A potential solution for household WEEE collection and segregation in India. *Waste Management & Research, 40*(7), 1047–1053. doi:10.1177/0734242X211052846 PMID:34726090

Song, Q., & Li, J. (2015). A review on human health consequences of metals exposure to ewaste in China. *Environmental Pollution, 196*, 450–461. doi:10.1016/j.envpol.2014.11.004 PMID:25468213

Suja, F., Abdul Rahman, R., Yusof, A., & Masdar, M. S. (2014). E-waste management scenarios in Malaysia. *Journal of Waste Management, 2014*, 1–7. doi:10.1155/2014/609169

Terazono, A., Murakami, S., Abe, N., Inanc, B., Moriguchi, Y., Sakai, S.-i., Kojima, M., Yoshida, A., Li, J., Yang, J., Wong, M. H., Jain, A., Kim, I.-S., Peralta, G. L., Lin, C.-C., Mungcharoen, T., & Williams, E. (2006). Current status and research on E-waste issues in Asia. *Journal of Material Cycles and Waste Management, 8*(1), 1–12. doi:10.1007/s10163-005-0147-0

The E-Waste Guide. (2017). *ewaste guide.info: a knowledge base for the sustainable recycling of e-Waste-Image gallery*. Available: https://ewasteguide.info/images_galleries

Velis, C. A., Wilson, D. C., Rocca, O., Smith, S. R., Mavropoulos, A., & Cheeseman, C. R. (2012). An analytical framework and tool ('InteRa') for integrating the informal recycling sector in waste and resource management systems in developing countries. *Waste Management & Research, 30*(9_suppl), 43–66. doi:10.1177/0734242X12454934 PMID:22993135

Voskergian, D., & Ishaq, I. (2023). *Smart E-waste Management System Utilizing Internet of Things and Deep Learning Approaches*. Academic Press.

Wang, F., Huisman, J., Stevels, A., & Balde, C. P. (2013). Enhancing e-waste estimates: Improving data quality by multivariate Input-Output Analysis. *Waste Management (New York, N.Y.), 33*(11), 2397–2407. doi:10.1016/j.wasman.2013.07.005 PMID:23899476

Wang, W., Tian, Y., Zhu, Q., & Zhong, Y. (2017). Barriers for household e-waste collection in China: Perspectives from formal collecting enterprises in Liaoning Province. *Journal of Cleaner Production, 153*, 299–308. doi:10.1016/j.jclepro.2017.03.202

Wang, Z., Zhang, B., & Guan, D. (2016). Take responsibility for electronic-waste disposal. *Nature, 536*(7614), 23–25. doi:10.1038/536023a PMID:27488785

Widmer, R., Oswald-Krapf, H., Sinha-Khetriwal, D., Schnellmann, M., & Böni, H. (2005). Global perspectives on e-waste. *Environmental Impact Assessment Review, 25*(5), 436–458. doi:10.1016/j.eiar.2005.04.001

Wong, M. H., Wu, S. C., Deng, W. J., Yu, X. Z., Luo, Q., Leung, A. O. W., Wong, C. S. C., Luksemburg, W. J., & Wong, A. S. (2007). Export of toxic chemicals – a review of the case of uncontrolled electronic-waste recycling. *Environmental Pollution, 149*(2), 131–140. doi:10.1016/j.envpol.2007.01.044 PMID:17412468

Zeng, X., Duan, H., Wang, F., & Li, J. (2017). Examining environmental management of ewaste: China's experience and lessons. *Renewable & Sustainable Energy Reviews, 72*, 1076–1082. doi:10.1016/j.rser.2016.10.015

Zeng, X., Mathews, J. A., & Li, J. (2018). *Urban Mining of E-Waste Is Becoming More Cost-Effective Than Virgin Mining*. Academic Press.

Zhang, S., & Forssberg, E. (1997). Mechanical separation-oriented characterization ofelectronic scrap. *Resources, Conservation and Recycling, 21*(4), 247–269. doi:10.1016/S0921-3449(97)00039-6

Zhou, E. P. (2022). Machine Learning For The Classification And Separation Of E-Waste. In *2022 IEEE MIT Undergraduate Research Technology Conference (URTC)* (pp. 1-5). IEEE. 10.1109/URTC56832.2022.10002242

Chapter 9
Achieving Green Sustainability in Computing Devices in Machine Learning and Deep Learning Techniques

S. Sharanya

Data Science and Business Systems, SRM Institute of Science and Technology, India

V. Vijayalakshmi

Data Science and Business Systems, SRM Institute of Science and Technology, India

R. Radha

Data Science and Business Systems, SRM Institute of Science and Technology, India

ABSTRACT

The accelerated growth in artificial intelligence, internet of devices, machine learning (ML), and deep learning at breakneck speed has attracted the attention of researchers in developing novel green solutions for reclaiming the green society. The intersection of these technologies with green sustainability will greatly impact the deployment of cutting-edge technologies with green solutions. Leveraging ML technologies to improve engineering techniques to reduce the toxins released in the environment in various forms is discussed in this work. The predominant area of focus is applying is developing green AI-based solutions with sustainability measures and metric in mind. The primary contribution of this work is the holistic analysis of the employment of green ML and deep learning techniques in fostering a sustainable environment. The potential scope of this research is to benefit the research community in developing novel ML and deep learning technologies for improving green sustainability.

DOI: 10.4018/979-8-3693-1552-1.ch009

1. GREEN COMPUTING AND SUSTAINABILITY

Green Computing (GC) or sustainable computing is gaining more popularity due to the detrimental effects of modern day Information and Communication Technologies (ICT). Through the term green computing was coined in early 1990s, it has received global acclamation just before a decade in almost all the sectors of engineering like manufacturing, automobiles and information technology (Murugesan, 2008). GC encompasses design, development, implementation and proper disposal of computing devices in sustainable fashion (Sarkar et al., 2021). The primary motor behind the initiatives of GC is to mitigate the carbon footprint of the computing devices by ensuring optimal usage. This deals with focusing on environment friendly aspects of cutting edge technologies by deploying techniques far efficient resource utilization at reduced computing cost (Jayalath et al., 2019). The canopy of GC extends to almost all computing resources like software, hardware, networks, data centres and other ICT tools. The recommend-able GC practices extends from optimised hardware design, low power consumption, using renewable power sources, efficient software, using virtual servers till e waste management (Rautela et al., 2021). These measures has significantly mitigated the energy consumption, decreased the device operational cost without compromising environmental sustainability. However the dark side of GC demands added investments on developing and deploying energy efficient hardware and tapping renewable energy power sources. Augmenting to this, the implementation of proper e-waste management require tremendous effort from all sectors of the corporates which hinder its widespread usage.

The 4 main pillars of accomplishing GC is shown in Fig 1(Wong et al., 2020). Practices of green usage focuses on reducing the energy consumption of computing devices in an environmental friendly manner. This will greatly reduce the carbon footprint. The next pillar focuses on proper disposal of e waste by encouraging the practice of using or refurbishing old computing devices. At a much higher level green design sheds light on designing environment friendly products without compromising the efficiency of the devices. Green manufacturing relies on producing energy efficient electronic components computing devices and other accessories which has mitigated negative impact on the environment.

Though the above taxonomy is very generic and suitable to almost all engineering disciplines, it has delineated green path for IT sector. From the perspective of computing technologies, GC kas extended its hands in three major domains namely sustainability at data centres, leveraging distributed computing and other ICT focused sustainable measures. The role of GC at data centers should never be undermined because they serve as the major contributor of carbon footprint among the computing sector. The measures such as using renewable energy sources, leveraging power management software at the servers, installing IT energy indicators, virtualization of servers, localised HVAC systems, implementing outsourcing and colocation services, usage of solid state devices and optimised storage capacity can greatly reduce the carbon footprints at data centers. Harnessing GC through distributed computing focuses on initiatives suggest virtualization at the client side, enabling thin clients, enhanced power management and improved network efficiency. There is an exhaustive list of measures that are equipped at other ICT based sustain-able development. Some major activities include leveraging cloud / edge/ fog computing, disposal and recycling of e waste, paper free environment, management of application energy portfolios, design and development of best practices for efficient and optimised resource management, unified communica-tion, conferencing collaboration telepresence and teleworking regulatory bodies with green procurement policies. Fig 2. Shows the carbon footprint of various computing devices.

Figure 1. Strategic pillars of green computing

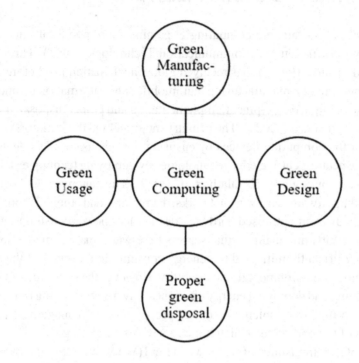

Figure 2. Carbon footprint of various computing devices

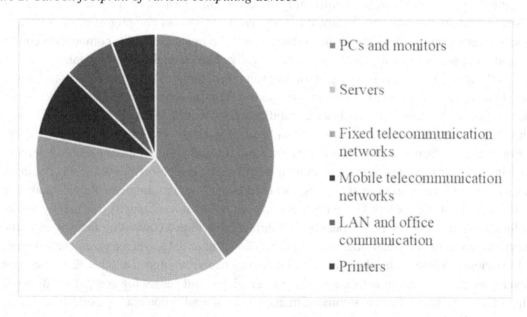

2. TOWARDS SUSTAINABLE AI

The field of Artificial Intelligence (AI) has witnessed a tremendous growth ask the technology finds its application in almost all fields. Green AI is a buzzword that is steadily gaining momentum among the corporate sectors. It is a bifurcation of AI to reduce the negative environmental impact of technologies which has caused the unfavourable climatic change across the globe. The canopy of green AI extends from designing and implementing models, systems, and algorithms that are more sustainable and environment friendly. This can be done by optimised energy usage, decreased greenhouse gas emissions into the atmosphere while promoting sustainable green practices in using the computing devices. Green AI unleashes much large potential to combat sustainability issues among the corporates, businesses and policy makers. It should not be forgotten that by enforcing the practices of green AI comes with its own challenges and limitations. It is well known that the learning process in AI is a power hungry task demands much larger resources of energy and computing devices. The backbone of AI is the data centers which has greater carbon footprint. In addition to this measures taken to ensure security and privacy of data adds on to the total carbon footprint. A more intelligent way would be to the benefits and potential challenges in enforcing green AI.

Green Machine Learning

Recent years has witnessed rampant growth in Machine Learning (ML) and Deep Learning (DL) techniques whose predictive power primarily depends on their learning capacities. They are widely used in large number of learning paradigms with different underlying architectures in versatile research areas like natural language processing, predictive analytics, computer vision, generative AI etc (Sharanya et al.,2022). The learning process in these techniques is done by hyper parameter tuning. This demands the deployment of intensive high performance computing environments augmented with optimization policies. The concern over the emission of carbon from the computing devices raised the awareness about the development of holistic and comprehensive ML models and algorithms but sensible trade off between the performance and green efficiency. The traditional ML algorithms are popularly known as Red ML models which are motivated by a single distinct performance index. On the other hand, Green ML focuses on constructing better learning paradigms with sound environment friendly practices (Silva et al., 2021). This is realised by including sustainability metrics as a part of the development of leraning models as shown in Fig3.

Figure 3. Evolution of green ML/DL techniques

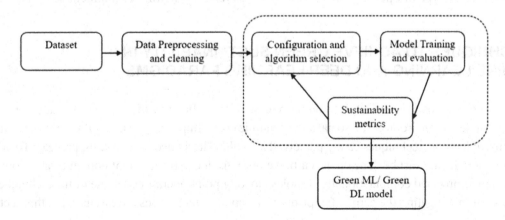

3. LANDSCAPE OF SUSTAINABILITY METRICS IN LEARNING PARADIGMS

The landscape of sustainability metrics in any domain primarily depends on four factors namely ecology, economics, politics and cultural background. From the perspective of Green AI, not many generic metrics have relevance to monitor the carbon footprints of computing devices. The sustainability metrics in Green AI focuses on gauging the optimal usage of computing resources and computational time. The predominant metrics that have a significant impact in mitigating the carbon emission of ML and DL models are discussed here:

- **Running time:** This is the time taken by the program to complete its execution. Though it cannot be a holistic measure of efficiency, but it has a positive correlation towards power consumption by computing devices. This metric can be used for the estimation of carbon footprints more precisely when augmented with details like power consumption of the underlying hardware (CPU/ GPU) and source of power (Baliga et al., 2010).
- **Resource utilization (CPU/ GPU) hours:** Quantising the running time of CPU and GPU is a direct measure of environmental impact. Measuring the running time of the processors in terms of the world clock time includes memory access time also (Scogland et al., 2010). On the other hand, measuring the real CPU or GPU time ignores the time taken for memory access operations.
- **Composition of floating point operations:** These operations are hardware dependent matrix which indirectly contributes towards carbon footprint. It is well known that floating point operations are computationally intensive which demands more power and more hardware resources (Wang et al., 2021). Incorporating good degree of optimization in terms of both memory and computation time can greatly reduce the carbon footprints.
- **Energy Consumption:** The energy consumption depends on the energy efficiency of the underlying hardware used in the computing devices. It is a very simple and direct metric to assess the carbon footprint of the computational hardware. It is easy to compute or approximate the actual carbon emissions by the computing device by knowing its energy consumption (Pandi et al., 2016). It is very easy for getting the energy consumption data of computing devices whereas the other supporting hardware such as electrical wires, switches, routers etc do not have a definite energy consumption which makes very difficult to measure the total carbon footprint.
- **Random Access Memory:** Deploying dynamic RAM devices place a crucial role in monitoring the energy consumption (Sofia et al., 2015). But the amount of energy consumption in RAM depends on the type of operation that is done such as reading, writing or maintenance.

4. TECHNIQUES TO ACHIEVE GREEN SUSTAINABILITY IN MACHINE LEARNING AND DEEP LEARNING PARADIGMS

the core value of green ML and DL techniques release of the efficiency of an array of evaluation criteria that are grounded on environmental, social and economic costs. Imparting sustainability in to the computation intensive learning techniques candy perceived as multi criteria decision making process. To acclaim efficiency more focus must be done on the carbon emissions due to the usage of non renewable sources of power, running time and computationally complex floating point operations. Some of the techniques that will be useful in mitigating the carbon footprint of ML and DL techniques are discussed in this section.

4.1 Granular Computing

The term granular computing encompasses a white horizon of knowledge that relies upon information granules which are represented in various computational domains like calculus, set theory, rough sets, fuzzy sets etc. These granules are then integrated to develop a methodological and developmental environment (Chu et al., 2023). Fig 4. Show the evolution of granular computing (Pedrycz et al., 2018).

Figure 4. Pollution of granular computing models

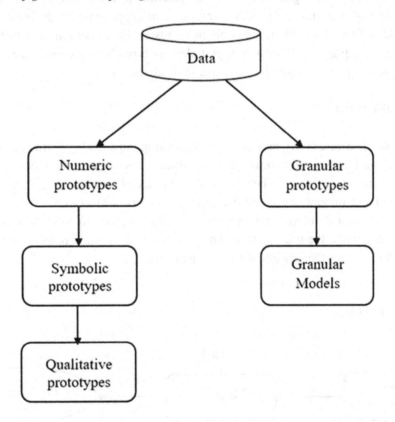

In the context of ML and DL, the granular computing and its representational information granules process tangible potential to implement green AI in multiple ways. As the world is moving towards developing explainable AI models, information granularity facilitates the development of interpretable models what good degree of explainability. Also, granular computing models aids in quantification of the underlying ML and DL models in establishing a mutually beneficial trade of between intensive computations and accuracy of the models. In addition to this, the granular computing helps in imparting privacy and confidentiality in the developed models by providing data abstraction and higher level communication. By this way, information granularity this exhibited at the lower level.

Precisely, information granules are chunks of knowledge which represents abstract data with well defined semantics. They actress functional modules in building interpretable and explainable models. These information granules are very useful in structuring semantically valid data. These granules are binded together based on their relationships such as distance, functionality, similarity etc. Therefore the information granules implicitly has multiple levels of abstraction which hides unnecessary details das aiding the development of explainable and interpretable models.

This knowledge based environment facilitates the design and further processing of the so developed information granules. In short, granular computing is all about a representation, development and interoperability off the information granules which are generally formalised as sets, fuzzy rules etc.

Granular Computing is a knowledge-based environment supporting the design and processing of information granules. Granular Computing is about representing, constructing, processing, and communicating information granules. Information granules are formalized as sets, fuzzy sets, rough sets, and probabilities, just to point at several among available alternatives.

4.2 Transfer Learning

The notion of transfer learning in ML and DL focus on reusing the already pre trained model on a previously unseen problem. The mission you see already gained knowledge from the previous task to generalise the new one. It has many potential benefits such as produced training time, improved performance and do not demand large amount of training data. In this strategy, a ML or DL model can be developed with little trading data as it develops on the previously trained model. In addition to this benefit, there is a significant production in the training time which saves the computing power and resources (Weiss et al., 2016). Fig 5 illustrates the process of transfer learning.

Figure 5. Transfer learning

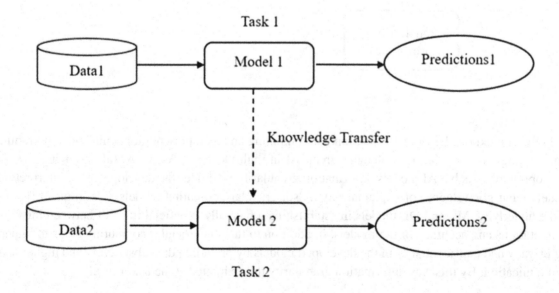

In this method, previous knowledge is occurred and applied to a new data to obtain new set of predictions. From the perspective of green ML, this results in producing the overall computation expenses has the models are reused. The new models are built as an adaptation of the previous pre trained models. Transfer learning is applied in scenarios where:

- There is very limited amount of training data in the target domain
- To improve the robustness of the already developed model
- to reduce the impact of cold which is widely prevalent in recommendation systems start problem

The primary challenge in the transfer learning life as when and how to transfer the knowledge efficiently into a new domain. In addition to this, bias in the data set may result in the development of poor transferred models. The data set mismatch will eventually lead to performance degradation as the new data set may violate the assumptions of the already trained model. Another predominant issue is the over footing and generalization of the models that are developing through transfer learning. During the process of transferring knowledge between different domains, the models may learn noise of the source data which may result in models with low generalization capacity.

In the process of knowledge transfer, some models made inadvertently lose their efficiency in predicting the original task which is commonly known as catastrophic forgetting. This kind of phenomenon can be observed when sequential retraining of the task overrides the task of being pre trained models. Ethical conduct this is another hurdle in developing transfer learning models as the fine tuned the model, inherits the sensitive information from source domain and use it for model development in the target domain.

4.3 Knowledge Distillation

Knowledge distillation process relies on transferring the obtain knowledge from a substantially large powerful model to low scale models, that are developed for real time applications. This can be thought as model compression technique. This technique can be applied to many types of ML and DL models, it is more popular in artificial neural networks, as they have multiple layers and tune several parameters.

The real challenge implementing large scale neural network models is its computational complexity. The process of knowledge distillation can be explored to limit the memory capacity and computational overhead of the deep neural networks (Goa et al., 2021). Fig 6 shows the process of knowledge distillation in neural networks among the parent and child models.

Figure 6. Knowledge distillation in neural networks

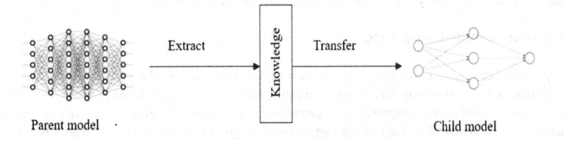

Parent model   Child model

It can be noted that knowledge distillation this hello custody transfer learning. However, in knowledge distillation the parent model is a highly sophisticated powerful learning model which has extremely high computational complexity and intensive operations. This model is generally referred to as teacher or parent model. Deploying the same complex parrot model in real time applications which involves the usage of handheld devices is computationally not feasible. These applications demands low power and low weight models which are generally referred to as student or child model. Hence the process of knowledge distillation can be seen as an significant step in achieving green AI.

4.4 Federated Learning

Federated learning or collaborative learning is an effective decentralized approach that is commonly deployed in training ML and DL models. The flexibility of this paradigm is that it do not demand a customised data exchange format between the local servers and clients. The core value of this strategy lies in training the edge devices on the raw data, thus increasing data privacy and reducing the computational load at the servers. The final model is then developed by leraning all the local updates. The primary motivation for developing federated learning models are:

- Federated leraning permit training the model locally, which prevents data privacy threats.
- This method ensures confidentiality as the updates to the global model can be done only by encrypted local models.
- This guarantees access to the heterogeneous type of data from various devices, at different geographic locations.

In general, federated learning stores a generic baseline model at the central server. Then, multiple copies of this baseline model is shared among the local clients, and they train the local data. Over a time, the encrypted local models can update the global model at the server. The updates from the local models are aggregated to make final updates to the global central model. After training the central model, it is again re-shared with other client's devices which are gain used for further iterations.

Two major types of federated leraning updates are possible: averaged federated learning and gradient-based federated learning. In the former method, the updates from local models are aggregated by averaging their changes. While in the later method, the gradient of the objective function of the model is estimated and then the updates are made. Fig 7 shows the learning procedure in Federated learning.

from the perspective of green AI, the Federated learning avoid unnecessary computations at the global server and cumulative update happened at the baseline server model. This eventually results in fewer number of computations and less resource usage. Hence, the carbon emissions of intensive ML and DL models are greatly reduced by Federated learning (Li et al. 2020).

4.5 Approximate Computing

Approximate computing has recently gained more popularity as it is perceived as the intersection of resource optimization made in both software as well as in hardware. It spawns over mitigating the carbon footprints by designing more efficient programming languages and by designing and developing new energy efficient hardware. It is obvious that, many computing applications give more accurate results than it is required, which can be seen as a wastage of time and resources. By leveraging this effect the approximate computing tries to produces the accuracy at very small scale to save energy and other computing resources.

Figure 7. Updates in federated learning

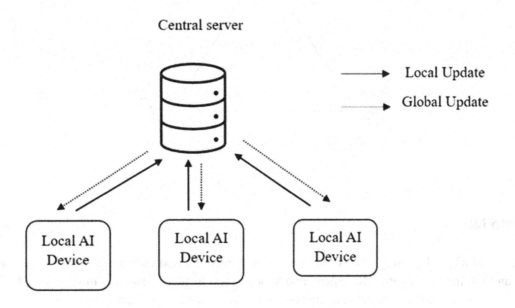

The development of various types of approximation techniques have been witnessed both in the hardware and software fronts. Some of the most popular techniques are dynamic procession knobs, unreliable hardware, loop perforations, neural network accelerators and numerical approximations (Han et al., 2013). The triad of approximation algorithms it can be better explained as follows and is shown as Fig 8:

- Software/algorithm level: These techniques are very simple like early stopping of a iterative algorithm, producing the number of loops, look perforations, code tuning techniques, employing early stopping and dropouts in artificial neural networks. Avoiding too much of decimals in floating point operations can also speed up the program execution at the software level.
- Architectural level: From the perspective of architecture, the approximations can be done by choosing appropriate low carbon system components. The literature witnesses the development of dedicated approximate accelerators application specific instruction processors to bring a significant reduction in the computing resources. This will eventually decrease the power consumption and processing time which directly leads to reduced carbon emissions.
- Circuit level: It can be seen that voltage scaling and induction of mathematical errors circuits are 2 common approximation techniques used to the circuit level. Politics scaling save energy and introduction of errors will save data propagation time.

Figure 8. Triad of approximate computing

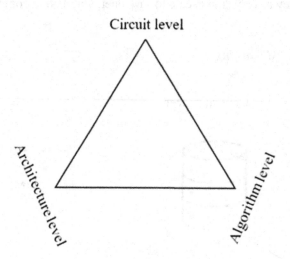

4.6 Auto ML

Automated Machine Learning (AUTO ML) envisions processes and methods to expand the horizons of ML and DL models beyond ML experts and to accelerate further research in this domain (He et al., 2021). The success of these models greatly rely on the following array of activities:

- Data preprocessing and cleaning
- Feature selection and feature engineering
- Choice of suitable model
- Hyper parameter tuning
- Model assessment.

Lot of efforts are necessary to design a complete ML and DL models. Automating the process of model development time and effort which eventually results in minimising the carbon footprint. Many AutoML platforms are available which makes the development of AI models more user friendly with better optimization. The AutoML platforms deploys optimal learning algorithm for a particular task by leveraging the following two techniques:

- Search for suitable neural network architecture: This aids the AutoML models to explorer and invasion new network architectures for the specific problem.
- Transfer learning: deploying already pre trade models on specific new data sets save unnecessary model training and hyper parameter optimizations. This also facilitates the real usage of already developed sophisticated models and harness their predictive power to suit to new problem domains.

The platform of AutoML performs wide range of tasks from handling raw data set 2 developing a complete architecture. More precisely, the research directions in Federated learning focus on three main areas namely developing energy efficient ML pipelines, optimised to search of energy efficient pipelines and optimised development of complete ML and DL algorithms using AutoML.

Development of Energy-Efficient ML Pipelines

Implementing ML and DL models to give energy efficient final output does not necessarily guarantee that the environment footprint of this process is sustainable. The usage of AutoML what's up long run solution which can greatly reduce the efforts taken for development of the models. It is always essential to investigate whether the carbon footprint of AutoML process is really energy efficient and is it worth of doing so. Another potential advantage of deploying AutoML apart from reducing the carbon footprint is designing dedicated pipeline spot energy constrained notes such as sensors mobile and edge devices which are equipped with battery power. A note for the way to save energy stop compress and accelerate the artificial neural network models that are used on handheld devices by reducing the number of floating point operations per second for every iteration in the neural network.

Developing Energy-Efficient AutoML Models

Another potential area which could be explored to reduce the carbon footprint of ML and DL models to optimise the underlying searching algorithms. Few notable methods that are adapted are:

i) **Warmstarting:** this mechanism integrates the knowledge that was gained in previous execution with the current iteration so that the optimization process need not be started from the scratch. This results in finding the right candidates in early iterations which greatly reduces the search space of the models (Ash et al., 2020). This method is a variation of the meta learning process which learns from the past experiences. A famous method for realising the warm starting procedure is to follow Hunstanton initial sequence of algorithms. The performance of these algorithms on the previously used data set is leveraged here to solve the new problem. Alternatively, auto-sklearn has a static portfolio which covers various types of use cases. The technique of warm starting is more popular in hyper parameter optimization and hence it has become a defacto standard in many AutoML platforms.

ii) **Zero Shot Auto ML:** This is perceived as extreme case of previously discussed warm starting. The zero short ML, recommends only one potential single pipeline as a candidate and this chosen pipeline was used by the system without any further evaluation. A more popular example of zero shot AutoML is Meta--Miner that obtains recommendations by forming ontology from the underlying data set and the algorithmic properties. Transfer learning and meta learning are also seen as different flavours of Zero shot AutoML. In these approaches the models either use an already existing pre trained model to develop solution for new computing problems (Ozturk et al., 2022). The data set Newton meter learning strategy demands the representation of meta features. Though these methods cannot be termed as completely energy efficient, but they play a crucial role in reducing the search space of the ML and DL algorithms.

5. TOWARDS SUSTAINABLE ML

The technological revolution has brought about a great change in developing Green solutions in almost all domains. Green ML has to be transparent about the obvious environmental impact. The developers of ML and DL models must realise the practice of providing a comprehensive summary of the efforts taken to reduce the carbon footprints by envisaging energy efficient solutions (Sharanya et al., 2022). The following are some of the important design considerations in moving forwards towards GreenML:

- **Design, Development and Evaluation:** This should focus on the key aspects of computations which embarks sustainability. A comprehensive list of measures like warmstarting, federated learning, multi-fidelity assessments, transfer learning etc must be enumerated. The developers must ensure that these best practices are adhered in developing a sustainable solution. Also, a proper quantification of the reduction in carbon footprint or improvement made in power efficiency must be estimated and mentioned during the design and development stage.
- **Resource Consumptions:** A brief note about the resources that are used for final assessment of the model like CPU or GPU hours, exploring parallelization, energy sources, the power of computing centres, fraction of the power that are generated from renewable energy sources etc. must be made. Also, formal assessments of carbon emissions and the reduction in the carbon footprint due to the adoption of sustainability measures must also be quantified 2 getter big picture of the overall implementation plan.

6. CHALLENGES IN IMPLEMENTING GREEN ML

Green ML and DL Offers manifold benefits to be environment. However the heterogeneous nature of these models which encompasses multifaceted tasks such as data collection, data transmission, communication, data cleaning and preprocessing, feature engineering, model selection, hyper parameter optimization and evaluation of the models posts a complex challenging ground that questions the portability and interoperability between these domains and devices. Throughout the life cycle of the development of ML and DL models Infosys data breach at multiple points. Leveraging crowd by solutions invite many type of intruder attacks like SQL injection, man in the middle attack, flooding attack and malware. As the models are scaled to handle large volume of heterogeneous data, the computing becomes more intense which escalates the total energy consumption. Using Federated learning, transfer learning and the other metrics which are discussed in this work must give a direct assessment of the carbon footprint. Also implementing distributed solutions has inherent challenges like threats related to privacy, confidentiality, latency, scalability, flexibility and energy consumption.

7. CONCLUSIONS AND FUTURE WORKS

The traditional ML and DL models and its development process do not primarily focus on reducing the carbon footprint or incorporating energy efficient methods as a part of its life cycle. Due to the climatic change and other detrimental environmental effects, the computing industry has now shifted its focus towards developing green solutions in almost all the fields. This work focuses on throwing light on the significant developments made in transforming the conventional ML and DL models into sustainable green solutions for the real time applications. The primary contribution of this work this is the assessment made on various strategies and techniques that can be leveraged during the process of model development without compromising environmental sustainability. Special focus has been given to incorporate energy efficient and resource optimised best practices during the development process. Augmenting to this the work also enumerates few research challenges, which may soon turn out to be potential motivation for future research directions. India future the development of ML and DL models must incorporate energy efficient and resource optimization methodologies as a part of its development process. These methods in addition to providing environmental benefits also churns out the cost and time taken for developing a new model from the scratch.

REFERENCES

Ash, J., & Adams, R. P. (2020). On warm-starting neural network training. *Advances in Neural Information Processing Systems, 33,* 3884–3894.

Baliga, J., Ayre, R. W., Hinton, K., & Tucker, R. S. (2010). Green cloud computing: Balancing energy in processing, storage, and transport. *Proceedings of the IEEE, 99*(1), 149–167. doi:10.1109/JPROC.2010.2060451

Chu, H., & Zhang, Y. (2023, August). A Green Granular Neural Network with Efficient Software-FPGA Co-designed Learning. In *IEEE 22nd International Conference on Cognitive Informatics and Cognitive Computing (ICCI* CC'2023).* IEEE.

Gou, J., Yu, B., Maybank, S. J., & Tao, D. (2021). Knowledge distillation: A survey. *International Journal of Computer Vision, 129*(6), 1789–1819. doi:10.1007/s11263-021-01453-z

Han, J., & Orshansky, M. (2013, May). Approximate computing: An emerging paradigm for energy-efficient design. In *2013 18th IEEE European Test Symposium (ETS)* (pp. 1-6). IEEE.

He, X., Zhao, K., & Chu, X. (2021). AutoML: A survey of the state-of-the-art. *Knowledge-Based Systems, 212,* 106622. doi:10.1016/j.knosys.2020.106622

Jayalath, J. M. T. I., Chathumali, E. J. A. P. C., Kothalawala, K. R. M., & Kuruwitaarachchi, N. (2019, March). Green cloud computing: A review on adoption of green-computing attributes and vendor specific implementations. In *2019 International Research Conference on Smart Computing and Systems Engineering (SCSE)* (pp. 158-164). IEEE. 10.23919/SCSE.2019.8842817

Li, L., Fan, Y., Tse, M., & Lin, K. Y. (2020). A review of applications in federated learning. *Computers & Industrial Engineering, 149,* 106854. doi:10.1016/j.cie.2020.106854

Murugesan, S. (2008). Harnessing green IT: Principles and practices. *IT Professional, 10*(1), 24–33. doi:10.1109/MITP.2008.10

Öztürk, E., Ferreira, F., Jomaa, H., Schmidt-Thieme, L., Grabocka, J., & Hutter, F. (2022, June). Zero-Shot AutoML with Pretrained Models. In *International Conference on Machine Learning* (pp. 17138-17155). PMLR.

Pandi, K. M., & Somasundaram, K. (2016). Energy efficient in virtual infrastructure and green cloud computing: A review. *Indian Journal of Science and Technology, 9*(11), 1–8. doi:10.17485/ijst/2016/v9i11/89399

Pedrycz, W. (2018). Granular computing for data analytics: a manifesto of human-centric computing. *IEEE/CAA Journal of Automatica Sinica, 5*(6), 1025-1034.

Rautela, R., Arya, S., Vishwakarma, S., Lee, J., Kim, K. H., & Kumar, S. (2021). E-waste management and its effects on the environment and human health. *The Science of the Total Environment, 773*, 145623. doi:10.1016/j.scitotenv.2021.145623 PMID:33592459

Sarkar, N. I., & Gul, S. (2021). Green computing and internet of things for smart cities: technologies, challenges, and implementation. *Green Computing in Smart Cities: Simulation and Techniques*, 35-50.

Scogland, T. R. W., Lin, H., & Feng, W. C. (2010). A first look at integrated GPUs for green high-performance computing. *Computer Science (Berlin, Germany), 25*(3-4), 125–134. doi:10.1007/s00450-010-0128-y

Selvaraj, S., Prabhu Kavin, B., Kavitha, C., & Lai, W. C. (2022). A Multiclass Fault Diagnosis Framework Using Context-Based Multilayered Bayesian Method for Centrifugal Pumps. *Electronics (Basel), 11*(23), 4014. doi:10.3390/electronics11234014

Sharanya, S., Venkataraman, R., & Murali, G. (2022). Predicting remaining useful life of turbofan engines using degradation signal based echo state network. *International Journal of Turbo & Jet-Engines*, (0).

Silva, G., Schulze, B., & Ferro, M. (2021). *Performance and energy efficiency analysis of machine learning algorithms towards green ai: a case study of decision tree algorithms* [Master's thesis]. National Lab. for Scientific Computing.

Sofia, A. S., & Kumar, P. G. (2015). Implementation of energy efficient green computing in cloud computing. *International Journal of Enterprise Network Management, 6*(3), 222–237. doi:10.1504/IJENM.2015.071135

Wang, X., Goyal, V., Yu, J., Bertacco, V., Boutros, A., Nurvitadhi, E., . . . Das, R. (2021, May). Compute-capable block RAMs for efficient deep learning acceleration on FPGAs. In *2021 IEEE 29th Annual International Symposium on Field-Programmable Custom Computing Machines (FCCM)* (pp. 88-96). IEEE. 10.1109/FCCM51124.2021.00018

Weiss, K., Khoshgoftaar, T. M., & Wang, D. (2016). A survey of transfer learning. *Journal of Big Data, 3*(1), 1–40. doi:10.1186/s40537-016-0043-6

Wong, T. H., Rogers, B. C., & Brown, R. R. (2020). Transforming cities through water-sensitive principles and practices. *One Earth, 3*(4), 436–447. doi:10.1016/j.oneear.2020.09.012

Chapter 10
Future Trends and Significant Solutions for Intelligent Computing Resource Management

Diya Biswas

Brainware University, India

Anuska Dutta

Brainware University, India

Shivnath Ghosh

Brainware University, India

Piyal Roy

Brainware University, India

ABSTRACT

Cloud providers place a high value on reducing energy consumption in cloud computing since it reduces operating costs and improves service sustainability. Cloud services are frequently replicated across providers to ensure high availability and dependability, which increases provider resource utilization and overhead. Finding the right balance between service replication and consolidation to lower energy usage and boost service uptime can be challenging for cloud resource management decision-makers. This chapter addresses this problem by presenting a ground-breaking technique known as "CRUZE," which is based on cuckoo optimization and considers energy efficiency, dependability, and comprehensive resource management in cloud computing, encompassing cooling systems, servers, networks, and storage. Using cloud resources, effectively illuminating and executing a range of jobs, CRUZE has significantly reduced energy usage by 20.1% while improving dependability and CPU utilization by effectively illustrating and executing a variety of workloads on cloud resources that have been allocated.

DOI: 10.4018/979-8-3693-1552-1.ch010

1. INTRODUCTION

The efficient management of computing resources is paramount. Now a days, the landscape of Intelligent Computing Resource Management (ICRM) is poised to undergo significant transformations to meet the growing demands of various industries and sectors. he rapid expansion of Internet-based technologies in recent times has prompted multinational corporations to install dispersed computing resources globally, including cloud computing and Cyber Physical Systems (CPS) (Al-Ansi et al., 2021). ICRM has become an increasingly important area of focus in today's rapidly changing technological environment. Computing hardware is point of contact in operations and performance in majority applications like artificial intelligence, edge computing, cloud computing, data analytics, and more.

Maintenance of pc hardware has turn out to be a prime challenge within the ever-evolving environment of pc technology. As we pass into the 21st century, the demand for electronic assets will increase at an exceptional charge. With the proliferation of information-in depth applications, the upward thrust of synthetic intelligence and system gaining knowledge of, the upward thrust of cloud computing as well as the Internet of Things, or IoT, the need to reveal computing a wisely it is far never incredible.

Intelligent computer resource management is being explored and advanced for a variety of reasons. Agencies must correctly and efficiently scale their computing resources as they expand their operations, which presents challenging conditions. This demands for automated solutions that can dynamically assign resources in accordance with current needs, optimizing performance and value at the same time.

Cost effectiveness: In a generation where computer assets account for a sizable amount of operating costs, cutting costs while maintaining or improving performance is a compelling goal. To achieve this equilibrium, intelligent aid control may be helpful.

Sustainability: Concerns about the environment are becoming more widespread. However, it also increases carbon footprints because wasteful resource allocation is no longer the most efficient. Smart resource management can result in a lesser ecological footprint and less strength usage.

Technological Advancements: The establishment of technologies like surface computing, 5G networks, and quantum computing increases intricacy to support in administration. Creative and adaptable solutions are needed to completely utilize these technologies. Data surge: To suppress the exponential expansion of data generation and consumption, robust aid control is indispensable. Smart answers can support in the effective processing and analysis of massive datasets.

2. EVOLUTION OF RESOURCE MANAGEMENT

The term "resource evaluation" refers to the process by which a researcher examines a potential source of information objectively and decides, after considering its veracity and correctness, whether or not it is appropriate for a given paper or project.This entails reviewing and evaluating the effectiveness and caliber of your resource management procedures, techniques, and equipment.

2.1 Traditional Resource Allocation Challenges

Traditional resource allocation challenges were characterized by static, inflexible, and manual processes that often led to resource underutilization, overallocation, and fragmentation. Intelligent computing resource management solutions aim to overcome these challenges through automation, machine learning, and data-driven decision-making, enabling organizations to optimize resource allocation and better adapt to evolving computing environments.

Resource allocation that is static: In the past, resources were frequently allocated in this way. This implies that resources remained devoted to that task after they were assigned to a particular task or application, even when the workload changed (Budde & Volz, 2019). This caused inefficient resource use and performance bottlenecks.

Underutilization: Using conventional resource allocation techniques frequently results in underutilization of computing resources (Papagianni et al., 2020). A server, for instance, might operate at barely 30% of its potential, wasting both energy and hardware resources.

Overallocation: On the other hand, overallocation of resources was also common. Allocating more resources than required for a task leads to increased costs and inefficiency, which is a concern for businesses and data centers.

Inflexibility: Traditional resource allocation models were often inflexible. They couldn't easily adapt to changing workloads or priorities. This resulted in poor performance during peak loads and inefficient resource usage during off-peak periods.

Resource Fragmentation: Over time, resources could become fragmented, with different applications and services each claiming a part of the available resources. This fragmentation could reduce overall system performance.

Lack of Predictive Capabilities: Traditional resource allocation lacked predictive capabilities. It could not anticipate resource demands based on historical data, trends, or machine learning models. This made it challenging to optimize resource allocation.

Manual Intervention: Many resource allocation processes were heavily reliant on manual intervention. IT administrators had to make decisions about resource allocation, which was both time-consuming and prone to human error.

Scalability: As computing infrastructures grew in complexity, traditional resource allocation methods struggled to scale. They were ill-suited for the dynamic, distributed, and highly virtualized environments that modern organizations require.

The Significance of Allocating Resources: The distribution of resources is a key factor in what propels organizational success. Organizations may maximize revenue, improve productivity, and optimize performance by making sure resources are allocated at the appropriate times and locations. It involves determining which areas are crucial in terms of resource requirements and distributing resources in a way that meets overall strategic goals and objectives. Effective resource allocation helps businesses maximize their available resources while reducing waste and unnecessary expenses. (Guazzone et al., 2011) dynamic allocation, containerization, cloud-native designs, automation, edge computing, and robust security measures. In order to fulfill the constantly changing demands of modern computing, these future trends and solutions are crucial (Liang et al., 2016). Adopting these advancements will be essential for maintaining competitiveness and making sure that the computer environment is frictionless, secure, and economical.

2.2 Emergence of Intelligent Resource Management

The emergence of intelligent resource management represents a significant change in the fields of technology and computing. As the demands on computing resources continue to grow in complexity and scale, traditional methods of resource allocation and optimization are proving inadequate. Artificial intelligence and machine learning are two cutting-edge technologies that are included into Intelligent Resource Management, and advanced algorithms to dynamically allocate and optimize computing resources in a more efficient and intelligent manner. This innovative approach enables organizations to meet the growing demands of their applications and services while minimizing costs and improving overall performance. In this dynamic landscape, we will explore future trends and significant solutions for Intelligent Computing Resource Management that promise to shape the future of technology and computing infrastructure.

The emergence of intelligent resource management is a significant development driven by advancements in technology, data analytics, and the increasing complexity of computing environments. This trend is poised to revolutionize how we allocate, optimize, and manage computing resources in a more efficient and automated manner. Let's explore the key drivers, trends, and solutions in the field of intelligent computing resource management.

Increased Complexity of Computing Environments

As organizations rely more on digital technologies, the complexity of their computing environments has surged. This complexity includes diverse hardware resources, cloud-based infrastructure, and heterogeneous software stacks.

To effectively manage these intricate ecosystems, intelligent solutions are required to make dynamic, data-driven decisions.

- **Making Decisions Based on Data:** The proliferation of big data and real-time data from multiple sources have enabled more informed decisions on resource allocation and management. Intelligent resource management systems can analyze this data to optimize resource utilization, predict future resource needs, and identify performance bottlenecks.
- **Artificial Intelligence and Machine Learning:** AI or ML techniques are being used to develop predictive and prescriptive algorithms that can adapt resource allocation in real-time based on historical data and current demands. These technologies enable systems to learn from past performance and make more efficient resource allocation decisions.
- **Automation and Orchestration:** Intelligent resource management systems can automate routine tasks like provisioning and scaling resources, which reduces the need for manual intervention. Orchestrating resources to meet specific application requirements or service level agreements (SLAs) becomes more efficient and responsive with intelligent management.
- **Edge Computing and The Cloud:** The adoption of cloud and edge computing has transformed resource management, with resources often being distributed across multiple data centers, cloud providers, and edge devices. Intelligent solutions can dynamically allocate resources across these distributed environments to optimize performance and cost-effectiveness.

- **Energy Efficiency and Sustainability:** There is a growing emphasis on energy efficiency and sustainability in data centers and computing environments. Intelligent resource management can help reduce energy consumption by optimizing resource utilization and placement (Negi et al., 2011).

- **Significant Solutions for Intelligent Computing Resource Management:** Predictive Analytics: Historical data and ML to predict future resource requirements, allowing for proactive allocation in this system.

- **Dynamic Scaling:** Intelligent resource management solutions can automatically adjust resource allocation based on traffic spikes or changes in workload, ensuring optimal performance and cost efficiency.

- **Application-Aware Management:** Some solutions are designed to be aware of the specific requirements of applications and allocate resources accordingly. For example, a database-intensive application may receive more CPU and memory resources when needed.

- **Hybrid and Multi-Cloud Management:** Intelligent systems can manage resources across various cloud providers and on-premises environments, optimizing costs and performance.

- Edge Resource Management: With the growth of edge computing, resource management solutions can efficiently allocate resources to edge devices, reducing latency and improving user experience.

- Green Computing: Resource management can also play a role in making computing environments more sustainable through maximizing the use of resources and energy.The emergence of intelligent resource management is driven by the need to efficiently allocate and optimize computing resources in increasingly complex and dynamic environments. Through data-driven decision-making, automation, and the application of AI and ML, these solutions are transforming how organizations manage their computing resources, making them more efficient, responsive, and sustainable.

2.3 Role of Automation and Orchestration

The launching of 5G (fifth generation) and its evolution (NextG), would herald the introduction of innovative, adaptable solutions that follow softwarization, openness, and division principles and the end of the era of inflexible, hardware-driven Radio Access Networks (RAN) architectures. This paradigm change, sometimes known as Open RAN, is characterized by unrestricted flexibility. It makes it possible to instantiate and control network functionalities across several network nodes. Traditionally, these functionalities were merged and executed in monolithic base stations. In IT, orchestration is often used for tasks like provisioning cloud resources, disaster recovery, and managing complex software deployments. In business, it can be used for supply chain management, customer onboarding, and order processing. The synergy of automation and orchestration is especially powerful. Automation handles the execution of individual tasks, while orchestration provides the framework for managing the end-to-end processes. This combination enables organizations to streamline operations, reduce manual intervention, and respond to dynamic business demands effectively.By automating and orchestrating processes, organizations can improve their overall efficiency, reduce operational costs, enhance reliability, and stay competitive in today's fast-paced, technology-driven world. These principles apply across various industries, from IT and manufacturing to healthcare and finance, and are crucial for staying agile and responsive to changing market conditions (Guazzone et al., 2011). Conflict mitigation: It's also important to make sure that different ML/AI models don't clash with one another and that, at all times, only one model is in charge of the same parameter or feature. Because of these factors, organizing intelligence about networks in the Open RAN is difficult and distinct, necessitating innovative, automated, and scalable methods. In this research, we introduce an automated smart orchestration system for the open radio access network (RAN) called OrchESTRAN, which aims to overcome these issues. Functioning as a rap in the non-RT RIC (Figure 1), Manage RAN complies with O-RAN guidelines and offers automated methods to: (i) Compile control requests from NOs; (ii) Select the most appropriate ML/AI models to achieve NOs' goals and avoid conflicts (Elgamal et al., 2023).

Figure 1. O-RAN reference architecture and interfaces (left). A tree graph of representation of an O-RAN network architecture as a tree graph (right) (D'Oro et al., 2022).

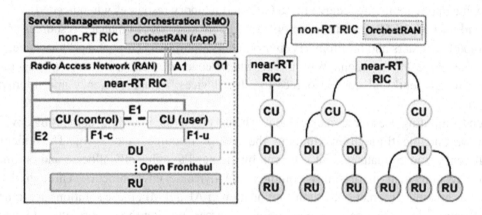

3. EDGE COMPUTING AND DECENTRALIZATION

When deep learning (DL) was initially proposed as a solution to problems, training data was gathered in order to facilitate captured data processing Due to the field's recent quick advancement, its applications have expanded to numerous industries, improving people's quality of life. However, gathering and sending such vast amounts of data into centralized storage facilities typically takes a lot of time, is inefficient, and raises privacy issues. Network capacity restrictions and other factors may contribute to significant latency. Furthermore, data privacy concerns arise from the possibility of personal data breaches associated with data transmission to a centralized computer resource. In particular, the growing awareness of data privacy in society has led to the promotion of legal restrictions like the General Data Protection Regulation (Asim & Abd El-Latif, 2021), which makes an explicitly defined framework even more impractical.

3.1 Edge Computing Architecture and Benefits

In today's rapidly evolving digital landscape, Edge computing has emerged as a result of the need for real-time data processing, low-latency applications, and distributed computing. Edge computing represents a significant shift from traditional centralized cloud architectures, as it brings data processing and computation closer to the data source, such as IoT devices, sensors, or end-user devices. This introduction provides an overview of edge computing architecture and its associated benefits. A decentralized computing paradigm known as "edge computing" shifts data processing and computation closer to the point of data generation or need, which is frequently at the "edge" of the network. Unlike conventional cloud computing, which processes data by sending it to a centrally located data center, edge computing processes data locally, minimizing latency and enhancing response times. In scenarios requiring real-time analytics, such driverless vehicles, industrial automation, remote monitoring, and augmented reality applications, this architecture is very important. T By exchanging sensor data about the environment while on the go, mobile crowdsensing (MCS) is a human-driven Internet of Things service that enables citizens to observe the phenomena of individual, community, or even societal value. Cloud-based con-textualized architectures are used in typical MCS service implementations. These architectures require a large amount of computational resources and generate a large amount of network traffic, both toward cloud-based MCS services and on mobile networks. Because mobile edge computing (MEC) allows for real-time data processing and aggregation close to data sources, it is a natural choice for distributing machine learning solutions by moving computation to the network edge. An MEC-based architecture also allows for significant performance improvements. Thus, less associated traffic will be in mobile channels, which will enable MCS. This research suggests an advanced computing architecture suitable for large-scale machine learning services by integrating key machine learning features into the reference MEC architecture. Apart from enhanced performance, the suggested architecture reduces privacy risks and gives citizens authority over the stream of shared sensor information. It is appropriate for both data analytics and real-time machine learning settings. It can integrate a lot of devices and allow for creative applications that need less network latency, which is in keeping with the 5G goal. Our analysis, based on numerous user traces, of a service overhang resulting from a distributed design and services modification at the network edge shows that this overhanging is controllable and insignificant when compared to the aforementioned benefits. The suggested architecture, when strengthened by international agreements, creates a setting for the establishment of an MCS market for the bartering and selling of both aggregated/processed information and raw sensor data.

3.2 Decentralized Resource Management Paradigms

Network and systems management has been gradually changing since the mid-1990s. It moved from centralized paradigms, in which all management processing occurs in a single management station, to distributed paradigms, in which management is dispersed over a potentially large number of nodes. A few of These concepts have been around for a while, and the SNMPv2 and CMIP protocols serve as examples of them. Nevertheless, a rush of new ones that have just lately emerged, based on distributed objects, mobile code, or intelligent agents appeared. This survey aims to categorize the main paradigms for network and systems management.to date, to assist systems and network administrators in creating a management application. The first section of the survey, provides a straightforward typology that is based on just one standard: The institutional model All paradigms are classified into four categories in this typology: centralized paradigms, weakly cooperative paradigms, heavily distributed hierarchical para-digms, and distributed hierarchical paradigms. An improved typology is developed in the survey's second section through the application of four criteria: delegation of authority granularity, semantic richness of the information model, level of specification for tasks, and the degree of computerized management. The selection of a management paradigm in a specific context is demonstrated in the study (Daigger, 2008). Due to the increasing popularity of the IoT, cloud-based solutions struggle to meet the demanding requirements of IoT applications. These requirements include rapid reaction times, heightened privacy and security, and the absence of computing power at the network's edge (Avasalcai & Dustdar, 2018). The core infrastructure of edge and fog concepts relies on computational resources distributed closer to the edge. However, the benefits of these paradigms are lost without sophisticated resource management strategies. This paper introduces a novel technique for allocating dispersed resources, aiming to facilitate the deployment and integration of various applications within an Internet of Things framework. The proposed algorithm involves two key steps: (i) mapping IoT applications at the network boundary and (ii) dynamically moving application components. It is assumed that the program is divided into multiple tasks, each representing a specific computation, before deployment due to the limited resources of edge devices. So, a model is created for an IoT application. Figure 2 presents a Directed Acyclic Graph (DAG), in which the vertices stand in for various tasks.

Figure 2. Directed acyclic graph (DAG) (Avasalcai & Dustdar, 2018)

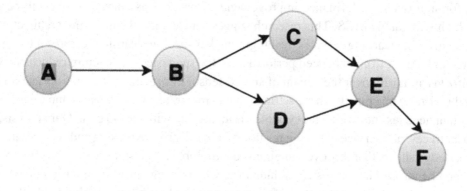

3.3 Distributed Ledger Technologies in Resource Allocation

The amount of data generated by edge grows drastically with the expanding Internet of Everything, resulting in greater requirements for network capacity. In the meantime, the network has to have less latency due to the advent of new applications. Edge computing, or the handling of data at a network's edge, is made possible by these two essential preconditions. When managing massive amounts of data, Edge Computing (EC) for cloud use guarantees service quality 1. The amount of useable data being generated (85 ZB) will be ten times greater than the amount that is saved or utilized (7.2 ZB) by 2021, predicts Cisco's Global Cloud Index 2. One technology that can help close this gap is EC. Simultaneously, huge data, processing capacity, and model advancements drive the growth of machine learning-based AI applications, particularly those involving deep learning algorithms. These days, apps are developed as a primary (Zhang et al., 2019).

3.4 Edge Intelligence for Real-Time Processing

The "smart" IoT is emerging as the next major force in the semiconductor business. Huge volumes of data from the physical world are available to us now. People are sensing us. These data must be minimized, examined, and used. These days, these data are transferred to the cloud, a central location, for evaluation. Data are processed in the cloud using well-established computing architectures to give services and analytics for consumers based on recognized company structures. This model is cloud-centric. not sustainable and will not be able to satisfy the demands of the intelligent Internet of Things, as Keshavarzi and van den Hoek have explained (Keshavarzi et al., 2020). Well, it is It is noteworthy because when it was launched in 1999, It was Proctor who defined IoT. According to Kevin Ashton of Gamble, "computers that observe, identify, and understand the world—without the limitations of human-entered data—are made possible by sensor technology." These PCs have communication through the Internet with one another. This explains a clever system, but because of its limitations The application produced passive (as opposed to smart) Internet of Things devices. As a result, the Internet of Things became a communication-centric concept where raw data gathered locally by IoT devices were sent to a central location by these IoT devices location of the analysis and processing. Smart IoT devices need a powerful engine to enable autonomy in decision making through small-system AI. Equally important is energy autonomy, particularly in times of scarcity and its supply fluctuations. This results in a unique class. of intelligent IoT gadgets that will have to rely on sporadic work in computers (Keshavarzi et al., 2020). Figure 3 shows that the AI for small systems engines will run on energy collected and must have a way to maintain their computational state if the energy supply runs out.

4. QUANTUM COMPUTING'S IMPACT ON RESOURCE MANAGEMENT

In the context of future trends and significant solutions for intelligent computing resource management, the integration of quantum computing fundamentals is a noteworthy aspect. Quantum computing can potentially revolutionize how computing resources are managed, offering new ways to optimize and solve complex problems. Here's an explanation of how quantum computing fundamentals can play a role in this domain:

4.1 Quantum Computing Fundamentals

Quantum computing fundamentals encompass the basic principles and concepts that underpin the operation of quantum computers, as described in the previous response. Integrating these fundamentals into intelligent computing resource management can have several implications:

- **Quantum-Assisted Optimization:** Quantum computers excel at solving optimization problems. Resource management often involves optimizing the allocation of resources like CPU, memory, and network bandwidth. Quantum algorithms, such as the Quantum Approximate Optimization Algorithm (QAOA), can be employed to find more efficient resource allocation strategies, leading to cost savings and improved performance.
- **Simulating Complex Systems:** Many resource management tasks involve simulating complex systems, such as traffic flow in networks or the behavior of financial markets. Quantum computers can simulate quantum systems more efficiently than classical computers. By using quantum simulators, resource managers can gain insights into system behavior, leading to better resource allocation decisions.

Figure 3. Energy-efficient local computing is essential to EI and to realizing the trillion-dollar ambition (Keshavarzi et al., 2020)

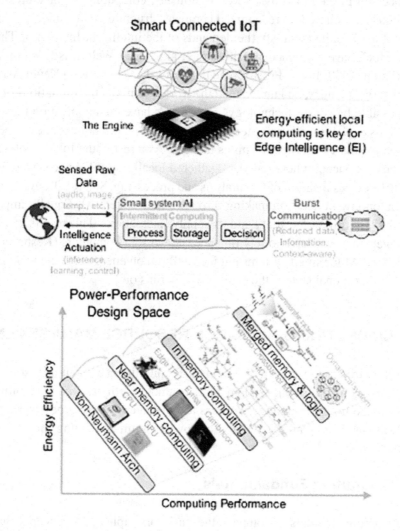

- **Enhanced Security:** Resource management in the context of cybersecurity and data protection can benefit from quantum cryptography. Quantum key distribution (QKD) provides theoretically unbreakable encryption methods. Quantum-safe resource management solutions can be developed to protect sensitive data and resources from future quantum attacks.
- **Parallel Processing:** Quantum computing leverages the concept of quantum parallelism, where multiple possibilities are explored simultaneously. This can be harnessed for tasks like load balancing and resource allocation, ensuring that computing resources are distributed optimally in real-time, even for dynamic workloads.
- **Quantum Machine Learning:** Quantum machine learning models can be used for intelligent resource management. These models can be trained and executed on quantum hardware to make predictions about resource needs and automatically allocate resources in real-time based on changing workloads and demand.

- **Quantum Algorithms for Analysis:** Resource managers can employ quantum algorithms for data analysis. For instance, Grover's algorithm can speed up the search process, helping to locate and identify resource bottlenecks or anomalies more quickly.
- **Energy Efficiency:** Quantum computing has the potential to reduce energy consumption. Resource management solutions that incorporate quantum algorithms can help minimize power usage, which is essential for environmentally responsible computing.

It is crucial to remember that the field of quantum computing is still in its infancy and that widely accessible, practical, large-scale quantum computers are not yet common. Therefore, the integration of quantum computing fundamentals into resource management solutions is a forward-looking trend that will become more significant as quantum technologies advance (Wang et al., 2021). As quantum hardware and software mature, the potential benefits for intelligent computing resource management will become increasingly evident. However, this integration will also bring new challenges, such as quantum error correction and the need for quantum-aware algorithms, which will need to be addressed to realize the full potential of quantum computing in resource.

4.2 Quantum Algorithms for Resource Optimization

Utilizing the foundations of quantum mechanics, quantum computing does specific computations notably more quickly than traditional computing. This makes it an ideal candidate for resource optimization problems that involve complex and vast search spaces. Quantum computers are inherently parallel, allowing them to explore multiple possibilities simultaneously. As quantum technologies mature, they will offer remarkable advantages for resource allocation and optimizationThe purpose of quantum algorithms is to efficiently solve problems by utilizing the unique characteristics of quantum computing. This method is applied in quantum analog computers such as D-Wave machines. Optimization problems like the traveling salesman problem or the job shop scheduling problem are suitable for its solution. Quantum annealers take advantage of quantum tunneling to find optimal solutions more quickly than classical algorithms. Variational quantum algorithms, like the Variational Quantum Eigensolver (VQE), can be adapted for resource optimization. They involve parameterized quantum circuits to find the best values for resource allocation and utilization, optimizing objectives such as cost or efficiency.

4.3 Quantum Machine Learning in Resource Allocation

Quantum computers leverage the principles of quantum mechanics to process information. Unlike classical computers, which use bits (0s and 1s), quantum computers use qubits, which can exist in multiple states simultaneously due to superposition. This parallelism allows quantum computers to handle certain types of problems much faster than classical computers. The Resource allocation in computing involves distributing computational resources like CPU, memory, and storage efficiently to meet the demands of various applications and workloads. This is a complex task that requires optimizing for factors like performance, cost, energy efficiency, and more. Quantum computers can simulate complex systems and environments more accurately than classical computers. In the context of resource allocation, this can help in simulating various resource allocation strategies and their impacts on the overall system, allowing for better-informed decisions. While QML holds great promise, it is important to note that practical quantum computers are still in their infancy, and there are several challenges to overcome, including error rates and scalability. Additionally, expertise in both quantum computing and machine learning is required to harness the full potential of QML.

4.4 Hybrid Quantum-Classical Resource Management

Quantum computing, while promising for certain types of problems, is not a panacea. Quantum computers are exceptionally well-suited for solving specific problems, such as factorization and optimization tasks, but they are not yet ready to replace classical computers entirely. They are sensitive to noise and decoherence, which limits their reliability and scalability. Classical computers, on the other hand, are well-established and reliable but struggle with certain complex problems that quantum computers excel in. Resource management in this context involves allocating and orchestrating computing resources effectively to tackle problems that are beyond the capabilities of classical computing alone. One significant trend is the development of quantum-inspired classical algorithms. These classical algorithms mimic the behavior of quantum algorithms to some extent, allowing classical computers to solve problems that were traditionally considered quantum-computing-only tasks. These hybrid algorithms can be used on classical hardware, making resource management more flexible. Cloud providers are exploring quantum cloud services, which offer access to both classical and quantum computing resources. Users can dynamically allocate these resources based on their computational needs. Efficient resource management in such cloud environments is essential to ensure cost-effective and timely execution of tasks. Quantum computing is inherently error-prone due to noise and decoherence. Managing and correcting errors is a crucial aspect of resource management in hybrid systems. Classical error-correcting codes are used in conjunction with quantum error-correcting codes to ensure the reliability of quantum computations. Resource management should be tailored to specific applications. For example, in drug discovery, hybrid quantum-classical systems can be used to simulate molecular structures and interactions efficiently. Resource allocation would be optimized for this particular use case.

5. AI-DRIVEN RESOURCE ALLOCATION

VLC, or visible light communication, offers a workable solution to the challenging requirements of new applications. Even though VLC offers orders of magnitude more bandwidth than the radio spectrum, making the most use of available resources is a significant task. The goal of making next-generation communication networks intelligent is becoming more and more pressing. AI-based technologies are taking the place of traditional handmade methods based on mathematical equations that assume previous understanding of the network in order to automate the administration of resources and adapt to changes within the network automatically. The creation of an intelligent resource allocation mechanism for an optically wires communication (OWC) system operating in an inconsistent network environment is explained in this article. The main goal is to maximize the total reward of the system, which is the total of the users' ratings. The Mixed Integer Linear Programming (MILP) model-based optimization scheme and the RL scheme's output are contrasted. Within the framework of contemporary computer landscapes characterized by growing resource demands and economic considerations, this paper introduces a novel framework that integrates AI for the creation of resource-efficient applications while maintaining a keen awareness of costs. The imperative to strike a harmonious equilibrium between application performance and expenses has never been more pressing, especially with the proliferation of cloud-based services. In response, our approach capitalizes on AI methodologies to dynamically analyze real-time application requisites, workload trends, and the availability of resources. Central to our methodology is the elevation of cost to a principal design determinant. We devise strategies that dynamically apportion resources, opt for suitable service tiers, and make necessary adjustments to application configurations. This duality of optimizing performance while curtailing expenditure underscores the essence of our approach. Rigorous simulations and empirical evaluations underscore the efficacy of our strategies across diverse scenarios, underscoring substantial cost reductions without compromising the quality of applications.

5.1 Machine Learning for Predictive Resource Planning

Machine Learning for Predictive Resource Planning involves the use of machine learning algorithms to predict future resource requirements based on historical data and real-time insights. This approach can be applied to various computing resources, including but not limited to CPU, memory, storage, and network bandwidth. The foundation of predictive resource planning is data. It involves collecting data on past resource utilization, workload patterns, and environmental factors that may influence resource demands. This data can come from various sources, such as system logs, monitoring tools, and user behavior. Feature engineering is the process of selecting and transforming relevant data attributes (features) that are used as input to machine learning models. In the context of resource planning, features could include time of day, user activity, application type, and more. Machine learning models are trained on historical data to learn patterns and relationships between resource consumption and input features. Popular machine learning algorithms such as regression, time series forecasting, and neural networks can be used to create predictive models. The field of machine learning is continually evolving, and future trends may involve the use of more advanced techniques like deep learning and reinforcement learning for even more accurate predictions. As edge computing becomes more prevalent, predictive resource planning will extend to edge devices, helping ensure efficient resource utilization in distributed environments. As the importance of resource planning grows, there will be a greater emphasis on making machine learning models more interpretable and explainable to gain user trust and regulatory compliance. Future systems may become increasingly autonomous, with machine learning models directly controlling resource allocation decisions based on predictions and user-defined policies.

5.2 Reinforcement Learning for Dynamic Resource Optimization

Reinforcement learning is a type of machine learning that focuses on decision-making in an environment to maximize a cumulative reward. RL models, often represented as Markov Decision Processes (MDPs), consist of an agent, a set of states, a set of actions, and a reward function. The agent interacts with the environment, taking actions, receiving feedback (rewards), and learning optimal policies to achieve long-term objectives. In the context of computing resource management, dynamic resource optimization refers to the allocation and reallocation of computing resources (e.g., CPU, memory, storage) in response to changing workloads and system conditions. This is essential for ensuring efficient resource utilization and maintaining system performance. In dynamic resource optimization, the system's current state could include factors like CPU utilization, memory usage, incoming request rate, and more. The reward function defines what the system aims to optimize. In resource management, this can be performance metrics (e.g., response time, throughput) and cost efficiency (e.g., minimizing operational costs). Balancing exploration (trying new resource allocation strategies) and exploitation (leveraging known, effective strategies) is a fundamental challenge in RL. In dynamic resource optimization, this translates to finding the right balance between optimizing for performance and resource cost.

5.3 Neural Network Models in Workload Prediction

Neural network models have become increasingly important in workload prediction and intelligent computing resource management. These models leverage artificial neural networks, which are inspired by the structure and functioning of the human brain, to make predictions and optimize resource allocation in various computing environments. Neural network models are data-driven, which means they learn from historical workload data and patterns. This is vital in workload prediction because traditional methods may not effectively capture complex and dynamic workload behaviors. By analyzing historical data, neural networks can identify trends and relationships that are difficult for human-designed algorithms to discover. Workload prediction often involves extracting meaningful features from the input data. Neural networks can automatically learn relevant features from raw data, reducing the need for manual feature engineering. This adaptability is especially useful when dealing with large and diverse datasets, such as system logs, sensor data, or network traffic. Deep neural networks, with multiple hidden layers, can capture hierarchical and abstract representations of data. In workload prediction, deep learning models such as recurrent neural networks (RNNs) and Long Short-Term Memory (LSTM) networks are effective at capturing temporal dependencies, which are crucial when predicting workloads with time-series characteristics. Once workload predictions are made, they can be used to optimize resource allocation. For example, cloud service providers can use workload predictions to allocate resources efficiently, scaling up or down as needed to meet demand. This can lead to cost savings and improved service quality. In data centers and edge computing environments, workload prediction is essential for optimizing energy consumption. Neural network models can help anticipate workload fluctuations and adjust resource allocation to minimize energy usage without sacrificing performance.

5.4 Explainable AI in Resource Management

Intelligent computing resource management is a critical aspect of modern IT and business operations, especially in the context of cloud computing, data centers, and other distributed computing environments. As we move into the future, several trends and solutions are emerging to enhance resource management, and one of these trends is the integration of Explainable AI (XAI). Cloud computing is becoming increasingly central to IT infrastructure. In the future, more organizations will rely on cloud resources to scale their operations. This means that efficient allocation and management of cloud resources will become even more critical. Many organizations are adopting multi-cloud strategies, using resources from various cloud providers to avoid vendor lock-in and increase redundancy. Managing resources across multiple clouds requires intelligent decision-making to optimize costs and performance, Edge computing is growing as a result of the proliferation of IoT devices and the requirement for low-latency processing. efficiently allocating resources at the cutting edge while upholding data security and integrity will be a significant challenge. Resource management systems, particularly in the context of cloud computing, distributed computing, and data centers, often involve complex algorithms and decision-making processes. These decisions impact the allocation of resources such as CPU, memory, storage, and network bandwidth. Explainable AI techniques help make these decisions more transparent and understandable. It enables stakeholders to have a clear understanding of why specific resource allocation decisions were made, which is vital for accountability. XAI can provide insights into resource allocation errors or inefficiencies by explaining how and why certain decisions were made. This information can be used to identify and rectify problems in resource management strategies, ultimately leading to more optimized and efficient resource usage. Resource management often requires human intervention, especially in cases of critical system failures or unexpected resource demands. Explainable AI systems can present explanations to human operators, helping them make informed decisions more effectively and quickly. In industries with strict regulations, such as healthcare or finance, it is essential to ensure that resource management decisions are compliant with legal requirements and ethical standards. XAI helps demonstrate that resource allocation adheres to established rules and guidelines, building trust among users and regulators. In an era of increasing concern for the environment, XAI can assist in resource management for sustainability. It can explain how resource allocation decisions impact energy consumption, helping organizations make greener choices and reduce their carbon footprint.

6. ENERGY-EFFICIENT RESOURCE MANAGEMENT

Energy-efficient resource management is about optimizing the use of computational resources while minimizing energy consumption. It involves a combination of hardware and software innovations, predictive analytics, and a broader shift toward sustainability. As technology evolves, these trends and solutions will play a crucial role in reducing energy consumption in the field of intelligent computing.

6.1 AI-Enabled Power Management Strategies

AI-Enabled Power Management Strategies represent a significant advancement in the field of intelligent computing resource management. These strategies leverage artificial intelligence to optimize the power consumption and performance of computer systems, contributing to more efficient and sustainable operations. Here's an explanation of this trend without points: In the realm of computing, energy consumption and performance are pivotal aspects that influence the overall efficiency and sustainability of systems. AI-Enabled Power Management Strategies encompass a range of innovative techniques that employ artificial intelligence to balance these two critical factors. One key facet of AI-Enabled Power Management is dynamic workload scheduling. Traditional power management strategies often use static configurations that do not adapt to changing workloads. AI, however, enables systems to analyze the real-time demands of applications and allocate power resources accordingly. By dynamically adjusting power allocation, AI-Enabled Power Management ensures that computing resources are utilized efficiently, reducing energy waste and enhancing overall system performance. Another noteworthy element of AI-Enabled Power Management is predictive analytics. Machine learning algorithms can learn from historical data and usage patterns to forecast future demands and energy requirements. This predictive capability enables systems to pre-emptively adjust power settings to meet upcoming needs, ultimately leading to better resource allocation and power efficiency. Furthermore, AI-Enabled Power Management can optimize hardware components. For instance, in data centers, AI can analyze temperature and usage patterns to dynamically control cooling systems. This not only saves energy but also prolongs the lifespan of hardware components. AI can also be used to optimize voltage and frequency levels in processors, striking a balance between performance and power consumption. Moreover, AI-Enabled

Power Management can contribute to sustainable practices. By reducing energy consumption, organizations can lower their carbon footprint and operating costs. Additionally, the ability to predict and respond to energy demands effectively can help prevent power shortages and grid congestion, which has broader societal and environmental implications.

6.2 Green Computing and Sustainable Resource Allocation

As the industry advances in generation and computing, savvy computing will turn out to be progressively basic (Turner et al., 2019). Two key zones in this matter are "green computing" and "sustainable asset appointment". These techniques are crucial to decrease the ecological effect of IT frameworks and make sure that assets are expended wisely and conscientiously. Green computing and sustainable asset appointment are basic components of savvy computing (Xu et al., 2018). They limit the environmental effect of IT frameworks, decrease energy usage, and guarantee efficient asset appointment, subsequently adding to a progressively supportable IT framework and ecologically cordial as the progression of innovation and ecological issues keep on in the coming years.

- **Green Computing:** Green Computing is a significant trend in intelligent computing resource management. It focuses on reducing the environmental footprint of information technology. The key principles of Green Computing include:
- **Energy Efficiency:** Green Computing promotes the use of energy-efficient hardware, software, and data centers. This involves optimizing power consumption and reducing electronic waste.
- **Virtualization:** Virtualization technologies enable the consolidation of multiple servers and applications on a single physical machine. This reduces the need for additional hardware, resulting in energy and resource savings.
- **Cloud Computing:** Cloud services allow for resource sharing and dynamic allocation, which can lead to more efficient use of computing resources. This reduces the environmental impact associated with maintaining dedicated hardware.
- **Recycling and E-Waste Management:** Green Computing also involves responsible disposal and recycling of electronic equipment to minimize the environmental impact of electronic waste.
- **Sustainable Resource Allocation:** Sustainable Resource Allocation is another crucial aspect of intelligent computing resource management. It focuses on ensuring that computing resources are allocated in a way that is both efficient and environmentally responsible. Key components of this trend include:
- **Resource Optimization:** Sustainable resource allocation aims to optimize the use of computing resources to minimize waste and enhance performance. This includes load balancing, resource allocation algorithms, and dynamic scaling.
- **Renewable Energy Sources:** Data centers and IT facilities can transition to renewable energy sources, such as solar and wind power, to reduce their carbon footprint and environmental impact.
- **Resource Monitoring and Management:** Real-time monitoring and management of computing resources help in identifying underutilized or overutilized resources, allowing for efficient allocation.
- **Sustainable Software Development:** Software plays a significant role in resource consumption. Sustainable software development practices include designing applications that are resource-efficient and minimize energy consumption.

- **Lifecycle Assessment:** Understanding the entire lifecycle of IT equipment and applications, from production to disposal, enables better decision-making in resource allocation and environmental impact reduction.

As the industry advances in generation and computing, savvy computing will turn out to be progressively basic. Two key zones in this matter are "green computing" and "sustainable asset appointment". These techniques are crucial to decrease the ecological effect of IT frameworks and make sure that assets are expended wisely and conscientiously. Green computing and sustainable asset appointment are basic components of savvy computing. They limit the environmental effect of IT frameworks, decrease energy usage, and guarantee efficient asset appointment, subsequently adding to a progressively supportable IT framework and ecologically cordial as the progression of innovation and ecological issues keep on in the coming years (Asim & Abd El-Latif, 2021; Aslanpour et al., 2020; Turner et al., 2019; Xu et al., 2018).

6.3 Using AI for Dynamic Voltage and Frequency Scaling (DVFS)

Social media sites like Instagram, stock markets, shopping websites, vacation portals, hospital administration, and government sectors now control the entire industry. The world is becoming more aware of how important it is to stay connected every day, and the only way this is feasible is with cloud-based enabled gadgets entering both the personal and professional spheres and allowing us to pick up where we left off with our work (Florence et al., 2016). The concepts behind cloud computing include "Pay as you go" and "Anything as a Service (AaaS)". For instance, as a national citizen, you can connect to a service provider and use the power that you use, only paying for what you use. On the other hand, any resource, including network platforms, software, and infrastructure, can be accessed through cloud services. Sophisticated DVFS systems, which can be based on integral derivative that is proportionate (PID) or artificial neural network, or ANN, technologies and architecture, have grown increasingly intricate over time (Figure 4). Dynamic Voltage and Frequency Scaling (DVFS) optimizes energy efficiency by dynamically altering the voltage and frequency of processors in response to workload demands. DVFS can intelligently fine-tune performance and power usage when combined with AI algorithms, enhancing overall system efficiency.

Figure 4. The terms quality of experience (QoE) and quality of service (QoS) (Khriji et al., 2022)

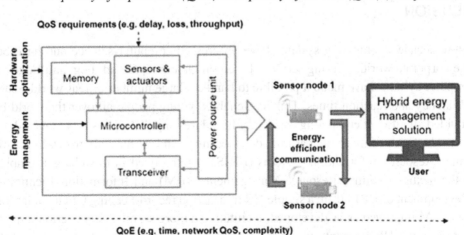

6.4 Renewable Energy Integration in Resource Management

Growing environmental degradation, constraints on fossil fuel availability, and an ever-increasing demand for electricity have resulted in the development of microgrids and the accumulation of clean energy resources (distributed generations). To schedule devices in the context of a function with objectives, constraints, and control parameters, an energy management and control unit (EMCU) is designed using algorithms based on genetics (GA), the optimization of binary particles swarms BPSO, and wind directed optimisation (WDO). By 2017, there will be a considerable increase in the number of nations producing more than 100 megawatt(MW) of renewable energy, according to forecasts. Promoting and developing renewable energy supply technologies is essential since conventional energy production has some unfavorable and irreversible externalities. It must be mentioned that economies of scale may be a significant factor in bringing down the cost of production per unit. Transmission and distribution costs, as well as technologies, do not differ much among the conventional and renewable energies Large numbers of renewable energy facilities are already connected to electrical grids. On top of those infrastructures, the energy-efficient Internet will be constructed (Aste et al., 2018). Multiple MGs coupled to an ADS provide a promising framework for integrating large-scale DERs under local conditions (Shayeghi & Shahryari, 2017). Additionally, this approach can improve financial gains, environmental friendliness, dependability, and system performance. An appropriate modern energy system that integrates DERs into electrical networks is the ADS, which has the ability to actively control the flow of electricity and DERs (Lv & Ai, 2016). Using hybrid energy systems, MGs can directly combine various energy resources and manage them as a type of hub for DGs, local loads, and ESSs. Large-scale intermittent RER integration into a multi-MGs-based ADS presents the following challenges for EM optimization: First and foremost, it is imperative to lessen the effects of RERs while still achieving even higher operational achievements. Second, RERs need to be successfully and appropriately employed. Thirdly, The first two issues can only be solved by taking into account the physical interactions that take place between machine guns and the ADN. Further, the suggested controller operates in a stationary reference frame in order to reduce the computational load related to tracking grid voltage phase angles and frame transformations. Even under the influence of unpredictable loads, the control strategy presented in this study is demonstrated to be reliable and stable in the face of load disruptions and uncertainty in microgrid parameters.

7. CONCLUSION

In the last few decades, computing systems have progressed computer science and have become the center of the corporate world, offering services based on Quantum, Cloud, Fog, and Edge computing. Current computer systems have made it possible to handle a large number of real-world issues that call for reduced latency and reaction times. This has. enabled young people all over the world build start-ups that provided significant computing capability for addressing difficult issues to expedite scientific advancement. The fields of machine learning, deep learning, and AI have all grown in popularity. The area of Human Resource Information Systems (HRIS) is introduced in this chapter, emphasizing its function at the nexus of Human Resource Management (HRM) and Information Technology (IT). It charts the development of HRIS from a simple HR transaction record-keeping tool to an important business partner. HRM improved as IT developed, facilitating a more seamless connection into HRIS. This development improved HRM's contribution to an organization's strategic goals by reducing the amount of time HR professionals spend on repetitive duties and increasing their emphasis on transformative actions. An introduction to HRIS and its different varieties—which are necessary for decision-making at all organizational levels—is given in this chapter. It avoids confusion by making a distinction between e-HRM and HRIS.

REFERENCES

Akyurt, İ. Z., Kuvvetli, Y., & Deveci, M. (2020). Enterprise resource planning in the age of industry 4.0: A general overview. *Logistics, 4*(0), 178–185.

Al-Ansi, A., Al-Ansi, A. M., Muthanna, A., Elgendy, I. A., & Koucheryavy, A. (2021). Survey on intelligence edge computing in 6G: Characteristics, challenges, potential use cases, and market drivers. *Future Internet, 13*(5), 118. doi:10.3390/fi13050118

Asim, M., & Abd El-Latif, A. A. (2021). Intelligent computational methods for multi-unmanned aerial vehicle-enabled autonomous mobile edge computing systems. *ISA Transactions*. PMID:34933773

Aslanpour, M. S., Gill, S. S., & Toosi, A. N. (2020). Performance evaluation metrics for cloud, fog and edge computing: A review, taxonomy, benchmarks and standards for future research. *Internet of Things : Engineering Cyber Physical Human Systems, 12*, 100273. doi:10.1016/j.iot.2020.100273

Aste, N., Buzzetti, M., Caputo, P., & Del Pero, C. (2018). Regional policies toward energy efficiency and renewable energy sources integration: Results of a wide monitoring campaign. *Sustainable Cities and Society, 36*, 215–224. doi:10.1016/j.scs.2017.10.005

Avasalcai, C., & Dustdar, S. (2018, October). Latency-aware decentralized resource management for IoT applications. In *Proceedings of the 8th International Conference on the Internet of Things* (pp. 1-4). 10.1145/3277593.3277637

Budde, F., & Volz, D. (2019). The next big thing? Quantum computing's potential impact on chemicals. *McKinsey*.

Buyya, R., Srirama, S. N., Casale, G., Calheiros, R., Simmhan, Y., Varghese, B., Gelenbe, E., Javadi, B., Vaquero, L. M., Netto, M. A. S., Toosi, A. N., Rodriguez, M. A., Llorente, I. M., Vimercati, S. D. C. D., Samarati, P., Milojicic, D., Varela, C., Bahsoon, R., Assuncao, M. D. D., ... Shen, H. (2018). A manifesto for future generation cloud computing: Research directions for the next decade. *ACM Computing Surveys, 51*(5), 1–38. doi:10.1145/3241737

Chen, J., Wei, Z., Li, S., & Cao, B. (2020). Artificial intelligence aided joint bit rate selection and radio resource allocation for adaptive video streaming over F-RANs. *IEEE Wireless Communications, 27*(2), 36–43. doi:10.1109/MWC.001.1900351

D'Oro, S., Bonati, L., Polese, M., & Melodia, T. (2022, May). Orchestran: Network automation through orchestrated intelligence in the open ran. In *IEEE INFOCOM 2022-IEEE Conference on Computer Communications* (pp. 270-279). IEEE.

Daigger, G. T. (2008, January). Aeesp Lecture: Evolving Urban Water and Residuals Management Paradigms: Water Reclamation and Reuse, Decentralization, Resource Recovery. In WEFTEC 2008 (pp. 1537-1565). Water Environment Federation.

Darwish, T. S., & Bakar, K. A. (2018). Fog based intelligent transportation big data analytics in the internet of vehicles environment: Motivations, architecture, challenges, and critical issues. *IEEE Access : Practical Innovations, Open Solutions*, 6, 15679–15701. doi:10.1109/ACCESS.2018.2815989

Dittakavi, R. S. S. (2023). AI-Optimized Cost-Aware Design Strategies for Resource-Efficient Applications. *Journal of Science and Technology*, 4(1), 1–10.

Elgamal, A. S., Aletri, O. Z., Yosuf, B. A., Qidan, A. A., El-Gorashi, T., & Elmirghani, J. M. (2023). AI-Driven Resource Allocation in Optical Wireless Communication Systems. *arXiv preprint arXiv:2304.03880*. doi:10.1109/ICTON59386.2023.10207473

Florence, A. P., Shanthi, V., & Simon, C. B. (2016). Energy conservation using dynamic voltage frequency scaling for computational cloud. *TheScientificWorldJournal*, 2016, 2016. doi:10.1155/2016/9328070 PMID:27239551

Gai, K., Qiu, M., Zhao, H., & Sun, X. (2017). Resource management in sustainable cyber-physical systems using heterogeneous cloud computing. *IEEE Transactions on Sustainable Computing*, 3(2), 60–72. doi:10.1109/TSUSC.2017.2723954

Gao, H., Huang, Z., Zhang, X., & Yang, H. (2023). Research and Design of a Decentralized Edge-Computing-Assisted LoRa Gateway. *Future Internet*, 15(6), 194. doi:10.3390/fi15060194

Gill, S. S., Xu, M., Ottaviani, C., Patros, P., Bahsoon, R., Shaghaghi, A., Golec, M., Stankovski, V., Wu, H., Abraham, A., Singh, M., Mehta, H., Ghosh, S. K., Baker, T., Parlikad, A. K., Lutfiyya, H., Kanhere, S. S., Sakellariou, R., Dustdar, S., ... Uhlig, S. (2022). AI for next generation computing: Emerging trends and future directions. *Internet of Things : Engineering Cyber Physical Human Systems*, 19, 100514. doi:10.1016/j.iot.2022.100514

Guazzone, M., Anglano, C., & Canonico, M. (2011, November). Energy-efficient resource management for cloud computing infrastructures. In *2011 IEEE Third International Conference on Cloud Computing Technology and Science* (pp. 424-431). IEEE. 10.1109/CloudCom.2011.63

Helmreich, R. L., Merritt, A. C., & Wilhelm, J. A. (2017). The evolution of crew resource management training in commercial aviation. In *Human error in aviation* (pp. 275–288). Routledge. doi:10.4324/9781315092898-15

Hewa, T., Gür, G., Kalla, A., Ylianttila, M., Bracken, A., & Liyanage, M. (2020). The role of blockchain in 6G: Challenges, opportunities and research directions. *2020 2nd 6G Wireless Summit (6G SUMMIT)*, 1-5.

Keshavarzi, A., Ni, K., Van Den Hoek, W., Datta, S., & Raychowdhury, A. (2020). Ferroelectronics for edge intelligence. *IEEE Micro*, 40(6), 33–48. doi:10.1109/MM.2020.3026667

Khan, W. Z., Ahmed, E., Hakak, S., Yaqoob, I., & Ahmed, A. (2019). Edge computing: A survey. *Future Generation Computer Systems*, 97, 219–235. doi:10.1016/j.future.2019.02.050

Khriji, S., Chéour, R., & Kanoun, O. (2022). Dynamic Voltage and Frequency Scaling and Duty-Cycling for Ultra Low-Power Wireless Sensor Nodes. *Electronics (Basel)*, *11*(24), 4071. doi:10.3390/electronics11244071

Krishnan, S. R., Nallakaruppan, M. K., Chengoden, R., Koppu, S., Iyapparaja, M., Sadhasivam, J., & Sethuraman, S. (2022). Smart water resource management using Artificial Intelligence—A review. *Sustainability (Basel)*, *14*(20), 13384. doi:10.3390/su142013384

Liang, L., Wang, W., Jia, Y., & Fu, S. (2016). A cluster-based energy-efficient resource management scheme for ultra-dense networks. *IEEE Access : Practical Innovations, Open Solutions*, *4*, 6823–6832. doi:10.1109/ACCESS.2016.2614517

Lv, T., & Ai, Q. (2016). Interactive energy management of networked microgrids-based active distribution system considering large-scale integration of renewable energy resources. *Applied Energy*, *163*, 408–422. doi:10.1016/j.apenergy.2015.10.179

Martinez, I., Hafid, A. S., & Jarray, A. (2020). Design, resource management, and evaluation of fog computing systems: A survey. *IEEE Internet of Things Journal*, *8*(4), 2494–2516. doi:10.1109/JIOT.2020.3022699

Morozov, V., Kalnichenko, O., & Mezentseva, O. O. O. M. (2020). The method of interaction modeling on basis of deep learning the neural networks in complex IT-projects. *International Journal of Computing*, *19*(1), 88–96. doi:10.47839/ijc.19.1.1697

Munir, H., Pervaiz, H., Hassan, S. A., Musavian, L., Ni, Q., Imran, M. A., & Tafazolli, R. (2018). Computationally intelligent techniques for resource management in mmwave small cell networks. *IEEE Wireless Communications*, *25*(4), 32–39. doi:10.1109/MWC.2018.1700400

Nabeeh, N. A., Smarandache, F., Abdel-Basset, M., El-Ghareeb, H. A., & Aboelfetouh, A. (2019). An integrated neutrosophic-topsis approach and its application to personnel selection: A new trend in brain processing and analysis. *IEEE Access : Practical Innovations, Open Solutions*, *7*, 29734–29744. doi:10.1109/ACCESS.2019.2899841

Negi, P. S., Negi, V., & Pandey, A. C. (2011). Impact of information technology on learning, teaching and human resource management in educational sector. *International Journal of Computer Science and Telecommunications*, *2*(4), 66–72.

Papagianni, C., Mangues-Bafalluy, J., Bermudez, P., Barmpounakis, S., De Vleeschauwer, D., Brenes, J., ... Pepe, T. (2020, June). 5Growth: AI-driven 5G for Automation in Vertical Industries. In *2020 European Conference on Networks and Communications (EuCNC)* (pp. 17-22). IEEE. 10.1109/EuCNC48522.2020.9200919

Ravi, G. S., Smith, K. N., Gokhale, P., & Chong, F. T. (2021, November). Quantum Computing in the Cloud: Analyzing job and machine characteristics. In *2021 IEEE International Symposium on Workload Characterization (IISWC)* (pp. 39-50). IEEE. 10.1109/IISWC53511.2021.00015

Rodoshi, R. T., Kim, T., & Choi, W. (2020). Resource management in cloud radio access network: Conventional and new approaches. *Sensors (Basel)*, *20*(9), 2708. doi:10.3390/s20092708 PMID:32397540

Samarasinghe, K. R., & Medis, A. (2020). Artificial intelligence based strategic human resource management (AISHRM) for industry 4.0. *Global Journal of Management and Business Research, 20*(2), 7–13. doi:10.34257/GJMBRGVOl20IS2PG7

Shaikh, P. H., Nor, N. B. M., Nallagownden, P., Elamvazuthi, I., & Ibrahim, T. (2014). A review on optimized control systems for building energy and comfort management of smart sustainable buildings. *Renewable & Sustainable Energy Reviews, 34*, 409–429. doi:10.1016/j.rser.2014.03.027

Shayeghi, H., & Shahryari, E. (2017). Integration and management technique of renewable energy resources in microgrid. *Energy Harvesting and Energy Efficiency: Technology, Methods, and Applications*, 393-421.

Sun, Y., Ochiai, H., & Esaki, H. (2021). Decentralized deep learning for multi-access edge computing: A survey on communication efficiency and trustworthiness. *IEEE Transactions on Artificial Intelligence, 3*(6), 963–972. doi:10.1109/TAI.2021.3133819

Thite, M., Kavanagh, M. J., & Johnson, R. D. (2012). Evolution of human resource management and human resource information systems. *Introduction To Human Resource Management*, 2-34.

Turner, C. J., Emmanouilidis, C., Tomiyama, T., Tiwari, A., & Roy, R. (2019). Intelligent decision support for maintenance: An overview and future trends. *International Journal of Computer Integrated Manufacturing, 32*(10), 936–959. doi:10.1080/0951192X.2019.1667033

Vhora, F., & Gandhi, J. (2020, March). A comprehensive survey on mobile edge computing: challenges, tools, applications. In *2020 fourth international conference on computing methodologies and communication (ICCMC)* (pp. 49-55). IEEE. 10.1109/ICCMC48092.2020.ICCMC-0009

Wang, D., Zhong, D., & Souri, A. (2021). Energy management solutions in the Internet of Things applications: Technical analysis and new research directions. *Cognitive Systems Research, 67*, 33–49. doi:10.1016/j.cogsys.2020.12.009

Wang, X., Han, Y., Leung, V. C., Niyato, D., Yan, X., & Chen, X. (2020). Convergence of edge computing and deep learning: A comprehensive survey. *IEEE Communications Surveys and Tutorials, 22*(2), 869–904. doi:10.1109/COMST.2020.2970550

Xu, L. D., Xu, E. L., & Li, L. (2018). Industry 4.0: State of the art and future trends. *International Journal of Production Research, 56*(8), 2941–2962. doi:10.1080/00207543.2018.1444806

Yabanci, O. (2019). From human resource management to intelligent human resource management: A conceptual perspective. *Human-Intelligent Systems Integration, 1*(2-4), 101–109. doi:10.1007/s42454-020-00007-x

Yuan, L., He, Q., Tan, S., Li, B., Yu, J., Chen, F., ... Yang, Y. (2021, April). Coopedge: A decentralized blockchain-based platform for cooperative edge computing. In *Proceedings of the Web Conference 2021* (pp. 2245-2257). 10.1145/3442381.3449994

Zhang, X., Wang, Y., Lu, S., Liu, L., & Shi, W. (2019, July). OpenEI: An open framework for edge intelligence. In *2019 IEEE 39th International Conference on Distributed Computing Systems (ICDCS)* (pp. 1840-1851). IEEE.

Chapter 11
Green Computing–Based Digital Waste Management and Resource Allocation for Distributed Fog Data Centers

N. Manikandan
SRM Institute of Science and Technology, India

R. Anto Arockia Rosaline
SRM Institute of Science and Technology, India

D. Vinod
SRM Institute of Science and Technology, India

P. Nancy
SRM Institute of Science and Technology, India

G. Premalatha
https://orcid.org/0000-0003-1031-5607
SRM Institute of Science and Technology, India

ABSTRACT

The term "green computing" describes the efficient use of resources in computing and IT/IS infrastructure. This study suggests a unique method for dispersed fog data centres' work scheduling and resource allocation based on digital waste management. Here, the bandwidth differential preemption evolution moving average method (BDPEMA) is used to control the network's digital waste while allocating resources. Reinforcement adversarial hierarchical group multi-objective cuckoo optimisation (RAHMCO) is used to schedule network tasks. In terms of resource sharing rate, energy efficiency, reaction time, quality of service, and makespan, experimental study is conducted. The proposed approaches have been evaluated in a simulated cloud environment. The proposed method outperformed the current rules when QoS features were considered. The proposed technique attained QoS of 66%, energy efficiency of 96%, resource sharing of 88%, response time of 45%, and makespan of 61%.

DOI: 10.4018/979-8-3693-1552-1.ch011

1. INTRODUCTION

Green computing embraces green infrastructure by studying as well as implementing global best practises in their design, manufacture, usage, and disposal (Mukta & Ahmed, 2020). Despite rise of African computer community, preliminary research revealed a lack of awareness of green computing. It is claimed that Africans continue to rely on information and communication technology (ICT) to sustain their way of life, with little concern given to the negative environmental effects of computers. Meanwhile, computing by both corporate and private users has been shown to increase global temperatures through the use of fossil fuel byproducts, worsen environment through production of hazardous chemicals, and reduce energy accessibility. Green IT, sometimes known as green computing, is study and practise of efficiently and successfully designing, manufacturing, and using computers, servers, screens, printers, storage devices, networking, and correspondence systems with zero or little environmental impact (Gupta, 2022). Green IT is also about utilizing IT to support, assist, and leverage other environmental projects, to raise green awareness. The primary goals are to maximise energy efficiency throughout item's lifetime, decrease use of hazardous chemicals, and enhance the recyclability or biodegradability of obsolete products as well as plant waste (Purnomo et al., 2021). Prasad et al. (2021) describes a method-driven engineering method to optimising architecture, energy utilisation, and operational costs of cloud auto-scaling infrastructure in order to provide greener computing environments with lower emissions from idle resources. In the absence of regular energy from public networks, many computing enterprises rely on fuel generators for power delivery. Because carbon is present in all hydrocarbon fuels, it is emitted as carbon dioxide (CO_2) after combustion. Non-combustible sources, such as sunlight, wind, nuclear, and hydropower (Das et al., 2022), on the other hand, lack the potential to convert hydrocarbons to CO_2, which is well recognised as a heat-trapping greenhouse gas. Scientists agree that the emission of greenhouse gases (GHGs) into the atmosphere has a negative impact on climate method. Nonetheless, because a cost-benefit analysis shows that benefits of ICT much outweigh the costs, steps for ecologically friendly ICT use in Sub-Saharan Africa must be put in place. Recycling computer equipment helps keep hazardous materials like lead and mercury out of landfills. Reuse can occur in a variety of ways. It is capable of carrying a load of obsolete equipment. Environmental management entails efforts such as assessing the current state of the planet, dealing with the direct and indirect effects of large-scale human operations such as agriculture, transportation, and assembly, and enlightening people's own choices in usage and behaviour. The current status of e-waste handling is quite terrible. By improving ewaste management, the globe will become greener and more environmentally friendly. Electronic devices built of several integrated circuit chips may cost less occasionally, but they consume enormous amounts of non-renewable resources. This is due to the fact that the chips are made of highly precious elements such as gold, lead, mercury, silicon, and so on. As a result, managing e-waste with the purpose of reusing these valuable metals is unquestionably beneficial. The development of e-wastes is currently expanding at an alarming rate, and the radiations from e-wastes can cause major genetic problems among personnel at collection sites as well as residents in surrounding areas (Gaharwar et al., 2022).

2. RELATED WORKS

Green computing has turned into a critical point in these new days. A few works have been completed with various methodologies in this field. Work Soesanto et al. (2023) surveyed the ongoing writing on green registering and its effects on supportable IT administrations with recognizing a center arrangement of standards to direct practical IT administration plan and influence centers to further develop client esteem, business esteem, and cultural worth. Bal et al. (2022) look at how power and cooling issues in supercomputing have developed from an optional worry to a fundamental plan basic. Joseph Williams (Das et al., 2022) tended to the part of green figuring that mirrors the structural technique and depicted building that design view's expectation's. Praveenchandar and Tamilarasi (2021) showed force of IT in imaginative ways to address mounting ecological problems as well as make IT frameworks and their utilization greener. He gave a comprehensive method for managing greening IT and featured how IT could help organizations in their natural drives and decrease their petroleum derivative side-effects. Saidi and Bardou (2023) explored the ongoing writing on green figuring and its impacts on practical IT administrations with recognizing a center arrangement of standards to direct reasonable IT administration plan and influence centers to further develop client esteem, business esteem, and cultural worth. Jamil et al. (2022) examined how power and cooling issues in supercomputing have developed from an optional worry to a fundamental plan basic. Rjoub et al. (2021) tended to the part of green registering that mirrors the building philosophy and portrayed fostering that engineering perspective's expectation's. Work introduced an extensive method for managing greening IT and featured how IT could help organizations in their ecological drives and lessen their non-renewable energy source side-effects (Sathiyamoorthi et al., 2021). Manikandan, Divya, and Janani (2022) gathered through data from 133 respondents in ventures that as of now work their in-house server cultivates and utilized the information to check the Greencomputing life cycle methodologies. He in like manner examined the Green figuring life cycle procedures to be applied for Greening server ranches in IT-based ventures. Researchers have proposed a variety of schemes for dynamic resources giving and management, which have been described in Potluri and Rao (2020). Numerous researchers have advanced answers for efficient resources distribution based on market mechanisms. One of the common market mechanisms in use at the moment is the CDA (Continuous Double Auction). It makes sure that the distribution of resources is very efficient and well-coordinated. Shu et al. (2021) demonstrates the efficiency of CDA-based online resource allocation. And to achieve efficient resource allocation in a cloud setting, Kumar et al. (2022) provides a cloud resource assignment strategy based on the CDA framework and Nash equilibrium. PSO was first suggested in Lepakshi and Prashanth (2020), where it was suggested as a method for optimisation. PSO was divided into two categories: discrete PSO and continuous PSO versions. PSO outperformed ACO or GA by doing many passes over search space as well as updating local best as well as global best solutions throughout every run.Authors added the idea of inertia weight to the first PSO in Liu (2022). The addition of inertia weight might hasten the convergence of PSO. Initially, it was suggested that inertia weight should fall between [0.9, 1.2], which can enhance PSO performance. Different inertia values allowed for more precise control over the solution search space. Lower numbers will confine the search in a certain location in the search space whereas higher values will cause overshooting. In Rugwiro et al. (2019), a Cost Aware Modified PSO (CA-PSO) was suggested. In Manikandan, Gobalakrishnan, and Pradeep (2022), writers use PSO to reduce the cost of all activities in a workflow while maintaining the deadline restrictions. The suggested PSO-based metaheuristic strategy is successful, but IC-PCP misses the application deadline.

3. BANDWIDTH DIFFERENTIAL PREEMPTION EVOLUTION MOVING AVERAGE ALGORITHM-BASED RESOURCE ALLOCATION

The design of cloud resource management method suggested in this study is shown in Figure 1. Register centre (RC), cloud environment monitor (CEM), infrastructure management (IM), and control centre (CC) are four parts of this resource management system. Following is a description of the four listed components. (i) RC: For connectivity and management, each physical server in a cloud data centre should register its data with RC. (ii) CEM: This component collects data about actual servers, such as host names and IP addresses, and keeps track of their operating statuses (started, running, shut down), as well as how much CPU, RAM, and disc storage they consume. (iii) IM: It is in charge of setting up and maintaining the virtualized infrastructures, which includes releasing VMs. (iv) CC: The computing centre is responsible for making the best resource allocation choice. Statuses as well as resource use for physical servers registered in RC are being watched by CEM. A new physical server's MAC address and IP address are registered to the RC as soon as it starts to use the cloud. When a user submits a service request to the cloud, CC receives request's resource needs. Based on the data gathered by CEM, CC decides how to allocate resources wisely. In order to manage actual servers as well as set up VMs, IM implements the allocation decision.

Figure 1. A framework of cloud resource management method

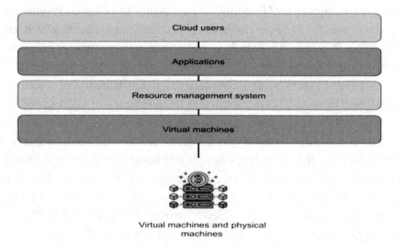

We use t to represent time and d to represent communication distance. Channel response of MP method in uth frequency subwindow is given by including N(u) Ref reflected rays by eqn (1)

$$h_u(d) = \alpha_{\text{LOS}}^{(u)}(d)\delta(t - t_{\text{LOS}})1_{\text{LOS}} + \sum_{q=1}^{N_{\text{Ref}}^{(i)}} \alpha_{\text{Ref}}^{(u,q)}(d)\delta(t - t_{\text{Ref}}^{(q)}) \tag{1}$$

where 1LOS is the pointer capability that is equivalent to 1 or 0 for presence of LOS way or not. For LOS way, α(i) LOS alludes to constriction, tLOS represents the postponement. For qth reflected way, α(i,q) Ref is constriction and t (q) Ref is postponement. In uth recurrence subwindow, middle recurrence is meant by fu . From one viewpoint, MP signal comprises of LOS and various reflected ways, when increases of sending as well as getting recieving wires are Gt = Gr = 0 dBi. Then again, with utilization of highgain recieving wires or extremely enormous recieving wire clusters, quantity of MP beams and defer spread decline, while absolute gotten signal power increments. Specifically, when increases of sending and getting recieving wires are Gt = Gr = 20 dBi, THz band transmission is profoundly directional as well as quantity of MP parts decreases to a modest number, i.e, Nu = 1.

Differential Preemption Evolution Model

Haphazardly select the underlying boundary esteems consistently on the stretches which x are lower and upper headed for every boundary. Every one of the singular vectors goes through change, recombination, and selection.In age G, every individual vector in the populace turns into an objective vector, where psdenotes the populace size. For every target vector, G Xi DEA applies a differential transformation activity to produce a freak individual, G Vi as indicated by the accompanying conditions by eqn (2)

$$V_i^G = X_j^G + F\left(X_k^G - X_q^G\right)$$

$$V_i^G = X_{best}^G + F\left(X_k^G - X_q^G\right)$$

$$V_i^G = X_{best}^G + F\left(X_j^G + X_k^G - X_q^G - X_r^G\right) \tag{2}$$

where, X_j^G, X_k^G, X_q^G, and X_r^G are randomly taken from population such that j, k, q, and r belong to $\{1,2,...,p_s\}$ and $j \neq k \neq q \neq r$; X_{best}^G is best individual at generation G by eqn (3)

$$U_{ji}^G = \begin{cases} V_{ji}^G, & \text{if rand}_j\left(0,1\right) \leq p_c \\ X_{ji}^G, & \text{otherwise,} \end{cases} \tag{3}$$

where $j=1,2,...,\rho$ denotes total number of design variables. randj(0, 1) function computes jth calculation of a uniform random number generator from interval [0, 1].

Boxing or grid method covered in this article is known as ATR (Average True Range) charts, which are frequently utilized in securities analysis with goal of determining the best value for the execution of the underlying process and disregarding any minor variations that may occur as a notional loss. Simply said, a small load factor during a specific session need not be seen as alarming if overall network channel is clean to handle the requirements. The suggested model is therefore built on creating Grids to manage the specific type of traffic reaching VMs. In order to forecast resource demand, the value of the first moving average is collected, and the value of the second moving average is measured at time 'T' of the 'a' user by eqn (4)

$$MAV_T^1\left(a\right) = \frac{x_T\left(a\right) + x_{T-1}\left(a\right) + ... + x_{T-(N-1)}\left(a\right)}{N}$$

$$MAV_T^2(a) = \frac{M_T^1(a) + M_{T-1}^1(a) + \ldots + M_{T-(N-1)}^1(a)}{N} \tag{4}$$

The likelihood of future province of Markov chain at every time is simply connected with condition of time, however not to condition of succession before time, which has no delayed consequence. Markov method is addressed by triples {S, π, P}, in which S addresses state space of irregular cycle as well as limited informational collection of arbitrary interaction. π is likelihood vector of chose starting state time, and P is likelihood move framework. Likelihood move lattice is gotten by recurrence assessment likelihood strategy, or by limiting squared aggregate blunder of likelihood vector about likelihood vector of present status and hypothetical state. Setting state worth of irregular cycle as S = {S1, S2, ..., Sn}, likelihood move normal for Markov chain not entirely settled by the restrictive likelihood, that is to say, likelihood of state Sj after k-time handling, when variable X is in state Si on time m. Whether information series is anticipated by Markov method needs χ 2 identification. Let fij be quantity of state I changes to state j, and Pij be likelihood of state I advances to state j. Measurement χ 2 is communicated as Condition (2), where, P•j is negligible likelihood of state j, which fulfills Condition (3) by eqn (5)

$$\chi^2 = 2\sum_{i=1}^m \sum_{j=1}^m f_{ij} \left| \lg \frac{P_{ij}}{P \cdot j} \right|$$

$$P_{\cdot j} = \frac{\sum_{i=1}^m f_{ij}}{\sum_{i=1}^m \sum_{j=1}^m f_{ij}} \tag{5}$$

If data sequence accords with $\chi^2 > \chi_a^2\left((m-1)^2\right)$ then Markov method is utilized to anticipate future pattern of information. In event that change likelihood of Markov bind from state Si to Sj in one stage is P (k) ij, then grid of state change likelihood in one stage is communicated as eqn (6).

$$P^{(1)} = \begin{bmatrix} P_{11} & \cdots & P_{1n} \\ \vdots & \ddots & \vdots \\ P_{n1} & \cdots & P_{nn} \end{bmatrix} \tag{6}$$

Assuming the arbitrary cycle is in I-th state at ongoing time, and times it moves to j-th state at following time is fij, then, at that point, Utilizing the technique for recurrence assessment likelihood, The likelihood Pij of state I changes to state j is determined by eqn (7).

$$P_{ij} = \frac{f_{ij}}{\sum_{j=1}^N f_{ij}} = P\{X = S_j \# X = S_i\} \tag{7}$$

Let π0 mean the underlying vector of stochastic interaction at time t, boundaries p1, p2, ..., pn signify the likelihood of each state around then. Then, underlying state vector is communicated as π0 = (p1, p2, ..., pn), and likelihood vector of arbitrary cycle at t = m is πm = π0P m. At point when the worth of m is sufficiently huge, the likelihood vector will watch out for a steady worth, which is communicated as Y = P πm × Si . As indicated by qualities of Markov cycle, future condition of stochastic interaction is anticipated by its authentic state. Anticipated worth Dt+1 is communicated as Recipe (6), which is inward result of state vector Xt+1 and typical worth of every state, where Xt+1 = (xt+1,1, xt+1,2, ..., xt+1,i ..., xt+1,N), on off chance that state is in I, worth of xt+1,i in framework is 1, and different factors of xt+1,j are set to nothing, where j is any state other than I by eqn (8)

$$D_{t+1} = X_{t+1}E_i = \sum_{i=1}^{N} x_{t+1,i}E_i \tag{8}$$

Targeting the direct result of the cloud resource allocation procedure is the goal of employing these metrics. Numerous session request management characteristics are taken into consideration for analysis and calculated for genuine range circumstances while keeping an eye on the key system components that have a substantial influence on service quality. The computation method is selected based on prioritisation of bandwidth or session requests or any other sort of services that might have an impact on the service quality, and as a result, the ATR method for selected metrics is carried out. Additionally, employing several sets of cloud resource scheduling activities to analyse difficulties as well as come to wise scheduling judgements in case of a complicated resource scheduling environment.

Reinforcement Adversarial Hierarchical Group Multi-Objective Cuckoo Optimization-Based Network Task Scheduling

Consider the following binary classification problem: z = (x, y) R d 1,+1. Minimising the fundamental cost function yields linear Q-learning classifier by eqn (9)

$$\mathcal{P}_n(\mathbf{w}) = \frac{\lambda}{2} \| \mathbf{w} \|^2 + \frac{1}{n} \sum_{i=1}^{n} \ell\left(y_i \mathbf{w}^{\top} \mathbf{x}_i\right) = \frac{1}{n} \sum_{i=1}^{n} \left(\frac{\lambda}{2} \| \mathbf{w} \|^2 + \ell\left(y_i \mathbf{w}^{\top} \mathbf{x}_i\right) \right) \tag{9}$$

Convex hull of a family of m vectors {uj} (j = 1, . . ., m; uj∈ R n), is set of all their convex combinations by eqn (11)

$$\bar{\mathcal{U}} = \left\{ \mathbf{u} \in \mathbb{R}^n \text{ such that } \mathbf{u} = \sum_{j=1}^{m} \alpha_j \mathbf{u}_j; \alpha_j \in \mathbb{R}^+ (\forall j): \sum_{j=1}^{m} \alpha_j = 1 \right\} \tag{10}$$

We define return and policy as follows. Return Gt is total discounted reward from time-step t by eqn (11)

$$G_t = R_{t+1} + \gamma R_{t+2} + \cdots = \sum_{k=0}^{\infty} \gamma^k R_{t+k+1} \tag{11}$$

State-esteem capability vπ(s) gives the drawn out worth of state s while following arrangement π. We embrace two brain organizations (NN), in particular the calculation NN as well as enemy NN. Result of calculation NN is a likelihood dispersion over the cost set for every client, while result from enemy NN is likelihood conveyances over financial plan set for all clients. Info to calculation NN or foe NN is chosen by learned procedure of other NN. More specifically, in every preparing cycle, cost succession as contribution to enemy NN and spending plan arrangement took care of into calculation NN are tested by result of calculation NN as well as foe NN. Cost set An as well as spending plan set B are thought to be known and limited in our detailing, which are assessed in light of verifiable follows if not given. We consider social government assistance expansion for single-type non-reused asset designation as well as evaluating. There are R units of asset supply altogether. Nusers show up after some time, every mentioning one unit of asset. Financial plan of client $i \in N$ is , indicating how much client will pay for getting one unit of the asset. At the point when client i shows up, the internet based calculation posts its cost pi for one unit of the asset. We accept the calculation doesn't know about the ongoing client's financial plan , yet knows posted costs and financial plans of past clients. Assuming that pi is no bigger than , client i acknowledges the cost and gets one unit of the asset. The piece of social government assistance because of tolerating client i is $(bi - pi) + pi = $.

We use Xi as irregular variable indicating regardless of whether client i is acknowledged. xi is acknowledged acknowledgment choice of irregular variable $:i = 1$ if $bi \geq pi$ and there exists accessible asset, and $xi = 0$, in any case. $xi = 1$ infers that one unit of asset is assigned to client i while $xi = 0$ demonstrates that user$i$ consumes no asset. Let $P(Xi = 1|X1 = x1, .., Xj = xj, 1 \leq j<i)$ indicate likelihood of client i being acknowledged molded on the acknowledged acknowledgment of past clients, which is likewise the likelihood at picking costs no bigger than bi . Additionally, $P(Xi = 0|X1 = x1, .., = xj, 1 \leq j<i)$ is the likelihood of not allotting asset to client i adapted on the acknowledged choices of past clients, i.e., the likelihood at picking costs bigger than bi . Terms connected with $P(Xi = 0|x1..j)$ are not displayed in accompanying definition and their slope is 0 by eqn (12)

$$\max\nolimits_{P(X_j = 1 | x_1 .. x_{i-1}), \forall j \in [i, N]} f = \sum\nolimits_{j=i}^{N} b_j P\left(X_j = 1 \# x_1 .. x_{i-1}\right)$$

subject to:

$$x_1 + x_2 + \ldots + x_{i-1} + \sum\nolimits_{j=i}^{N} P\left(X_j = 1 \# x_1 \ldots x_{i-1}\right) \leq R$$

$$P(X_j = 1 | x_1 \ldots x_{i-1}) \in [0,1], \forall j \in [i, N] \tag{12}$$

The goal capability (1) is the normal social government assistance accomplished by the calculation adapted on the choices before client i. (1a) is the asset requirement, which limits the normal asset utilization by the quantity of accessible assets, because of the randomization idea of our calculation NN. The contingent likelihood in light of past acknowledged choices would guarantee optimality of total grouping, that clients with biggest R spending plans are accepted. Above issue is a straight program, major areas of strength for where holds. We can loosen up imperative (1a) by presenting Lagrangian multiplier λ, and get accompanying Lagrangian capability by eqn (13)

$$\mathcal{L}\left(P\left(X_j = 1 | x_1 .. x_{i-1}\right), \lambda\right) = \sum\nolimits_{j=i}^{N} b_j P\left(X_j = 1 | x_1 \ldots x_{i-1}\right)$$
$$+ \lambda\left(R - x_1 - .. - x_{i-1} - \sum\nolimits_{j=i}^{N} P\left(X_j = 1 | x_1 \ldots x_{i-1}\right)\right) \tag{13}$$
$$= \sum\nolimits_{j=i}^{N} \left(b_j - \lambda\right) P\left(X_j = 1 | x_1 \ldots x_{i-1}\right) + \lambda\left(R - x_1 - .. - x_{i-1}\right)$$

Dual function is then by eqn (14)

$$\mathcal{G}(\lambda) = \max_{P(X_j = 1\#x_1.x_{i-1}),} \mathcal{L}\left(P(X_j = 1\#x_1 \ldots x_{i-1}), \lambda\right)$$

$$OPT = \mathcal{G}(\lambda^*) = \max_{P(X_j = 1\#x_1 \ldots x_{i-1}), \forall j \in [i,N]} \mathcal{L}\left(P(X_j = 1\#x_1 \ldots x_{i-1}), \lambda^*\right)$$

$$= \max_{\substack{P(X_j = 1\#x_1, x_{i-1}) \\ \forall j \in [i,N]}} \sum_{j=i}^{N}(b_i - \lambda^*)P(X_j = 1\#x_1..x_{i-1}) + \lambda^*(R - x_1 - \ldots - x_{i-1}) \tag{14}$$

Given $\lambda*, *(R - x1 - .. - xi-1)$ is a constant, and solving primal issue is equivalent to solving by eqn (15)

$$\max_{P(X_j = 1\#x_1, x_{i-1}), \forall j \in [i,N]} \sum_{j=i}^{N}(b_i - \lambda^*)P(X_j = 1\#x_1 \ldots x_{i-1}) \tag{15}$$

subject to:

$$P(X_j = 1| x1 \ldots x_{i-1}) \in [0,1], \forall j \in [i,N] \tag{16}$$

Cuckoo search calculation is a multitude insight based calculation. This calculation was propelled by regular way of behaving of cuckoos. In this calculation, each egg in a home shows the competitor arrangement. As a general rule, every cuckoo can lay simply a solitary egg into a home in remarkable shape though every home can has various eggs addressing a bunch of arrangement. The primary goal of CS is to make new arrangements that will supplant the most horrendously terrible arrangements in the ongoing home populace. The bit by bit process is made sense of beneath; Stage 1: Introduction Stage In this segment, a populace of home (arrangement) is made haphazardly. Stage 2: Producing New Cuckoo Stage In this part, we create new cuckoos utilizing demand flights. Then we work out goal capability of every answer for figuring out nature of arrangements. Stage 3: Wellness Assessment Stage After arrangement age, we compute the wellness capability, and afterward pick the best solution.

fitness= least goal capabilities.

Stage 4: Updation Stage After wellness computation, the new arrangement is made utilizing Eq. (17).

$$I_i^{New} = I_i^{(t+1)} = I_i^{(t)} + \alpha \oplus \text{Levy}(\lambda) \tag{17}$$

To deal with the undertaking to distribute virtual machine a legitimate planning is required. To remember this insightful some cycle "I" has been created. The emphasess "I" has been the underlying populace by eqn (18)

$$I = (NI_{max} - NI_{min}) \times \text{Rand}[0,l] + NI_{min} \tag{18}$$

This number "I" recommends new cycles number to be produced by parent iterator at laying stage. Every parent iterator is permitted to change request for a restricted undertakings with the limitations of control stream reliance. This predetermined number is known as layingRadius (LR). Every parent emphasis has processed LRVM and LRt values. LRVM and LRt are processed for every emphasis by utilizing following equ. 10 and equ.fdgrespectively. LRVM = [γ x current Emphasis I/Absolute of all Cycle I] by eqn (19)

$$x(var_{hivm} \ var_{lowrm})$$

$$x(var_{hit} \ var_{low}t) \tag{19}$$

The migration cost (MC) is first parameter of the fitness function. The amount of money spent to complete the process is referred to as the migration cost. Migration cost is kept to a minimum in this schedule. Overall migration cost is evaluated utilizing Movement factor (MF) and the Cost factor (CF). Eq (20) can be used to calculate MC.

$$MC = \frac{M^F + C^F}{2}$$

$$M^F = \frac{1}{PM} \left[\sum_{i=1}^{VM_i} \left(\frac{\text{Number of movements}}{\text{Total } VM} \right) \right]$$

$$C^F = \sum_{i=1}^{VM_i} \left(\frac{\text{Cost to run} \times \text{Memory of task}}{VM \times PM} \right) \tag{20}$$

The second fitness function parameter is energy consumption (EC). The amount of energy expended during the job scheduling process in a CC environment. EC value must be as low as possible for our objective function. Eq. (21) is used to calculate the EC.

$$EC = \frac{1}{PM_i \times VM_j} \left[\sum_{i=1}^{PM} \sum_{j=1}^{VM} Q_{ij} Y_{max} + \left(i - K_{ij} \right) \lambda_{ij} Y_{max} \right] \tag{21}$$

Where

$$Q_{ij}, Y_{max} \in [0,1]$$

$$\lambda_{ij} = \frac{1}{2} \left[\left(\frac{CPU \text{ utilized}_{ij}}{CPU_{ij}} \right) + \left(\frac{MU_{ij}}{M_{ij}} \right) \right]$$

Memory utilisation is the fitness function's third parameter. Memory used for job scheduling in a CC environment. The MU value for our fitness function must be as low as possible. Eq. (14) gives the MC calculation by eqn (22)

$$memory = \frac{1}{PM \times VM}\left[\sum_{i=1}^{PM}\sum_{j=1}^{VM}\frac{1}{2}\left(\frac{CPU\ utilized_{ij}}{CPU_{ij}} + \frac{MU_{ij}}{M_{ij}}\right)\right] \qquad (22)$$

Credit is the fitness function's fourth parameter. The credit is determined by implementation time. Because task was performed inside deadline, method assigned credit to VM, which is rising by 1. At initially, we assign each virtual computer a credit of 0. Eq. (15 is used to calculate the credit.

$$CR = \frac{CR}{P + CR} \qquad (23)$$

The objective function's final parameter is a penalty. Punishment is also affected by execution time. Job extending deadline indicates that method allocates a penalty to VM, resulting in a decrease of - 1. Eq. (24) is used to calculate the penalty.

$$P = \frac{P}{P + CR} \qquad (24)$$

Objective function of our research is defined in Eq. (7). We can see from the phrasing that issue is one of task scheduling. Solving such issues with a mathematical programming method will take a long time for a huge task. Figure-2 depicts the proposed network task scheduling flowchart.

Figure 2. Flowchart for proposed network task scheduling method

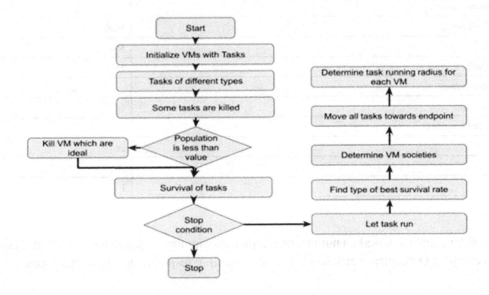

4. RESULTS AND DISCUSSION

In this section, we examine outcomes of suggested estimation-based task booking approach. We realised our proposed endeavour job booking utilising Java (jdk 1.6) with cloudSim gadgets, movement of trials were carried out on a PC with Windows 7 Working system at 2 GHz twofold centre PC with 4 GB Crush running a 64-digit edition of Windows 2007. Central idea of our suggested method is VM booking with CHSA estimate. We allocate N number of tasks as well as M number of assets at each stage of the process. To organise the work in terms of cost, memory consumption, energy consumption, credit, punishment, and wellness capacity. We used three different plan arrangements in this work. The presentation is then evaluated based on three subcategories, such as (1) execution with 5 genuine machines and 14 VMs, (2) execution with 10 real machines and 26 VM, and (3) execution with 10 genuine machines and 31 VM. Here 4 and 5 such endeavours were utilized for planning cycles to designate in VMs as well as every task contains 3 sub tasks. Total number of micro processor used is equivalent to amount to number of VMs utilized. Trial and error specifications like server ranch, VMs, and clients' interest defined in Table1.

Table 1. Experimentation configurations

Data Center Setup	
Number of hosts	01
Data center	01
VMs configuration	
MIPS	1000
Number of VMs	5 – 50
Bandwidth	1000
VM memory	512
Nuner of PEs per VM	1
Workload environment	
Numbers of cloudlets	1000
Data Set	Montage
Number of users	15
ACO parameters setup	
Number of ants (m)	10
Number of iterations (max)	100
a	.3
ρ	.4
β	1

Table-1 shows analysis based on number of samples and number of users. Parametric analysis is carried out in terms of QoS, energy efficiency, resource sharing, response time, and makespan.

Table 2. Comparative analysis for various parameters

Cases	Techniques	QoS	Energy Efficiency	Resource Sharing Rate	Response Time	Makespan
Number of VMs	**DRL**	56	91	77	36	51
	LFU_DRM	61	93	79	38	53
	BDPEMA_RAHMCO	63	95	81	39	55
Number of Tasks	**DRL**	64	93	83	42	56
	LFU_DRM	65	94	85	43	59
	BDPEMA_RAHMCO	66	96	88	45	61

Figure 3. Comparative based on number of VMs

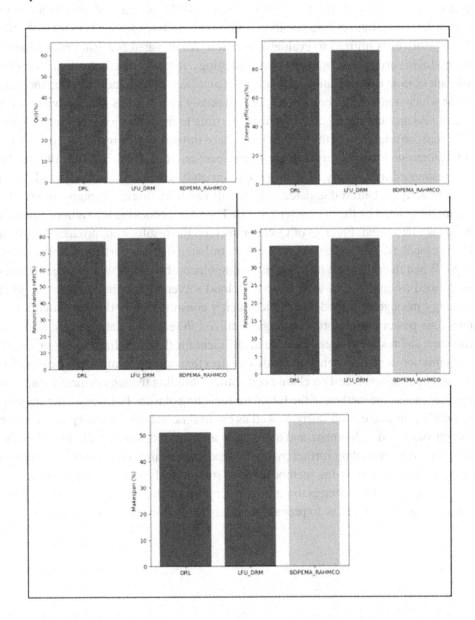

The above figure- 2 (a) – (e) shows comparative based on number of VMs. Here proposed technique QoS of 63%, energy efficiency 95%, resource sharing of 81%, response time 39%, and makespan 55%; existing DRL QoS 56%, energy efficiency 91%, resource sharing 77%, response time 36%, and makespan 51%, LFU_DRM QoS 61%, energy efficiency 93%, resource sharing 79%, response time 38%, and makespan 53%. The waste level progressively dropped till it reached 9 cm at time = 8, as seen from time axis (y-axis). Rubbish level stayed steady until time = 12 after reaching a threshold distance of 4 cm at time = 9. A notice was sent to collector during this time, and they collected and dumped e-waste from smart bin after that. After trash was collected, level of rubbish rose as well as measured at 28 cm at time = 13.

From above figure 3 (a)- (e) shows comparative based on number of samples. Here proposed technique QoS 66%, energy efficiency 96%, resource sharing 88%, response time 45%, and makespan 61%; existing DRL QoS 64%, energy efficiency 93%, resource sharing 83%, response time 42%, and makespan 56%, LFU_DRM attained QoS 65%, energy efficiency 94%, resource sharing of 85%, response time of 43%, and makespan of 59%.t measures the percentage of successfully recognised objects among all of the items the system detected. Ratio of true positives (things successfully detected) to total of true positives as well as false positives is utilized to evaluate precision. An elevated precision rating means that the system is successful in correctly recognising pertinent things. It means that the system doesn't frequently mistake irrelevant objects for target item. In our case, a precision of 95% indicates that there is a reasonably low rate at which non-smartphone goods are mistakenly identified as smartphones by the system.

We also take into account the algorithm's response time to incoming jobs as a second performance parameter. The real consideration of the request takes place throughout the response time. In other words, we may claim that resource availability directly affects reaction time. The scheduling of tasks has a direct impact on the resources' availability. If jobs are appropriately scheduled, resources will automatically become available early or before due dates, which will result in shorter response times. Results and performance are influenced by the parameter's value. Therefore, choosing the right parameters is also a crucial issue to take into mind. In terms of QoS parameters, the results were produced utilising various sets of VMs and cloudlets. The processor and memory utilisation in our particular work are instances of resource usage. When the other two techniques are combined, the proportion of one method is always greater. It is defined as energy unit used by every cloud server to distribute resources. Management employs an energy management model to reduce energy consumption in this particular job. Although there are numerous power consumption strategies utilized in real-time data centres, such as dynamic voltage, frequency, and resource sleep, they are insufficient for the virtualized environment. The gas is cleaned by the emission control system, which also verifies that only clean air is released. Additionally, it provides a range of de-dusters based on client need while upholding the requirement for approved emissions. Pyrolysis is seen as the method of the future for recycling plastic. To improve the characterisation, monitoring of emissions and ensure safety as well as environmental sustainability of e-waste pyrolysis, ongoing experimentation, development, and more study are required. Future work on our research paper will focus on a few areas, including further pyrolysis process optimisation, improving data-driven decision making by utilising cutting-edge methods, waste stream analysis for a simple recycling process, managing and controlling solar batteries to improve their performance as well as lengthen their lifespan, and streamlining recycling workflow to prepare recyclers for various recycling methods.

Figure 4. Comparative based on number of samples

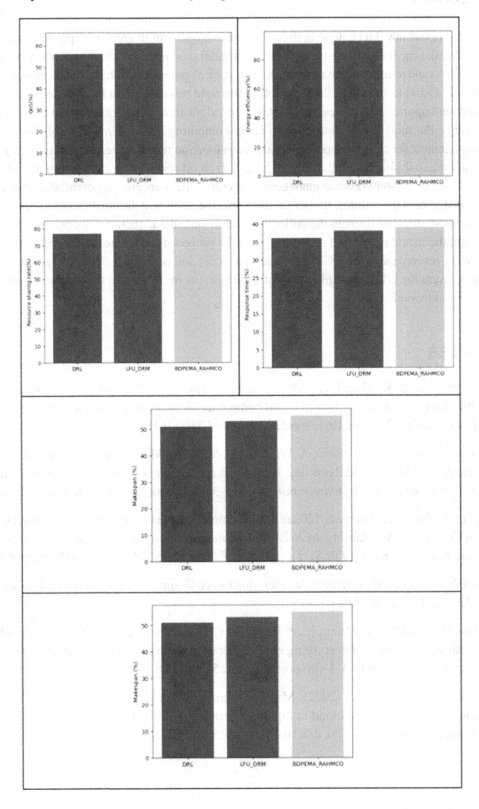

5. CONCLUSION

This research propose novel technique in digital waste management in green computing based resource allocation and task scheduling using bandwidth differential preemption evolution moving average algorithm (BDPEMA) and reinforcement adversarial hierarchical group multi-objective cuckoo optimization (RAHMCO). It assists in matching the right job to the right resource. Results and analyses also reveal that the suggested approach beat previously established algorithms. The optimisation process may also be used to make the most use of cutting-edge cloud computing components like IoT. Future work in resource management for cloud computing would focus more on renewable energy, automating resource management, and analysing user behaviour to speed up cloud services. In conclusion, the adoption of our cloud- as well as IoT-based waste management method offers enormous potential to transform current waste management procedures. It offers significant cost savings, a diminished environmental effect, and increased sustainability through the collection of real-time data, operational optimisation, resource allocation, and the manufacturing of recycled goods.the proposed technique attained QoS 66%, energy efficiency 96%, resource sharing 88%, response time 45%, and makespan 61%. To secure widespread adoption and assure the effective deployment of such systems in the future, more research and security issues must be addressed.

REFERENCES

Bal, P. K., Mohapatra, S. K., Das, T. K., Srinivasan, K., & Hu, Y. C. (2022). A joint resource allocation, security with efficient task scheduling in cloud computing using hybrid machine learning techniques. *Sensors (Basel), 22(3)*, 1242. doi:10.3390/s22031242 PMID:35161987

Das, K., Saha, S., Chowdhury, S., Reza, A. W., Paul, S., & Arefin, M. S. (2022, October). A Sustainable E-waste Management System and Recycling Trade for Bangladesh in Green IT. In *International Conference on Intelligent Computing & Optimization* (pp. 351-360). Cham: Springer International Publishing.

Gaharwar, G., Pandya, J., & Pandya, S. (2022, July). Comprehensive Framework for Assessing Organization's Green Computing Adaptability. In *2022 1st International Conference on Sustainable Technology for Power and Energy Systems (STPES)* (pp. 1-6). IEEE. 10.1109/STPES54845.2022.10006640

Gupta, C. (2022). E-Waste Management-Accelerating Green Computing. *International Journal of Early Childhood Special Education, 14*(5).

Jamil, B., Ijaz, H., Shojafar, M., Munir, K., & Buyya, R. (2022). Resource allocation and task scheduling in fog computing and internet of everything environments: A taxonomy, review, and future directions. *ACM Computing Surveys, 54*(11s), 1–38. doi:10.1145/3513002

Kumar, Y., Kaul, S., & Hu, Y. C. (2022). Machine learning for energy-resource allocation, workflow scheduling and live migration in cloud computing: State-of-the-art survey. *Sustainable Computing : Informatics and Systems, 36*, 100780. doi:10.1016/j.suscom.2022.100780

Lepakshi, V. A., & Prashanth, C. S. R. (2020, March). Efficient resource allocation with score for reliable task scheduling in cloud computing systems. In *2020 2nd International Conference on Innovative Mechanisms for Industry Applications (ICIMIA)* (pp. 6-12). IEEE. 10.1109/ICIMIA48430.2020.9074914

Liu, H. (2022). Research on cloud computing adaptive task scheduling based on ant colony algorithm. *Optik (Stuttgart)*, *258*, 168677. doi:10.1016/j.ijleo.2022.168677

Manikandan, N., Divya, P., & Janani, S. (2022). BWFSO: Hybrid Black-widow and Fish swarm optimization Algorithm for resource allocation and task scheduling in cloud computing. *Materials Today: Proceedings*, *62*, 4903–4908. doi:10.1016/j.matpr.2022.03.535

Manikandan, N., Gobalakrishnan, N., & Pradeep, K. (2022). Bee optimization based random double adaptive whale optimization model for task scheduling in cloud computing environment. *Computer Communications*, *187*, 35–44. doi:10.1016/j.comcom.2022.01.016

Mukta, T. A., & Ahmed, I. (2020). Review on E-waste management strategies for implementing green computing. *Int. J. Comput. Appl*, *177*, 45–52.

Potluri, S., & Rao, K. S. (2020). Optimization model for QoS based task scheduling in cloud computing environment. *Indonesian Journal of Electrical Engineering and Computer Science*, *18*(2), 1081–1088. doi:10.11591/ijeecs.v18.i2.pp1081-1088

Prasad, K. D., Murthy, P. K., Gireesh, C. H., Prasad, M., & Sravani, K. (2021). Prioritization of e-waste management strategies towards green computing using AHP-QFD approach. *Proc. Eng*, *3*, 33–40.

Praveenchandar, J., & Tamilarasi, A. (2021). Dynamic resource allocation with optimized task scheduling and improved power management in cloud computing. *Journal of Ambient Intelligence and Humanized Computing*, *12*(3), 4147–4159. doi:10.1007/s12652-020-01794-6

Purnomo, A., Anam, F., Afia, N., Septianto, A., & Mufliq, A. (2021, August). Four decades of the green computing study: a bibliometric overview. In *2021 International Conference on Information Management and Technology (ICIMTech)* (Vol. 1, pp. 795-800). IEEE. 10.1109/ICIMTech53080.2021.9535069

Rjoub, G., Bentahar, J., Abdel Wahab, O., & Saleh Bataineh, A. (2021). Deep and reinforcement learning for automated task scheduling in large-scale cloud computing systems. *Concurrency and Computation*, *33*(23), e5919. doi:10.1002/cpe.5919

Rugwiro, U., Gu, C., & Ding, W. (2019). Task scheduling and resource allocation based on ant-colony optimization and deep reinforcement learning. *Journal of Internet Technology*, *20*(5), 1463–1475.

Saidi, K., & Bardou, D. (2023). Task scheduling and VM placement to resource allocation in Cloud computing: Challenges and opportunities. *Cluster Computing*, *26*(5), 3069–3087. doi:10.1007/s10586-023-04098-4

Sathiyamoorthi, V., Keerthika, P., Suresh, P., Zhang, Z. J., Rao, A. P., & Logeswaran, K. (2021). Adaptive fault tolerant resource allocation scheme for cloud computing environments. *Journal of Organizational and End User Computing*, *33*(5), 135–152. doi:10.4018/JOEUC.20210901.oa7

Shu, W., Cai, K., & Xiong, N. N. (2021). Research on strong agile response task scheduling optimization enhancement with optimal resource usage in green cloud computing. *Future Generation Computer Systems*, *124*, 12–20. doi:10.1016/j.future.2021.05.012

Soesanto, H., Maarif, M. S., Anwar, S., & Yurianto, Y. (2023). Current Trend, Future Direction, and Enablers of e-Waste Management: Bibliometric Analysis and Literature Review. *Polish Journal of Environmental Studies*, *32*(4), 3455–3465. doi:10.15244/pjoes/163607

Chapter 12
Sustainable Waste Management OOA-Enhanced MobileNetV2-TC Model for Trash Image Classification

B. Manjunatha
New Horizon College of Engineering, India

K. Dinesh Kumar
https://orcid.org/0000-0003-0843-1561
Amrita Vishwa Vidyapeetham, India

Sam Goundar
https://orcid.org/0000-0001-6465-1097
RMIT University, India

Balasubramanian Prabhu Kavin
https://orcid.org/0000-0001-6939-4683
SRM Institute of Science and Technology, India

Gan Hong Seng
XJTLU Entrepreneur College, Xi'an Jiaotong-Liverpool University, China

ABSTRACT

E-waste is an invisible, indirect waste that contaminates natural resources like the air, water, and soil, endangering the ecosystem, people, and animals. Long-term waste accumulation and contamination can harm the resources found in the environment. Since traditional waste management systems are very inefficient and the number of people living in urban areas is increasing, waste management systems in these areas face challenges. However, by combining a variety of sensors with deep learning (DL) models, waste resources can be used effectively. For this chapter, firstly, the Trashnet dataset with 2527 images in six classes and the VN-trash dataset, which comprises three classes and 5904 images, are collected. Then the collected images are preprocessed using truncated gaussian filter. After that, pre-trained convolutional neural network (CNN) models (Resnet20 and VGG19) are applied to the images in order to extract features. In order to enhance the predictive performance, this study then creates a MobileNetV2 model for trash classification (TC) called MNetV2-TC.

DOI: 10.4018/979-8-3693-1552-1.ch012

1. INTRODUCTION

In the rapidly urbanising world of today, waste creation is an inevitable byproduct of modern living. The sheer amount of items that are thrown away, commonly known as "trash," from homes to businesses, presents a serious threat to the sustainability of the environment (Zhang, Yang, Zhang et al, 2021). Fundamentally, trash is any variety of items that are either no longer considered necessary or have passed their prime and are no longer useful (Gupta, 2020). The term "trash" encompasses a wide range of materials, from plastic bottles, food packaging, and paper used in households to industrial byproducts like electronic waste and construction debris (Masand et al., 2021). The problem is not only the sheer amount of waste generated, but also the variety of materials involved, each of which needs a different disposal strategy to reduce its negative effects on the environment. Keeping up with this ever-increasing waste mountain has become increasingly important as urbanisation and consumption patterns change (Vo, Vo, & Le, 2019).

Effective waste management techniques are receiving more attention as a result of the pressing need to address the growing waste crisis (Tiyajamorn et al., 2019). A crucial element of efficient waste management is the precise and methodical classification of waste. Appropriate trash classification is important because of how it affects resource recovery, recycling, and environmental preservation (Yu, 2020). Different waste types require different disposal techniques, and improper waste segregation can lead to pollution of the environment, increased use of landfills, and lost recycling opportunities (Ozkaya & Seyfi, 2019). Communities can improve the efficiency of their waste management procedures, lessen their impact on the environment, and get closer to a more sustainable future by using sophisticated classification techniques to understand the composition of waste (Mao et al., 2021).

Waste sorting has historically been a labour-intensive procedure that relies on hand labour to separate various materials (He et al., 2020). But with the advent of DL, an artificial intelligence subfield motivated by the composition and operations of the human brain, the trash classification industry has undergone significant transformation (Yang et al., 2020). DL algorithms that have demonstrated remarkable performance in tasks related to image classification and recognition are convolutional neural networks (CNNs) (Zhang, Zhang, Mu et al, 2021). These algorithms are highly accurate at differentiating between different materials when applied to trash images, which makes them indispensable tools for automating the waste sorting process (Gupta et al., 2022). Effective trash management techniques are becoming more and more necessary as communities deal with the fallout from inappropriate waste disposal. Classifying and managing various types of trash through the use of advanced technologies, especially DL, is one promising solution that is soon to be realised (Mythili & Anbarasi, 2022).

1.1. Motivation

The pressing need for efficient waste management is the driving force behind the classification of trash images. Waste production both increases in quantity and variety as urbanisation picks up speed. Conventional waste sorting techniques are error-prone and labor-intensive. Using DL to power image classification expedites the process by identifying different materials automatically. This improves resource recovery, encourages appropriate recycling, lowers environmental pollution, and boosts the effectiveness of waste management systems. Trash image classification ultimately aims to address the growing difficulties associated with contemporary waste disposal by developing a more technologically sophisticated and sustainable method.

1.2. Main Contributions

- The main contribution is the use of a Truncated Gaussian Filter in image pre-processing as a noise reduction method.
- VGG19 and ResNet20 are examples of pre-trained DL models, are used to extract features. This stage entails extracting pertinent information from the pictures to make classification easier later on.
- Based on the MobileNetV2 architecture, the study presents a novel trash classification model called MNetV2-TC. The purpose of this model is to improve predictive performance in tasks involving trash classification.
- The Orca Optimisation Algorithm (OOA), a novel meta-heuristic algorithm, is used to tune hyperparameters while creating the MNetV2-TC model. By optimising the model's parameters, this technique seeks to increase the model's efficacy.
- Trashnet and VN-Trash are two different datasets used to assess the suggested MNetV2-TC model respectively.
- On the Trashnet dataset, 98.87% accuracy was attained and on the VN-Trash dataset, it was even higher at 99.63%, indicating the model's outstanding performance.

1.3. Organization of Work

The remaining sections of the study are organised like shadows: Part 2 summarises relevant literature, Part 3 gives a brief description of the suggested model, Part 4 shows the analysis and validation results, and Section 5 concludes with a summary.

2. RELATED WORKS

In Lilhore et al. (2023), a clever trash classification system for sustainable development was created by combining a hybrid CNN-LSTM with transfer learning. A hybrid model that blended convolutional neural networks (CNN) and long short-term memory (LSTM) was used to separate the trash into categories for recyclables and organic materials. The model was able to predict and classify waste categories by utilising ImageNet's advantages and transfer learning (TL). In addition, an enhanced procedure for data augmentation was employed to address the problems of data sampling and overfitting. The TrashNet dataset which was divided into classes for organic waste (17,005) and recyclable waste (10,025) was the sample used for an experimental analysis aimed at assessing the model's efficacy. Tested using Python in combination with multiple existing CNN models (ResNet-34, ResNet-50, VGG-16, and AlexNet), the suggested hybrid model was evaluated using testing and training loss, accuracy, recall, precision, and other performance-measuring metrics. An adaptive moment estimator (AME) optimisation algorithm was employed to create each model across a range of epochs. With a small quantity of loss modelling for testing, validation, and training, the AME optimisation yielded the best accuracy and optimisation for the suggested approach. With a precision of 95.45%, the suggested model exhibited the highest accuracy.

Girsang et al. (2023) aimed to employ DL to enhance and expedite the waste sorting procedure. Two types of datasets, organic and inorganic, were collected and divided into three sections for the purpose of conducting the research:testing, training, and validation. Hyperparameter testing and preprocessing were used to identify the optimal learning objectives. MobileNet, VGG16, and Xception were the models employed in the investigation. Interestingly, the accuracy of the MobileNet model was 93.35%.

An independent waste classification system Transformative Vision using a Multilayer Hybrid Convolution Neural Network (VT-MLH-CNN) was introduced by Alrayes et al. (2023). Reducing classification time and increasing trash classification accuracy were the objectives of the strategy. The steps in the process included gathering data, extracting features, and normalising data. The recommended model performed better when the number of connections and network modules were changed. The outcomes of the simulation demonstrated that the recommended method had a more straightforward network model and worked better in trash categorization accuracy than some existing attempts. With a classification accuracy of up to 95.8%, the efficacy of the suggested approach was proven by multiple tests on the TrashNet dataset.

AA-ResNet, a feature extraction network's adaptive attention mechanism, was suggested by Luo and Hu (2023) for image classification. The model consisted of a pre-trained generator, a complimentary network, and a pattern-directed feature extraction network. The idea enhanced the capacity for feature representation by effectively utilising both local and global picture data. To reduce overfitting, a specially designed categorization helped train the entire model as a loss function to deal with multitasking challenges. Experiments showed that the method works well for picture classification on datasets like Caltech-101, Caltech-256, and Cifar-10.

The goal of Abu-Qdais et al. (2023) was to evaluate both conventional and deep learning models for machine learning to produce a computerised waste classification model. Trashnet datasets that were generated and made available to the models underwent testing and training on members of the public. The outcomes showed that in terms of prediction ability, the Convolutional Neural Network (CNN) for deep machine learning performed better than the standard machine learning models, like Support Vector Machine (SVM) and Random Forest (RF). Tests using a local garbage dataset combined with the Trashnet dataset produced results for SVM, RF, and CNN of 62.5%, 72.0%, and 92.7% accuracy, respectively.

For the purpose of trash identification and classification, an improved MobileNetV2 DL model was introduced by Jin et al. (2023). An attention mechanism was incorporated into the MobileNetV2 model's the initial and final convolution layers to increase the accuracy of recognition. To enhance the model's generalisation ability, transfer learning was applied with pre-trained weight parameters. Principal component analysis (PCA) lowered the final fully connected layer's dimension for real-time operation on an external gadget. outcomes of experiments, based on the "Huawei Cloud" datasets, demonstrated 30.1% model volume compression compared to the standard MobileNetV2 model, an average Raspberry Pi 4B CPU inference time of 600 ms and an accuracy of 90.7% in trash classification. A garbage sorting prototype that was created and produced to evaluate the model's performance in actual trash identification was able to achieve an average garbage classification accuracy of 89.26%.

A system for detecting junk pictures using DL was showcased by Li and Liu (2023). Dropout was used to prevent overfitting, the adagrad adaptive technique was used to debug deep neural network parameters, and ReLU was selected to address the gradient dispersion of neural network training. The method first extracted shape features from the trash images in the dataset after extracting edge, colour, and texture features. As a result, the processing of trash image data was centralised. A modified probability density function was used to classify the image based on a number of garbage-related characteristics. The result was a comprehensive garbage image identification system.

2.1. Research Gaps

The proposed MNetV2-TC model aims to fill in some research gaps, even though the previously mentioned studies have made significant progress in trash classification using DL techniques. To increase accuracy, a number of studies concentrate on hybrid models, such as Vision Transformer-based models or CNN-LSTM with transfer learning. Nonetheless, there is a dearth of focus on the particular difficulties associated with noise reduction in image pre-processing, an issue that the suggested model addresses by employing a modulated Gaussian filter. The Orca Optimisation Algorithm is also incorporated into the suggested MNetV2-TC model, offering an innovative method for hyperparameter tuning. The examination of innovative optimisation algorithms in the context of trash classification and a thorough assessment of noise reduction methods are two areas that still require research.

3. PROPOSED METHODOLOGY

Figure 1 shows the proposed MNetV2-TC model's work flow with the techniques used in this paper.

Figure 1. Block diagram

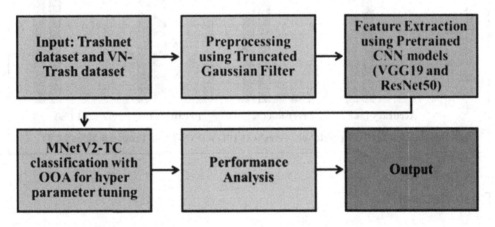

3.1. Dataset Description

A. Trashnet Dataset

The first data set is known as Trashnet (Vo, Vo, & Le, 2019), was gathered using mobile devices and comprises 2527 images divided into six categories: trash, paper, cardboard, glass, plastic, and metal. The subject of the photos was set against an artificially lit white background provided by the sun or a room. Table 1 displayed the statistics of the images for each class, and Figure 2 displays multiple samples from each class in this dataset.

Table 1. The Trashnet dataset's statistics

S. No	Classes	No. of Images
1	Glass	501
2	Metal	410
3	Cardboard	403
4	Trash	137
5	Plastic	482
6	Paper	594

Figure 2. Samples of Trashnet dataset

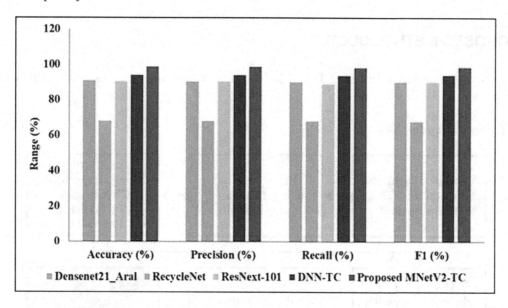

B. VN-Trash Dataset

Regarding the garbage classification issue, this study additionally gathers another dataset known as the VN-trash dataset. Three categories of wastes are included in this dataset: medical, inorganic, and organic wastes. In addition to taking pictures outside, the 5904 photos in this dataset were gathered through web crawling. Table 2 displays the VN-trash dataset's statistics and description. Also, Figure 3 shows multiple samples from each class in VNtrash. The experimental datasets, which include the Trashnet and VN-trash datasets, are split into 60%, 20%, and 20% sections in this section for the training, testing sets, and validation, as well, to allow both datasets to be used in the same experimental setting. Both of the experimental datasets' respective numbers of images, training, validation, and testing sets is shown in Table 3.

Table 2. The VN-trash dataset's statistics

S. No	Class	Objects	No. of Images
1	Organic	Juice peel, plants, paper, cardboard, seeds, etc...	2015
2	Inorganic	Bone, eggshell, glass, plastic, etc...	2003
3	Medical	Swabs, scalpels, tissue, etc...	1886

Figure 3. Samples of VN-trash dataset

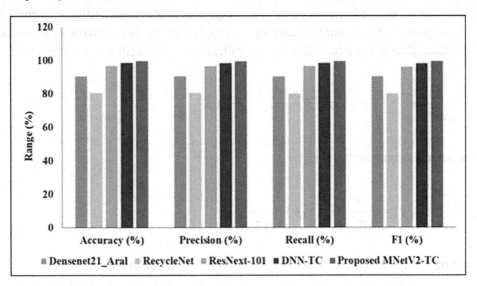

Table 3. No of images across two datasets in training, validation, and testing sets

No. of Images	VN-Trash Dataset	Trashnet Dataset
Training set	3772	1516
Validation set	1061	506
Testing set	1071	505

3.2. Preprocessing Using Truncated Gaussian Filter

Gaussian half-filters for image analysis are reliable and easy to use. Even in situations when there is a great deal of visual noise, these directed filters operate brilliantly because they precisely reduce noise and identify contours by pointing the pixel surrounding each one in the correct direction. A common and extensively used filter type that is comparable to, if not directly inspired by, orientated half-filters is called a "steerable filter." The 2D Gaussian used in these screens is full and isotropic, meaning that the gradient's amplitude is determined by the energy in the filter's maximal response direction (Magnier & Hayat, 2023). Freeman and Adelson's research indicates that the basic isotropic derivatives of the Gaussian with regard to the x and y directions can be linearly combined to create the 2D directional Gaussian first derivative $G_{\sigma,\theta,}$ angled steering θ.

$$G_{\sigma,\theta}(x,y) = cos(\theta) \cdot \frac{\partial G_\sigma}{\partial x}(x,y) + sin(\theta) \cdot \frac{\partial G_\sigma}{\partial y}(x,y), \tag{1}$$

where the coordinates of the pixel are indicated as (x,y), and σ symbolises the Gaussian standard deviation G_σ.

Frequently, in digital pictures, more than one contour will pass through a single pixel. Take a pixel at the intersection of multiple directions at a corner, for example. Half-filters are an efficient way to estimate these directions. These filters are useful and effective when used in conjunction with other image processing methods for tasks like corner detection, image restoration using partial differential equations (PDE), and descriptors. Mehrotra and Nichani proposed applying these truncated filters' responses directly to corner detection. Examining a two-dimensional isotropic Gaussian filter G the standard deviation σ:

$$G_\sigma(x,y) = \frac{1}{2\pi\sigma^2} \cdot e^{-\frac{x^2+y^2}{2\sigma^2}}, \; with \; \sigma \in R_+^* \; and \; (x,y) \in R^2, \tag{2}$$

its initial derivative is computed using:

$$G_\sigma'(x,y) = \frac{\partial G_\sigma}{\partial x}(x,y) = -\frac{x}{\sigma^2} \cdot G_\sigma(x,y) = \frac{-x}{2\pi\sigma^4} \cdot e^{-\frac{x^2+y^2}{2\sigma^2}}. \tag{3}$$

Consequently, the truncated Gaussian derivative HG' can be written as:

$$HG_\sigma'(x,y) = H(y) \cdot \left(-\frac{x}{2\pi\sigma^4} \cdot e^{-\frac{x^2+y^2}{2\sigma^2}} \right) \tag{4}$$

where H signifies the step function of Heaviside:

$$H(s) = \{1, \; if \; s>0; \; 0, \; elsewhere. \tag{5}$$

3.3. Pre-Trained CNN for Feature Extraction

This section introduces feature extraction using a CNN that has already been trained. Three layers make up CNNs: fully connected, pooling, and convolutional layers. The convolutional and pooling layers are the most important layers. To extract features, a convolution layer convolves a section of the image with multiple filters. The features in an input image can be interpreted by a CNN more precisely because of its higher layer count. The convolution's output mapping is compressed by the pooling layer (Ali et al., 2022). VGG19 and ResNet50 were the study employed two pre-trained networks. This paper uses two pre-trained models, which are explained in the following section.

3.3.1. VGG19

A convolutional neural network with 19 layers is called the VGG neural network. In 2014, Simonyan and Zisserman trained and developed it at Oxford University. Their research paper "Very Deep CNN for Large-Scale Image Recognition" from 2015 contains all the information. The VGG19 network was additionally trained with over One million pictures from the ImageNet database. Of course, you could import the model from ImageNet that includes training weights. Its pre-trained network can classify up to 1000 items. Bright images with a 224 x 224 pixel resolution were used to train the network, as Figure 4 illustrates.

Figure 4. VGG19 architecture

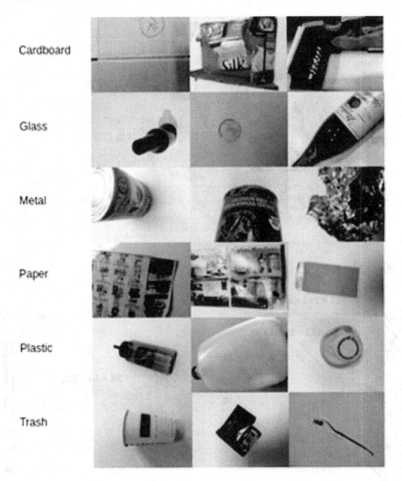

3.3.2. ResNet50

Convolutional neural network with 50 layers is called ResNet50. Similar to VGG19, it was trained on 224x224 pixel coloured images and is capable of classifying up to 1000 objects. Furthermore, the over one million images in the ImageNet collection were used to train this model. The model was created and trained by Microsoft in 2015, and the results of the model's performance are published in "Deep Residual Learning for Image Recognition."

A residual block from a ResNet is shown in Figure 5. By establishing short cuts that perform identity mapping, stack layers perform residual network mapping, as seen in the figure (x). The residual function of the stacked layers F is increased by their outputs (x).

When the deep network was being trained via backpropagation, the shallow layers were affected by an error gradient that was found. As one went deeper into the levels, this error got smaller and smaller until it finally disappeared. In really deep networks, The term "gradient vanishing problem" describes this phenomenon. Relative learning can be used to solve the issue, as shown in Figures 5 and 6.

Figure 5. Residual block of ResNet

Figure 6. Initial residual unit

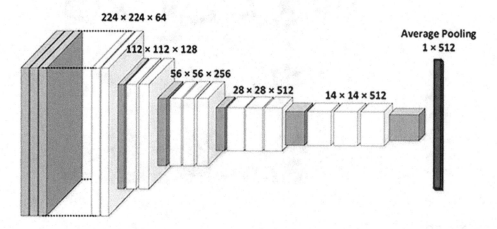

Figure 6 displays the initial residual branch, also known as unit l, in the residual network. The figure shows weights, batch normalisation (*BN*), and a corrected linear unit (*ReLU*). To calculate a residual unit's input and output, the following formulas were applied:

$$y_l = h(x_l) + F(x_l + W_l) \tag{6}$$

$$x_{l+1} = f(y_l) \tag{7}$$

where $h(x_l)$ symbolises the mapping of identities, F symbolises the residual function, x_l symbolises the source, and W_l symbolises the coefficient of weight. The mapping of identities, indicated by $h(x_l) = x_l$, is the ResNet architecture's cornerstone. The networks for which the residual networks were created with 34, 50, 101, and 152 layer counts. In this investigation, ResNet50 was used. There are fifty layers in the network.

3.4. Image Classification With MobileNetV2

To address the classification problem, MobileNetV2, deep Neural Network technology has been applied. TensorFlow was used to load ImageNet's pretrained weights. Subsequently, the base layers are defrosted to prevent any damage to previously acquired features (Nagrath et al., 2021). Subsequently, incorporating new trainable layers, These strata are instructed using the gathered dataset to ascertain the characteristics required for categorising trash photos. The weights are then saved after the model has been adjusted. By using pre-trained models, pre-biased weights can be utilised without erasing previously acquired features and avoid needless computational costs.

3.4.1. Building Blocks of MobileNetV2

Convolutional neural networks serve as the foundation for MobileNetV2, a DL model that utilises the subsequent layers and features.

A. Convolutional Layer

The core component of a convolutional neural network is this layer. A third function is created by convolution, which is the mathematical fusion of two functions. In order to make feature extraction from images easier, it uses a sliding window mechanism. Feature map creation is made easier as a result. The output C is obtained by convolutioning two functional matrices, namely the convolutional kernel B and the input image matrix as:

$$C(T) = (A * B)(x) = \int_{-\infty}^{\infty} A(T) \times B(T - x) dT \tag{8}$$

Figure 7 illustrates how the two matrices are convolutionally operated.

B. Pooling Layer

Utilising the pooling operations enables a reduction in computations faster by reducing the input matrix's size without compromising a lot of features. It is possible to use a variety of pooling operations, some of which are listed below:

i.) Max Pooling: It employs the highest value found in the chosen area where the kernel is at that moment to determine the value of the cell's matrix output.

ii.) Average Pooling: All values currently within the current region of the kernel are averaged to determine the output for that cell's matrix value. The average-pooling operation is depicted in Figure 8.

Figure 7. Convolutional operation

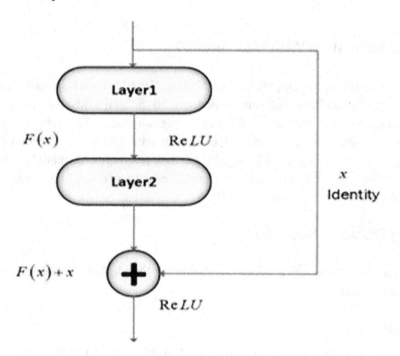

Figure 8. Standard-pooling procedure

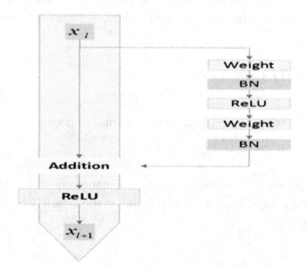

C. Dropout Layer

By eliminating arbitrary biassed neurons in the model, this lessens the likelihood of overfitting during training. Both visible and hidden layers may contain these neurons. A neuron's probability of the dropout ratio can be changed to change who gets dropped.

D. Non-Linear Layer

Typically, these layers follow convolutional layers in order. The non-linear functions most frequently utilised are the sigmoid function, tanh functions, and different types of Rectified Linear Units (ReLU), such as Leaky, Noisy, Exponential, and so forth. The following list contains several types of non-linear functions and the formulas for them.

Sigmoid: $\sigma(x) = \dfrac{1}{1 + e^{-x}}$ (9)

Leaky Relu: $f(x) = max(0.1x, x)$ (10)

Tanh: $f(x) = tanh(x)$ (11)

Maxout: $f(x) = max\left(w_1^T x + b_1, w_2^T x + b_2\right)$ (12)

Relu: $f(x) = max(0, x)$ (13)

ELU: $f(x) = \{xx \geq 0\alpha(ex - 1)x < 0$ (14)

E. Fully Connected Layer

These layers fully connect to activations layers and are appended to the model. These layers aid in binary or multi-class classification of the provided images. One activation function that is utilised in these layers is called SoftMax, and it yields the probability result for the anticipated output classes.

F. Linear Bottlenecks

Since it is not possible to reduce multiple matrix multiplications to just one numerical calculation, neural networks apply ReLU6 and other non-linear activation functions, which enable the easy removal of multiple discrepancies. This allows for the multilayer neural network's construction. Considering that Values less than 0 are rejected by the ReLU activation function. The measurements of the network expand as more channels are added to compensate for information loss. The block's layers are compressed to create a residual block that has been reversed, and the opposite procedure is completed as previously mentioned. For the skip connections, this is carried out at the linkage point; it might affect network functionality. The linear bottleneck concept was introduced as a solution to this, which provides a linear output for the final convolution of the leftover block prior to adding it to the initial activation.

3.4.2. Hyper Parameter Tuning Using OOA

In this work, OOA is used to perform hyperparameter tuning in the MNetV2-TC model.

A. Orcas in Nature

A cunning strategy employed by certain Antarctic orca species drives seals away from patches of floating ice by the animals' ability to hunt in packs and generate massive waves (Golilarz et al., 2020). When this unusual performance was initially noticed in the late 1970s, it was thought to be an isolated incident. Since then, numerous comparable observations have been reported by marine researchers (Baswaraju et al., 2023). This is repeated by the orcas in various directions when the seal jumps off the ice and into the sea.

B. OOA

The ice floe is washed by orcas using varying energies and directions in the Orca algorithm. Each orca's energy determines whether or not It has the ability to cleanse the ice floe the seal (X_- s). The ice floe's area can be decreased by the orca by washing it. Orcas are positioned in the radius R-delineated area and are oriented randomly during this process (Thirumalraj et al., 2023). When orcas approach an ice floe from a distance, they move more quickly and exert more energy in washing it. However, if an orca collides with an ice floe, they lose less energy.

The orca can strike the ice floe farther away and approach the seal with greater force. The orcas' energy was utilised to calculate their new location in order to simulate this behaviour. With the suggested algorithm, The following formulas can be used to calculate each orca's energy:

$$e_i = F_i - F_s \tag{15}$$

Where, F_i and F_s indicate the fitness of X_i and X_s respectively. Next, the energy vector is sorted $[e]_{N \times 1}$}, and identify the most and least energetic orcas. It uses the following equation to calculate the normalised energy value by sorting the energy vector:

$$E_i = \frac{e_i - e_{min}}{e_{max} - e_{min}} \tag{16}$$

The orca approaches the ice floe and lies in the area that is highlighted, halfway between the two circles. The radius of the first circle is R, and the radius of the second circle is $(R - d)$:

$$d_i = E_i \times R \tag{17}$$

Every iteration will see the removal of P percent of the orca population, and random generators will be generated in the search space (Khang, 2023). The algorithm is able to break free from local minima thanks to this process. Algorithm 1 shows the pseudo code for the suggested algorithm.

Algorithm 1. The proposed OOA algorithm's pseudocode.

```
Create the orca population from scratch (N+1) in the search area at random
Define the initial radius (R = R₀) of ice floe
Determine each orca's fitness level and arrange them
while (the stopping requirement is not met) do
●          • assign the best response as the Seal (Xₛ)
●          Compute the normalized energy of each orca (Eᵢ) by Eq. (15) and (16)
●          Using their energy in the designated area, direct the orcas towards
the seal and ice floe by Eq. (17)
●          To create the next iteration, remove the worst orcas (P%) and gener-
ate them randomly.
end while
```

4. RESULTS AND DISCUSSIONS

4.1. Experimental Setup

The Pytorch framework is a free Python DL library, was used to conduct the experimental procedures, which were implemented in Python 3.7. With an Intel Core i7-4790K (4.0 GHz × 8 cores), 16 GB of RAM, and a GeForce GTX 1080 graphics card, the operating system is Ubuntu 16.04 LTS.

4.2. Performance Metrics

The four primary analytical metrics that were developed to evaluate the built-in classification system's effectiveness using the ILPD dataset were false positive (FP), false negative (TN), true positive (TP), and false negative (FN).

The efficacy of a classification model is evaluated using the ratio of accurate assumptions to all assumptions made (ACC):

$$Accuracy = \frac{TP + TN}{TP + FP + TN + FN} \tag{18}$$

Precision (PR), measures how many positively detected examples there are compared to all positive examples:

$$Precision = \frac{TP}{TP + FP} \tag{19}$$

Out of every positive instance, the percentage of cases classified as positively is known as the Recall (RC) or sensitivity.

$$Recall = \frac{TP}{TP + FP} \tag{20}$$

An integrated metric that combines PR and RC is represented by a single numerical value, the F1-score (F1):

$$F1 = \frac{Precision * Recall}{Precision + Recall} \tag{21}$$

Table 4. Performance analysis of different classes of two experimental datasets

Datasets	Classes	ACC (%)	PR (%)	RC (%)	F1 (%)
Trashnet Dataset	Glass	97.56	97.42	97.34	97.25
	Paper	97.88	97.65	97.09	97.38
	Cardboard	98.23	98.18	97.73	98.01
	Plastic	98.76	97.45	98.34	98.63
	Metal	98.87	98.68	97.72	97.42
	Trash	98.59	97.59	98.81	97.66
VN-Trash Dataset	Organic	97.63	98.79	99.29	98.05
	Inorganic	98.24	99.35	97.92	99.27
	Medical	99.63	98.14	98.45	98.22

Table 4 and Figure 9 Trashnet and VN-Trash dataset evaluation results are shown below. The Trashnet dataset's classification results for the different classes are shown below: Glass obtained 97.56% ACC, 97.42% PR, 97.34% RC, and 97.25% F1 score. The paper displayed a 97.88% ACC, a 97.65% PR, a 97.09% RC, and a 97.38% F1 score. Cardboard showed 98.23% ACC, 98.18% PR, 97.73% RC, and a 98.01% F1 score. With an F1 score of 98.63%, RC of 98.34%, ACC of 98.76%, and PR of 97.45%, Plastic performed exceptionally well. Metal demonstrated 98.87% ACC, 98.68% PR, 97.72% RC, and a 97.42% F1 score. Last but not least, Trash produced results including 98.59% ACC, 97.59% PR, 98.81% RC, and 97.66% F1 score. Now let's look at the VN-Trash dataset. Here, Organic showed 97.63% ACC, 98.79% PR, 99.29% RC, and 98.05% F1 score. With a 99.27% F1 score, 98.24% ACC, 99.35% PR, and 97.92% RC, Inorganic performed the best. With a 99.63% ACC rate, 98.14% PR rate, 98.45% RC rate, and 98.22% F1 score, Medical showed the highest performance. These metrics shed light on how well the corresponding datasets perform in terms of classification across various classes.

The performance metrics for various models of Trashnet dataset are presented in table 5 and figure 10. The Densenet21_Aral model achieved an ACC of 91.01%, PR of 90.42%, RC of 90.23%, and an F1 score of 90.13%. The RecycleNet model demonstrated an ACC of 68.23%, PR of 68.12%, RC of 68.02%, and an F1 score of 67.88%. The ResNext-101 model exhibited an ACC of 90.42%, PR of 90.31%, RC of 89.23%, and an F1 score of 90.23%. The DNN-TC model showcased high performance with an ACC of 94.12%, PR of 94.09%, RC of 93.86%, and an F1 score of 94.11%. With an F1 score of 98.54%, an ACC of 98.87%, a PR of 98.68%, and an RC of 98.34%, the Proposed MNetV2-TC model outperformed the other models. These metrics give a thorough summary of the classification performance of each model, with the Proposed MNetV2-TC model outperforming all other models in each and every statistic evaluated.

Figure 9. Analysis of various classes of both datasets

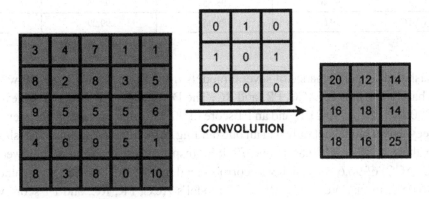

Table 5. Trashnet dataset result analysis

Models	ACC (%)	PR (%)	RC (%)	F1 (%)
Densenet21_Aral	91.01	90.42	90.23	90.13
RecycleNet	68.23	68.12	68.02	67.88
ResNext-101	90.42	90.31	89.23	90.23
DNN-TC	94.12	94.09	93.86	94.11
Proposed MNetV2-TC	98.87	98.68	98.34	98.54

Figure 10. Trashnet dataset graphical validation

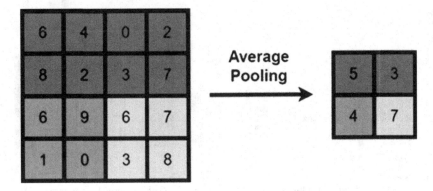

Table 6. VN-trash dataset result analysis

Models	ACC (%)	PR (%)	RC (%)	F1 (%)
Densenet21_Aral	90.45	90.34	90.21	90.08
RecycleNet	80.54	80.46	79.86	79.98
ResNext-101	96.63	96.54	96.42	96.05
DNN-TC	98.46	98.35	98.26	98.08
Proposed MNetV2-TC	99.63	99.35	99.29	99.27

The VN-Trash dataset performance of several models from Table 6 and Figure 11, with an analysis of each model based on F1 score, ACC, PR, and RC. The Densenet21_Aral model achieved an ACC of 90.45%, PR of 90.34%, RC of 90.21%, and an F1 score of 90.08%, demonstrating excellent performance overall. The RecycleNet model not only had an outstanding ACC of 80.54%, but it also showed PR, RC, and F1 score values of 80.46%, 79.86%, and 79.98%, respectively. ResNext-101 showed a high classification ACC (ACC) of 96.63%, which was corroborated by PR, RC, and F1 score values of 96.54%, 96.42%, and 96.05%, in that order. The DNN-TC model's ACC, PR, RC, and F1 score values are reported as 98.46%, 98.35%, and 98.08%, in that order. Finally, the Proposed MNetV2-TC model yielded remarkable results with an ACC of 99.63%, PR of 99.35%, RC of 99.29%, and an F1 score of 99.27%. Together, these metrics provide a thorough evaluation of the classification performance of each model, emphasising the Proposed MNetV2-TC model's outstanding results across the board. When OOA is used in MNetV2-TC for hyperparameter tweaking, the recommended model's ACC rate is higher than that of other models.

Figure 11. Graphical analysis of VN-trash dataset

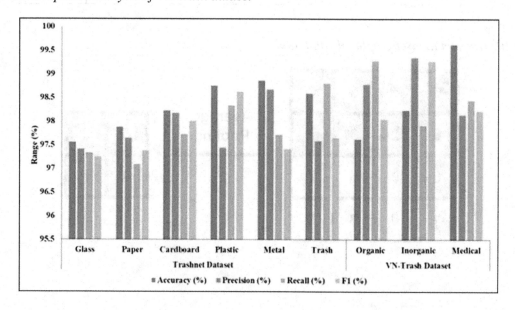

5. CONCLUSION

In summary, this research highlights the detrimental consequences of waste accumulation on the environment, animals, and human health while addressing the crucial problem of ineffective waste management systems in urban settings. Acknowledging the obstacles presented by the expanding population and the constraints of conventional waste management, the study suggests a unique strategy that integrates diverse sensors and DL models to optimise the use of waste resources. During the experimentation phase, datasets such as The VN-trash dataset contains photos of organic, inorganic, and medical waste. from Vietnam, and the Trashnet dataset are gathered and preprocessed. Pre-trained DL models—VGG19 and Resnet20—are used to extract features. Next, a MobileNetV2 model for trash classification called MNetV2-TC is presented in the study. It uses the Orca optimisation algorithm (OOA) to adjust its hyperparameters. Experiments conducted on the VN-trash dataset show that MNetV2-TC outperforms cutting-edge techniques in trash classification. The suggested model performs better predictably when the OOA is used to adjust the hyperparameters. This is a positive step towards trash classification systems that are more precise and effective. The suggested MNetV2-TC model displays impressive results on two different datasets: the Trashnet dataset shows an accuracy of 98.87%, and the VN-Trash dataset shows an even higher accuracy of 99.63%. In order to evaluate MNetV2-TC's robustness and generalizability, it is essential to expand its evaluation to a variety of datasets and real-world scenarios in future work.

REFERENCES

Abu-Qdais, H., Shatnawi, N., & Esra'a, A. A. (2023). *Intelligent solid waste classification system using combination of image processing and machine learning models.* Academic Press.

Ali, M. A., & PP, F. RSalama Abd Elminaam, D. (2022). A Feature Selection Based on Improved Artificial Hummingbird Algorithm Using Random Opposition-Based Learning for Solving Waste Classification Problem. *Mathematics, 10*(15), 2675. doi:10.3390/math10152675

Alrayes, F. S., Asiri, M. M., Maashi, M. S., Nour, M. K., Rizwanullah, M., Osman, A. E., Drar, S., & Zamani, A. S. (2023). Waste classification using vision transformer based on multilayer hybrid convolution neural network. *Urban Climate, 49*, 101483. doi:10.1016/j.uclim.2023.101483

Baswaraju, S., Maheswari, V. U., Chennam, K. K., Thirumalraj, A., Kantipudi, M. P., & Aluvalu, R. (2023). Future Food Production Prediction Using AROA Based Hybrid Deep Learning Model in Agri-Sector. *Human-Centric Intelligent Systems*, 1-16.

Girsang, A. S., Pratama, H., & Santo Agustinus, L. P. (2023). Classification Organic and Inorganic Waste with Convolutional Neural Network Using Deep Learning. *International Journal of Intelligent Systems and Applications in Engineering, 11*(2), 343–348.

Golilarz, N. A., Gao, H., Addeh, A., & Pirasteh, S. (2020, December). ORCA optimization algorithm: a new meta-heuristic tool for complex optimization problems. In *2020 17th International Computer Conference on Wavelet Active Media Technology and Information Processing (ICCWAMTIP)* (pp. 198-204). IEEE. 10.1109/ICCWAMTIP51612.2020.9317473

Gupta, H. (2020). *Trash Image Classification System using Machine Learning and Deep Learning Algorithms* [Doctoral dissertation]. National College of Ireland.

Gupta, T., Joshi, R., Mukhopadhyay, D., Sachdeva, K., Jain, N., Virmani, D., & Garcia-Hernandez, L. (2022). A deep learning approach based hardware solution to categorise garbage in environment. *Complex & Intelligent Systems*, *8*(2), 1–24. doi:10.1007/s40747-021-00529-0

He, Y., Gu, Q., & Shi, M. (2020). *Trash Classification Using Convolutional Neural Networks Project Category: Computer Vision*. Academic Press.

Jin, S., Yang, Z., Królczykg, G., Liu, X., Gardoni, P., & Li, Z. (2023). Garbage detection and classification using a new deep learning-based machine vision system as a tool for sustainable waste recycling. *Waste Management (New York, N.Y.)*, *162*, 123–130. doi:10.1016/j.wasman.2023.02.014 PMID:36989995

Khang, A. (Ed.). (2023). *AI and IoT-Based Technologies for Precision Medicine*. IGI Global. doi:10.4018/979-8-3693-0876-9

Li, Y., & Liu, W. (2023). Deep learning-based garbage image recognition algorithm. *Applied Nanoscience*, *13*(2), 1415–1424. doi:10.1007/s13204-021-02068-z

Lilhore, U. K., Simaiya, S., Dalal, S., & Damaševičius, R. (2023). A smart waste classification model using hybrid CNN-LSTM with transfer learning for sustainable environment. *Multimedia Tools and Applications*, 1–25. doi:10.1007/s11042-023-16677-z

Luo, J., & Hu, D. (2023). An Image Classification Method Based on Adaptive Attention Mechanism and Feature Extraction Network. *Computational Intelligence and Neuroscience*, *2023*, 2023. doi:10.1155/2023/4305594 PMID:36844695

Magnier, B., & Hayat, K. (2023). Revisiting Mehrotra and Nichani's Corner Detection Method for Improvement with Truncated Anisotropic Gaussian Filtering. *Sensors (Basel)*, *23*(20), 8653. doi:10.3390/s23208653 PMID:37896745

Mao, W. L., Chen, W. C., Wang, C. T., & Lin, Y. H. (2021). Recycling waste classification using optimized convolutional neural network. *Resources, Conservation and Recycling*, *164*, 105132. doi:10.1016/j.resconrec.2020.105132

Masand, A., Chauhan, S., Jangid, M., Kumar, R., & Roy, S. (2021). Scrapnet: An efficient approach to trash classification. *IEEE Access: Practical Innovations, Open Solutions*, *9*, 130947–130958. doi:10.1109/ACCESS.2021.3111230

Mythili, T., & Anbarasi, A. (2022). A concatenation of deep and texture features for medicinal trash image classification using EnSegNet-DNN-based transfer learning. *Materials Today: Proceedings*, *62*, 4691–4698. doi:10.1016/j.matpr.2022.03.129

Nagrath, P., Jain, R., Madan, A., Arora, R., Kataria, P., & Hemanth, J. (2021). SSDMNV2: A real time DNN-based face mask detection system using single shot multibox detector and MobileNetV2. *Sustainable Cities and Society*, *66*, 102692. doi:10.1016/j.scs.2020.102692 PMID:33425664

Ozkaya, U., & Seyfi, L. (2019). Fine-tuning models comparisons on garbage classification for recyclability. *arXiv preprint arXiv:1908.04393*.

Thirumalraj, A., Asha, V., & Kavin, B. P. (2023). An Improved Hunter-Prey Optimizer-Based DenseNet Model for Classification of Hyper-Spectral Images. In *AI and IoT-Based Technologies for Precision Medicine* (pp. 76–96). IGI Global. doi:10.4018/979-8-3693-0876-9.ch005

Tiyajamorn, P., Lorprasertkul, P., Assabumrungrat, R., Poomarin, W., & Chancharoen, R. (2019, November). Automatic trash classification using convolutional neural network machine learning. In *2019 IEEE International Conference on Cybernetics and Intelligent Systems (CIS) and IEEE Conference on Robotics, Automation and Mechatronics (RAM)* (pp. 71-76). IEEE. 10.1109/CIS-RAM47153.2019.9095775

Vo, A. H., Vo, M. T., & Le, T. (2019). A novel framework for trash classification using deep transfer learning. *IEEE Access : Practical Innovations, Open Solutions*, 7, 178631–178639. doi:10.1109/ACCESS.2019.2959033

Yang, H. Q., Lu, T., Guo, W. J., Chang, S., & Song, J. F. (2020, January). RecycleTrashNet: Strengthening Training Efficiency for Trash Classification Via Composite Pooling. In *Proceedings of the 2020 2nd International Conference on Advanced Control, Automation and Artificial Intelligence (ACAAI 2020), Wuhan, China* (pp. 11-13). 10.12783/dtetr/acaai2020/34190

Yu, Y. (2020, August). A computer vision based detection system for trash bins identification during trash classification. *Journal of Physics: Conference Series*, 1617(1), 012015. doi:10.1088/1742-6596/1617/1/012015

Zhang, Q., Yang, Q., Zhang, X., Bao, Q., Su, J., & Liu, X. (2021). Waste image classification based on transfer learning and convolutional neural network. *Waste Management (New York, N.Y.)*, 135, 150–157. doi:10.1016/j.wasman.2021.08.038 PMID:34509053

Zhang, Q., Zhang, X., Mu, X., Wang, Z., Tian, R., Wang, X., & Liu, X. (2021). Recyclable waste image recognition based on deep learning. *Resources, Conservation and Recycling*, 171, 105636. doi:10.1016/j.resconrec.2021.105636

Chapter 13
Computational Intelligence for Green Cloud Computing and Digital Waste Management

Sana Dahmani
Independent Researcher, Germany

ABSTRACT

The intersection of environmental protection, Sustainable Development Goals, and the role of information technology (IT) is to foster it. Artificial intelligence addresses environmental challenges, offering solutions such as emissions reduction, cost savings, legal compliance, HR attraction, optimized investments, and waste management. The "three Rs" of green IT emphasize sustainable hardware management through reuse, refurbishment, and recycling. Green IT aligns with broad sustainability and ESG standards and is fostered with innovative AI solutions to integrate and optimize AI-based green computing algorithms which address environmental impacts and data utilization strategies, optimizing energy consumption, and mitigating digital waste and carbon footprint. The conclusion advocates increased use of AI-built technologies for the adoption of renewable energy sources, energy-efficient hardware, hardware optimization, and exploration of external cloud solutions for a more sustainable future.

I. INTRODUCTION

Digital transformation acts as a stimulus for sustainability through the promotion of innovation, enhancement of resource efficiency and enablement of eco-conscious decision-making. Integrating artificial intelligence (AI) and machine learning (ML) facilitates data analysis for the purpose of identifying inefficiencies and optimizing resource deployment, and reducing energy consumption within buildings, factories, and transportation. Digital platforms support circular economy principles through the minimization of waste through product reuse and sharing. Blockchain enhances supply chain transparency for ethical sourcing, a better ebergy efficiency and the development of smart cities (Parmentola et al., 2021). Digital tools empower individuals to make environmentally conscious choices and suypport organiza-

DOI: 10.4018/979-8-3693-1552-1.ch013

tions to provide optimal sustainable solutions. Smart grid technologies optimize energy distribution and integrate renewable sources, thus, reducing dependancy on fossil fuels. Precision agriculture, enabled by sensors and AI, minimizes resource use through the real-time environmental monitoring which also provides crucial data for informed decision-making, and smart city initiatives optimize urban planning for reduced environmental footprints. To sum up, digital transformation contributes to sustainable development, harmonizing technological advancement with environmental stewardship.

According to (UNDP, 2023), digital transformation hastens sustainable and inclusive development. To harness the immense hidden and identified potential of digital technologies for development while ensuring inclusivity and sustainability, a conscious endeavor must be prepared to guide technology development and national digital transformation. The United Nations Development Programme (UNDP) advocates for an inclusive digital transformation which addresses the needs of the most vulnerable groups, covering the destitute, women, and people with disabilities. The core goal is to prevent the exacerbation of existing inequalities and empower underrepresented groups, fostering meaningful participation and gender equality. Furthermore, the emphasis is on safeguarding individuals from the adverse effects of digital technologies and promoting the use of responsible, rights-based, and open digital technology. Drawing on over a decade of experience in supporting developing countries with digital transformation, the UNDP has responded to the accelerated digitalization prompted by the COVID-19 pandemic. Employing a whole-of-society, inclusive, and rights-based approach, the UNDP advises countries on digital strategies, governance frameworks, and the implementation of digital public infrastructure. Insights gained from country-level digital interventions highlight three key ways in which digital transformation propels sustainable and inclusive development: enabling economic development, building resilience, and facilitating climate adaptation. In the contemporary technological landscape, IT/OT integration and digital transformation are pivotal paradigms that converge Information Technology (IT) and Operational Technology (OT). Historically functioning in isolation, IT and OT integration aims to dismantle silos, fostering connectivity to enhance communication, streamline processes, and unlock the full potential of data and technologies. Digital transformation involves leveraging various technologies to reshape business processes, organizational structures, and customer experiences, achieving efficiency, innovation, and growth. Key sub-topics include data-driven decision-making, customer experience transformation, process automation, business model innovation, digital culture and leadership, cybersecurity, digital talent development, and the integration of emerging technologies. These interconnected elements collectively shape organizations' trajectory in the ongoing digital transformation journey. Digital transformation and cloud computing, influential forces in today's business landscape, involve leveraging digital technologies to fundamentally alter business operations and provide value to customers. Cloud computing, delivering various computing services over the Internet on a pay-as-you-go basis, enhances agility, innovation, and scalability, reduces costs, improves collaboration, and offers robust security measures. This integration fosters an improved customer experience by providing global accessibility to data and applications. Various strategies including cloud-first, hybrid cloud, multi-cloud, and cloud-native, can be adopted based on specific business needs and goals. The implementation of digital transformation and cloud computing involves assessing the current IT infrastructure, developing a cloud strategy, selecting a well-fitting provider, migrating IT infrastructure to the cloud, and managing the cloud environment to ensure security, reliability, and scalability. While the process is complex, successfully embracing these transformations, appoints businesses for future success (Andhra Pradesh Grama Sachivalayam, 2019). The concept of the "Green Cloud" encapsulates the prospective environmental benefits offered by cloud computing, a paradigm that migrates IT services to the internet. Anticipated benefits include a notable 38%

reduction in data center energy consumption by 2020, achieved through measures such as consolidation, optimizing power usage efficiency, advancing recycling practices, minimizing emissions, and reducing water usage for cooling. Acknowledging the significant energy consumption linked to data storage, the Storage Networking Industry Association (SNIA) advocates for the adoption of energy-efficient storage technologies and architectures (Kamiya & Nagpal, 2015).

II. STATE-OF-THE-ART: GREEN COMPUTING

Green Computing is often referred to as sustainable computing or sustainable IT (Information Technology) while its wider spectrum is known as ICT (Information and Communications Technology) sustainability aimed at incorporating an environmentally conscious approach for computing and IT information technology. Green Computing encompasses the strategic development and elevation of computer chips, systems, networks, and software in order to maximize efficiency by means of the enhancement of energy utilization of optimal energy usage and detrimental environmental impacts. Green Computing has been a focus point of multiple researchers. As a matter of fact, its target consists in mitigating the negative impact generated by technology, hardware and software by leveraging the optimization and scale up of the green computing potential and aptitudes of technologies. It strives at minimizing the technology's environmental footprint and augment energy efficiency along the supply chain ranging from cradle to grave, i.e. from raw material to end of life management. The core focus areas of green computing initiatives embody the energy efficiency optimization for hardware and software, energy-intensive data centers, the sustainable materials usage and electronics manufacturing process, the integration of circular economy standards highlighted by material recycling and electronic device lifespan prolongation and the promotion of company sustainability reporting transparency. Within a dynamic green computing landscape, this IT responsible focus is poised to play a crucial role in the mitigation of the environmental impact of technology industry especially with the continuous growth in computing power and both data consumption and generation. This green evolution aimed at fostering a more sustainable computing ecosystem mandates ongoing innovations in hardware, software, and data management (NVIDIA, 2022).

What Is Digital Waste?

In this work, we break down the Digital waste into 1. Data waste and 2. E-waste, short for electronic waste. Data waste covers superfluous online data, which range from outdated photos and unessential messages to advertising emails and internet downloads. The latter collaborative web-based activity contributes substantially to environmental pollution, through the emission which contribute significantly to environmental pollution, emitting significant amounts of CO_2 deemed to be harmful for the planet. In the same context, we define the digital carbon / CO_2 footprint emerging with the continuous progress and spread of high-technology industries. Indeed, as of October 2023 the Internet users worldwide were estimated to 5.3 billion covering 65.7% of the global population and of this total, 61.4% are classified as social media users (Ani Petrosyan, 2023). Moreover, the number of connected Internet of Things (IoT) devices worldwide increased between 2019 and 2023 from 8.6 to 13.14 billion and is prone to doubling from 15.1 billion in 2020 to more than 29 billion IoT devices in the horizons of 2030 (Vailshery, 2023). Data storage increases the demand for data centers, indeed, their number has increased. A data center inventory was published by CBRE Group Inc (CBRE) and showed an increase in megawatts (MW) of total inventory between Q1 2022 to Q1 2023: Data center supply grew year-over-year in Frankfurt, London, Amsterdam and Paris (FLAP) as well as in North America, led by Virginia where it increased with 19.5% . The growth in Latin America-mainly throughout Brazil, Mexico, Chile and Colombia was led by Brazil with an inventory reaching 127% between 2020 and 2022. Finally, in Asia-Pacific, inventory growth was led by Tokyo, Sydney, and Singapore exceeding half a gigawatt each (CBRE, 2023).

Even if digital transformation and cutting-edge technologies contributed to the promotion of sustainability through efficient resource management, precision agriculture, supply chain transparency and smart cities etc., the data generation, use and transfer of digital devices and related infrastructure, showed a considerable digital footprint. The latter is easier to compute and track at the individual level than at larger scales, as a matter of fact, there is a lack and even absence of global statistics derived from direct measurements of the energy consumption resulting from digital activities, not even at the national scale. Their status-quo shows estimated measurements derived either through extrapolating measurements conducted on specific samples (such as a group of data centers) or through the application of models which assumptions are often not clearly disclosed. In this context, a peer reviewed study with a high level of granularity and clear calculation formula carried out by (Andrae & Edler, 2015) with a worldwide digital technology scope model. They demonstrated that considering the worldwide electricity mix, the proportion of greenhouse gas (GHG) emissions linked to the Digital era is poised to rise from 2.5% in 2013 to 4% in 2020, amounting to 2.1 gigatons. Figure 1 describes the recorded and expected GHG emissions of the digital products.

Figure 1. World digital GHG emissions of in Gigatons of CO2eq (Andrae & Edler, 2015, The Schift Project, 2019)

GHG emissions in GtCO$_2$eq	2015	2020	2025	CAGR 2015/2020	CAGR 2020/2025
Expected - 2015	1.4	1.7	2.5	4%	8%
Worst - 2015	2.3	3.6	7.6	9.4%	16%
Expected updated	1.5	2.3	3.6	9.2%	9.9%

Each individual search query, streamed video, and all types of cloud computing, when iterated billions of times, contribute to the growing global energy demand and, consequently, the high CO_2 emissions. With its big data sizes, video streaming constitutes the major contributor to the digital footprint. In comparison to them, activities such as the use of a search engine or the transfer of text-only emails have a minimal impact (The Shift Project, 2019). The growing emphasis on reducing the environmental impact of data centers and cloud services contrasts with the lack of attention given to data waste. As organizations increasingly aim to leverage data, the issue of data waste emerges as a significant concern. Five strategies are introduced in order to combat data waste while concurrently gaining better control over the corporate data landscape. The focus is on addressing the unsustainable storage practices of accumulating vast amounts of unnecessary data, which contributes to energy consumption, resource depletion, and e-waste.

(Roosa Säntti, 2022) recommends mapping the existing data landscape, assessing its environmental impact, implementing a managed lifecycle for data, avoiding unnecessary data ingestion and minimizing data duplication. By actively managing data waste, organizations not only contribute to sustainable development goals (SDGs) covering climate mitigation, resource management, industry innovation and infrastructure; responsible consumption and production; but also enhance their overall data mastery. Therefore, involving people in embracing a sustainable data vision and strategy for successful outcomes is crucial for an effective collaboration. Several approaches are recommended to further mitigate data waste and advance data master such as relying on shared data, procuring data from external providers, limiting data movement, and adopting energy-saving storage media. According to Capgemini's Research Institute, the global e-waste generated in 2019 hedged 53.6 million tons and only 17.4% of it was documented to be collected or recycled while only 89% of the organizations recycles less than 10% of their hardware waste. The International Energy Agency claimed that data centers constituted nearly 1% of the world's energy demand in 2019. All told, the technology industry is projected to have an approximate contribution of 14% of the global emissions by 2040, which represents a significant increase from the current 3%. In addition to the energy demand arising from electronic device usage, the production and disposal of these devices have significant environmental consequences. The environmental impact of manufacturing these devices is Similar to, if not surpassing, the carbon cost incurred during their usage (Evangelidis and Davies, 2021).

The surge in e-waste, driven by technological growth and consumption, poses health risks through hazardous substances. Improper disposal, including burning, harms the environment and exposes millions, especially in informal recycling, to dangers. Inadequate regulations and infrastructure, particularly in low- and middle-income countries, amplify risks. Illegal e-waste movement persists, threatening global health. Urgent efforts are needed for responsible e-waste management (WHO, 2023). The growing demand for electronic devices, driven by technology and cloud computing, contributes significantly to e-waste. Sustainable design, extended device lifespans, and efficient e-waste management are crucial for mitigating environmental and health risks. Prioritizing energy efficiency and responsible data management in cloud computing and AI operations is essential for overall sustainability. At its core, data waste refers to the missed potential of extracting value from data or overspending on the acquisition, storage, and utilization of data across various scales of systems. Table 1 describes the data waste generated at each step of the Data Management Lifecycle. Several metrics can track this data usage, and as a consequence, the waste it generates.

Global data creation soared to record-breaking levels in 2020, projected to surpass 180 zettabytes by 2025, signaling an era of unprecedented information abundance. Accelerating data growth rates, exemplified by a 29% annual increase in mobile broadband data usage in OECD countries from 2017 to 2021, highlight the expanding internet-connected population and growing reliance on digital services. Despite a growth deceleration in 2022, organizations strive to optimize data storage costs by increasing data utilization rates, measuring actively used stored data. The increasing amount of data contributes to the escalation of data storage costs including both hardware and software ; power, cooling and workforce expenses. Tackeling those issues relies on the sustaianble data lifecycle management throught its compression, alternative data destruction methods, leveraging volume discounts to lower disposed data, cloud optimization and audits conduction. Moreover, the growing environmental impact of e-waste takes upward the disposal costs.

Table 1. Data management lifecycle and data waste

Data Management Cycle		Data Waste Aspect
Data Collection	Data is collected from several various sources including sensors, transactions, manual input etc.	Collecting unnecessary or duplicate data is the cause of wasted storage space and processing resources of data within the computing system
Data Storage	Data is stored in a secure and accessible location, such as a database or cloud storage.	Storing data inefficiently through practices such as using redundant copies or outdated storage technologies is prone to trigger increased costs and make challenging data retrieval
Data Processing	Data is cleaned, transformed, and analyzed to extract meaningful insights, information and correlations through its mining	Improper data cleaning and transformation can lead to inaccurate or incomplete data, rendering it unusable, meaningless or unaccurate for analysis and interpretation
Data Usage	Data is used for various purposes such as research, decision-making or reporting	Failing to utilize data effectively or using outdated data can lead to missed opportunities and poor decision-making processes
Data Archiving	Data which is no longer actively used gets archived for long-term storage or to later serve as historical reference	Archiving data without proper metadata or suitable data retrieval mechanisms can lead to inaccessibility and loss of valuable information
Data Disposal	Data which has reached the end of its utility is disposed of safely to protect privacy and restrict data leakage	Disposing of data in insecurely can expose sensitive information and lead to environmental hazard

Recent researches pointed the environmental impact of extensive data storage practices of Big Data (Lucivero, 2022). As a matter of fact, the study showed an ethical dilemma in respect of the environmental footprint of large-scale data storage, emphasizing the limited nature of big data resources and underscoring the relevance of joint efforts from stakeholders in addressing environmental dimensions and the ethical considerations in building the future policies. AI play a pivptal role in digital waste managment through the automation of data flows throughout the data lifecycle management, the storage optimization and waste minimization through demand prediction, anomaly detection, dynamic resource allocation and unceseccary components identification.

AI enables predictive maintenance, reducing e-waste by forecasting equipment failures and extending equipment lifespan. AI-driven data analytics identifies waste patterns, monitors energy consumption, detects inefficiencies, and supports circular economy initiatives like product repair and recycling, contributing to waste reduction. According to (Nokia company, 2023), from the energy generated in power plants, only 90% makes it to the network, incurring a 10% loss during transmission. Within this remaining energy, around 80% is used by the radio access, while the rest is distributed to transport, core, and Operations support system(OSS). Passive auxiliary components, such as air conditioning and power systems, contribute to 30% of the network's energy consumption (equivalent to 35% of the original energy), leaving 70% (or 65% of the original energy) for the direct consumption by the network elements. In this context. 78% of telcos rely on AI energy solutions to optimize network energy.

III. AI SOLUTIONS FOR GREEN COMPUTING: GREEN CLOUD COMPUTING, ML, DEEP LEARNING

1. Green AI Solutions Concepts

AI is instrumental in minimizing data waste by optimizing processes like collection, storage, and processing. Algorithms filter irrelevant information, compress data, and automate tasks, freeing resources for more complex assignments. Leveraging extensive datasets, AI extracts valuable insights for informed decision-making. It ensures data governance, compliance, and quality by classifying, tracking, and rectifying errors. This maximizes data value, enabling informed, data-driven decisions for business growth and innovation.

Companies embrace green computing practices through several which aim at:

- **Providing environmental-friendly solutions:** Reduce carbon footprints and waste, and manage resources in a sustainable way.
- **The implementation of energy-efficient components**: such as central processing units (CPUs), servers, peripherals, power systems, and other IT equipment. Their commitment extends to minimizing resource consumption and adopting proper electronic waste disposal methods.
- **Incorporation of green computing into operational frameworks:** is a pivotal element of Environmental, Social, and Governance (ESG) initiatives, emphasizing the integration of sustainable and ethical business practices. This aligns with broader business sustainability endeavors that seek to establish companies on a trajectory of sustained success through conscientious corporate management and strategic approaches.
- **Enhancing energy efficiency in data centers:** is pivotal for green computing, given that these centers are substantial energy consumers, constituting about 2% of global electricity consumption. **The mitigation of the data centers** environmental footprint requires the adoption of diverse strategies to boost energy efficiency.
  - **Hardware Optimization:** Harnesses energy-efficient servers characterized with higher power supply efficiency (PSE) ratings which significantly decrease energy consumption. It virtualizes multiple servers onto a single physical one, thus, consolidating resources and decreasing energy usage. The implementation of energy-efficient storage technologies, including as solid-state drives (SSDs), supports lower power consumption.
  - **Cooling Optimization:** Uses free cooling systems in order to leverage ambient air for cooling and, thus, reduce or even eliminate the need for energy-intensive mechanical cooling. The utilization of liquid cooling systems is more efficient than traditional air cooling, as it enhances cooling performance and curtails energy consumptio through precision cooling systems whicg target cooling to specific areas optimize energy usage.
  - **Data Center Management Practices:** Implement power management software and strategies such as power capping and load balancing to optimize power usage. Utilizing Data Center Infrastructure Management (DCIM) tools to monitor and manage infrastructure helps identify and address inefficiencies. Optimizing workload placement and scheduling aligns workloads with the most energy-efficient resources to redue consumption.
  - **Renewable Energy Integration:** Install on-site renewable energy sources-such as solar panels or wind turbine for the reduction of reliance on grid electricity procure renewable energy

credits (RECs) or directly source renewable energy offsets data center carbon emissions. Power purchase agreements (PPAs) also known as electricity power agreement-a prolonged contract between customer and electricity generator- based on renewable energy providers are used to guarantee a steady power supply for data centers.

    ○ **Predictive Maintenance and Anomaly Detection:** Data collected from sensors is analyzed and harnessed for a proactive systems upkeeping and downtime reduction thus enhancing their reliability.

    ○ **Resource Optimization:** consists in optimal allocation of data centers in the strategic locations and optimized energy utilization to decrease the lowers overall energy consumption.

    ○ **Data Storage Optimization:** Thanks to AI, redundant or outdated data are identified and removed to reduce storage requirements and the interconnected energy consumption in order to optimize data management, data storage costs and enhance data security.

- **E-waste Management and Resource Recovery:** AI supports e-waste management by proficiently identifying, classifying, and sorting the different e-waste types and components. This facilitates efficient recycling and resource recovery, thus, mitigating the environmental impact of e-waste disposal and endorsing circular economy practices.

- **Smart Grid Integration and Renewable Energy Management:** AI facilitates the integration of renewable energy sources into the grid and help optimize energy distribution and storage. Smart grid management enhances the utilization of renewable energy, reduces reliance on fossil fuels and advocates for sustainable energy consumption.

- **AI-Powered Sustainability Reporting and Analytics:** AI analyzes and visualizes data related to energy consumption, resource usage, and carbon emissions. Those analytics and reporting empower businesses and organizations to monitor sustainability progress; identify improvement areas, and make informed decisions, fostering a reduced environmental impact and the accomplishment of sustainability goals.

- **Green Building Design and Optimization:** AI optimizes building design and operations to minimize energy consumption and maximize efficiency. This encompasses the optimization of HVAC systems, smart lighting control based on occupancy, and the wise use of natural daylight, contributing to environmentally conscious building practices.

- **Data Waste Management:** In the context of machine learning, the adage "Garbage in, garbage out" is a widely acknowledged principle. AI-experts are well aware that the caliber of the data significantly influences the outcome's quality. However, the term "garbage" encompasses a broad and growing spectrum within data science, including deficiently labeled or inaccurate data which mirrors inherent human biases, and incomplete datasets. To echo Tolstoy, exemplary datasets share common characteristics, whereas subpar datasets each possess their distinctive and lamentable shortcomings (S&P, 2019).

AI minimizes data waste by optimizing the complete data lifecycle covering collection, storage, processing, and decision-making. During collection, AI filters out irrelevant data, ensures efficient storage and eliminates spam and irrelevant content. Automatic classification and real-time analysis ensure only actionable information is stored for timely decisions. In storage, AI removes redundancies and reduces the footprint and predictive storage to optimize allocation. Archiving and purging identify outdated data in order to minimize storage and energy use. AI automates processing tasks, freeing resources for complex analysis. Anomaly detection technology improves data quality while summarization/visualization offers quick insights. Predictive analytics forecasts trends assess risks and enhance data-driven processes, thus, reducing waste and improving efficiency for productivity gains.

2. AI Solutions Concepts for E-Waste Management

Artificial Intelligence (AI) is pioneering transformative solutions to tackle the emerging and growing issue of e-waste including data centers and further hardware implied in the use of any AI project, identified with substantial environmental and health hazards due to hazardous components. AI-driven technologies have revamped the e-waste management processes covering collection, transportation, segregation, treatment and storage and/or recycling. AI-based interventions in e-waste management is shown in serveral aspects as shown in Figure 2.

Figure 2. How AI contributes to the sustainable e-waste management?

Automated E-waste Identification and Sorting	• AI-driven image recognition systems help identofy and classify several e-waste types including: computers, smartphones, TVs and batteries • Automated identification optimizes sorting and ensures optimal recycling or treatment facilities
Optimized E-waste Collection and Transportation	• Route optimization algorithms developed to identify the most optimal and efficient locations to minimize transportation distance, time, resource consumption and cost. • AI can predict e-waste generation patterns to enable an optimal collection scheduling and resource allocation
Precise E-waste Disassembly and Processing	• Robots help disassemble e-waste and separate valuable and hazardous materials
Material Recovery and Recycling	• AI-driven sensor systems help identify and classify materials and conduct an efficient recycling process
Automated Condition Evaluation	• AI-driven advanced image analysis has the ability of evaluating hardware and detect defects, scratches and damage which determine soundness of refurbishment through sensor data analysis
Quality Assurance and Anomaly Detection	• AI supports assessing the state of refurbished hardware quality and predict costs associated with recycling several hardware types
Predictive Maintenance of E-waste Processing Equipment	• Help monitor e-waste processing equipment performance and identify potential issues before breakages and failures occur among which downtime rediction and equipment optimization
Regulatory Compliance and Reporting	• AI aids supports enhancing compliance and mitigating envrionemntal risk by means of real-time data for analysis and regulatory reporting
E-waste Lifecycle Management Platforms	• AI-powered platforms help integrate and manage data across the e-waste lifecycle and optimization resource management
Circular Economy Applications	• AI fosters circular economy initiatives by optimizing e-waste collection, refurbishment, and reuse. AI systems help identify items suitable for repair
Consumer Education and Engagement	• Chatbots and virtual assistants provide personalized e-waste disposal guidance chatbots and virtual assistants provide personalized e-waste disposal guidance
E-waste Data Analytics and Policy Insights	• AI goes through and alayzes e-waste big data

3. AI-Based Algorithms for the Implementation of Green Solutions

AI algorithms play a pivotal role in enhancing energy efficiency across different sectors including buildings, transportation, manufacturing and energy grids. Those algorithms help conduct data analysis for trends and anomalies recogniction to mitigate environmental impact. AI algorithms support refining

HVAC systems, lighting, and further building operations. Through the inspection of real-time data, AI helps adjust temperature settings, brightness levels and ventilation to minimize consumption use while ensuring user comfort. In the field of industrial equipment, AI –analysis powered powered predictive maintenance harnesses data to anticipate potential failures, thus, effectively decreasing downtime and maintenance costs by up to 40%.

AI's role extends to optimizing smart grids, enhancing energy supply and demand management to minimize waste. It enhances grid stability and reduces energy losses by up to 10%. Furthermore, AI algorithms help optimize transportation systems by enhancing traffic flow, reducing congestion and improving fuel efficiency, showcasing potential fuel consumption reductions of up to 20%. Lastly, AI-driven analyses of household energy consumption patterns enable personalized recommendations for energy reduction, facilitating behavior change and decreasing household energy consumption by up to 15% (Utilities One, 2023). AI can support the journey towards Net Zero. In effect, (Hassan et al., 2023) quantified the correlation of AI-enhanced solutions to the three of renewable energy efficiency, emission reduction and environmental sustainability.

3.1 Building Energy Management Systems (BEMS)

Energy systems play a pivotal role in the infrastructure of buildings through the entailment of considerable expenses and exertion of a crucial influence on the prosperity of businesses and services. Achieving optimal energy utilization is made possible through the implementation of computer-based systems like Building Energy Management Systems (BEMS), which offer precise and automated control over energy-related systems. Commonly used to automate diverse functions, including energy management, Building Management Systems (BMS) establish connectivity between building components and a central campus (Sayed & Gabbar, 2018). AI algorithms, including DRL, LSTM, GANs, SVMs, Decision Trees, Ensemble Methods, and various learning techniques, significantly enhance building energy efficiency. They optimize consumption, predict demand, simulate BEMS configurations, and detect anomalies. Collaborative methods like Federated Learning and MARL improve efficiency, while techniques such as Transfer Learning and MBRL accelerate green building initiatives. DQNs, DDQNs, PPO, and DDPG optimize HVAC parameters, ensuring real-time management. HRL optimizes energy management at various levels, and FRL addresses data privacy for collaborative optimization.

3.2 Smart Grid Optimization

- **Demand Forecasting:** is based on algorithms which predict future energy demand based on historical data, weather patterns etc. and helps meeting power demand without oversupplying or undersupplying the grid. Various AI algorithms have been developed for different applications, including artificial neural networks (ANN) and statistical methodologies which use historical energy demand, weather data, and calendar information commonly introduced as inputs (Verwiebe et al., 2021). CNN hybrid models integrating Gated Recurrent Unit (GRU) and Long Short-Term Memory (LSTM) showed respectively efficient optimization strategy for energy storage at a large scale within a smart grid to enhances predictive performance and computational capabilities (Li, 2023) and demand forecasting in supply chain management (Rastogi and A.K, 2023). convolutional neural networks (CNNs) is used for pattern extraction from time series data, autoassociators for anomaly detection, support vector machines (SVM) for energy demand forecasting, decision trees for interpretable demand prediction, random forests for accurate demand prediction, Bayesian networks for probabilistic energy demand forecasts, particle filtering to tackle real-time energy demand, ARIMA (AutoRegressive Integrated Moving Average) for forecasting with strong seasonal patterns and trends, linear regression for simple demand prediction with limited data, decision trees for demand prediction with outlier robustness and moving averages for short-term forecasting, exponential smoothing for smoothing noise in data.By leveraging AI, organizations can develop sustainable IT infrastructure and reduce their environmental footprint.

- **Distributed Energy Resource (DER) Management:** DER management algorithms optimize distributed energy resources like solar panels and wind turbines for maximum grid contribution and stability. They serve as esteemed tools for green computing, thus, contributing to sustainable practices through the enhancmeent of energy efficiency and the mitigation of environmental impact. Strategies include demand response for consumption optimization, load shifting, and peak load management. Renewable energy integration involves forecasting, and energy storage optimization controls cycles for grid support. Grid integration maintains stability, and electric vehicle management optimizes charging for grid-friendly use. Microgrid control prioritizes loads and manages transitions. Optimization methods use machine learning and predictive analytics, adjusting DER operation to changing conditions. Various AI algorithms are emerging to enhance green computing by optimizing the efficiency and sustainability of distributed energy resources (DER). Notable algorithms, such as MADDPG, DQN, PPO, GANs, GBT, SVM, GA, TS, and PSO, play key roles in decentralized policy learning, optimal policy determination, direct policy optimization, data generation, output prediction, type classification, optimal placement and configuration, schedule and control optimization, and collaboration coordination for DER management. For instance, MADDPG facilitates coordinated DER-grid interaction, DQN determines optimal DER allocation, and PPO enables real-time energy management. GANs enhance DER scheduling through data generation, GBT predicts DER output, and SVM classifies DER types. GA optimizes DER placement, TS refines schedules and control strategies, while PSO enhances collaboration.

4. Optimizing Energy Efficiency

- **Model Quantization**: Model quantization, is a machine learning technique poised to reduce deep learning models' memory footprint and computational demands. They represent parameters with fewer bits, which helps enhance energy efficiency, especially for models deployed on resource-constrained edge devices. Furthermore, size and complexity reduction can be achieved through AI algorithms such as key quantization algorithms including post-training quantization (PTQ) and quantization-aware training (QAT). Various AI algorithms, including KD, pruning, low-rank approximation, hardware-aware quantization, dynamic quantization, joint quantization and pruning, NAS, and sparsity-inducing regularization, play a crucial role in achieving these improvements. The implementation of these techniques is vital for sustainable computing, particularly in the widespread integration of AI in resource-limited environments.

- **Hardware-Aware Optimization:** Optimizing AI model deployment involves tailoring the architecture and training process for specific hardware, utilizing accelerators like GPUs or TPUs, and minimizing memory and computational bottlenecks for energy-efficient Green Training. Green Training further leverages renewable energy sources and infrastructure optimizations, such as cloud providers with renewable energy options and hyperparameter tuning, to reduce overall energy consumption and training time. Various AI algorithms and techniques, including mapping algorithms to hardware, designing efficient neural network architectures, model compression, hardware emulation, and testing for real-world optimization, contribute to enhancing energy efficiency in resource-constrained environments. These strategies enable the development of sustainable AI solutions that minimize energy consumption, aligning with green computing practices.

- **Model Compression:** AI algorithms contribute to model compression, reducing size without sacrificing performance, using techniques like pruning, quantization, and knowledge distillation. This approach, crucial for enhancing energy efficiency and deploying AI models on resource-limited devices, employs various methods, including post-training quantization, quantization-aware training, knowledge distillation, pruning, low-rank approximation, hardware-aware quantization, dynamic quantization, joint quantization and pruning, network architecture search (NAS), parameter-sharing, matrix factorization, sparsity-inducing regularization, weight normalization, gradient-based optimization, and reinforcement learning-based model compression. Implementation of these methods significantly reduces model size and complexity, leading to notable improvements in energy efficiency, inference time, and resource utilization. These advancements support sustainable computing practices and promote the widespread integration of AI in edge devices and resource-constrained environments.

- **Efficient Model Deployment:** Focuses on the improvement of the approach through which the model is deployed and put into service. This optimization involves techniques such as batching, which enables the processing of several inputs at once; catching, which avoids redundant computations ; and efficient handling of the data transfer protocol-aimed at transmitting the data deom a server to a client through a computer nework- in order to improve energy savings. **Table 2** presents the related algorithms.

Table 2. AI algorithms for efficient model deployment solution

AI Algorithm	Description	Application Domains
Model Partitioning	Divides a large AI model into smaller, independently executable modules	Efficient deployment on hardware with limited memory or processing power
Model Specialization	Tailors the model architecture and parameters to a specific task or domain	Improves accuracy and efficiency compared to a general-purpose model
Federated Learning	Enables collaborative training of AI models without sharing raw data between participants	Preserves data privacy and reduces the need for centralized data storage and processing
Hardware-Aware Optimization (HAO)	Considers the specific characteristics of the hardware platform to optimize performance and energy efficiency	Algorithm selection, parameterization, data partitioning, memory optimization, instruction-level optimization
Quantized Neural Networks (QNNs)	Represent weights and activations of neural networks using lower precision data types	Reduces model size and memory footprint while maintaining accuracy
Low-Rank Approximation	Approximates weights and activations of a neural network using lower-rank matrices	Reduces model size without compromising accuracy
Knowledge Graph Embedding	Represents entities and relationships in a knowledge graph as low-dimensional vectors	Enables efficient storage, retrieval, and reasoning over large knowledge graphs
One-Shot Learning (OSL)	Learns from a single example per class	Tasks with limited or expensive data
Few-Shot Learning (FSL)	Learns from a small number of examples per class	Rapid adaptation to new tasks, domain adaptation

- **Hardware-Software Co-design (HSDC):** Consists in the hatrdware and software design for the sake of performnace optimization, energy efficiency of AI applications; taking into account hardware capabilities and constraints during the design process and vice-versa, thus, saving energy. There are AI Algorithms for HSDC Solutions such as algorithms, including machine learning (ML), reinforcement learning (RL), neural architecture search (NAS), graph neural networks (GNNs), generative adversarial networks (GANs), transfer learning, automated design space exploration, predictive maintenance, anomaly detection, and energy-aware resource allocation, are developed to enhance energy efficiency in hardware and software systems. ML models aid in performance modeling, RL optimizes hardware management, NAS focuses on hardware-aware model design, GNNs analyze hardware-software interaction, GANs explore hardware design, transfer learning optimizes hardware-software synergy, and automated design space exploration ensures system optimization. These algorithms collectively contribute to improved energy efficiency by optimizing resource utilization, predicting energy consumption, designing efficient architectures, analyzing system behavior, and managing energy resources effectively.
- **AI-assisted Resource Optimization:** AI plays a vital role in optimizing resource allocation and energy conservation across diverse technical domains like cloud computing, data centers, and power grids. Employing sophisticated algorithms, such as Long Short-Term Memory (LSTM) networks for time series prediction, Gradient Boosting Trees (GBT) for accurate predictions and regression, and Support Vector Machines (SVM) for classification and regression, AI-driven solutions find applications in forecasting energy demand in smart grids, detecting anomalies in power systems, ensuring predictive maintenance for IoT devices, optimizing demand response for electric vehicles, routing energy efficiently in data centers, and analyzing power system stability. The strategic use of these AI algorithms empowers intelligent resource management systems to make informed decisions, resulting in optimized resource utilization, reduced energy consumption, and enhanced overall system stability.

5. Reducing Waste and Emissions

Model Predictive Control (MPC) is a machine learning algorithm optimizing manufacturing processes by predicting and minimizing waste through real-time sensor data analysis. Reinforcement Learning (RL) contributes to waste reduction by training robots for accurate waste sorting. Supervised and Unsupervised Learning automate waste sorting and detect inefficiencies. Natural Language Processing (NLP) extracts insights from text data to identify waste trends and develop reduction strategies. Computer Vision analyzes waste streams through image and video processing for automated sorting. AI-driven waste patterns are foracasted thanks to predictive analytics which help foresee potentail failures. Digital Twins hinging in simulations based on data entries, simulates waste management processes and contributes to the optimization of facilities targetting reduced waste. AI intervenes at the supply chain level through its optimization by the reduction of waste and emissions inventory and the sound management of resources.

6. Developing Renewable Energy Resources

Solar panels, wind turbines, and other renewable sources can be AI-empowered and operate driven by algorithms prone to optimizing design by assessing large datasets to increase efficiency and reduce costs. Optimization covers optimal locations for facilities for photovoltaic panels, geothermal reservoirs and wind turbines etc. AI algorithms help predict energy demand thus supporting a sound management of power generation and distribution. Even though several challenges related to privacy and security were highlighted in addition to the "black box" nature of certain AI techniques, AI showed contribution to reliability and resilience of smart grids (Omitaomu and Niu, 2021).

7. ESG-Related AI Algorithms

AI and environmental, social, and governance (ESG) correlation is revolutionizing sustainable development by investing cutting-edge technologies to enhance ESG-aligned decisions and ensure compliance. AI algorithms are harenessed across a wide spectrum of ESG applications ranging from climate change mitigation to biodiversity conservation, environmental and social impact assessment, and sustainable resource allocation. AI intercedes at those segments through ensuring regulatory compliance and reporting, developing ESG-focused investment strategies, optimizing portfolios and assist stakeholder engagement regarding ESG investing. In fact, machine learning and NLP algorithms enhance corporate governance and ESG reporting, while optimization algorithms develop ESG-focused investment strategies. AI is transforming ESG investing by enabling more informed and impactful decision-making. Machine learning and computer vision algorithms support the monitoring of environmental data, while predictive analytics and optimization algorithms help develop climate change mitigation strategies thanks to their modelling and forecast abilities. Natural language processing (NLP) and sentiment analysis algorithms assess social impacts and help investigate and extract key information from documents for the sake of compliance.

IV. GREEN CLOUD COMPUTING AND DIGITAL WASTE MANAGEMENT

The advent and development of cloud computing has revolutionized business operations thanks to its resource optimization. Virtualization enables resource efficiency through the reduction of both energy consumption and waste. Cloud applications support remote work and help decrease commuting emissions. Cloud computing is achieved through the virtualization supports cost reduction for cloud providers, through allowing their users a flexible resource scaling. Services are delivered over the Internet network, thus, providing on-demand access without the need for physical infrastructure management. Users who interact with cloud services, send requests over the Internet to a provider's data center for further processing whike cloud computing offers advantages such as scalability, cost-effectiveness, flexibility, reliability, and agility in orderto play a more significant role in value delivery and operational landscapes (Andhra Pradesh Grama Sachivalayam, 2019). Even though providers are harnessing energy-efficient data centers through leveraging renewables and virtualization, several challenges are still encountered including substantial energy consumption and e-waste. Thus, the selection of providers with sustainable practices, optimizing resource usage, promoting remote work, and advocating for sustainable data centers are crucial. Green cloud initiatives aim to maximize eco-friendliness by optimizing applications and incorporating renewables, aligning with sustainability and cost-saving goals. Green cloud refers to the environmentally conscious approach to cloud computing for the mitigation of energy demands and addressing environmental concerns and potential issues. Within a data center, several active servers require constant power and cooling which are significant contributor to data consumption.

The carbon footprint of a cloud provider is based on four key pillars. Firstly, the location of the data center which plays a crucial role as it embodies significant factors including climate particularities, innovative solutions such as underground facilities or recycling waste heat for nearby homes and land dynamics which increases the costs of disaster managements on some high-risk regions. Secondly, the type of energy used is pivotal, with a shift towards green energy and the utilization of battery banks to store renewable energy and further cope with the discontinuities associated with the use of renewable energies. Thirdly, employing modern hardware and infrastructure involves energy-saving strategies such as dynamic voltage and frequency scaling (DVFS) and utilizing efficient data-storage devices like SSDs. This optimization enhances efficiency and reduces overall energy demand. Lastly, workload shifting strategies, like those seen in **Txture**, help avoid peak times, ultimately reducing network traffic and contributing to the green cloud initiative (Txture).

In 2022, global data center electricity consumption, excluding cryptocurrency mining, ranged from 240 to 340 TWh, constituting 1-1.3% of total global electricity demand. Cloud computing, leveraging economies of scale, dynamic resource allocation, and energy-efficient practices, proves effective in reducing energy consumption compared to traditional on-premises IT infrastructure. Energy-efficient cloud-based software contributes to business energy savings and reduces e-waste. Many cloud providers integrate renewable energy sources, enhancing sustainability. Despite its green advantages, cloud waste may occur due to inefficient resource allocation, leading to increased costs and reduced efficiency. While offering unparalleled computational resources, cloud computing's environmental footprint poses sustainability challenges, demanding careful consideration of energy consumption, e-waste, water usage, and broader environmental implications (IEA, 2023).

Addressing these challenges opens the doors for sustaianble resource maanagement by including a shift to clean clean energy sources including renewable energies, in order to mitigate the carbon footprint. At the data center scale, impelementing energy-efficient technologies can help reduce energy consumption and its associated emissions. Moreover, embracing circular economy principles such as hardware repurposing through refurbishment and recycling and extending the lifespan of IT resources are crucial solutions to minimize e-waste. Leveraging big data and AI for sustainability involves optimizing energy usage, identifying inefficiencies, and developing sustainable solutions tailored to the unique demands of cloud computing; and reaching those sustainability targets governments and businesses to advocate for clean energy adoption, investments in research and development for energy-efficient hardware and software, implementation of sustainable practices within cloud operations, and elevation of the public awareness through campaigns and educational initiatives, and fostering cloud computing responsible energy consumption (PVcase, 2023).

Several algorithms which may be used in the cloud to ensure several aspects of green IT. In terms of Task Scheduling Algorithms for Green Cloud Computing, (Zong, 2020) suggested s a taskscheduling algorithm that combines Genetic Algorithm (GA) and Ant Colony Optimization (ACO) which showed the ability of the algorithm to achieve energy optimization. Moreover, green ML showed efficiency in tackling cloud computing energy efficiency (Hassan et al., 2020). According to the survey conducted by (Demirci, 2015), Hybrid (Reinforcement + Supervised) learning models were harnessed for the adjustement of CPU frequency and the allocation of power-aware server. Supervised models were used in intelligent and power-aware scheduling poised to turn off the unutilized servers, load prediction to allocate resources optimally and worload classification and consolidation. Unsupervised learning showed utilizations in forecasting unused machines to sleep while reinforcement was used in spatially-aware load in order to decrease cooling expances.Due to the crisis aspect of climate change, the urgent role of AI in tackling this issue is be addressed. As a matter of fact, even tough AI proved itself to be efficient in climate modeling, smart grid design and further sound resource management applications ; AI demonstrated serious environmenatl issues. According to a study conducted by the University of Massachusetts Amherst, the carbon footprint of training big language models is considerable and is equivalent to about 300,000 kg of carbon dioxide emissions. Tech professionals have expressed concern about the larger carbon impact of big tech's AI deployment infrastructure, which goes beyond model training (Dhar, 2020).

CONCLUSION

A comprehensive strategy to lessen the environmental impact of IT operations is known as "green computing." This entails techniques to improve energy efficiency, such as the application of virtualization technologies, the use of energy-efficient components, and cooling system optimization. Furthermore, efforts like recycling, remanufacturing, and prolonging equipment lifespan—as well as decreasing raw material utilization in IT equipment production—help conserve resources. Waste reduction aims to reduce electronic waste (e-waste) as much as possible by using equipment for longer periods of time, disposing of it properly, and designing products that prioritize being recyclable. Green computing practices include, but are not limited to, the use of energy-efficient hardware selection, virtualization of servers to reduce energy consumption, adoption of cloud computing for centralized resources, data deduplication for efficient storage, use of green cooling systems such as free cooling and liquid cooling, and responsible e-waste disposal through recycling.

In addition to its obvious environmental advantages, green computing lowers costs through improved operational efficiency and lower energy costs. This makes it a desirable tactic for companies looking to draw in clients and staff who care about the environment Robust data waste management practices are crucial for lowering storage costs, enhancing data security, ensuring compliance with data privacy laws, and improving data quality. Data waste management involves the identification, classification, and disposal of outdated or superfluous data within an organization. Thorough data discovery and inventory procedures, value-and sensitivity-based data classification, and prioritizing data according to its importance to key smart objectives are some of the key best practices. The evaluation of data quality is crucial for the identification and amendment of the errors, while optimally managing data archiving and policies prone to responsibly managing inactive or outdated data.

In the context of AI's capabilities in data analysis, optimization, and automation, the e-waste management sector can significantly enhance efficiency, diminish environmental impact, and contribute to fostering a more circular economy. Green algorithms, designed with an ecological focus, optimize energy efficiency and minimize the environmental footprint associated with artificial intelligence models. The key areas of focus include increasing energy efficiency, waste management, sustainable resource management, and reducing carbon footprint. Green computing's primary objective is the reduction of energy consumption, leading to decreased energy expenses for organizations and a diminished carbon footprint, particularly concerning IT assets. Beyond financial benefits, the adoption of green computing practices helps organizations adhere to regulatory requirements, potentially providing a competitive advantage in customer relations and investor appeal.

In order to mitigate environmental impact, it is advised to minimize video streaming, which accounts for 75% of global data traffic. The production, energy consumption, and operation of devices such as (smartphones, laptops, TVs) across internet networks, data centers, and servers/routers, showed a significant contribution to the overall energy footprint during streaming, thus, it is recommended to resort to downloading which is a less energy-intensive alternative. Furthemore, playing songs including audio files instead of streaming videos on platforms like YouTube and choosing lower video resolutions and disabling auto-play features are able to decrease energy consumption. The extension of the lifespan of electronic devices, such as smartphones and televisions, is crucial as well as the proper disposal of old devices, the regular cleaning of email boxes to reduce and further minimize data storage, and prioritizing local data storage over cloud usage are recommended practices. At the individual level, using WLAN networks instead of mobile networks showed a better energy-efficient, let alone for integrating renewable energy sources in the energy mix (Kamiya, 2020 ; SRF, 2020 ; MyClimate, 2023).

Engaging in an hour of video conferencing results in the emission of 1kg of CO_2, but this impact can be drastically reduced through opting alternatively to audio-only communication, therefore, leading to a 96% carbon reduction. The majority of electronic devices' lifetime carbon emissions, amounting to 80%, stem from their production, and the prolongation of the lifespan of a computer from 4 to 6 years, as suggested by a University of Edinburgh study, could prevent approximately 190kg of carbon emissions. Furthermore, streaming online content constitutes a significant portion, 58-60%, of internet traffic as generating 300 million tonnes of CO_2 annually which accounts approximately for 1% of the global emissions, according to research from the Shift Project. In this context, considering that 1 tonne of CO_2 aligns with the average emissions produced by one passenger on a round-trip flight from Paris to New York is to be considered (Evangelidis and Davies, 2021).

REFERENCES

Andrae, A., & Edler, T. (2015). On global electricity usage of communication technology: Trends to 2030. *Challenges*, *6*(1), 117–157. doi:10.3390/challe6010117

CBRE. (2023). *Global data center trends 2023*. Retrieved October 5, 2023, from https://www.cbre.com/insights/reports/global-data-center-trends-2023

Demirci, M. (2015). A Survey of Machine Learning Applications for Energy-Efficient Resource Management in Cloud Computing Environments. In *2015 IEEE 14th International Conference on Machine Learning and Applications (ICMLA)* (pp. 205). 10.1109/ICMLA.2015.205

Dhar, P. (2020). The carbon impact of artificial intelligence. *Nature Machine Intelligence*, *2*(8), 423–425. doi:10.1038/s42256-020-0219-9

Evangelidis, H., & Davies, R. (2023, November 29). *Are you aware of your digital carbon footprint?* Retrieved October 5, 2023, from https://www.capgemini.com/gb-en/insights/expert-perspectives/are-you-aware-of-your-digital-carbon-footprint/

Grama Sachivalayam Andhra Pradesh. (2019). *Digital Assistant Training Handbook*. Author.

Hassan, M. B., Saeed, R. A., Khalifa, O., Ali, E. S., Mokhtar, R. A., & Hashim, A. A. (2022). Green Machine Learning for Green Cloud Energy Efficiency. In *2022 IEEE 2nd International Maghreb Meeting of the Conference on Sciences and Techniques of Automatic Control and Computer Engineering (MI-STA)* (pp. 288-294). 10.1109/MI-STA54861.2022.9837531

HassanQ.SameenA. Z.SalmanH. M.Al-JibooryA.JaszczurM. (2023, May 25). The role of renewable energy and artificial intelligence towards environmental sustainability and net zero. Research Square. doi:10.21203/rs.3.rs-2970234/v1

International Energy Agency (IEA). (2023). *Data centres and data transmission networks*. Retrieved from https://www.iea.org/energy-system/buildings/data-centres-and-data-transmission-networks

Kamiya, M. (2020). *The carbon footprint of streaming video on Netflix*. Retrieved October 5, 2023, from https://www.carbonbrief.org/factcheck-what-is-the-carbon-footprint-of-streaming-video-on-netflix/

Kamiya & Shakti Nagpal. (2015). *Green Cloud Computing Resource Managing Policies: A Survey*. Academic Press.

Li, X. (2023). CNN-GRU model based on attention mechanism for large-scale energy storage optimization in smart grid. *Frontiers in Energy Research*, *11*, 1228256. Advance online publication. doi:10.3389/fenrg.2023.1228256

Lucivero, F. (2019). Big Data, Big Waste? A Reflection on the Environmental Sustainability of Big Data Initiatives. *Science and Engineering Ethics*, *26*(2), 1009–1030. doi:10.1007/s11948-019-00171-7 PMID:31893331

MyClimate. (2023). *What is a digital carbon footprint?* Retrieved October 5, 2023, from https://www.myclimate.org/en/information/faq/faq-detail/what-is-a-digital-carbon-footprint/: https://www.myclimate.org/en/information/faq/faq-detail/what-is-a-digital-carbon-footprint/

N. P. M K., Rastogi, & A. K. (2023). Demand Forecasting in Supply Chain Management using CNN-LSTM Hybrid Model. In *14th International Conference on Computing Communication and Networking Technologies (ICCCNT)* (pp. 1-5). 10.1109/ICCCNT56998.2023.10307665

Nokia. (2023). *Energy efficiency*. Retrieved October 5, 2023, from https://www.nokia.com/networks/bss-oss/ava/energy-efficiency/?did=D000000007BR&gad_source=1&gclid=CjwKCAiAvdCrBhBREiwAX6-6Uu8DkwDjjBpM81xs6olZKW-poLxSFWvpyW6fg5ISSpj6nsBsyQn9UhoCgMoQAvD_BwE

NVIDIA. (2022). *What is green computing?* Retrieved from https://blogs.nvidia.com/blog/what-is-green-computing/

OECD. (2023). *Data usage per mobile broadband user*. Retrieved October 5, 2023, from https://www.oecd.org/digital/broadband/broadband-statistics-update.htm

Omitaomu, O. A., & Niu, H. (2021). Artificial Intelligence Techniques in Smart Grid: A Survey. *Smart Cities*, *4*(2), 548–568. doi:10.3390/smartcities4020029

Parmentola, A., Petrillo, A., Tutore, I., & De Felice, F. (2021). Is blockchain able to enhance environmental sustainability? A systematic review and research agenda from the perspective of Sustainable Development Goals (SDGs). *Business Strategy and the Environment*. Advance online publication. doi:10.1002/bse.2882

Petrosyan, A. (2023). *Worldwide digital population 2023*. Retrieved October 5, 2023, from https://www.statista.com/statistics/1044012/us-digital-audience/

PVcase. (2023). *The environmental impact of computing problems and possible solutions.* Retrieved from https://pvcase.com/blog/the-environmental-impact-of-computing-problems-and-possible-solutions/#:~:text=Cloud%20computing%20is%20only%20as,of%20e%2Dwaste%20is%20recycled

Säntti, R. (2022). *Five ways to battle data waste.* Capgemini. Retrieved from https://www.capgemini.com/insights/expert-perspectives/five-ways-to-battle-data-waste/

Sayed, K., & Gabbar, H. (2018). *Building Energy Management Systems (BEMS).* doi:10.1002/9781119422099.ch2

S&P Global. (2019). *Avoiding garbage in machine learning shell.* Retrieved from https://www.spglobal.com/en/research-insights/articles/avoiding-garbage-in-machine-learning-shell

SRF. (2020). *Co2-Fussabdruck im Internet: Surfe ich das Klima kaputt?* Retrieved October 5, 2023, from https://www.srf.ch/kultur/gesellschaft-religion/co2-fussabdruck-im-internet-surfe-ich-das-klima-kaputt

Taylor, R. (2023). *Worldwide data created 2023.* Retrieved October 5, 2023, from https://www.statista.com/statistics/871513/worldwide-data-created/: https://www.statista.com/statistics/871513/worldwide-data-created/

The Shift Project. (2019). *Lean ICT Report.* Retrieved October 5, 2023, from https://theshiftproject.org/en/home/

Txture. (2022). *Txture Cloud Transformation YouTube channel.* Retrieved from https://www.youtube.com/@txturecloudtransformation

United Nations Development Programme (UNDP). (2023). *Three ways digital transformation accelerates sustainable and inclusive development.* Retrieved from https://www.undp.org/blog/three-ways-digital-transformation-accelerates-sustainable-and-inclusive-development

Utilities One. (2023). *Demystifying the impact of AI on energy management.* Retrieved October 5, 2023, from https://utilitiesone.com/demystifying-the-impact-of-ai-on-energy-management: https://utilitiesone.com/demystifying-the-impact-of-ai-on-energy-management

Vailshery, A. (2023). *Internet of Things (IoT) Connected Devices Worldwide 2022-2030.* Retrieved October 5, 2023, from https://www.statista.com/statistics/1183457/iot-connected-devices-worldwide/

Verwiebe, P. A., Seim, S., Burges, S., Schulz, L., & Müller-Kirchenbauer, J. (2021). Modeling Energy Demand—A Systematic Literature Review. *Energies, 14*(23), 7859. doi:10.3390/en14237859

World Health Organization (WHO). (2023). *Electronic waste (e-waste) fact sheet.* Retrieved from https://www.who.int/news-room/fact-sheets/detail/electronic-waste-(e-waste)

Zong, Z. (2020). An Improvement of Task Scheduling Algorithms for Green Cloud Computing. In *2020 15th International Conference on Computer Science & Education (ICCSE)* (pp. 654-657). 10.1109/ICCSE49874.2020.9201785

Chapter 14

Efficient Resource Management in Green Computing Based on ISHOA Task Scheduling With Secure ChaCha20–Poly1305 Authenticated Encryption– Based Data Transmission

B. Santosh Kumar

New Horizon College of Engineering, India

K. A. Jayasheel Kumar

New Horizon College of Engineering, India

Balasubramanian Prabhu Kavin

iD https://orcid.org/0000-0001-6939-4683

SRM Institute of Science and Technology, India

Gan Hong Seng

XJTLU Entrepreneur College, Xi'an Jiaotong-Liverpool University, China

ABSTRACT

One of the hottest new technologies that allows users to handle a broad range of resources and massive amounts of data in the cloud is green computing resource management. One of the biggest obstacles is task scheduling, and poor management leads to a decrease in productivity. The task must be efficiently scheduled to ensure optimal resource utilisation and minimal execution time. Given this, this study suggests a fresh method for efficient task scheduling in a green computing environment that also offers improved security. There is a development of an enhanced spotted hyena optimizer (ISHO). Finding the ideal or almost ideal subset with a straightforward structure to minimise the specified fitness function is a strong point of the SHO. It enhances a switch strategy in the spotted hyena's position updating mechanism and generates random positions in place of the violated spotted hyenas in order to support the proposed ISHO's exploration characteristics. Second, the data is encrypted using the ChaCha20–Poly1305 authenticated encryption algorithm, ensuring secure data transmission.

DOI: 10.4018/979-8-3693-1552-1.ch014

1. INTRODUCTION

In the field of green computing, resource management is essential, especially when it comes to task scheduling. Reducing the environmental effects of computer systems is the aim of green computing, and one of the most important ways to do this is through efficient resource management. Optimising resource utilisation becomes essential when it comes to task scheduling, which is the process of allocating computing resources to different tasks. Green computing places a strong emphasis on resource efficiency to lower energy use and environmental impact. Green Computing aims to optimise resource allocation to minimise energy consumption and maximise efficiency through intelligent task scheduling algorithms. This entails taking into account elements like the distribution of the workload, patterns of power consumption, and the utilisation of hardware that uses less energy. Green computing improves computer system performance and makes the computing environment more environmentally friendly and sustainable by utilising cutting-edge resource management strategies. Task scheduling needs to incorporate resource management techniques more and more as technology develops in order to promote environmentally conscious and more ecologically friendly computing practises.

A key component of computer science and information technology, task scheduling is essential to the effective and efficient use of computing resources. The fundamental idea behind task scheduling is to distribute computational tasks wisely among available resources, like processors, to maximise throughput, minimise completion times, or save energy usage. Efficient task scheduling is crucial for maximising system performance and resource utilisation, regardless of whether the system is in a distributed system, cloud computing environment, or standalone computer. The difficulties in handling different workloads, different priorities, and resource limitations require the creation of complex algorithms and plans to coordinate task completion. Scheduling tasks is very important in the context of emerging paradigms such as Green Computing, where the focus is on minimising environmental impact, in addition to its impact on the overall efficiency and responsiveness of computing systems. Task scheduling is essential in establishing the sustainability, scalability, and performance of computing systems in a variety of domains as technology develops.

By converting data into a safe and unintelligible format, encryption plays a critical role in protecting data privacy by averting unwanted access or interception. Encryption is a vital instrument in the digital age, as it shields private data from potential breaches and threats, especially in the context of massive data transmissions and storage. Fundamentally, encryption is the process of encrypting data using keys and algorithms so that anyone without the correct decryption key cannot decode it. This makes sure that information is rendered effectively useless even in the event that it is intercepted during transmission or unauthorised individuals gain access to storage systems. Encryption is used in many different contexts when it comes to protecting personal information. One method of securing communication channels is end-to-end encryption, which encrypts data at the source and only permits decryption at the end of the intended recipient. In platforms like email services, messaging apps, and others where privacy is critical, this is especially important. Further protecting information kept on devices or servers is data at rest encryption.

1.1. Main Contributions

- **Improved Spotted Hyena Optimizer (ISHO):** In order to schedule tasks efficiently in a green computing environment, the paper presents an improved version of the Spotted Hyena Optimizer

(SHO), named ISHO. Finding ideal or almost ideal subsets to minimise fitness functions is a well-known capability of SHO.

- **Security Enhancement with ChaCha20–Poly1305:** The suggested method uses the authenticated encryption algorithm ChaCha20–Poly1305, which prioritises security. This guarantees safe data transfer, protecting the confidentiality and integrity of the data being handled.

1.2. Organization of work

The remaining sections of the study are organised like shadows: Section 2 summarises relevant literature; Section 3 gives a brief description of the suggested model; Section 4 shows the conclusions and analysis of validation; and Section 5 concludes with a summary.

2. RELATED WORKS

Chandrashekar et al. (2023) sought to provide the best possible task scheduling algorithm, which was evaluated and contrasted with other algorithms already in use in terms of cost, makespan, and efficiency. The algorithm known as Hybrid Weighted Ant Colony Optimisation, or HWACO, improved upon the previously accessible Ant Colony Optimisation Algorithm, was used in the paper in order to clarify and resolve the problem with scheduling. The results obtained from utilising the suggested HWACO were more advantageous; the goal of achieving convergence quickly was achieved. Therefore, the suggested algorithm is an optimal task scheduling algorithm because it outperformed other conventional algorithms like MIN-MIN Algorithm (MM), First-Come-First-Serve (FCFS), Modified-Transfer-Function-Based Binary Particle Swarm Optimisation (MTF-BPSO), Ant Colony Optimisation (ACO), and Quantum-Based Avian Navigation Optimizer Algorithm (QANA).

Zhang, Chu, Song, Wang, and Pan (2022) provided a work scheduling approach based on the advanced Phasmatodea Population Evolution (APPE) regarding diverse cloud environments. Through the optimisation of the nearest optimal solutions' convergent evolution, the algorithm decreased the time required to find solutions. Next, in order to keep the algorithm from going into local optimisation, a restart strategy was added, and its exploration and development capabilities were balanced. In addition, the evaluation function was designed to identify optimal solutions by taking into account length of time, cost of resources, and level of load balancing. The APPE algorithm performed better than comparable algorithms when tested on thirty benchmark functions. Concurrently, the algorithm resolved the issue of task scheduling within the cloud computing setting. Comparing this method to other algorithms, the convergence time was faster and the resource usage was higher.

Malathi and Priyadarsini (2023) researched the benefits of using heuristic approaches to create a cloud-based load balancer algorithm. There were two significant developments made to load balancing methods. Better application of the hybrid technique resulted in outstanding results in terms of virtual machine resource consumption and maximum turnaround time. The optimizer lion was created as the initial contribution to assist in load balancing by creating the ideal virtual machine parameter selection. To improve the selection process, It was possible to create two selection probabilities, such as the virtual machine selection probability and the task scheduling probability. The virtual machine's properties and the task itself both influenced the fitness criteria used by the lion optimizer. Through the modification of global search criteria relevant to the lion optimizer, the second contribution involved the creation of a genetic algorithm. Experimental results proved the efficacy of the algorithm based on a hybrid lion for genetics.

Yadav and Mishra (2023) concentrated on developing a more advanced and refined ordinal optimisation technique in order to minimise the size of to maximise the makespan, search the area where the best scheduling can occur in the least amount of time. To satisfy the present need for an ideal schedule for minimum makespan, enhanced reiterative ordinal optimisation was used. By carefully assigning the load to a highly promising schedule and using horse race circumstances as selection criteria, this optimisation attained low overhead. Optimal schedules were generated by combining the recommended ordinal optimisation technique with linear regression in order to minimise makespan. Moreover, the suggested formula (obtained through linear regression) predicted any dynamic workload in the future for the goal of the shortest makespan period.

Kallapu et al. (2023) utilised the attribute-aware encryption technique driven by blockchain to create a real-time secure cloud communication method. Data owners could give data users fine-grained search permissions by using attribute-based encryption technology. In order to provide safe an increase in blockchain keyword searches and access to encrypted data, the suggested solution integrated accessible encryption technology. A helpful comparison of attribute-based encryption algorithms developed recently was made available by the study. The two primary components of the access control strategy were a linear secret-sharing system and two types of access trees. It was decided to use Type-A bilinear pairing parameter and set the base field of the elliptic curve to 512b. With this method, keywords were encrypted using attribute-based encryption and stored on a distant server. Moreover, the blockchain contained both the ciphertext and the encrypted data blockchain. The system's ability to generate keys, build trapdoors, and retrieve keywords was measured quantitatively.

Yang et al. (2023) suggested a productive attribute-based encryption system for multi-cloud environments that includes data security classification. They improved the security of outsourcing data by dividing dividing an information security tier for the data owner and keeping it with various cloud service providers. Additionally, their plan could reduce the computing load on the data user by utilising the cloud to fully enable outsourcing decryption, which would allow for granular access control determined by Ciphertext-Policy Attribute-Based Encryption (CP-ABE). The security analysis's findings demonstrated the effectiveness of their plan in thwarting selective-attribute plaintext attacks and safeguarding data privacy.

Alsubai suggested a novel approach to task scheduling using swarms that included a security measure (Alsubai et al., 2023) to encrypt cloud data during task scheduling and distribute tasks as efficiently as possible utilising the resources at hand. It might integrate the Chameleon Swarm Algorithm (CSA) and the Moth Swarm Algorithm (MSA) for task scheduling. Polymorphic Advanced Encryption Standard (P-AES) could be used for information security for cloud-scheduled tasks. This approach offered a novel perspective on the optimal application of swarm intelligence cloud task scheduling algorithms. By taking advantage of the advantages of the employed algorithms, the approach was able to offer safe and efficient task scheduling by combining P-AES with MSA and CSA. Throughput, execution time, latency, speed, and bandwidth usage, average waiting time, cost, makespan, and response time were among the metrics used in the study to assess how well the suggested strategy performed. There were many different tasks used in the simulation, ranging from 1000 to 5000.

2.1. Research Gaps

While the existing models present novel techniques to job scheduling, load balancing, and secure communication in cloud systems, there remains a significant research gap in integrating these disparate components. There is, in particular, a scarcity of comprehensive studies that address job scheduling optimisation, load balancing efficiency, and security concerns in cloud computing. Existing research focuses on specific areas such as scheduling algorithms, load balancing strategies, or encryption methods; nevertheless, there is a need for holistic approaches that integrate these parts to create a coherent and resilient solution for secure and efficient cloud computing environments.

3. PROPOSED METHODOLOGY

Figure 1 shows the workflow of the proposed model.

Figure 1. Block diagram

3.1. System Design

Scheduling in cloud computing can be separated into two categories: jobs and tasks. Among them, Hadoop is the scheduling technique that is most frequently applied to task scheduling, with the primary goal being system performance enhancement. Using a mapping of tasks and resources, task scheduling assigns resources to task applications. Currently, assigning jobs to processors in a way that maximises makespan, minimises cost, and achieves high efficiency is a crucial aspect of task scheduling (Zhang, Chu, Song, Wang, & Pan, 2022). To overcome the limitations of the existing studies of task scheduling algorithms for cloud computing, a novel approach is presented here. First, an evaluation model and task scheduling system are developed in order to weigh the advantages and disadvantages of allocating tasks according to makespan, cost, and load level. The ISHO-based task scheduling approach is then used to modify the task scheduling model. The framework for evaluation is used to determine the best work scheduling strategy

The virtual machine receives the task that the user submits to the cloud by the task scheduling method after being added the task manager added it to the queue. Every task is distinct and does not have any precedence. Figure 2 a Framework for Scheduling Tasks in Cloud Computing Systems (Khang, 2023).

This article's virtual machine uses the task queue to process tasks in a serial fashion. Let n and m are the two attributes of the task; m is the number of tasks (Baswaraju et al., 2023). In machine language instructions (MI), a len is the task's length, and the unit is millions. Four characteristics make up a virtual machine: n, MIPS, RAM, and bandwidth. The number of virtual machines (VMs) is denoted by n, and MIPS stands for mean processor performance of single-word fixed-point instructions. Memory is represented by RAM, and bandwidth by bandwidth.

Figure 2. Task scheduling model for cloud computing

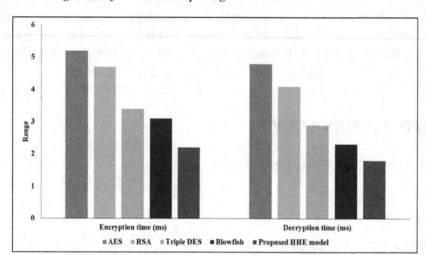

A. Evaluation Model

Made-to-order, profit, waiting time, cost, and completion time are the main metrics used to evaluate how well cloud computing task planning is working. From the standpoints of expense and degree of imbalance, and makespan, this paper examines task scheduling performance in detail. It go into more detail about these performance indicators below (Thirumalraj & Rajesh, 2023).

Let Task $=\{T_1, T_2, \ldots, T_m\}$, $(50 \leq m \leq 500)$. Task denotes the task queue that users of the cloud have submitted, and m denotes the total number of tasks. Let $T\_length = \{len_1, len_2, \ldots, len_m\}$. len_i represents the duration of the i-th task. $VM = \{VM_1, VM_2, \ldots, VM_n\}$. VM_j denotes the j-th virtual machine, and n denotes its quantity. $ESC = \{ESC_{ij}\}_{m*n}$, $ESC_{ij} = 1$ symbolises the reality that task i is carried out on the VMj, otherwise $ESC_{ij} = 0$. $ETC = \{ETC_{ij}\}_{m*n}$ represents the anticipated completion time, or the task's processing time i on the VMj, which the subsequent formula uses to calculate:

$$E_{TC}ij = \frac{len_i}{M_{PPS}^j} \tag{1}$$

M_{IPS} indicates the speed at which the VMj.

- **Makespan**

Makespan is a crucial metric for assessing how effective cloud-based task scheduling is. The formula that follows is used to calculate the makespan, which is the task's completion time and represents the overall operating time of all VMs:

$$\text{Makespan} = \max_{j}\left(\sum_{i=1}^{m} ETC_{ij}*ESC_{ij}\right) \tag{2}$$

- **Cost**

The following costs are computed per hour in accordance with VM specifications: USD0.12, USD0.13, USD0.17, USD0.48, USD0.52, and USD0.96 are all included. Use the following formula to determine the virtual machine's cost.

$$\text{Cost} = \sum_{j=1}^{n}\left(\text{cost}_j *\left(\sum_{i=1}^{m} ETC_{ij}*ESC_{ij}\right)\right) \tag{3}$$

cos_j is the jth virtual machine's hourly rate; Memory, bandwidth, and MIPS all affect its resource cost in a heterogeneous environment (Thirumalraj & Balasubramanian, n.d.); Virtual machine memory, or RAM, and bandwidth refer to the VM's respective bandwidths.

- **Load**

$$\text{Load} = \sqrt{\varphi*\frac{\sum_{j=1}^{n}\text{load}_j*VL_j}{n*\text{Makespan}}} \tag{4}$$

φ symbolises the extent of the system's imbalance. n shows the quantity of virtual machines, load j represents, as indicated by the following formula (Thirumalraj et al., 2023), the extent to which RAM, bandwidth, and MIPS each affect the virtual machine:

$$\varphi = \sqrt{\frac{\sum_{j=1}^{n}\left(VL_j - \overline{VL_j}\right)^2}{n}} \tag{5}$$

$$\text{load}_j = \zeta*MIPS + \delta*RAM + \eta*\text{bandwidth} \tag{6}$$

$$VL_j = \sum_{i=1}^{m} ETC_{ij}*ESC_{ij} \tag{7}$$

$$\overline{VL_j} = \frac{\sum_{j=1}^{n}VL_j}{n} \tag{8}$$

Here, VL_j indicates the duration of the $VMi, \overline{VL_j}$ shows the VM's average running time, load j is connected to bandwidth, RAM, and MIPS, ζ, δ, η are each of the trio of weight values. One obtains the objective function formula as follows using the performance indicators mentioned above:

$$\text{fitness} = \text{Makespan} * \text{Cost} * \text{Load} \tag{9}$$

3.2. ISHO

3.2.1 Spotted Hyena Algorithm

In SHA (El-Ela et al., 2020), The spotted hyenas adjust their locations in relation to the desired outcome. This conduct is carried out as:

$$\vec{D}_{SH} = \left| \vec{C} \vec{X}_T - \vec{X}_{SH} \right| \tag{10}$$

$$\vec{X}_{SH} = \vec{X}_T - \vec{A} \vec{D}_{SH} \tag{11}$$

where D_{SH} symbolises the distance between the location of the spotted hyena and (X_{SH}) and the prey's intended location (X_T). A and C are these the revised co-efficient parameters:

$$\vec{A} = 2 \vec{a} r - \vec{a} \tag{12}$$

$$\vec{C} = 2r \tag{13}$$

Where, the vector a, that symbolized in Eq. (14), is linearly decreased from 5 to 0 as:

$$\vec{a} = 5 \cdot \left(1 - \frac{\text{iter}}{\text{Max}_{\text{iter}}} \right) \tag{14}$$

Where, r is a figure chosen at random. Subsequently, a designated set of a number (N) of the finest hyenas (C_H) is consisted as follows:

$$\vec{C}_H = \vec{X}_{SH} + X_{SH+1} + X_{SH+2} + \ldots\ldots X_{SH+N} \tag{15}$$

where, N refers to a certain quantity of hyenas, which is assessed as:

$$N = \text{count}_{n1} \left(X_{SH}, X_{SH+1}, X_{SH+2}, \ldots\ldots \overrightarrow{(X}_{SH} + \vec{M}) \right) \tag{16}$$

where, n1 is the quantity of spotted hyenas. M is a vector that is random within range. [0.5, 1]. The hyenas then use Equation (17) to update their positions so they can attack their tracked preys.

$$\vec{X}_{SHnew} = \frac{\vec{C}_H}{N} \tag{17}$$

Following an update to the spotted hyenas' positions, the following boundary checks are made:

$$X_{SHnew}(d) = \begin{cases} X(d)^{max} & if \ X_{SHnew}(d) > X(d)^{max} \\ X_{SHnew}(d) & if \ X(d)^{min} > X_{SHnew}(d) > X(d)^{max} \\ X(d)^{min} & if \ X_{SHnew}(d) < X(d)^{min} \end{cases} \quad d = 1,2\ldots Dim \tag{18}$$

The superscripts 'min' and 'max' denote the lower or upper bounds, respectively. Dim denotes the dimension under consideration.

3.2.2. Improved Spotted Hyena Algorithm (ISHA) for the Considered Problem

Within the fundamental SHA, the elite group of hyenas $\left(\vec{C}_H\right)$ ought to be chosen for every iteration. What happens if there isn't a workable answer in the starting population is a crucial question that comes up here. This might have happened if some of the loads weren't supplied or if the initial hyenas in the load flow didn't get convergence. This will result in the best hyenas in the group being empty. $\left(\vec{C}_H\right)$ and will prevent the SHA's updating mechanism from working. First, a switch strategy is suggested to update the positions of the spotted hyenas in order to avoid this block:

$$X_{SHnew} = \overrightarrow{X_{SH}} + \varepsilon \vec{R}\left(1 - \frac{iter}{Max_{iter}}\right) \tag{19}$$

Where, ε is very small value ($\varepsilon = 0.001$) and R is an arbitrary vector in the interval [0.5,1]. To enhance the exploration features for looking for new regions, another change is suggested. With this update, the infringed spotted hyenas will be placed in new, random locations as follows:

$$X_{SHnew}(d) = \begin{cases} X(d)^{max} & if \ X_{SHnew}(d) > X(d)^{max} \\ X_{SH}^{*}(d) & if \ X(d)^{min} > X_{SHnew}(d) > X(d)^{max} \\ X(d)^{min} & if \ X_{SHnew}(d) < X(d)^{min} \\ d = 1,2,\ldots\ldots Dim \end{cases} \tag{20}$$

Where, $X_{SH}*(d)$ is a random value that falls inside each control variable's acceptable range(d).

3.3. ChaCha20-Poly1305 Authenticated Encryption

3.3.1. ChaCha20

A. Algorithm

The Salsa20 algorithm is the foundation of the ChaCha20 cypher. The algorithm first creates an initial matrix (I_M) 256-bit key, four documentation-defined constants, a 96-bit nonce, and a 32-bit initial counter. The I_M 512 bits in size, and it is represented by a 4 by 4 matrix with 32-bit unsigned integers assigned to each matrix entry. Additionally, the setup of I_M is written in little-endian. Here is an example of the matrix illustration (21).

$$I_M = \begin{bmatrix} 61707865 & 3320646e & 79622d32 & 6b206574 \\ Key_0 & Key_1 & Key_2 & Key_3 \\ Key & Key_5 & Key_6 & Key_7 \\ Counter & Nonce_0 & Nonce_1 & Nonce_2 \end{bmatrix} \tag{21}$$

Algorithm 1 (Serrano et al., 2022) displays the suite of cyphers ChaCha20. First, the values of an operation matrix (O_M) are initialised I_M. It apply ten iterations to the internal condition by the ChaCha20 algorithm O_M, where eight quarter-round function applications make up each double-round. The rounds come to an end once all states have been updated O_M. The state that the input took is preserved in the same location as the outcome of the QR operation. In the end, the keystream is produced by appending the I_M and O_M. Stated differently, the decryption procedure is just the same encryption operation employing the same keystream and cypher stream. To obtain the following plaintext block, additionally, for each block processed, the counter value is increased by one I_M.

Algorithm 1. *The cypher suite algorithm ChaCha20*

    Require: $K \in (0,1)^{256}$, $N \in (0,1)^{96}$, $C \in (0,1)^{32}$, $PT \in (0,1)^*$
    Ensure: CT = ChaCha20 (K,N,C,PT)
        $I_M \leftarrow$ Init(K,N,C) $\triangleright$ The first matrix is set up as (1).
            for $x \leftarrow 0$ to ($[P/512] - 1$) do
        $O_M \leftarrow I_M$
            for $y \leftarrow 0$ to 9 do
        $O_M[0,4,8,12] \leftarrow$ QR($O_M[0,4,8,12]$)
        $O_M[1,5,9,13] \leftarrow$ QR($O_M[1,5,9,13]$)
        $O_M[2,6,10,14] \leftarrow$ QR($O_M[2,6,10,14]$)
        $O_M[3,7,11,15] \leftarrow$ QR($O_M[3,7,11,15]$)
        $O_M[0,5,10,15] \leftarrow$ QR($O_M[0,5,10,15]$)
        $O_M[1,6,11,12] \leftarrow$ QR($O_M[1,6,11,12]$)
        $O_M[2,7,8,13] \leftarrow$ QR($O_M[2,7,8,13]$)
        $O_M[3,4,9,14] \leftarrow$ QR($O_M[3,4,9,14]$)
    end for
        $S \leftarrow$ Serialize($O_M + I_M$)

for $z \leftarrow 0$ to 511 do

$\quad CT[512x+z] \leftarrow PT[512x+z] \oplus S[z]$

end for

$\quad I_M[12] \leftarrow I_M[12] + 1 \triangleright$ Change the counter's value in the original matrix.

end for

return CT

The QR function is included in both the column and diagonal rounds $(A,B,C,D)=$QR (a,b,c,d). It has the following effects on the state. A 32-bit word is added using a carry-less method, as indicated by the QR algorithm.

$$\begin{cases} x = a+b; & y = \left(d \oplus x\right) \lll 16; \\ w = x+y; & z = \left(b \oplus w\right) \lll 12; \\ A = x+f; & D = \left(y \oplus A\right) \lll 8; \\ C = w+D; & B = \left(z \oplus C\right) \lll 7; \end{cases} \tag{22}$$

B. Hardware Implementation

The ChaCha20 primitive's implementation is seen in Figure 3. The two components of the ChaCha20 implementation are separated. Initially a 256-bit key, a 96-bit nonce, and a 32-bit counter are required for a block function that is indicated in blue to generate an I_M in the original regulations. The conditions of the I_M are arranged using the little-endian system. The initial states of the Initial regs are then obtained by the operation regs. A 32-bit register makes up each of the 16 states that all matrices have. Using the QR modules, a red-highlighted Finite State Machine (FSM) completes the 20 rounds. One, two, and four quarter-round operations can be performed in parallel with the round by the ChaCha20 primitive. On the other hand, the ChaCha20 algorithm shows that paralleling the diagonal and column rounds is not possible. As a result, four quarter-round operations are the maximum. Each column's and the diagonal round's quarter-round operators' input is managed by the FSM. Only the outcomes of the preceding rounds determine each round's O_M. Furthermore, the Next and Init signals designate the subsequent 512 bits to be processed and the beginning of the process, respectively. The Matrix Adder receives the final matrix after 20 rounds (F_M), incorporating the Operation and Initial Regulations. Secondly, a green-highlighted Crypto Function takes the F_M produced by the Block function. An XOR operation is used to create the 512-bit Cypher Text on F_M as well as the Plain Text. A legitimate signal is produced upon completion of the crypto-function.

Figure 4: The operation of a QR code. The inputs a, b, c, and d represent the stages of an operation. The QR operation uses four Add-Rotate-XOR (ARX) cells. A wire permutation is used in place of to minimise the delay of the ARX cell, the rotate operation highlighted in red is used instead of combinational logic.

Figure 3. The primitive architecture of ChaCha20

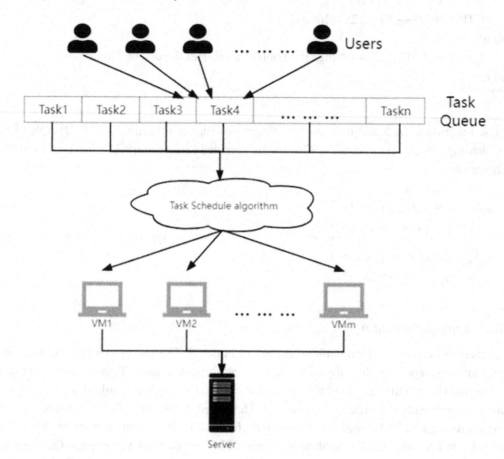

Figure 4. Operation in quarters

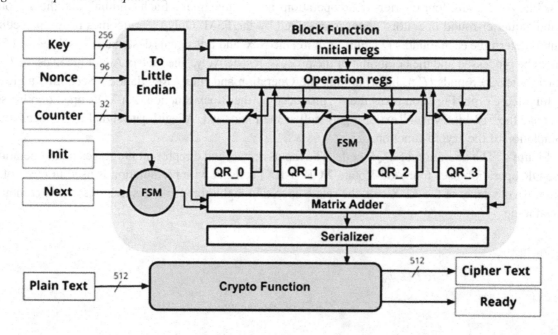

3.3.2. Poly1305

A. Algorithm

As a one-time authenticator, the Poly1305 algorithm creates a 16-byte message authentication code (MAC) via a message of any length and a 32-byte one-time key. Algorithm 2 displays the authenticator algorithm for Poly1305. First of all, there are two sections to the key, indicated by s and r, respectively. The pair (s,r) ought to be distinct and erratic for every Poly1305 algorithm call. But still, the r and s can be produced in a pseudorandom manner. Moreover, r may possess a steady, but it must be changed. Pad1305 divides the message's arbitrary length in half based on its length q fpieces totaling 16 bytes. Little-endian format is used to read the arbitrary-length message, and the r is secured. The clamp function then clears some bits of r, such that $\bar{r} = r_0 + r_1 + r_2 + r_3$ -where $r_0 \in \{0,1,2,\ldots,2^{28}-1\}$, $r_1/2^{32} \in \{0,4,8,\ldots,2^{28}-4\}$, $r_1/2^{64} \in \{0,4,8,\ldots,2^{28}-4\}$, and $r_1/2^{96} \in \{0,4,8,\ldots,2^{28}-4\}$. After initialising the accumulator h, the polynomial outcome (24), is added. Additionally, a 128-bit truncation of the MAC occurs.

$$T = \left(\sum_{i=1}^{q} m_i r^{-q-i+1} \mathrm{mod} 2^{130} - 5 \right) + s\,\mathrm{mod}2^{128} \qquad (23)$$

Algorithm 2. *The Poly1305 authenticator algorithm*

    Require: $K \in (0,1)^{256}$, $M \in (0,1)^L$, $L \in \mathbb{N}$
    Ensure: $MAC = $ Poly1305(K,M,L)
        $(r,s) \leftarrow K$
        $m = (m_1,\ldots,m_q) \leftarrow$ Pad1305(L)
        $\bar{r} \leftarrow$ Clamp(r)
        $h \leftarrow 0$
            for $i \leftarrow 0$ to $(q-1)$ do
        $h \leftarrow h +$ Polynomial$(m,\bar{r},q,i)$
    end for
        $MAC \leftarrow h + s\,\mathrm{mod}2^{128}$
return MAC

B. Hardware Implementation

The Poly1305 core implementation is split into two sections in Figure 5. Using the 256-bit key, a PBlock that is indicated in blue first creates an initial r and s. Next, the Multi-Multiplier and Accumulator (MulAcc) replicates the polynomial given in (23), operating the 128-bit of Block. A 32-bit accumulator and 32-bit unsigned multiplier make up the MulAcc implementation. With four MulAcc integrated into its architecture, A block's 128 bits can be processed by the Poly1305 primitive within a single cycle. It's over by the documentation that there is no set length for the message. As a result, 128 bits are processed by the core at each stage. Block len is a signal that indicates how many bytes are in each Block. Every message block utilised for authentication is managed by a red-highlighted FSM, who also oversees the MulAcc interaction. A new message block is indicated by the signals Next and Init and the new message process, respectively. The finish MulAcc operation is initiated when the final section the message is presented, signalled by the Finish signal. The data is therefore transferred to the Final Block. The data accumulation is then sent to the Final Block (shown in green), which generates the MAC, in the second step. The MAC can be determined by adding the data accumulation that each MulAcc produces. Moreover, the sum's result is truncated to 128 bits. Lastly, the generation of the MAC is indicated by the signal Ready.

Figure 5. Poly1305 primitive architecture

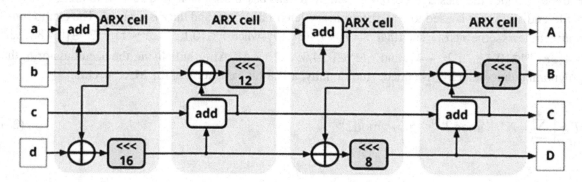

The experiments of the proposed method are evaluated with various components and their specifications are depicted in Table 1.

Table 1. Experimental configuration.

Component	Specifications
Operating system	Windows (X86 ultimate) 64-bit OS
Processor	Intel Pentium CPU G2030 @ 3.00 GHZ
System	64 Bit OS system
RAM	4GB
Hard disk	1 TB

4. RESULTS AND DISCUSSIONS

4.1. Validation Analysis

- **Peak Signal-to-Noise Ratio (PSNR):** Compares the encrypted signal to the original data to determine its quality.

$$PSNR = 10\log_{10}\left(\frac{I^2}{MSE}\right) \tag{24}$$

- **Bit Error Rate (BER):** The number of bit mistakes in encrypted data compared to original data.

$$BER = \frac{\text{No. of bits in Errors}}{\text{Total Number of Bits}} \tag{25}$$

- **Correlation Coefficient (CC):** The linear relationship between the original and encrypted data is measured.

$$CC = \frac{\sum_{i=1}^{N}\left(l_i - d(I)\right) - \left(m_i - d(m)\right)}{\sqrt{\sum_{i=1}^{N}\left(l_i - d(l)\right)^2 - \left(m_i - d(m)\right)^2}} \tag{26}$$

In the above result measurements, one of the image's measurements is represented by the sign N, and the maximum pixel value is represented by the symbol M. The input data and encrypted pictures are represented by A and B, while the periodicity coefficients are indicated by c_1 and c_2.

- **Structural Similarity Index (SSI):** Evaluates the structural similarity between the original and encrypted data.

$$SSI = \frac{\left(2\text{mean}\left(A \times B\right) + c_1\right)\left(2\text{con}\left(A*B\right) + c_2\right)}{\left(\text{mean}\left(A^2\right) + \text{mean}\left(B^2\right) + c_1\right)\left(\text{con}\left(A^2\right) + \text{con}\left(B^2\right) + c_2\right)} \tag{27}$$

Table 2. Task scheduling analysis of proposed ISHO model

Models	Makespan (Time Units)	Throughput (Tasks/Sec)	Degree of Imbalance (%)
ACO	150	30	20
FFO	100	26	15
MBO	120	19	10
CSO	130	12	12
Proposed ISHO model	80	8	5

Various optimisation methods have been examined in the context of task scheduling from table 2 and figures 6 and 7, based on performance criteria such as makespan (measured in time units), throughput (expressed in tasks per second), and the degree of imbalance. The Ant Colony Optimisation (ACO) model had a time period of 150 units, a throughput of 30 jobs per second, and a degree of imbalance of 20%. The Fruit Fly Optimisation (FFO) model revealed a 100-unit makespan, a throughput of 26 operations per second, and a 15% degree of imbalance. The MBO model produced a makespan of 120 units, a throughput of 19 tasks per second, and a 10% degree of imbalance. The makespan of the Cuckoo Search Optimisation (CSO) model was 130 units, the throughput was 12 jobs per second, and the degree of imbalance was 12%. Notably, the proposed Improved Spotted Hyenna Optimisation (ISHO) model outperformed the others with a makespan of 80 units, a throughput of 8 jobs per second, and a 5% degree of imbalance, suggesting its efficiency in task scheduling.

Figure 6. Makespan analysis

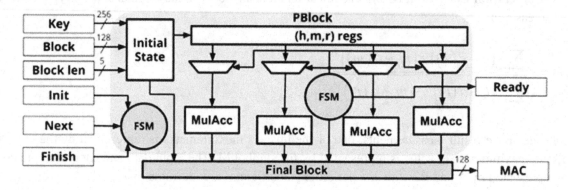

Figure 7. Throughput and degree of imbalance analysis

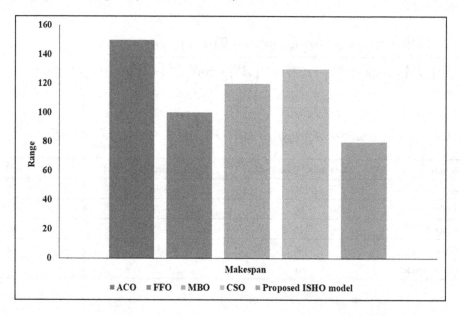

Table 3. Analysis of proposed HHE encryption model with other existing models

Models	PSNR	BER	SSI	CC
AES	30	0.3	0.93	0.91
RSA	34	0.7	0.92	0.95
Triple DES	38	0.9	0.93	0.94
Blowfish	40	0.5	0.94	0.92
Proposed HHE model	43	0.4	0.97	0.98

The evaluation of several encryption models in table 3 and figures 8 and 9 reveals distinct performance characteristics, with measures like Peak Signal-to-Noise Ratio (PSNR), Bit Error Rate (BER), Structural Similarity Index (SSI), and Cross-Correlation (CC) providing useful insights. The Advanced Encryption Standard (AES) achieves a PSNR of 30 dB, a BER of 0.3, an SSI of 0.93, and a CC of 0.91 among the examined encryption algorithms. Despite a slightly higher BER of 0.7 and a comparable SSI of 0.92, the RSA encryption model outperforms, particularly in PSNR of 34 dB and CC of 0.95. The PSNR of the Triple Data Encryption Standard (Triple DES) is 38 dB, the BER is 0.9, the SSI is 0.93, and the CC is 0.94. The PSNR of the Blowfish encryption model is 40 dB, the BER is 0.5, the SSI is 0.94, and the CC is 0.92. Notably, the suggested Hierarchical Hybrid Encryption (HHE) model beats all competitors with a higher PSNR of 43 dB, a lower BER of 0.4, and remarkable SSI and CC values of 0.97 and 0.98, respectively. These findings highlight the HHE model's efficiency in balancing high-quality encryption and robust security, establishing it as a promising solution for secure data transit and storage applications.

Figure 8. PSNR evaluation

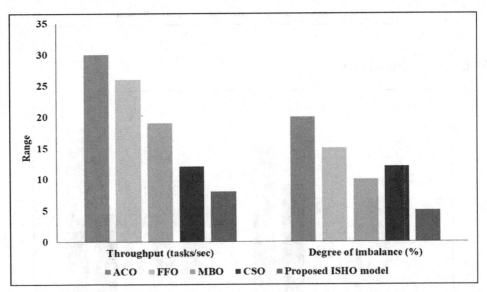

Table 4. Execution time analysis

Models	Encryption Time (ms)	Decryption Time (ms)
AES	5.2	4.8
RSA	4.7	4.1
Triple DES	3.4	2.9
Blowfish	3.1	2.3
Proposed HHE model	2.2	1.8

Figure 9. SSI and CC analysis

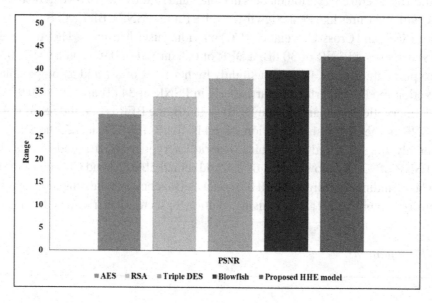

Figure 10. Execution time analysis

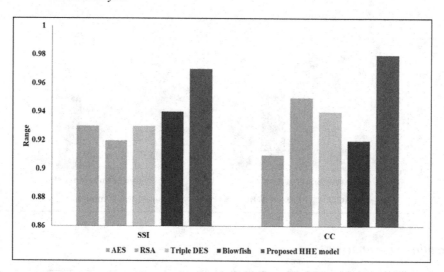

The encryption and decryption timings are critical in evaluating the computational efficiency of various encryption models from table 4 and figure 10. The Advanced Encryption Standard (AES) requires 5.2 milliseconds for encryption and 4.8 milliseconds for decryption across the tested devices. With encryption and decryption times of 4.7 and 4.1 milliseconds, respectively, the RSA encryption model performs marginally better. The Triple Data Encryption Standard (Triple DES) is more efficient, with encryption and decryption times lowered to 3.4 and 2.9 milliseconds, respectively. With encryption and decryption speeds of 3.1 and 2.3 milliseconds, respectively, the Blowfish encryption paradigm improves computing speed even further. Notably, the suggested Hierarchical Hybrid Encryption (HHE) model is highly efficient, with encryption and decryption times reduced to 2.2 and 1.8 milliseconds, respectively.

These findings underscore the HHE model's computing capability, establishing it as an ideal solution for circumstances where fast encryption and decryption operations are critical for preserving operational efficiency and responsiveness.

5. CONCLUSION

To conclude, the new method for efficient work scheduling and improved security in green computing that makes use of the ChaCha20–Poly1305 authenticated encryption algorithm and the Improved Spotted Hyena Optimizer (ISHO) shows promise in resolving the issues of resource management and security in cloud environments. With the improved exploration features of the ISHO integrated, task scheduling becomes more efficient, resulting in better use of resources and shorter execution times. By ensuring a secure data transmission process, the ChaCha20–Poly1305 encryption algorithm adds a crucial layer of protection to the sensitive data being processed in the green computing environment. The proposed approach is compared with various other task scheduling-based approaches for various performance metrics and achieves 43 dB of PSNR. The suggested ISHO can be further improved and optimised in the future to increase its effectiveness and suitability for various cloud computing scenarios. The ChaCha20-Poly1305 encryption algorithm's security features might also be looked into more in order to adjust to changing cybersecurity threats.

REFERENCES

Alsubai, S., Garg, H., & Alqahtani, A. (2023). A Novel Hybrid MSA-CSA Algorithm for Cloud Computing Task Scheduling Problems. *Symmetry*, *15*(10), 1931. doi:10.3390/sym15101931

Baswaraju, S., Maheswari, V. U., Chennam, K. K., Thirumalraj, A., Kantipudi, M. P., & Aluvalu, R. (2023). Future Food Production Prediction Using AROA Based Hybrid Deep Learning Model in Agri-Sector. *Human-Centric Intelligent Systems*, 1-16.

Chandrashekar, C., Krishnadoss, P., Kedalu Poornachary, V., Ananthakrishnan, B., & Rangasamy, K. (2023). HWACOA scheduler: Hybrid weighted ant colony optimization algorithm for task scheduling in cloud computing. *Applied Sciences (Basel, Switzerland)*, *13*(6), 3433. doi:10.3390/app13063433

El-Ela, A. A. A., El-Sehiemy, R. A., Shaheen, A. M., & Kotb, N. (2020). Optimal allocation of DGs with network reconfiguration using improved spotted hyena algorithm. *WSEAS Transactions on Power Systems*, *15*, 60–67. doi:10.37394/232016.2020.15.7

Kallapu, B., Dodmane, R., Thota, S., & Sahu, A. K. (2023). Enhancing Cloud Communication Security: A Blockchain-Powered Framework with Attribute-Aware Encryption. *Electronics (Basel)*, *12*(18), 3890. doi:10.3390/electronics12183890

Khang, A. (Ed.). (2023). *AI and IoT-Based Technologies for Precision Medicine*. IGI Global. doi:10.4018/979-8-3693-0876-9

Malathi, K., & Priyadarsini, K. (2023). Hybrid lion–GA optimization algorithm-based task scheduling approach in cloud computing. *Applied Nanoscience*, *13*(3), 2601–2610. doi:10.1007/s13204-021-02336-y

Serrano, R., Duran, C., Sarmiento, M., Pham, C. K., & Hoang, T. T. (2022). ChaCha20–Poly1305 Authenticated Encryption with Additional Data for Transport Layer Security 1.3. *Cryptography*, *6*(2), 30. doi:10.3390/cryptography6020030

Thirumalraj, A., Asha, V., & Kavin, B. P. (2023). An Improved Hunter-Prey Optimizer-Based DenseNet Model for Classification of Hyper-Spectral Images. In *AI and IoT-Based Technologies for Precision Medicine* (pp. 76–96). IGI Global. doi:10.4018/979-8-3693-0876-9.ch005

Thirumalraj, A., & Balasubramanian, P. K. (n.d.). *Designing a Modified Grey Wolf Optimizer Based Cyclegan Model for Eeg Mi Classification in Bci*. Academic Press.

Thirumalraj, A., & Rajesh, T. (2023). *An Improved ARO Model for Task Offloading in Vehicular Cloud Computing in VANET*. Academic Press.

Yadav, M., & Mishra, A. (2023). An enhanced ordinal optimization with lower scheduling overhead based novel approach for task scheduling in cloud computing environment. *Journal of Cloud Computing (Heidelberg, Germany)*, *12*(1), 1–14. doi:10.1186/s13677-023-00392-z

Yang, G., Li, P., Xiao, K., He, Y., Xu, G., Wang, C., & Chen, X. (2023). An Efficient Attribute-Based Encryption Scheme with Data Security Classification in the Multi-Cloud Environment. *Electronics (Basel)*, *12*(20), 4237. doi:10.3390/electronics12204237

Zhang, A. N., Chu, S. C., Song, P. C., Wang, H., & Pan, J. S. (2022). Task scheduling in cloud computing environment using advanced phasmatodea population evolution algorithms. *Electronics (Basel)*, *11*(9), 1451. doi:10.3390/electronics11091451

Zhang, A. N., Chu, S. C., Song, P. C., Wang, H., & Pan, J. S. (2022). Task scheduling in cloud computing environment using advanced phasmatodea population evolution algorithms. *Electronics (Basel)*, *11*(9), 1451. doi:10.3390/electronics11091451

Chapter 15
Dual–CNN–Based Waste Classification System Using IoT and HDS Algorithm

A. V. Kalpana
Ⓘ https://orcid.org/0000-0003-2289-4968
SRM Institute of Science and Technology, India

S. Suchitra
Ⓘ https://orcid.org/0000-0001-7127-0479
SRM Institute of Science and Technology, India

Ram Prasath
Ⓘ https://orcid.org/0000-0002-2667-9184
SRM Instiute of Science and Technology, India

K. Arthi
SRM Institute of Science and Technology, India

J. Shobana
Ⓘ https://orcid.org/0000-0001-9754-2604
SRM Institute of Science and Technology, India

T. Nadana Ravishankar
Ⓘ https://orcid.org/0000-0002-9854-5150
SRM Institute of Science and Technology, India

ABSTRACT

Efficient waste management is crucial in today's environmental landscape, necessitating comprehensive approaches involving recycling, landfill practices, and cutting-edge technological integration. The proposed approach introduces a sophisticated waste management system, harnessing dual or twofold convolutional neural networks (D-CNN or TF-CNN) and a histogram density segmentation (HDS) algorithm. This intelligent system equips users with the means to enact essential safety protocols while handling waste materials. Notably, this research presents groundbreaking contributions: Firstly, a geometrically designed smart trash box, incorporating ultrasonic and load measurement sensors controlled by a microcontroller, aimed at optimizing waste containment and collection. Secondly, an intelligent method leverages deep learning for the precise classification of digestible and indigestible waste through image processing. Lastly, a cutting-edge real-time waste monitoring system, employing short-range Bluetooth and long-range IoT technology through a dedicated Android application was proposed.

DOI: 10.4018/979-8-3693-1552-1.ch015

INTRODUCTION

Waste management is the term used to describe the necessary tasks and operations that must be completed from the beginning to the end. Waste may be solid, liquid, or gaseous. Numerous procedures are used to handle various waste kinds, such as residential, industrial, and biological garbage. Glasses, cardboard, plastics, papers, and bio garbage are examples of household waste. No waste can be recycled or divided into materials or biological categories when it comes to home activities. As per EUROSTAT 423 million tonne, or 56 percent, of the residential garbage formed in the European Union held in the year 2016 was reprocessed and recycled. In 2016, landfills within the European Union handled 179 million tons of domestically produced waste, constituting 24 percent of the overall total. The reports underscore the importance of efficient household waste management to enhance recycling processes. Integrating modern technology into the waste management system could yield significant benefits. Globally, civic solid waste production amounts to 2.01 billion tons annually, with at least 33 percent not environmentally safely managed. Globally, people produce an average daily waste per capita ranging from 0.11 to 4.54 kg, indicating significant variability. Notably, although high-income countries account for only 16 percent of the world's population, they contribute more than 34 percent (683 million tons) to the total global waste output.

Worldwide, trash production is anticipated to reach 3.40 billion loads of tonne by 2050, outpacing population growth by more than double that amount. Generally, more incomes are associated with more garbage being produced. The quantity of waste formed by a person in a single day is anticipated to climb by 19% in high-income nations by 2050, compared to at least 40% in low-income and middle-income nations. Waste generation is lower at lower income levels, but as income levels increase, it rises more rapidly in low-income communities than in high-income groups. By 2050, the projected total amount of waste generated in low-income nations is expected to be over three times greater than the quantity illustrated in Figure 1 (World Bank, 2012). The Middle East and North Africa region, responsible for 6% of the global waste, ranks second to last, with Eastern Asia and the Pacific leading at 23%. Middle East of Africa, Sub-Saharan region of Africa, North of Africa, and South of Asia are projected to have the quickest rates of rise in trash creation. By 2050, the total amount of rubbish produced in these parts of the world is expected to more than triple, double, and quadruple, respectively. Currently, almost 50% of these areas' trash is discarded outside. Urgent action is necessary due to the high rate of waste increase, which will significantly affect public health, economic development, and the environment.

Waste collection shows a key part in waste management, and the associated charges vary considerably depends on income levels. In high-income and upper-middle-class countries, there is often nearly universal garbage collection coverage. Waste is collected to the amount of 48% in cities in low-income countries; outside of urban areas, the rate drops to 26%. Over 44% of the world's waste is poised in Sub-Saharan region of Africa, whereas at least 90% is poised together in Europe, North America as well as South America and Central Asia. Currently, the most dangerous problem is waste streamline, regardless of its source—domestic or industrial. The length of the consumer product's life cycle, from the opinion of manufacture to the point of clearance, determines the size of the waste streamline. Furthermore, in a few years, the garbage streamlining, consumption, and disposal practices of our society will result in a "waste tsunami." We dispose of an enormous 2 billion tons of trash annually (Kaza et al., n.d.). The World Bank reports that trash creation has been increasing annually and is predicted to reach 3.40 billion loads of tonne, or 70% more, by the year 2050—more than twice than the pace of evolution in population (World Bank, 2012). This is primarily caused by a number of variables, including population expansion, the spread of urbanization, economic expansion, and changing consumer purchasing and lifestyle patterns.

Figure 1. Regional trash production estimates (in million tons/year) (World Bank, 2012)

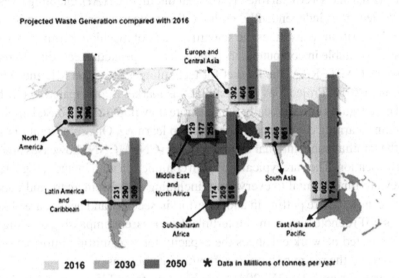

Figure 1 displays the annual trash produced by region worldwide together with projections for the near future. Waste creation is predicted to increase by two or three times in the majority of emerging nations, especially in the sectors of South East Asia as well as Sub-Saharan region of Africa (World Bank, 2012) as shown in Figure 2.

Figure 2. Annual garbage production and projections for it in the coming years across the globe (World Bank, 2012)

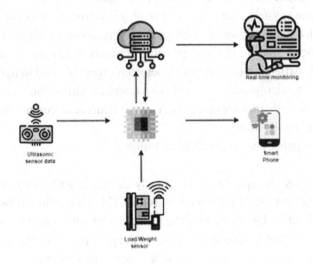

Machine learning (ML) plays a crucial role in Artificial Intelligence (AI), enabling a system to acquire knowledge and make decisions independently, without direct supervision. The examination of specific statistical models and algorithms constitutes the scientific study of machine learning. Because ML offers the best characteristics available in computers, its popularity has peaked. According to recent data from Tractica, also known as CAPTERRA, the market for technologies based on ML and AI experienced a growth of $1.4 billion in 2016. Projections suggest that this market is anticipated to reach $59.8 billion by the year 2025. These figures unequivocally demonstrate how popular ML-based applications are. In a similar form, machine learning cannot exist without deep learning. One important discussion of deep neural networks is the familiar convolutional neural network (CNN). CNN showcases amazing advancements in picture identification. They are typically used in conjunction with image categorization to assess visual imagery. They are fundamental to everything, including self-driving cars and Facebook's photo tagging. Behind the scenes, they're putting in a lot of effort in security and healthcare alike. Conversely, the Internet of Things (IoT) denotes a system of interlinked devices, encompassing both digital and analog entities. This interconnected network enhances the capacity for transmitting information throughout a network, extending beyond the realm of human-computer interactions.

According to estimates from (FUSON, 2020), there are up to 127 innovative IoT devices linked to municipal networks each second on middling. 328 million pieces are connected every month as a consequence of the speedy expansion. These statistics demonstrate how IoT will play a big role in contemporary computing. The Internet of Things business market has grown along with its daily rise in popularity. According to data, the smart home sector alone will account for $151.2 billion of the worldwide IoT industry in 2023. The IoT arcade as a whole can be alienated into a numerous areas, including the banking and financial transaction sector, the health market, and the equipment category for education and training. Researchers predict that in the coming years, every device—whether digital or analog—will be part of the IoT. STATISTA (2020) projects that by 2023, the IoT market is expected to reach $1.1 trillion. These figures highlight the important role that IoT plays in the digital age.

By implementing a smart system, the suggested method allows users to take the appropriate safety measures on the system for handling waste. This paper offers several insights to the area of waste administration in the interest of finding the best possible aqueous solutions. A smart approach to apply deep learning for waste classification to distinguish between digestible and indigestible waste e; A smart garbage can is built using a microprocessor, load sensor and an ultrasonic sensor. A clever method of keeping an eye on waste in real time utilizing short-range Bluetooth connectivity and long-range IoT technology via an Android application.

Our contribution to the paper can be précised as follows:

- Creation of a smart trash receptacle, meticulously designed with integrated ultrasonic and load measurement sensors governed by a microcontroller. This innovation aims to revolutionize waste containment and collection by optimizing the process through precise sensor-based monitoring.
- Implementing an intelligent approach that utilizes deep learning for the accurate classification of digestible and indigestible wastes through image processing techniques.
- Proposed a sophisticated real-time waste monitoring system that utilizes short-range Bluetooth and long-range IoT technology within a dedicated Android application.

The rest of this research article is structured as follows. Section 2 delivers a review of related work, focusing on various AI techniques applied to Waste Classification Systems. Section 3 gives proposed method for accurately classifying the waste as digestible and indigestible wastes. Section 4 discusses on Performance evaluation for the proposed system and our conclusion is specified in Section 5

RELATED WORKS

Recently, cutting-edge technologies such as computer vision, IoT, and machine learning have played crucial roles in identifying and categorizing waste across diverse waste management domains. Computer vision, a rapidly advancing field, employs algorithms that enable computers to comprehend and classify digital images akin to human capabilities. Deep learning is a subfield of artificial intelligence that uses ANNs, which mimic brain activity, to learn features. Activities in garbage sorting are made possible by machine learning structures, which consist of several layers of a predetermined size. This study aims to explore cutting-edge waste management in addition to classification initiatives in smart metropolises. While significant study has focused on waste collection and processing, limited attention has been given to bin devices, which hold substantial potential in addressing waste management challenges. Previous studies have extensively covered garbage management systems, yet the specific exploration of bin devices remains relatively scarce despite their capacity to revolutionize waste controlling practices.

Dugdhe et al. (2016) showcased an innovative system integrating RFID-enabled trash bins, ultrasonic measurement tools, and gas sensors. This system also incorporated a garbage collection schedule for vehicles, utilizing mathematical algorithms to calculate the most efficient routes between full bins and those emitting hazardous gasses. Catania and Ventura (2014) employed a Smart-M3 platform as well as intelligent monitoring to develop an efficient garbage collection system. Their methodology comprises two main phases: firstly, a continuous measurement process that monitors, stores, and transmits trash levels within compartments; secondly, leveraging this data to optimize and refine the routes for garbage collection.

Bueno-Delgado et al. (2019) proposed a garbage disposal system designed for rural regions that makes use of LoRaWAN technology and optimises the routes used for garbage collection. However, despite the implementation of IoT, the method lacked consistent connectivity and optimization capabilities across the rubbish bins involved. Misra, Das, Chakrabortty, and Das (2018) proposed a virtualized IoT-centered waste management system that anticipates trash bin spillage beforehand. This system utilizes ultrasonic sensors to gauge the waste level in bins, transmitting this data over the internet to a database for analysis and recommendation. Subsequently, this information is leveraged to map garbage containers and determine whether collection is necessary. This distinctive strategy aims to forecast future events while drawing insights from the current waste status.

A trash can linked to an Internet of Things database was presented by Hussain et al. (2020); Comprehensive pollution projections are generated by the information system once it assesses the state of the bin. Urban trash management with the "Smart Bin" concept was demonstrated by Sharma et al. (2015). This network of sensor-enabled smart bins, connected via cellular connections to the city's waste management system, produces a significant amount of data. In an effort to encourage users, Abd Wahab et al. (2014) created a technologically advanced Recycle dustbin fitted onto a trash-type detection system to award opinions centred on the bulk or sort of waste dumped. While this method addresses garbage sorting issues by deducting user points for inappropriate waste disposal, accurately sensing the garbage type remains unpredictable and subject to change. Consequently, this approach has not contributed to the enhancement of the waste collection system. Liu et al. (2010) developed a Bayesian statistics-drive framework for intelligent garbage categorization, utilizing a straightforward information processing method. Despite its robust mathematical foundation, this technology cannot achieve full automation as it relies on features obtained from human intervention.

Bobulski and Kubanek (2019) developed a method for trash categorization that makes use of CNN methods and computational imaging. Their research primarily concentrated on detecting polyethylene. Additionally, the authors conducted numerous experiments aimed at polyester, polypropylene and high-density polyethylene are the substances being identified. Sreelakshmi et al. (2019) utilised Capsule Neural Network (Capsule-Net) as part of their strategy for managing garbage, allowing for the identification of items other than plastic. They conducted research on two public datasets, achieving accuracy rates of 96.3% and 95.7%. Moreover, they thoroughly integrated and tested the system across various hardware devices.

Huiyu (2019) presented an inventive classification model employing deep learning mechanisms to distinguish different types of waste. They also applied the system for garbage recycling. In a separate study, Adedeji and Wang (2019) proposed a widely-used technique for trash recognition utilizing deep learning. The model's use in the recycling waste classification was also emphasized by the authors. Misra, Das, Chakrabortty, and Das (2018) explored an innovative system for detecting and categorising electronic waste (e-waste). Their model incorporated a CNN for classification and an RCNN for identification across multiple types of e-waste. The authors reported detection as well as classification accuracy ranging from 90- 97 percent.

Samann (2017) introduced an impactful method for automated and robust waste management. A intelligent garbage can with many gas detectors and an ultrasonic detector was unveiled by the writers. They also suggested a smartphone app and cloud server for waste visualisation in instantaneous fashion. Malapur and Pattanshetti (2017) presented an intelligent trash can that is affordable and made for optimal waste management. Through IoT integration, the authors included gadgets like a sensor for ultrasound, an Arduino micro, and a module for the GSM network. Through the usage of the GPS module, the organisation was able to notify the user through a text message if the waste level reached a certain threshold. The device also had a memory card to send users an audio message and a PIR motion sensor. According to the authors, the suggested system performed satisfactorily.

Balaji (2017) created an intelligent garbage can that can detect waste levels. For operation, the device required both Wi-Fi as well as a web server. The authors gauged the trash can's level of fullness using an infrared distance measuring device. The gathered information and outcomes were then sent via an Android application through the web server. Bai et al. (2018) developed an intelligent waste management system based on the IoT to address and minimize food waste. The authors have centralized control over every component by using mesh technology. In order to collect and analyze data on food spoilage, their model included a router and a server. After a successful trial period, the technology reduced food waste by 33 percent. Muthugala et al. (2020) unveiled a rubbish pick-up robot with ground navigation capabilities. The authors claimed that by using deep learning processes, the suggested architecture was able to effectively detect junk. The prototype's accuracy rate in detecting waste was 95%.

To improve monitoring and management systems, numerous researchers have used a variety of approaches, such as methods based on machine learning (Khan, Abbas, Atta et al, 2020; Zahra et al., 2021), deep learning using transfer learning approaches (Ghazal et al., 2022; Ihnaini et al., 2021; Khan, Abbas, Khan et al, 2020), and computational intelligence approaches (Abbas et al., 2020; Khan et al.,

2021; Khan, Abbas, Ditta et al, 2020). The collection system was the main focus of the research that was described in this paper. It placed emphasis on the creation of intelligent trash cans that can produce data and manage waste effectively. In turn, this data helps the collecting system by enabling changes to the position and capacity of the compartments. This approach, which focuses on route optimization, continuously seeks to lower fuel consumption, transportation material utilization, and human resource time as well as costs. Additional research projects focus on sorting waste materials into different recycling groups. Reuse, recycling, and categorization are by definition the best waste management techniques. This study's technique incorporates knowledge from solutions based on the literature. Although numerous studies have proposed Internet of Things (IoT) systems for managing waste, no one has yet provided a detailed blueprint for such a system that makes use of the jargon associated with the reinforcement learning paradigm.

PROPOSED METHODOLOGY

The suggested approach serves as a blueprint for smart waste bins, as outlined in module 1. This module facilitates real-time data monitoring through the IoT In module 2, the waste sorting process is implemented, integrating dual-convolutional neural networks (dual CNN) and a histogram density-based segmentation algorithm. When two mathematical models of structure are combined, they produce outstanding results in waste management. Sorting waste into certain groups facilitates the identification of recyclable materials. When it comes to waste categorization, machine learning algorithms are unrivalled. The possibility of reducing the use of recyclable materials drives the author to use deep learning for garbage sorting and waste tracking to differentiate recyclable waste. This article primarily divided them into digestible and indigestible waste types. In this article, we have distinguished between two primary categories of waste: digestible and indigestible. We have utilized refined models to classify garbage instead of having enough data to do so. The utilization of deep learning technology for waste classification facilitates the extraction of waste categories from photographs. Trash boxes' layout makes it possible for several sensors to collect data and transmit readings for monitoring.

Implementation of Trash Box

Public trash cans perform the dual roles of intelligent entry points and trash receptacles. The following data is sent by the trash can to the system's central database: the weight and percentage measurement of the waste can. Each garbage can is assigned a unique detection number and is integrated into the Internet of Things (IoT) network for connectivity.

The design and development technique of the innovative smart garbage box architecture are covered in this section. The schematic diagram of the garbage box we built is displayed in Figure 3. The overall process in this setup is controlled by a microcontroller called "ESP8266" NodeMCU, as depicted in the image. This microcontroller is outfitted with an ultrasonic sensor that gauges the fill level of the trash box. The top of our created device is where the ultrasonic sensor is placed. The intensity of emptiness in the garbage can is verified by the transmission and reception of ultrasound signals. The microcontroller receives the calculated value. Additionally, a load measuring sensor is positioned at the surface's base. The waste's weight in kilograms is determined by this sensor. In the context of time, the load measurement sensor operates based on the accumulating load of garbage. The load value increases proportionally with the periodic growth in the amount of rubbish in the trash box. The microcontroller is then set to the modified value. The created Android application obtains the values from the load measurement sensor plus ultrasonic transducer. Using an application developed using Android, operators can speedily see the heaviness of the garbage and the amount of space that is currently empty in the bin. In the presence of an internet connection, the pertinent data will be transmitted to the cloud server. Subsequently, operators can monitor the real-time data using an Android application.

Figure 3. Implementation of Trash Box

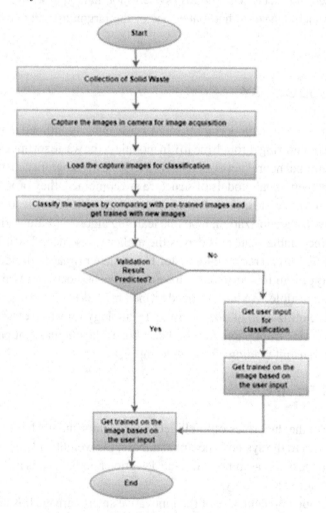

Figure 4 illustrates the flow diagram and Figure 5 illustrates the hardware system model for the recommended solid waste sorting scheme. The classification process consists of two phases. In the initial phase, solid waste is collected, and images are captured using a camera. A proximity sensor activates the camera module when waste is detected, capturing images with the Raspberry Pi Camera Module v2. The second phase involves waste classification through a dedicated module. The image cataloguing module authenticates captured images, categorizes solid waste, and directs them to the appropriate bin.

The system conveys the category of collected trash through an automatic pre-recorded voice message. A pre-programmed automatic control unit transmits information stored in the cloud. The system can seek input from sanitary workers for unidentified waste particles, facilitating self-training for future predictions without human intervention. The waste sorting model, which has been pre-trained, is based on CNN architecture. Comprehensive explanations of its structure and functioning can be found in sections such as "Building a Machine Learning Model," "Training and Evaluation," and "Evaluation Standards."

Figure 4. Workflow of the proposed system

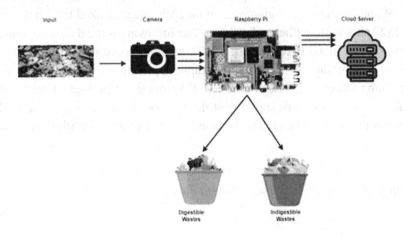

Figure 5. Hardware model for the proposed system

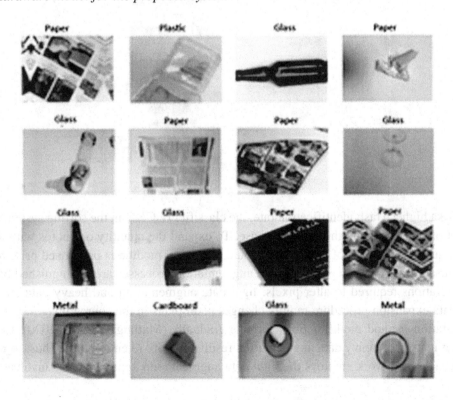

Dataset

Finding recyclable materials is crucial for the survival of civilization and humanity. Furthermore, it is indisputable that the global ecosystem is valuable and necessary for a planet that can support human habitation. Our goal in this effort is to categorize the recycling of items including cardboard, paper, glass, and metal. The models were trained using the TrashNet (Aral et al., 2018) dataset. Paper, glass, plastic, metal, carton, and waste subcategories are included in this dataset as shown in Figure 6.

This collection is made up of pictures of trash shot against a white background. The variances in the dataset are a result of the various lighting and exposure settings chosen for each image. The initial dataset has a size of approximately 3.5 gigabytes, with each image resized to dimensions of 512×384 pixels. There are 2527 photos altogether in TrashNet. The following is the dataset's content: Paper (594), glass (501), rubbish (137), metal (410), plastic (482), and cardboard (403). In the data augmentation process, the test database contains about 1000 photographs depicting both organic and ecological garbage. Meanwhile, the training folder comprises around 13,000 images of biological waste and ten thousand photos of recyclable waste. To significantly boost the variety of data available for model training, the suggested pre-processing model is created and trained using data augmentation, eliminating the necessity for extra data.

Figure 6. Illustration of TrashNet data from different classes

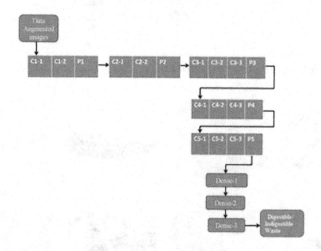

To achieve a higher Trash identification rate, the classifier of CNN in the sorting system requires an extensive set of images during the training phase. To expand the quantity of photos without compromising pixel quality in images of trash, the data augmentation module is positioned prior to the CNN classification segment. Two categories of data augmentation processes are distinguished based on the number of iterations required to alter pixels: light data augmentation and heavy data augmentation. The heavy data expansion procedure rapidly changes pixel plans in the source images, while the light data augmentation method modifies pixel formats gradually (Muthugala et al., 2020). Compared to light data augmentation, heavy data augmentation results in more retention loss. Thus, to increase the number of images, this work employs the light data augmentation approach. This involves a randomized flip transformation with a 0.5 probability ratio for left and right flips and an affine translation of pixels with scaling factors of 0.5, 0.7, and 0.9. The random flip transformation creates two augmented images (representing left and right), whereas the affine conversion generates three enhanced images at scales of 0.5, 0.7, and 0.9. Consequently, each source image in this work yields five photos following the data augmentation procedure. In this research, 16% of the photos are allocated for testing, 14% for validation, and 70% for training.

Edge Preserving Image Fusion

The Trash dataset provided the trash images used in this study for the digestible and indigestible garbage cases. Since there is no noise in the photos in this dataset, no de-noising technique or algorithm was employed in this research. As part of the pre-processing technique, this research proposes the use of the edge-conserving image fusion procedure to improve the pixels in trash images. Fusion entails combining the pixels of two consecutive photographs to generate an enhanced image (Kaur et al. 2021). Traditional image fusion techniques include pixel-level and area-level approaches, which combine every pixel in two images to generate an enhanced image. In a trash detection system, edge pixels carry more significance than non-edge pixels. Therefore, it is crucial to retain these edge pixels during the fusion process. To ensure edge preservation, this work recommends the use of an edge-preserving image fusion technique.

The 'Canny' edge detector (Wu et al., 2019) is utilized on together source trash images, denoted as $S1(i, j)$ and $S2(i, j)$. This application is done to identify edges in the images, as explained below.

$$E_1(i,j) = canny(S_1(i,j)) \tag{1}$$

$$E_2(i,j) = canny(S_2(i,j)) \tag{2}$$

$E_1(i, j)$ and $E_2(i, j)$ are added to create the fused image $F_i(i,j)$, as explained in Equation (1) and Equation (2).

$$F_i(i,j) = E_1(i,j) + K* E_2(i,j) \tag{3}$$

where $E_1(i,j)$ as well as $E_2(i,j)$ are the edge-detected source images, respectively, and $F_i(i,j)$ as stated in Equation (3) corresponds to the fused picture. In addition, K represents the kappa factor that is involved.

The fused image can only be obtained by means of the Kappa factor. The following procedures are used to find the Kappa factor.

Step 1: As mentioned below, get the Euclidean Mean as stated in Equation (4) from the fused image in addition to source image 1.

$$EM_1 = \frac{1}{2} \|F(i, j) - S_1(i, j)\|^2 \tag{4}$$

Step 2: As previously mentioned, calculate the Euclidean Mean value, as defined in Equation (5), between the fused image and source image 2.

$$EM_2 = \frac{1}{2}\|F(i,j) - S_2(i,j)\|^2 \tag{5}$$

Step 3: Equation (6) and Equation (7) is used to find the Kappa factor if $EM_1 \geq EM_2$.

$$K = EM_1 \tag{6}$$

else

$$K = EM_2 \tag{7}$$

The Kappa factor has a value in the range of 0.5 and 0.7. The garbage pixel intensities are used to calculate the Kappa factor.

Proposed CNN Waste Classifier

A sophisticated feed-forward network is the convolution neural network (CNN). It can resolve a variety of issues that were previously unresolved. CNN utilizes picture feature extraction to function. CNNs are frequently employed for image classification owing to their excellent accuracy. Image classification involves analyzing an image to identify the specific category it belongs to. In our proposed method, we categorized garbage into two primary groups: digestible waste and indigestible waste. Paper, cardboard, glass, metal, and other materials can all be considered indigestible garbage.

As evidenced by several decades' worth of research, deep learning effectively contributes to better classification outcomes when compared to machine learning techniques. CNN has been utilized to identify certain Regions of Interest (RoI) within the image. Consequently, the CNN classifier is used in this work to categorize garbage images as either digestible garbage or indigestible garbage images. With the use of convolutional layers, the current CNN architectures create features on their own and use those features to obtain classification results. The classification rate is low even though this method conducts the classification procedure significantly. More internal features are needed in the suggested work in order to increase the categorization rate. Hence, the Dual Mode CNN (DM-CNN) is proposed as a technique for classifying garbage images. In this approach, Mode-1 CNN is employed to extract features from the data-augmented image, and subsequently, Mode-2 CNN classifies the features extracted by Mode-1 CNN.

This research study discusses the modified VGG-16, wherein all internal components are configured in parallel to employ convolutional layers for constructing internal features. The architecture introduced in this study is now denoted as the proposed VGG-16 configuration. As a result, there is a notable reduction in the volume of feature generation through convolutional layers. Subsequently, the Mode-2 CNN (Yolo-v2) is utilized for the classification of these generated features. In our approach, the features extracted by Mode-1 CNN act as the input for Mode-2 CNN. Leveraging the internally generated features from Mode-1 CNN, Mode-2 CNN further refines and enhances these features. The augmentation contributes to an improved classification rate for the suggested trash detection system. Both CNN modes follow the sequence of Convolutional, Activation, Pooling and Dense layers. The Convolutional layer calculates inner linear features $C(i,j)$ from the input image, as expressed in Equation (8). This computation involves the Convolution process between the kernels of the Convolutional layers, denoted as $K(p,q)$, and the input data-augmented image, denoted as $D(i,j)$.

$$C(i,j) = \sum_{p}\sum_{q} K(p,q) * D(i-p, j-q) \qquad (8)$$

The pooling layer serves to reduce the computational cost associated with the internally generated features, which are the outputs of the convolutional layer. It operates by reducing the computational dimensionality of the output patterns from each convolutional layer. In this study, Max pooling is utilized to address overfitting concerns in the pooling layer output, although Average pooling is another type of pooling layer.

The pooling and convolutional layers generate output features in a straightforward pattern. However, for implementing complex applications, linear patterns are insufficient. Therefore, it is crucial to transform these linear patterns into non-linear patterns, a task accomplished by the activation layer. Various activation functions, such as the Linear Rectification Unit (ReLU) and Sigmoid, are available. This study utilizes the ReLU activation function to mitigate gradient-related challenges in the layer learning process. Furthermore, the dense layers embody fully connected neural networks, establishing connections between the output nodes of the current layer and the nodes in the preceding layer.

The VGG-CNN diagram, illustrated in Figure 7, serves as the foundation for the architectural schematic of the proposed Mode-1 CNN, depicted in Figure 8. The architectural diagram of Mode-1 CNN adapts the VGG-CNN architecture by reconfiguring the design of the internal modules from a serial to a parallel configuration. Both architectures consist of five internal architectural modules, labeled as Modules 1 through 5, and both accept data-augmented images as input. Modules 1 and 2 consist of two Convolutional layers (C) and a single Pooling layer (P). Modules 3, 4, and 5 each include three Convolutional layers (C) along with a Pooling layer (P), as depicted in Figure 7. After passing through each of the three Dense layers, the output from Module 5 is classified as either digestible or indigestible garbage. The data-augmented image is fed into the proposed VGG-16 architecture diagram, and it is simultaneously passed to Modules 1 and 2. The internal feature matrix is created by integrating the results of Modules 1 and 2. A simultaneous pass of this within created characteristic matrix is made to Modules 3, 4, and 5. Three thick layers are passed through before the feature integrator integrates the outputs to these 3 architectural internal modules and generates a feature matrix (Feature Vector-FV). As illustrated in Figure 8, the output from the 3rd dense layer provides the categorization results, distinguishing between digestible and indigestible garbage. The internal layers of this proposed VGG-16 design are structured using hyper parameters.

The following is a list of the hyper parameters that this architecture uses.

- Total count of CNN kernels
- Dimensions of every kernels
- Total count of PL
- Dimension of PL
- Walk size.

Total count of CNN in the CNN1-1, CNN1-2, CNN2-1, and CNN2-2 designs in the planned VGG-16 is established at 128. Additionally, 256 kernels are fixed in the designs of CNN 3-2, CNN3-1, and CNN3-3; 512 kernels are fixed in the designs of CNN4-1, CNN4-2, and CNN4-3; and 512 kernels are fixed in the designs of CNN5-2, CNN5-1, and CNN5-3. In this proposed system, hyper parameters are chosen and fixed according to the category accuracy. In this proposed approach, the hyperparameter values are kept constant once the classification accuracy reaches the optimal level.

Figure 7. Existing VGG-16 architecture for waste classification

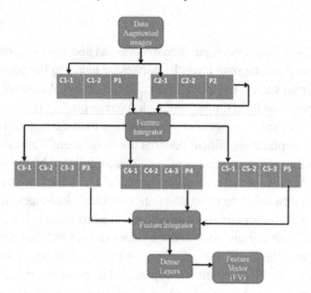

Figure 8. Proposed VGG-16 architecture for waste classification

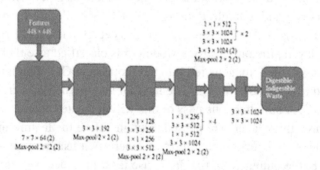

As per Tanvir Ahmad et al. (2020), the features obtained are categorized through Mode-1 CNN, and the YOLO-v2 is utilized as the mode-2 CNN for further processing. The YOLO-v2 architecture, employed in this research to classify digestible and indigestible wastes, is illustrated in Figure 9. A matrix with 448 rows and 448 columns contains the elements from the Mode-1 CNN architecture. To obtain classification results, the feature matrix undergoes processing through the sets of Convolutional (C) and Pooling (P) layers. The number of filters within each of the Convolutional layers in the YOLO-v2 architecture is increased linearly. This is done to achieve significant classification outcomes in a shorter duration for detection. The percentage of this architecture's learning is 10^{-4} over 150 epochs with a 20-lot size are employed in this architecture's training mode. The garbage Region of Interest (RoI) is now segmented using the YOLO v2 categorized Glioma brain picture. This work proposes the following garbage segmentation using Histogram Density Segmentation (HDS) algorithm to segment the RoI of the garbage in the categorized appearance.

Figure 9. YOLO-v2 architecture for classifying digestible and indigestible waste

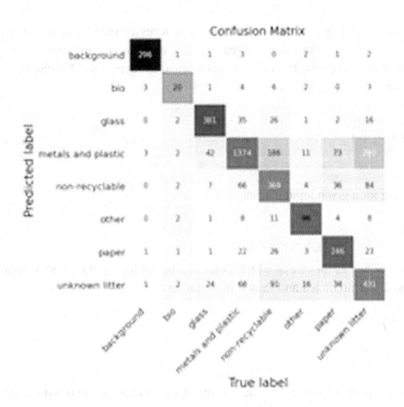

Algorithm: The HDS algorithm for statistical concentration

Input: Garbage picture
Output: Digestible and Indigestible Garbage RoI
  Start;
    Step 1: For every pixel in trash picture I, find the histogram count (h_i).
    Step 2: Calculate the average Key Factor (KF) using the computed distribution count and the formula.

$$KF = \frac{1}{N}\sum_i h_i$$

    N represent quantity of calculations made on the histogram.
    Step 3: Using *KF*, divide the waste image into two separate parts as instructed in the adhering to standards.
$I_1 = I \geq KF$
$I_2 = I < KF$
    Step 4: Figure out the PDF of the two computed images by using the formulas provided below.
$p1 = \mathrm{pdf}(I_1)$
$p2 = \mathrm{pdf}(I_2)$
    Step 5: With the use of the computed PDF p1 and p2, the garbage ROI may be ascertained using the equation below.
$p1 \leq GarbageRoI \leq p2$ (14)
  End;

EXPERIMENTAL EVALUATION

Performance and accuracy rate of the proposed system is accessed using multiple attributes like Mean and absolute error (MAE) rate, Precision, F-Measure, Recall, accuracy and specificity.

MAE: This attribute computes the accuracy rate tagged to regressor. This can be computed as depicted in Equation(9)

$$MAE = \frac{1}{n} \sum_{i-1}^{N} (x_i - x_i) \qquad (9)$$

x_i - Denotes right tag for a given input dataset
x_j - Prediction rate
N - Total samples

Precision: It is defined as the rate at which Total number of inputs classified accurately to the total number of false positives predicted and shown in Equation (10).

$$Precision = \frac{TP}{TP + FP} \qquad (10)$$

Recall: This measure is defined as the ratio of total number classes identified accurately to the total false negatives and shown in Equation (11).

$$Recall = \frac{TP}{TP + FN} \qquad (11)$$

F-Measure: F-beta-measure is mean of harmonic values between recalls, precision recall. This value ranges from 0 to 1. If f-measure is closer to 1 then the accuracy is poor and if the value is near to 0 then accuracy is positive and shown in Equation (12).

$$FMeasure = 2 \frac{precision \times recall}{precision + recall} \qquad (12)$$

Accuracy: There are two types of accuracies 1) accuracy with training dataset and 2) Actual accuracy associated with actual data to be validated. This normally means the perfection rate of predicted value and shown in Equation (13).

$$Accuracy = \frac{(TP + TN)}{(TP + FP + TN + FN)} \qquad (13)$$

TP - True-Positive

TN - True-Negative

FP - False-Positive

FN - False Negative

Sensitivity: Rate at which true positives are accurately identified by a model is termed as sensitivity and shown in Equation (14).

$$Sensitivity = \frac{TP}{TP + FN} \tag{14}$$

Specificity: Similar to sensitivity this parameter is the rate at which True negatives are identified. It is computed as stated in Equation (15)

$$Specificity = \frac{TN}{FP + TN} \tag{15}$$

4.1 Results and Discussions

The experimental setup utilizes a dedicated cloud server featuring an Intel(R) Xeon(R) CPU E5-2630 v4 @ 2.20 GHz with a TDP rating of 85 W (thermal design power). The operating system used in this research is Red Hat Enterprise Linux Server 7.4 x86_64. The system is equipped with two Tesla K40m GPUs, each having a TGP (Total Graphics Power) rating of 235 W. The InceptionNet and Xception models exhibit a computational cost of approximately 16 ms per frame. When the server operates at full capacity, it can generate around 76 frames per second. Consequently, the system can support a minimum of 4 industrial stations at a real-time frame rate of up to 19 frames per second.

The confusion matrix as shown in Figure 10 shows that the model primarily misidentified glass with plastic and metal. Owing to its transparency and reflection, glass is thought to be the hardest substance to classify—even by humans—without specifics like weight and surface characteristics.

Figure 10. Confusion matrix of TrashNet dataset

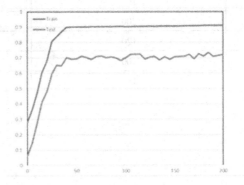

A number of indicators, including the confusion matrix and the Receiver Operating Characteristic (ROC) curve, are used to assess the effectiveness of the proposed system, which is shown in Figure 11.

Figure 11. ROC curve

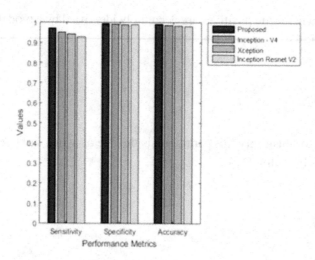

Performance and accuracy rate of the proposed system is accessed using multiple attributes like MAE - Mean and absolute error rate, Precision, Recall, F-Measure, accuracy and specificity.

Table 1 demonstrate how the proposed trained model outperforms more intricately modified models. The outcomes were significantly influenced by hyper-parameter tuning and data augmentation strategies.

Table 1. Epochs and accuracy results on test data, compared with Sarala et al. (2023), using the Trash-Net dataset

Model	Epochs	Accuracy
Inception ResnetV2	100	89.13%
Xception	100	80.25%
Xception	150	82.07%
CNN	100	78.07%
Inception-V4	200	89.33%
Inception-V4	120	94.14%
Dual-CNN+HDS (Proposed)	100	95.23%
Dual-CNN+HDS (Proposed)	**150**	**96.12%**

Efficiency of the Proposed System

Effectiveness of our method is evaluated with the other existing techniques like Inception V4, Xception and Inception V2 and the results are presented in Table 2.

Table 2. Efficiency comparison of proposed system

Measures (%)	Proposed	Inception V4	Xception	Inception V2
Accuracy	99.1	98.2	98.5	98.3
F-Measure	97.2	93.1	95.2	94.5
Recall	98.5	93.2	96.2	93.2
Precision	98.2	93.4	94.1	93.2
Sensitivity	97.3	93.1	93.7	92.1
Specificity	99.4	99.1	92.2	94.6

Our proposed approach has achieved 98.2% accuracy in comparison with other existing algorithms 98% for Inception V4, 98.6% for Xception and 98.4% for Inception V2, that are considered minimum in comparison with the recommended approach as shown in Figure 12 and Figure 13.

Figure 12. Comparison of sensitivity, specificity, and accuracy

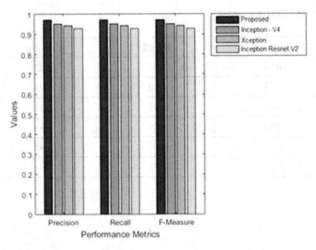

In this research methodology, the obtained results in recall, F-measure, precision, and sensitivity demonstrate considerable potential for improvement. The specificity value, reaching 99.5%, exceeds that of alternative approaches. Figure 12 provides a visual representation of the assessment of sensitivity, specificity, and accuracy among different classifiers. Likewise, Figure 13 offers a graphical distinction of F-measure, recall, and precision employing diverse classifiers.

Figure 13. Comparison of precision, recall, and F-measure

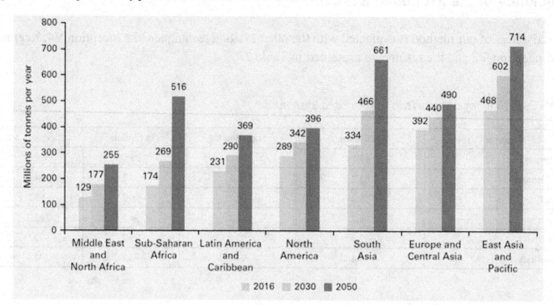

Table 3 displays the classification report for the Trash Dataset on test data, presenting Precision, Recall, F1-Score, and Support values for each category. The results indicate that cardboard is notably more predictable compared to the other types of trashes.

Table 3. Classification report on test data

Category	Precision	Recall	F1-Score	Support
Cardboard	0.97	0.96	0.97	103
Glass	0.93	0.91	0.93	124
Metal	0.92	0.91	0.92	102
Paper	0.94	0.97	0.95	139
Plastic	0.93	0.92	0.92	123
Trash	0.83	0.82	0.83	36

CONCLUSION

In conclusion, the integration of deep learning and IoT in waste management, as explored in this study, presents a transformative approach towards creating a sustainable and efficient waste handling infrastructure. The fusion of Dual Neural Systems and the Separation of Density Histograms algorithm offers a sophisticated waste management system, enabling users to implement crucial safety protocols. The innovative contributions, including a geometrically designed smart trash box, intelligent waste classification through deep learning, and a real-time monitoring system, underscore the adaptability and robustness of the proposed solution. The system's noteworthy classification accuracy of 96.12% underscores its potential for broad application, extending beyond waste management facilities to household settings. These results highlight the promising trajectory of advanced technologies in waste management, paving the way for a more environmentally conscious and effective waste management paradigm. In future research, it is crucial to extend the exploration of advanced technologies in waste management, with a particular emphasis on scaling the proposed smart waste management system for urban implementation. Further refinement of deep learning algorithms for waste classification is warranted, with a focus on continuous training and improvement to enhance accuracy, especially in managing diverse waste streams. Additionally, the incorporation of predictive analytics holds promise for proactive waste management, utilizing historical data to forecast overflows, optimize collection routes, and reduce environmental impact.

REFERENCES

Abbas, S., Khan, M. A., Falcon-Morales, L. E., Rehman, A., Saeed, Y., Zareei, M., Zeb, A., & Mohamed, E. M. (2020). Modeling, simulation and optimization of power plant energy sustainability for IoT enabled smart cities empowered with deep extreme learning machine. *IEEE Access : Practical Innovations, Open Solutions*, 8, 39982–39997. doi:10.1109/ACCESS.2020.2976452

Abd Wahab, M. H., Kadir, A. A., Tomari, M. R., & Jabbar, M. H. (2014). Smart Recycle Bin: A Conceptual Approach of Smart Waste Management with Integrated Web Based System. *2014 International Conference on IT Convergence and Security (ICITCS)*, 1-4. 10.1109/ICITCS.2014.7021812

Adedeji, O., & Wang, Z. (2019). Intelligent waste classification system using deep learning convolutional neural network. *Procedia Manufacturing*, 35, 607–612. doi:10.1016/j.promfg.2019.05.086

Aral, R. A., Keskin, S. R., & Kaya, M. (2018). Classification of trashnet dataset based on deep learning models. *Proceedings of the 2018 IEEE International Conference on Big Data (Big Data)*, 2058–206.

Bai, J., Lian, S., Liu, Z., Wang, K., & Liu, D. (2018). Deep learning based robot for automatically picking up garbage on the grass. *IEEE Transactions on Consumer Electronics*, 64(3), 382–389. doi:10.1109/TCE.2018.2859629

Balaji, D. (2017). *Smart trash can using internet of things*. Academic Press.

Bobulski, J., & Kubanek, M. (2019). Waste Classification System Using Image Processing and Convolutional Neural Networks. *International Conference on Artificial Neural Networks*, 350-361. 10.1007/978-3-030-20518-8_30

Bueno-Delgado, M.-V., Romero-G'azquez, J.-L., Jim'enez, P., & Pav'on-Mariño, P. (2019). Optimal path planning for selective waste collection in smart cities. *Sensors (Basel)*, 19(9), 1973. doi:10.3390/s19091973 PMID:31035549

Catania, V., & Ventura, D. (2014). An Approach for Monitoring and Smart Planning of Urban Solid Waste Management Using Smart-M3 Platform. *Proceedings of the 15th Conference of Open Innovations Association FRUCT*, 24–31. 10.1109/FRUCT.2014.6872422

Chatterjee, S., Nizamani, F. A., & Nürnberger, A. (n.d.). Classification of brain tumours in MR images using deep spatio-spatial models. *Scientific Reports*, 12.

Dugdhe, S., Shelar, P., Jire, S., & Apte, A. (2016). Efficient waste collection system. *2016 International Conference on Internet of Things and Applications (IOTA),* 143-147, 10.1109/IOTA.2016.7562711

Ghazal, T.M., Abbas, S., Munir, S., Khan, M.A., Ahmad, M., Issa, G.F., Zahra, S.B., Khan, M.A., & Hasan, M.K. (2022). *Alzheimer Disease Detection Empowered with Transfer Learning.* Academic Press.

Huiyu, L. (2019). Automatic Classifications and Recognition for Recycled Garbage by Utilizing Deep Learning Technology. *7th International Conference on Information Technology: IoT and Smart City,* 1–4.

Hussain, A., Draz, U., Ali, T., Tariq, S., Irfan, M., Glowacz, A., Antonino Daviu, J. A., Yasin, S., & Rahman, S. (2020). Waste Management and Prediction of Air Pollutants Using IoT and Machine Learning Approach. *Energies, 13*(15), 3930. doi:10.3390/en13153930

Ihnaini, B., Khan, M. A., Khan, T. A., Abbas, S., Daoud, M. S., Ahmad, M., & Khan, M. A. (2021). A Smart Healthcare Recommendation System for Multidisciplinary Diabetes Patients with Data Fusion Based on Deep Ensemble Learning. *Computational Intelligence and Neuroscience, 2021,* 2021. doi:10.1155/2021/4243700 PMID:34567101

Kaza, S., Yao, L., Bhada-Tata, P., & Van Woerden, F. (n.d.). *What a Waste 2.0: A Global Snapshot of Solid Waste Management to 2050.* World Bank. https://openknowledge.worldbank.org/h andle/10986/30317

Khan, A. H., Khan, M. A., Abbas, S., Siddiqui, S. Y., Saeed, M. A., Alfayad, M., & Elmitwally, N. S. (2021). Simulation, Modeling, and Optimization of Intelligent Kidney Disease Prediction Empowered with Computational Intelligence Approaches. Academic Press.

Khan, M.A., Abbas, S., Atta, A., Ditta, A., Alquhayz, H., Khan, M.F., & Naqvi, R.A. (2020). *Intelligent cloud based heart disease prediction system empowered with supervised machine learning.* Academic Press.

Khan, M.A., Abbas, S., Khan, K.M., Al Ghamdi, M.A., & Rehman, A. (2020). *Intelligent forecasting model of COVID-19 novel coronavirus outbreak empowered with deep extreme learning.* Academic Press.

Khan, T. AAbbas, SDitta, AKhan, M. AAlquhayz, HFatima, AKhan, M. F. (2020). IoMT-Based Smart Monitoring Hierarchical Fuzzy Inference System for Diagnosis of COVID-19 machine. *Computers, Materials & Continua, 65*(3), 1329–1342.

Liu, C., Sharan, L., Adelson, E. H., & Rosenholtz, R. (2010). Exploring features in a bayesian framework for material recognition. In *Computer Vision and Pattern Recognition (CVPR).* IEEE. 10.1109/CVPR.2010.5540207

Malapur, B. S., & Pattanshetti, V. R. (2017). IoT based waste management: An application to smart city. *IEEE International Conference on Energy, Communication, Data Analytics and Soft Computing (ICECDS),* 2476-2486. 10.1109/ICECDS.2017.8389897

Misra, D., Das, G., Chakrabortty, T., & Das, D. (2018). An IoT-based waste management system monitored by cloud. *Journal of Material Cycles and Waste Management, 20*(3), 1–9. doi:10.1007/s10163-018-0720-y

Muthugala, M. V. J., Samarakoon, S. B. P., & Elara, M. R. (2020). Tradeoff between Area Coverage and Energy Usage of a Self-Reconfigurable Floor Cleaning Robot based on User Preference. *IEEE Access : Practical Innovations, Open Solutions, 8,* 76267–76275. doi:10.1109/ACCESS.2020.2988977

Samann, F. E. (2017). The design and implementation of smart trash bins. *Academic Journal of Nawroz University*, *6*(3), 141–148. doi:10.25007/ajnu.v6n3a103

Sarala, B., Sumathy, G., Kalpana, A. V., & Jasmine Hephzipah, J. (2023). Glioma brain tumor detection using dual convolutional neural networks and histogram density segmentation algorithm. *Biomedical Signal Processing and Control*, *85*, 104859. doi:10.1016/j.bspc.2023.104859

Sharma, N., Singha, N., & Dutta, T. (2015). Smart Bin Implementation for Smart Cities. *International Journal of Scientific and Engineering Research*, *6*(9), 787–791.

Sreelakshmi, K., Akarsh, S., Vinayakumar, R., & Soman, K.P. (2019). *Capsule Networks and Visualization for Segregation of Plastic and Non-Plastic Wastes*. IEEE.

World Bank. (2012). *What a Waste: A Global Review of Solid Waste Management Urban Development Series Knowledge Papers*. Academic Press.

Zahra, S.B., Khan, M.A., Abbas, S., Khan, K.M., Al-Ghamdi, M.A., & Almotiri, S.H. (2021). *Marker-Based and Marker-Less Motion Capturing Video Data: Person and Activity Identification Comparison Based on Machine Learning Approaches*. Academic Press.

Chapter 16
Intelligent Healthcare Provisioning in Fog Using Grey Wolf Optimization

Rajalakshmi Shenbaga Moorthy
ⓘD https://orcid.org/0000-0003-3297-6123
Sri Ramachandra Institute of Higher Education and Research, India

K. S. Arikumar
VIT-AP University, India

Sahaya Beni Prathiba
Vellore Institute of Technology, Chennai, India

P. Pabitha
Anna University, India

ABSTRACT

The increasing population rate plays a vital role in bringing challenges in the provisioning of health care. The data was initially collected and kept in the cloud, where the machine learning algorithm was run on the data and decisions were then transmitted back to the client device. This incurs a significant delay for transferring the data and getting back the result. Thus, in this chapter, fog layer is introduced between device layer and cloud layer for processing the sensor data. The introduction of fog layer tends to minimize the delay incurred by the cloud, as analyzing the health data is close to the device that generates the data. For conducting the best analytics on health data received from sensors, the grey wolf optimization (GWO)-based k-nearest neighbor (K-NN) is proposed. GWO K-NN is integrated in the fog nodes, which is close to the device generating the health data, thereby providing timely decisions. The proposed GWO K-NN works on the fitness of accuracy and misclassification rate of K-NN, and it models the hunting behavior of wolves.

DOI: 10.4018/979-8-3693-1552-1.ch016

INTRODUCTION

The World wide web, a marvel of human innovation, has dominated a number of industries, including enterprise, academia, governance, information science, electrical engineering, networking, and telecommunications, among many others (Chitra and Jayalakshmi, 2021). In the current technological period, the domains of computer science, electronics, and telecommunication are being integrated to create a new technical field called "Internet of Things (IoT)" (Miraz et al., 2015 and Paul et al., 2018). The Web of Things is a great advance in web 2.0 that is bringing daily fluctuations to everyone's lives, regardless of whether they're aware of it or not. The IoT makes an effort to collect and transmit data so that analytics may be performed on it and possibly produce useful insights. The development of IoT has given researchers the opportunity to come up with a workable solution for widespread healthcare applications. A person in a far-off place can haphazardly find a trustworthy solution at a reasonable price. The fast development of three sectors, including cloud technology, mobile applications, and wearables, has changed the way that health care is traditionally provided in favor of more advanced, ubiquitous health care (Arikumar et al., 2022 and Chen et al., 2017).

Smart health care system is essential as it does restrict the person to utilize the benefits of health care from anywhere particularly during pandemic conditions. Various IoT devices embedded in the human body generates data periodically thereby massive amount of data is collected which needs to be analyzed carefully for potential insights (Manogaran et al., 2017 and MA et al., 2018). Traditionally, the analytics process often happens in the centralized cloud computing that may incur delay which is crucial. As, timely decision is essential in any health care application, in this paper Fog Computing is used which can perform analytics at the edge servers thereby minimizing the delay.

Wearable device data will be analyzed by machine learning techniques to produce insights (Arikumar and Natarajan 2020). Several machine learning techniques are available for processing the data, including supervised algorithms that may process data with class labels, unsupervised algorithms that try to cluster the data, and semi-supervised algorithms that operate on a reward or penalty basis (Magoulas, 1999 and Obermeyer, 2016). The hyperparameters have a significant impact on how well the machine learning model performs. The optimal value chosen for the hyperparameter prevents the machine learning model from overfitting. Choosing optimal value for the hyperparameter is often a serious research issue. In this study, the sensitive health data are analyzed using the non-parametric technique K-Nearest Neighbor (K-NN). The value of K in K-NN has an impact on changing the true positives to true negatives which is really a serious challenge. Consequently, in order to improve the decision, it is crucial to identify the ideal value of K.

Metaheuristic algorithms are one such way of finding the optimal values for NP-Hard problems (Dietterich, 2000). Hyperparameters of machine learning algorithms can be found using metaheuristic algorithms. Particle swarm optimization, ant colony optimization, Firefly optimization technique, etc. are a few examples of metaheuristic algorithms. There is room for improvement because no algorithm is effective for all real-world challenges. Thus, in this research the authors had used Grey Wolf Optimization to find the optimal value for K in K-NN. Exploration and Exploitation are the two main factors of metaheuristic algorithms which prevents from falling in local optima. For the purpose of determining the ideal value of K, GWO makes use of the positions of the three best wolves.

Following is a list of the paper's primary contributions:

- Design of Fog layer to bring the analytics closer to edge devices for providing timely decisions for health care applications
- Analytics in the Fog layer is designed using Grey Wolf Optimization (GWO) based K-Nearest Neighbor (K-NN)
- The proposed Fog based architecture for provisioning Health Care as a Service is compared with cloud-based provisioning in terms of network usage, latency and energy consumption
- • Other machine learning algorithms currently in use are contrasted with the proposed GWO K-NN for performance indicators like accuracy, precision, recall, and root mean square error.
- • Other metaheuristic algorithms like the Genetic Algorithm (GA) and Particle Swarm Optimization (PSO)are contrasted with the proposed GWO K-NN
- Further, the proposed GWO K-NN is compared with state-of-the-art method Firefly algorithm with fuzzy clustering

The rest of the chapter is structured as follows: The introduction to health care as a service and the use of machine learning are both covered in depth in Section 1. Section 2 gives the detailed study on various existing methodologies available in Fog based provisioning of healthcare, and the application of various meta heuristic algorithms with machine learning algorithms. Section 3 details the proposed architecture together with the proposed GWO K-NN for performing analytics. Section 4 gives the description about detailed experimentation and comparison carried out with various mechanisms. Finally, the research is concluded in Section 5 with future scope.

RELATED WORKS

This section details the various existing mechanisms applied for analyzing health data gathered through sensors and the emergence of fog computing as a middle tier between cloud and device layer. Also, the study has been extended to learn various mechanisms available for hyperparameter optimization of the value of K in K-NN.

Survey on Fog Computing

Fog computing had been used for monitoring the patients suffering with chronic diseases. As the cloud incurs delay, while transferring the data gathered from sensors embedded in the patient's body to cloud and further the data needs to travel from cloud to hospital. Tri-tier architecture was designed for latency-sensitive, context-aware health applications that used cloud, fog and sensors. Data analysis and aggregation was done in Fog tier. The work was distributed by using a task scheduling mechanism. Once each processor finished execution, data were aggregated which included schema mapping, duplicate detection and data fusion. Simulation was carried to measure latency and network usage for 5 different configurations. It was observed that when there were 64 monitoring devices, Fog achieved 8.47 ms, whereas the cloud achieved 3225.91 ms latency. Similarly, cloud consumed 1102 KB bandwidth, whereas Fog consumed 189 kb of bandwidth (Awasthi and Goel 2021). Though cloud offers potential benefits which is effectively utilized by many industrial professionals, the serious drawback that it incurs is delay, which leads to generation of Fog Computing. Fog Computing was proven to be efficient for applications like health-care which requires immediate response (Kaur et al., 2021). FETCH, a framework that uses deep learning mechanisms in Fog computing for efficient monitoring of healthcare activities. The designed framework was integrated with automated monitoring which makes it most suitable for health-care activities. Once the data was gathered from various sensors, preprocessing was done Using Singular Value Decomposition, Principal Component Analysis (PCA), and Set Partitioning in Hierarchical Trees (SPIHT). An ensemble model based on deep learning was fed the preprocessed data. Cleveland heart disease dataset was utilized to evaluate the model's effectiveness. It was observed that when Fog is integrated with Cloud, 16.89 kbps bandwidth gets consumed for transferring and processing heartbeat data packets (Verma et al., 2022). Fog Computing along with Cloud Computing had been utilized for minimizing latency in health care applications. Random Forest based classifier was used for segregation of the patient data. The classifier was trained with real data gathered through IoT Sensors. The designed model had a latency reduction of 95% (Kishor et al., 2021). Fog Computing integrated with cloud computing had been used for timely prediction of heart disease. In addition, optimized Cascaded Convolution Neural Network (CCNN) were used for enhancing analytics. Galactic Swarm Optimization (GSO) was used for tuning hyperparameters of CCNN. The designed GSO-CCNN improved precision by 3.7% than Particle Swarm Optimization based CCNN (PSO-CCNN) (Raju et al., 2022). Three layers—the cloud layer, the fog layer, and the sensor network layer—were used in the construction of the fog-based health monitoring framework. The sensor layer is in charge of gathering information like blood pressure, temperature, and pulse rate. The devices in the fog layer receive the data that has been collected. In the Fog layer, a large number of distributed nodes known as gateways carry out a variety of tasks include storing the data obtained from the sensor layer and analyzing the data to produce useful judgements. In order to provide timely decisions, analytics using machine learning algorithms are integrated in Fog Layer. In addition to that, the data gets encrypted and the encrypted data is transferred to the cloud (El-hadad et al., 2022). HealthFog was designed with the aim to integrate deep learning algorithms in edge computing for providing timely decisions for heart disease. The deep learning architecture was used to classify the data instances generated from sensors in the body area network. The data often retrieved from oximeters of ECG devices is in graphical format, which needs to be preprocessed before analyzing. The ensemble deep learning architecture was trained with cleveland dataset and was tested with real data gathered from sensor (Tuli et al., 2020).

Survey on Metaheuristic Algorithms

The Sine cosine algorithm had been integrated with K-NN (K=1) (SCAK-NN) for detecting the phishing attack. The designed SCAK-NN intends to select the nearest optimal instance for a new instance in minimal time. SCAK-NN achieved the highest accuracy of 97.18% than Decision tree (95.88) and Naive Bayes (92.98) (Moorthy and Pabitha, 2020). Modified Grey Wolf Optimization (MGWO) has been used for optimal intrusion detection by selecting the optimal subset of features with K-NN as a classifier. Since K-NN performs well in low dimensional dataset, MGWO was used to extract the relevant set of features for optimal classification (Seth and Chandra, 2018). Fast tuning had been designed to optimize the hyperparameters of K-NN. The method was based on the principle of grid search. The hyperparameters such as k, t (number of topmost neighbors) which was used to predict the category of document in case of document summarization was determined using Fast tuning-based grid search (Ghawi and Pfeffer, 2019). Enhanced grey wolf optimization was used for optimal selection of important information from the health dataset. After choosing the crucial features, the performance is assessed by machine learning classification using the specified features as input (Chakraborty et al., 2019) The centers of hidden neurons in RBFNN were located using K-Means based on Particle Swarm Optimization. Numerous metaheuristic algorithms, including the genetic algorithm (GA), the sine cosine algorithm (SCA), and the whale optimization algorithm (WOA), were used to find the centers because the performance of RBFNN depends on the optimal value of the centers of hidden neurons. It was found that PSO based K-Means achieved high levels of performance (Moorthy and Pabitha, 2021).

From the study, it was observed that the existing mechanism of analyzing data in the cloud incurs huge delay, with maximum network usage and energy consumption. Also, it is evident from the study that the performance of conventional K-NN is severely affected by K. Thus, the proposed mechanism of integrating Fog in between device layer and cloud layer solves the problem of delay in making decision and also Grey Wolf optimization technique of finding the value of K in K-NN, increases accuracy and reduces the misclassification rate.

PROPOSED SYSTEM

The proposed Fog based Architecture for provisioning Health Care as a Service (FAHAAS) is shown in Figure 1. It has three layers, with the device layer at the bottom, which contains a spectrum of IoT devices and sensors. The purpose of the device layer is to acquire information from various sensors and Web of Things devices which are integrated with users and transmit it to the fog nodes, a middle layer. The suggested algorithm known GWO-K-NN is incorporated into the Fog Layer's quick information processing, and the completed result is made transparent to healthcare practitioners. The centralized cloud, the top layer in the proposed FAHAAS, receives the information after it has been processed by the fog node.

Device Layer

Many IoT sensors and devices have been successfully implemented in the medical industry for the on-demand and ubiquitous delivery of health care as a service. There is a network of sensors linked to the patient's body called the Body Area Network (BAN). The sensors in BAN collect data from the patient's body while monitoring the performance of several components. The acquired data is transmitted to the processing node so that it can make the best decisions feasible depending on the value which the sensor generated. As an instance, a patient-implanted Electrocardiogram probe aims to communicate ECG signals to the processing node. A message or alarm is sent to the victim's smartphone by the processing node using the values provided by the ECG sensor.

Other sensors, including blood pressure, temperature and biometric sensors are now frequently employed in healthcare institutions and enable for the transmission of patient data even over long distances. Furthermore, the utilization of sensor-enabled gadgets, including that of smartwatches and smartphones, can gather information from the patient's body and relay it to processing nodes for proactive decisions. This investigation concentrates on data received through ECG sensors and seeks to figure out whether a person suffers heart disease or not.

Figure 1. Fog-based architecture for provisioning healthcare as a service

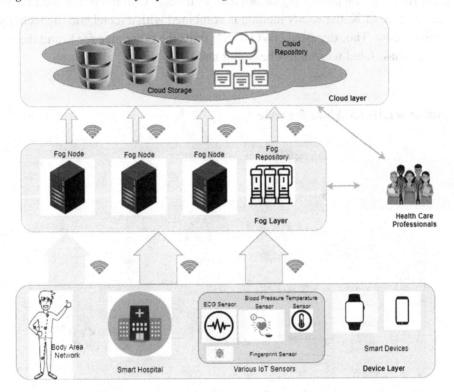

Fog Layer

For the processing of information gathered through various devices at the device layer, the fog layer has one or more fog nodes. With the objective of expediting the decision-making process, the processing capability is coupled with Fog Node. It is vital to have a useful result because the patient's life depends on the accuracy of the health care data. Any processing lag will make the circumstance worse and lead to significant loss. As a consequence, in this article Fog Nodes are employed to process distributed, instead of centralized, health care data. The patient's information is transmitted to centralized cloud-based storage, where processing takes place, in the traditional method of data storage and processing in the cloud. Such centralized processing causes significant end-to-end delays, consumes more network bandwidth, and clogs the network as a result of greater traffic. To address the aforementioned issues with traditional cloud computing, the globally dispersed Fog nodes are used to manage medical data effectively closer to the patient. By processing the information at the location where it is present, each Fog node helps to prevent network congestion and delay. Importantly, it produces rapid results, preserving the patient's life. Each node in the Fog Layer is integrated with GWO K-NN for performing processing of healthcare data as shown in Figure 2.

K-Nearest Neighbor is a lazy learning method that delays classifying until it sees the test instance, which is why Grey Wolf Optimization was integrated with it. The value of K in the K-NN model has a significant impact on the classification. Figure 3 represents the impact of K in the K-NN classifier. From the figure it is evident that the K in K-NN has a direct impact on the predicted class which in turn has direct influence on accuracy and misclassification. Therefore, it is very essential for the classifier to produce optimal results while processing health data as it deals with the life of the patient. In order to discover the ideal value of K, the K-NN method is combined with a metaheuristic technique known as Grey Wolf Optimization. The suggested GWO K-NN finds the best value for K and then assigns the new instance the best class label when it reaches a fog node.

Figure 2. Integration of GWO K-NN in fog node

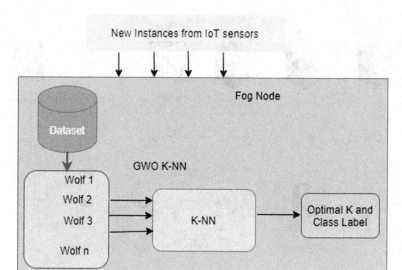

The Grey Wolf Optimization is a metaheuristic algorithm (Mirjalili et al., 2014) inspired by the hunting behavior of wolves. The wolves have a hierarchy of leadership in which the roles are held by four different types of grey-colored wolves, including alpha α, beta β, delta δ, and omega ω. Each wolf does the operation including hunting, prey searching, prey encircling and prey attacking. Alpha wolf is the leader wolf that is responsible for making decisions. Alpha conveys its decision to the entire pack. The other wolves in the pack always obey the alpha wolf's words. Beta wolf comes next in the hierarchy which helps the alpha wolf for making effective decisions. Delta wolves are termed to be subordinate wolves which command the lowest level omega wolves and report to alpha and beta wolves. When correlating GWO with real world problem, α is considered as the best candidate solution. β and δ wolves are considered to be the second and third best solution in pack. The GWO includes processes such as prey encircling, hunting, exploitation (attacking the prey) and exploration (searching for the prey). List of symbols used are specified in Table 1.

Prey Encircling

While encircling the prey, the position of the grey wolf for the next iteration is computed based on its current position and distance between itself to prey which is represented in Eq (1) and (2).

$$\overrightarrow{dis} \leftarrow \left| \overrightarrow{C} * \overrightarrow{P_{prey}}(t) - \overrightarrow{P}(t) \right| \tag{1}$$

Table 1. List of notations

Symbols	Description
$\overrightarrow{dis}$	Distance between the Prey and the wolf
$\overrightarrow{P_{prey}(t)}$	Position of the prey at time t
$\overrightarrow{P(t)}$	Position of the wolf at time t
$\vec{A}$	Coefficient vector
$\vec{C}$	Coefficient vector
$\overrightarrow{rand_1}$, $\overrightarrow{rand_2}$	Random value
$\overrightarrow{dis_\alpha}$	Distance between the alpha and omega wolves
$\overrightarrow{dis_\beta}$	Distance between the beta and omega wolves
$\overrightarrow{dis_\gamma}$	Distance between the delta and omega woves
t_p	True Positive
t_n	True Negative
f_p	False Positive
f_n	False Negative
TI	Test Instance
I_i	ith instance
D	Dataset

Figure 3. Impact of K in K-nearest neighbor

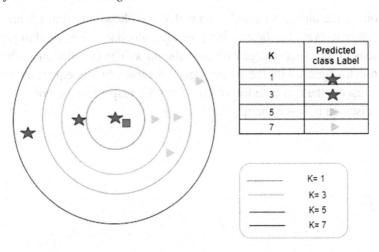

$$\overrightarrow{P(t+1)} \leftarrow \overrightarrow{P(t)} - \vec{A} * \overrightarrow{dis} \tag{2}$$

The coefficient vectors $\vec{A}$ *and* $\vec{C}$ are computed using Eq (3) and (4) respectively.

$$\vec{A} \leftarrow 2 * \vec{a} * \overrightarrow{rand_1} - \vec{a} \tag{3}$$

$$\vec{C} \leftarrow 2 * \overrightarrow{rand_2} \tag{4}$$

The diagrammatic representation of prey encircling is shown in Figure 4.

Figure 4. Encircling of prey

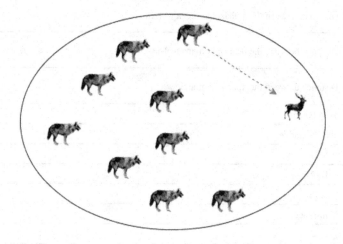

Hunting

The optimal positions of the alpha, beta, and delta wolves, as shown in Figure 5, are used to mimic the hunting behavior of grey wolves. The best wolf's position, which is represented in Eq (11), will be used to update the positions of the other omega wolves in the pack. The calculation of the distance between the alpha, beta, and delta agents with the omega agent is shown by the equations (5), (6), and (7). Eq (8), (9) and (10) represents the computation of position with respect to the position of alpha, beta and delta agent and the distance between them.

$$\overrightarrow{dis_\alpha} \leftarrow \left| \overrightarrow{C_1} * \overrightarrow{P_\alpha} - \vec{P} \right| \tag{5}$$

$$\overrightarrow{dis_\beta} \leftarrow \left| \overrightarrow{C_2} * \overrightarrow{P_\beta} - \vec{P} \right| \tag{6}$$

$$\overrightarrow{dis_\gamma} \leftarrow \left| \overrightarrow{C_3} * \overrightarrow{P_\gamma} - \overrightarrow{P} \right| \tag{7}$$

$$\overrightarrow{P_1} \leftarrow \overrightarrow{P_\alpha} - \overrightarrow{A_1} * \overrightarrow{dis_\alpha} \tag{8}$$

$$\overrightarrow{P_2} \leftarrow \overrightarrow{P_\beta} - \overrightarrow{A_2} * \overrightarrow{dis_\beta} \tag{9}$$

$$\overrightarrow{P_3} \leftarrow \overrightarrow{P_\gamma} - \overrightarrow{A_3} * \overrightarrow{dis_\gamma} \tag{10}$$

$$\overrightarrow{P(t+1)} \leftarrow \frac{\overrightarrow{P_1} + \overrightarrow{P_2} + \overrightarrow{P_3}}{3} \tag{11}$$

Figure 5. Hunting behavior of wolf pack

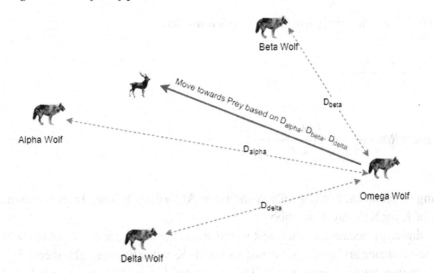

Attacking a Prey (Exploitation)

The attack on the prey causes the grey wolves to cease their search. Over the span of an iteration, $\vec{a}$ is decreased from 2 to 0. This leads to change in the $\vec{A}$. Based on the value of $\vec{A}$ and if $\left\| \vec{A} \right\| < 1$, then the wolf moves towards the prey. This happens when $\vec{A} \in [-1, 1]$.

Hunt for Prey (Exploration)

When the value of $\left\|\vec{A}\right\|$ is greater than 1, then the wolves usually go in different directions looking for prey. This happens when $\vec{A} \in [-1,1]$. In addition, $\vec{C}$ promotes exploration, as it adds weight to the position of the prey as specified in (1).

Fitness Function

Each wolf is represented with only one dimension i.e., the value of K in K-NN. The multi-objective function evaluated in terms of accuracy and misclassification rate is the fitness function used to assess the wolf. At each iteration, GWO evaluates the fitness of the wolf based on the accuracy and misclassification rate of K-NN which is depicted in Eq (12). The proportion of instances that are correctly classified to all instances is a measure of accuracy and it is represented in Eq (13). The percentage of instances that are incorrectly classified relative to the total number of instances is known as the misclassification rate and it is represented in Eq (14).

$$Maximize\ Fit(W_i) \leftarrow w_1 * Accuracy + w_2 * MisclassificationRate \tag{12}$$

$$Accuracy \leftarrow \frac{t_p + t_n}{t_p + t_n + f_p + f_n} \tag{13}$$

$$MisclassificationRate \leftarrow \frac{f_n + f_p}{t_p + t_n + f_p + f_n} \tag{14}$$

The working of proposed GWO K-NN is shown in Algorithm 1. The algorithm intends to find the optimal value of K for K-Nearest Neighbor.

The cosine similarity between the new test instance and the instances in the dataset will be computed when the new test instance arrives at the fog node after the K- value has been obtained using GWO K-NN. K- Neighbors are then found using Eq (15). The class label for the new instance is based primarily on the class labels of the neighbors, indicating whether or not heart disease is present.

$$CosineSimilarity\left(TI, I_i \in D\right) \leftarrow \frac{TI.I_i}{\|TI\| * \|I_i\|} \tag{15}$$

Cloud Layer

Though, there are numerous advantages of using Fog nodes, there exists some limitations including limited storage and processing overhead. As a result, fog nodes periodically relay data to the cloud. High-end servers are embedded into the cloud layer. The health care professionals can able to access the health data both from cloud as well as Fog. Thus, if fog nodes are not accessible, then the data can be retrieved effectively from the cloud.

Algorithm 1. GWO K-NN

Input:	Dataset $D \leftarrow \{I_1, I_2, \ldots, I_N\}$, TI
Output:	Optimal value of K
	Initialize $Pack \leftarrow \{wolf_1, wolf_2, \ldots, wolf_n\}$
	Initialize $a \leftarrow 2$
	Generate $\vec{A}$ and $\vec{C}$ randomly
	For each $wolf_i \in Pack$
	Compute Fitness using Eq (12)
	End For
	$P_\alpha \leftarrow FirstBestWolf(Max(Fitness))$
	$P_\beta \leftarrow SecondBestWolf(Max(Fitness))$
	$P_\alpha \leftarrow ThirdBestWolf(Max(Fitness))$
	While $t \leq MaxIteration$
	For each $Wolf_i \in Pack$
	If $wolf_i \neq \alpha$ && $wolf_i \neq \beta$ then
	Update the position of the omega wolf as in Eq (11)
	End If
	End For
	Compute $a \leftarrow 2 - \dfrac{2-0}{MaxIteration} * t$
	Compute $\vec{A}$ as in Eq (3)
	Compute $\vec{C}$ as in Eq (4)
	For each $wolf_i \in Pack$
	Compute Fitness as in Eq (12)
	End For
	$P_\alpha \leftarrow FirstBestWolf(Max(Fitness))$
	$P_\beta \leftarrow SecondBestWolf(Max(Fitness))$
	$P_\alpha \leftarrow ThirdBestWolf(Max(Fitness))$
	End While
	Return P_α and Class Label

EXPERIMENTAL RESULTS

The experimentation of suggested FAHAAS is done in a system with configuration of 16GB RAM on an X64 based processor. In order to compare the proposed Fog based provisioning of Health care as a service with traditional cloud environment, simulation has been carried out in Java. Five workloads had been generated each comprising a different number of requests. Also, to validate the proposed GWO K-NN integrated in Fog node, the heart dataset from UCI repository (Janosi et al. 1988) had been used.

The proposed GWO K-NN had been compared with conventional K-NN, Decision Tree and Random Forest. Additionally, the suggested GWO K-NN has been assessed alongside K-NNs based on genetic algorithms (GA K-NN) and particle swarm optimization (PSO K-NN). Also, the proposed GWO K-NN has been compared with the method specified in (Moorthy and Pabitha, 2021).

Comparison of Latency

Latency measures the time taken for sending the request and receiving the response. The simulation has been carried out by varying the number of health care requests from 100 to 500 and each time latency incurred in cloud and Fog environments has been analyzed. Figure 6 represents the comparison of latency in Fog and cloud. It is observed that the latency incurred in Fog is minimum as the processing and storage of data is nearer to Device. Whereas in the case of cloud, the latency is high as the storage and processing happens at the remote. Figure 6 clearly shows that when the number of requests rises, latency rises as well. In all the cases, the delay incurred in Fog nodes is less than in the cloud. When the number of requests is 200 the latency incurred in Fog is 0.35 which is 0.45 in cloud. This shows 28.57% reduction in latency while using Fog environment. Also, when the number of requests is 500, the latency is 51.51% less in Fog than Cloud.

Figure 6. Comparison of latency

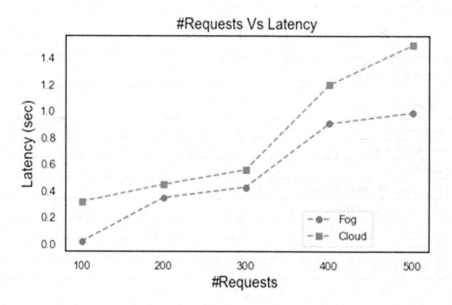

Comparison of Network Usage

The next level of comparison is made to measure the network usage in kilobytes while using Fog environment and cloud environment for provisioning health care as a service. From the Figure 7, it is evident that the network usage in cloud for satisfying health care requests is very high when compared to the Fog nodes. The time it takes to send a request and receive a response in fog is the shortest since the fog nodes are placed closer to the device that created the data as opposed to farther away. As a result, there is less network usage than cloud. When the number of health care requests is 100, the network usage in the cloud is 86.11% higher than Fog. When the number of health requests is 300, the network usage in cloud is 45129.43 kb, which is 87.52% greater than Fog.

Figure 7. Comparison of network usage

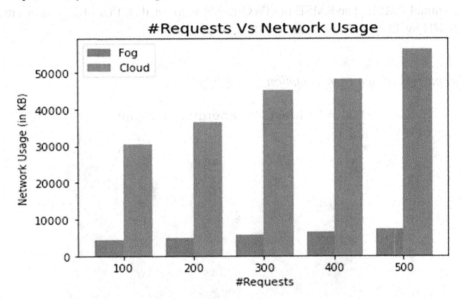

Comparison of Energy Consumption

Energy consumption is the next metric used to validate the necessity of Fog nodes for processing and storing the user data. The figure 8 clearly shows that energy usage grows in tandem with an increase in request volume. The energy consumed by the fog node is less than the cloud which is shown in the Figure 8. For example, when the number of health requests is 400, the energy consumption in Fog node is 42.59% less than Cloud. The reason behind is that since the Fog nodes are situated closer to the Device layer from where the IoT devices generate data, also it takes minimal time for processing the request, the energy consumption is minimum in Fog than in Cloud.

Comparison of GWO K-NN With Traditional Machine Learning Algorithms

Accuracy, precision, recall, and root mean square error are the metrics used to assess the proposed GWO K-NN. Figure 9 compares the metrics obtained by the proposed GWO K-NN algorithm to those obtained by other widely used algorithms including Decision Tree, Random Forest, and K-NN. It is clear from Figure 9 that GWO K-NN is more accurate than other algorithms. The accuracy of the suggested GWO K-NN is high since it uses exploration and exploitation approaches to determine the ideal value of K to locate K such neighbors. GWO K-NN outperforms Decision Tree, Random Forest, and K-NN in terms of accuracy by 7.46%, 8.5%, and 6.1%, correspondingly. Similarly, the recall and precision values are compared with the other conventional algorithms. The recall of GWO K-NN is 7.29%, 2.41% and 8.29% greater than Decision tree, Random Forest and K-NN respectively as shown in Figure 9. Precision is a metric that compares the proportion of accurately anticipated positive instances to all positive instances. As shown in Figure 9, the precision of GWO K-NN is higher than conventional algorithms. The precision for GWO K-NN is 6.94%, 0.08% and 8.29% higher than Decision tree, Random Forest and K-NN respectively. Next, Root Mean Squared Error (RMSE) is compared with various algorithms which is shown in Figure. It is clear from Figure 9 that GWO K-NN outperforms other conventional algorithms in terms of minimal RMSE. The RMSE of GWO K-NN is lower than that of Decision Tree, Random Forest, and K-NN by 45.47%, 30.64%, and 18.05%, respectively.

Figure 8. Comparison of energy consumption

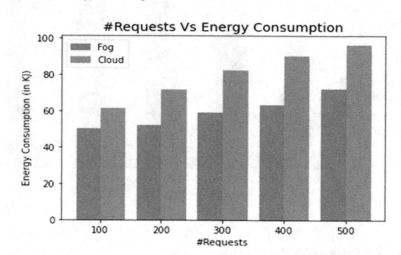

Comparison of GWO K-NN With Other Metaheuristic Algorithms

In terms of accuracy, the developed GWO K-NN is contrasted with other standard metaheuristic algorithms as PSO based K-NN and GA based K-NN. In comparison to PSO-based K-NN and GA-based K-NN, the presented GWO K-NN performs more accurately. The GWO K-NN offers a higher degree of exploration and exploitation to search in the solution space by using alpha, beta, and delta wolves to compute the position of additional wolves that aim to move towards the prey. The genetic algorithm which relies on the selection, mutation and cross over operator may result in overcrowding of individuals which actually does not take to the solution. Particle Swarm Optimization on the other hand uses global best value in addition to local best value, and suffers with a problem of stagnation and premature convergence. Thus, the resulting K-value is not optimal enough to boost the performance of K-NN. As a result, accuracy is lower in PSO K-NN and GA K-NN than GWN K-NN. From the figure 10, it is evident that the accuracy of GWO K-NN is 3.2%, 8.1% higher than PSO K-NN and GA K-NN respectively.

Figure 9. Comparison of performance metrics

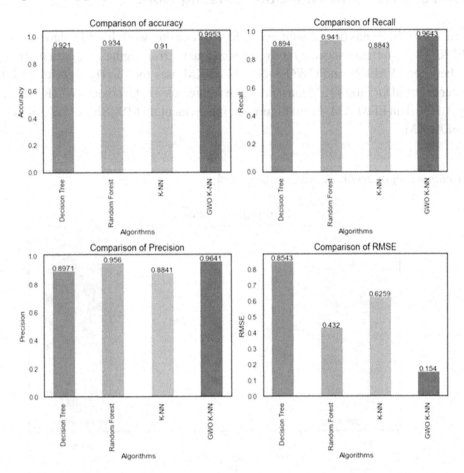

Figure 10. Comparison of accuracy with other metaheuristic algorithms

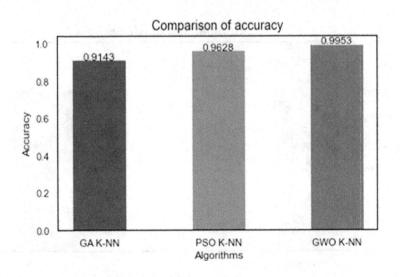

Comparison of GWO K-NN With Other Existing Work

The proposed GWO K-NN has been compared with other existing work FireFly algorithms with Fuzzy K-Means (FFAFKM) in terms of accuracy and root mean square error. Figure 11 represents the comparison of accuracy between FFAFKM and GWO K-NN. FFAFKM uses the Firefly algorithm for finding the optimal cluster centroid for Fuzzy K-Means. From the figure, it is evident that, GWO K-NN improves the accuracy by 1.10% than FFAFKM. From Figure 12, it is evident that RMSE of GWO K-NN is 23.64% better than FFAFKM.

Figure 11. Comparison of accuracy with existing work

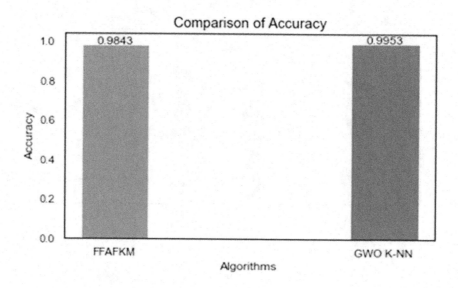

Figure 12. Comparison of RMSE with existing work

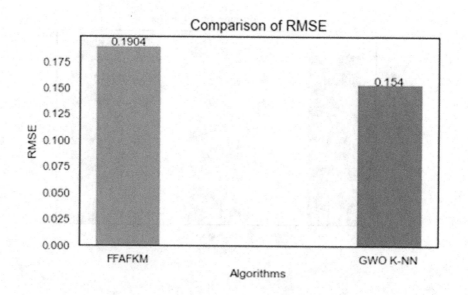

CONCLUSION

Timely decision is essential in terms of pervasive health care which involves gathering data through IoT devices and processing the data in Fog layer / Cloud layer. When data is analyzed on the cloud, a significant amount of time is required for processing the data and communicating the decision. Thus, in this paper, Fog layer was introduced between cloud and device layer for providing timely decisions. Also, GWO K-NN was used in Fog layer which enhances the analytics process. The proposed GWO K-NN alleviates the problem of conventional K-NN, whose accuracy depends on optimal value of K. The proposed GWO K-NN on Fog is evaluated against conventional machine learning techniques, metaheuristic algorithms, and cutting-edge techniques on the heart dataset. GWO K-NN performs better than other algorithms across every circumstance. Additionally, the effectiveness of GWO K-NN on Fog is compared to that of cloud, and from the results of the experiments, it can be seen that the latter has higher energy consumption, network utilization, and latency.

Conflict of interest

The authors declare there is no conflict of interest.

REFERENCES

Arikumar, K. S., & Natarajan, V. (2020). FIoT: a QoS-aware fog-IoT framework to minimize latency in IoT applications via fog offloading. In *Evolution in Computational Intelligence: Frontiers in Intelligent Computing: Theory and Applications (FICTA 2020)* (Vol. 1, pp. 551–559). Springer Singapore.

Arikumar, K. S., Prathiba, S. B., Alazab, M., Gadekallu, T. R., Pandya, S., Khan, J. M., & Moorthy, R. S. (2022). FL-PMI: Federated learning-based person movement identification through wearable devices in smart healthcare systems. *Sensors (Basel)*, 22(4), 1377. doi:10.3390/s22041377 PMID:35214282

Awasthi, A., & Goel, N. (2021). Phishing Website Prediction: A Comparison of Machine Learning Techniques. *Data Intelligence and Cognitive Informatics Proceedings of ICDICI, 2020*, 637–650.

Chakraborty, C., Kishor, A., & Rodrigues, J. J. (2022). Novel Enhanced-Grey Wolf Optimization hybrid machine learning technique for biomedical data computation. *Computers & Electrical Engineering*, 99, 107778. doi:10.1016/j.compeleceng.2022.107778

Chen, M., Ma, Y., Li, Y., Wu, D., Zhang, Y., & Youn, C. H. (2017). Wearable 2.0: Enabling human-cloud integration in next generation healthcare systems. *IEEE Communications Magazine*, 55(1), 54–61. doi:10.1109/MCOM.2017.1600410CM

Chitra, S., & Jayalakshmi, V. (2021). A review of healthcare applications on Internet of Things. *Computer Networks, Big Data and IoT. Proceedings of ICCBI, 2020*, 227–237.

Dietterich, T. G. (2000, June). Ensemble methods in machine learning. In *International workshop on multiple classifier systems* (pp. 1-15). Springer Berlin Heidelberg.

Elhadad, A., Alanazi, F., Taloba, A. I., & Abozeid, A. (2022). Fog computing service in the healthcare monitoring system for managing the real-time notification. *Journal of Healthcare Engineering, 2022,* 2022. doi:10.1155/2022/5337733 PMID:35340260

Ghawi, R., & Pfeffer, J. (2019). Efficient hyperparameter tuning with grid search for text categorization using kNN approach with BM25 similarity. *Open Computer Science, 9*(1), 160–180. doi:10.1515/comp-2019-0011

Kaur, J., Verma, R., Alharbe, N. R., Agrawal, A., & Khan, R. A. (2021). Importance of fog computing in healthcare 4.0. *Fog Computing for Healthcare 4.0 Environments: Technical, Societal, and Future Implications,* 79-101.

Kishor, A., Chakraborty, C., & Jeberson, W. (2021). Intelligent healthcare data segregation using fog computing with internet of things and machine learning. *International Journal of Engineering Systems Modelling and Simulation, 12*(2-3), 188–194. doi:10.1504/IJESMS.2021.115533

Ma, X., Wang, Z., Zhou, S., Wen, H., & Zhang, Y. (2018, June). Intelligent healthcare systems assisted by data analytics and mobile computing. In 2018 14th International Wireless Communications & Mobile Computing Conference (IWCMC) (pp. 1317-1322). IEEE. doi:10.1109/IWCMC.2018.8450377

Magoulas, G. D., & Prentza, A. (1999). Machine learning in medical applications. In *Advanced course on artificial intelligence* (pp. 300–307). Springer Berlin Heidelberg.

Manogaran, G., Thota, C., Lopez, D., & Sundarasekar, R. (2017). Big data security intelligence for healthcare industry 4.0. *Cybersecurity for Industry 4.0: Analysis for Design and Manufacturing,* 103-126.

Miraz, M. H., Ali, M., Excell, P. S., & Picking, R. (2015). A review on Internet of Things (IoT), Internet of everything (IoE) and Internet of nano things (IoNT). *2015 Internet Technologies and Applications (ITA),* 219-224.

Mirjalili, S., Mirjalili, S. M., & Lewis, A. (2014). Grey wolf optimizer. *Advances in Engineering Software, 69,* 46–61. doi:10.1016/j.advengsoft.2013.12.007

Moorthy, R. S., & Pabitha, P. (2020). Optimal detection of phising attack using SCA based K-NN. *Procedia Computer Science, 171,* 1716–1725. doi:10.1016/j.procs.2020.04.184

Moorthy, R. S., & Pabitha, P. (2021). Prediction of Parkinson's disease using improved radial basis function neural network. *Computers, Materials & Continua, 68*(3), 3101–3119. doi:10.32604/cmc.2021.016489

Moorthy, R. S., & Pabitha, P. (2021). Design of Wireless Sensor Networks Using Fog Computing for the Optimal Provisioning of Analytics as a Service. In *Machine Learning and Deep Learning Techniques in Wireless and Mobile Networking Systems* (pp. 153–173). CRC Press. doi:10.1201/9781003107477-9

Obermeyer, Z., & Emanuel, E. J. (2016). Predicting the future—Big data, machine learning, and clinical medicine. *The New England Journal of Medicine, 375*(13), 1216–1219. doi:10.1056/NEJMp1606181 PMID:27682033

Paul, A., Pinjari, H., Hong, W. H., Seo, H. C., & Rho, S. (2018). Fog computing-based IoT for health monitoring system. *Journal of Sensors, 2018,* 2018. doi:10.1155/2018/1386470

Raju, K. B., Dara, S., Vidyarthi, A., Gupta, V. M., & Khan, B. (2022). Smart heart disease prediction system with IoT and fog computing sectors enabled by cascaded deep learning model. *Computational Intelligence and Neuroscience, 2022*, 2022. doi:10.1155/2022/1070697 PMID:35047027

Seth, J. K., & Chandra, S. (2018). MIDS: Metaheuristic based intrusion detection system for cloud using k-NN and MGWO. In *Advances in Computing and Data Sciences: Second International Conference, ICACDS 2018, Dehradun, India, April 20-21, 2018, Revised Selected Papers, Part I 2* (pp. 411-420). Springer Singapore.

Tuli, S., Basumatary, N., Gill, S. S., Kahani, M., Arya, R. C., Wander, G. S., & Buyya, R. (2020). HealthFog: An ensemble deep learning based Smart Healthcare System for Automatic Diagnosis of Heart Diseases in integrated IoT and fog computing environments. *Future Generation Computer Systems, 104*, 187–200. doi:10.1016/j.future.2019.10.043

Verma, P., Tiwari, R., Hong, W. C., Upadhyay, S., & Yeh, Y. H. (2022). FETCH: A deep learning-based fog computing and IoT integrated environment for healthcare monitoring and diagnosis. *IEEE Access : Practical Innovations, Open Solutions, 10*, 12548–12563. doi:10.1109/ACCESS.2022.3143793

Chapter 17
Fog Computing–Based Framework and Solutions for Intelligent Systems:
Enabling Autonomy in Vehicles

Shashi

Department of Computer Application, CCS University Meerut, Meerut, India

M. Dhanalakshmi

ⓘ https://orcid.org/0000-0003-3920-6949

Department of Computer Science and Engineering, New Horizon College of Engineering, Bangalore, India

K. Tamilarasi

School of Computer Science and Engineering, Vellore Institute of Technology, Chennai, India & VIT University, India

S. Saravanan

Department of Mechanical Engineering, Bannari Amman Institute of Technology, Sathyamangalam, India

G. Sujatha

Department of Networking and Communications, SRM Institute of Science and Technology, Chennai, India

Sampath Boopathi

ⓘ https://orcid.org/0000-0002-2065-6539

Department of Mechanical Engineering, Muthayammal Engineering College, Namakkal, India

ABSTRACT

The automotive industry is increasingly focusing on autonomous vehicles, leading to a need for intelligent systems that enable safe and efficient self-driving. Fog computing is a promising paradigm for real-time data processing and communication in autonomous vehicles. This chapter presents a comprehensive framework and solutions for integrating fog computing into intelligent vehicle systems, enabling autonomous features, low-latency data processing, reliable communication, and enhanced decision-making capabilities. By offloading computational tasks to nearby fog nodes, this framework optimizes resource utilization, reduces network congestion, and enhances vehicle autonomy. The chapter discusses various use cases, architectures, communication protocols, and security considerations within fog computing, ultimately contributing to the evolution of intelligent and autonomous vehicles.

DOI: 10.4018/979-8-3693-1552-1.ch017

INTRODUCTION

The automotive industry is experiencing a transformation due to advancements in autonomous vehicle technologies, transforming transportation, safety, and mobility. This transformation improves road safety, reduces traffic congestion, and enhances travel efficiency. The integration of artificial intelligence, sensor systems, machine learning, and connectivity drives vehicles towards higher levels of autonomy, bringing self-driving cars closer to a tangible reality. The concept of autonomous vehicles traces its roots back to the early 20th century, with visionary ideas and prototypes that laid the foundation for today's breakthroughs (Ahangar et al., 2021; Kato et al., 2018). However, it is the recent convergence of computational power, data availability, and algorithmic innovation that has catapulted autonomous vehicles into the forefront of technological advancement. From advanced driver assistance systems (ADAS) to fully autonomous cars capable of navigating complex urban environments, the journey towards vehicular autonomy has been marked by incremental achievements, profound challenges, and a wealth of interdisciplinary research (Butt et al., 2022; Hakak et al., 2022).

This comprehensive research exploration delves into the multifaceted landscape of autonomous vehicles. It aims to provide a holistic understanding of the underlying technologies, research trends, challenges, and societal implications associated with the pursuit of vehicular autonomy. By examining the progression from basic driver assistance features to highly autonomous systems, this exploration seeks to shed light on the intricate interplay between hardware and software, regulation and ethics, human-machine interaction, and the transformative potential for various industries (Fernandes & Nunes, 2012; Hakak et al., 2022). Autonomous vehicles rely on an array of sensors, including cameras, lidar, radar, and ultrasonic sensors, to perceive their surroundings. The fusion of data from these sensors and the subsequent interpretation of the environment form the cornerstone of autonomous navigation and decision-making. Cutting-edge machine learning techniques, such as deep learning, reinforcement learning, and probabilistic modeling, empower vehicles to learn from data and adapt to diverse driving scenarios. These algorithms enable the recognition of objects, prediction of behavior, and optimization of driving trajectories (Cui et al., 2018; Hoermann et al., 2017). Autonomous vehicles must not only perceive their environment but also make real-time decisions to navigate safely and efficiently. Control and planning algorithms determine how the vehicle should move and interact with its surroundings, considering factors like traffic, pedestrians, and road conditions.

Ensuring the safety of autonomous vehicles is paramount. Rigorous testing, simulation, and validation processes are essential to building trust in these systems and minimizing the risk of accidents in complex real-world environments. The introduction of autonomous vehicles raises ethical dilemmas, such as how vehicles should prioritize the safety of occupants versus pedestrians in potential collision scenarios (Barik et al., 2018; Wu et al., 2017). Additionally, legal frameworks and regulations must evolve to accommodate the unique challenges posed by autonomous driving. The success of autonomous vehicles hinges on how well humans can interact with and trust these systems. Understanding user perceptions, preferences, and behaviors is crucial for designing user-friendly interfaces and fostering public acceptance. The advent of autonomous vehicles will have far-reaching impacts on various industries, including transportation, logistics, urban planning, and more. Exploring these effects is essential for anticipating economic shifts and making informed policy decisions (Aljumah et al., 2021; Barik et al., 2018).

The collaborative nature of autonomous vehicle advancement requires a critical examination of the state of the art, addressing challenges, and envisioning a future where autonomous vehicles coexist harmoniously with society. Engineers, researchers, policymakers, ethicists, and the general public all play pivotal roles in shaping the technology's trajectory. This contributes to the ongoing dialogue surrounding this transformative journey. Intelligent systems have revolutionized technology, transforming interactions in smart cities, industrial automation, healthcare, and transportation (Aljumah et al., 2021; Ijaz et al., 2021). Their integration leads to efficiency, accuracy, and convenience, revolutionizing various industries. However, this integration has also given rise to new challenges, particularly in terms of data processing, real-time decision-making, and the demand for higher computational capabilities. At the heart of these challenges lies the concept of "Fog Computing," a paradigm that addresses the limitations of traditional cloud computing and edge computing by bridging the gap between the two. Fog computing extends the cloud's reach to the edge of the network, enabling data processing and analytics to take place closer to the data source (Alajali et al., 2019; Singh & Singh, 2023). This approach not only reduces latency but also enhances the efficiency of intelligent systems, making them more responsive, reliable, and capable of handling the immense data volumes generated by the Internet of Things (IoT) and other sensor-based technologies (Hema et al., 2023; Karthik et al., 2023; Reddy et al., 2023; Samikannu et al., 2022).

This comprehensive exploration delves into the Fog Computing-Based Framework and Solutions for Intelligent Systems, aiming to provide a thorough understanding of the principles, applications, challenges, and potential of this innovative paradigm. Examining fog computing and intelligent systems' fusion reveals transformative impacts on industries, individuals, and society. Traditional cloud computing offers resources but faces limitations in real-time data processing (Hema et al., 2023; Rahamathunnisa et al., 2023). Edge computing, on the other hand, alleviates some of these challenges by performing computations closer to the data source, reducing latency and bandwidth usage (Padhy et al., 2023; Rani et al., 2023). Fog computing builds upon these concepts, creating a distributed architecture that extends from the cloud to the edge, encompassing a diverse range of devices, sensors, and communication endpoints.

Intelligent systems, driven by artificial intelligence, machine learning, and advanced analytics, are powering an array of applications, from autonomous vehicles to smart healthcare devices. However, the success of these systems hinges on their ability to process and act upon data in real time. Fog computing offers a compelling solution by providing a dynamic, decentralized infrastructure that can accommodate the demanding processing requirements of intelligent applications (Lin et al., 2020; Tripathy et al., 2022). The fusion of fog computing with intelligent systems has far-reaching implications across industries. In manufacturing and industrial automation, real-time monitoring and predictive maintenance become feasible, optimizing production processes and minimizing downtime. In healthcare, remote patient monitoring and real-time diagnostics enhance patient care and enable timely interventions. Smart cities benefit from improved traffic management, energy efficiency, and public safety through the integration of fog-enabled intelligent systems (Pareek et al., 2021; Vilela et al., 2020).

Implementing fog computing-based solutions for intelligent systems is not without its challenges. Managing the complexity of distributed architectures, ensuring data security and privacy, and orchestrating seamless communication between fog nodes and cloud services require innovative approaches. Researchers and practitioners are actively exploring techniques such as edge intelligence, dynamic resource allocation, and adaptive networking protocols to overcome these hurdles (Syamala et al., 2023; Venkateswaran et al., 2023). The convergence of fog computing and intelligent systems marks a critical juncture in the evolution of technology. As the number of connected devices and sensors continues to grow, the demand for efficient, responsive, and scalable computing infrastructure intensifies. Fog computing offers a middle ground that balances the benefits of cloud computing's extensive resources with the low latency and real-time capabilities of edge computing.

Background and Motivation

The automotive industry has been witnessing a transformative shift towards the development of autonomous vehicles, driven by the promise of increased safety, efficiency, and convenience. As vehicles become more autonomous, they rely on a complex interplay of sensors, data processing, decision-making algorithms, and communication systems. However, the realization of full autonomy presents challenges related to real-time data processing, low-latency communication, and the need for localized decision-making. Traditional cloud computing architectures, while useful for many applications, might fall short in meeting the stringent requirements of autonomous vehicles due to latency issues and the sheer volume of data generated.

This chapter addresses these challenges by exploring the integration of fog computing into the context of intelligent vehicle systems. Fog computing, which extends cloud capabilities to the edge of the network, offers a solution that can bridge the gap between centralized cloud computing and the real-time demands of autonomous vehicles. By distributing computational tasks closer to the source of data, fog computing enables efficient data processing, reduced latency, and enhanced vehicle autonomy.

Objectives

The primary objectives of this chapter are as follows:

- **Provide a Comprehensive Understanding:** The chapter aims to provide a clear and comprehensive understanding of the concepts of fog computing and its relevance to intelligent vehicle systems. It delves into the key principles, benefits, and challenges associated with fog computing in the context of autonomous vehicles.
- **Present a Framework for Integration:** A key objective is to present a well-defined framework for integrating fog computing into intelligent vehicle systems. This framework outlines the architectural considerations, data offloading strategies, real-time processing techniques, and communication protocols required to enable seamless fog-based solutions.
- **Highlight the Role in Vehicle Autonomy:** The chapter explores how fog computing can empower vehicle autonomy by enabling real-time decision-making at the edge, enhancing sensor data fusion, and supporting adaptive learning algorithms. It emphasizes how fog computing can contribute to making vehicles smarter and more capable of navigating complex environments.
- **Discuss Use Cases and Applications:** To provide practical insights, the chapter discusses various use cases and applications of fog computing in the automotive domain. It examines scenarios such as advanced driver assistance systems (ADAS), traffic management, and vehicle-to-infrastructure (V2I) communication, demonstrating the versatility of fog-based solutions.
- **Address Security Considerations:** Recognizing the importance of security, the chapter also addresses security considerations unique to fog-based vehicle systems. It explores potential threats, vulnerabilities, and strategies for ensuring secure data transmission and access control (Anitha et al., 2023; Boopathi, 2023b; Karthik et al., 2023).

By achieving these objectives, this chapter aims to contribute to the body of knowledge surrounding the integration of fog computing into intelligent vehicle systems, fostering advancements in autonomous driving technology and enhancing the overall safety and efficiency of future vehicles.

Scope and Organization

The scope of this chapter encompasses a comprehensive exploration of fog computing's integration into intelligent vehicle systems to enable autonomy. The focus will be on addressing the challenges of real-time data processing, communication, and decision-making in the context of autonomous vehicles (Kirsanova et al., 2021; Zhang et al., 2022). The chapter will provide insights into the principles, framework, applications, and security considerations related to fog computing, all within the context of enhancing vehicle autonomy.

Autonomous Vehicles and Intelligent Systems: This section lays the foundation by discussing the evolution of autonomous vehicles. It traces the technological advancements and milestones that have led to the current state of autonomous driving. The section then delves into the pivotal role of intelligent systems in achieving vehicle autonomy. It highlights how artificial intelligence, machine learning, and sensor fusion contribute to the decision-making processes necessary for safe and efficient autonomous driving.

Evolution of Autonomous Vehicles: This subsection explores the historical development of autonomous vehicles, from early experimental prototypes to the cutting-edge autonomous technologies of today. It outlines key breakthroughs, challenges faced, and the societal impact of autonomous driving. The evolution of autonomous vehicles sets the stage for understanding the complex requirements and capabilities that modern intelligent systems must possess.

Role of Intelligent Systems in Autonomy: Here, the focus shifts to the significance of intelligent systems in realizing autonomous vehicles. The subsection emphasizes the central role of AI algorithms, deep learning models, and sensor technologies in enabling vehicles to perceive their surroundings, interpret complex scenarios, and make informed decisions. Discussions will revolve around how these systems interact with real-time data streams and contribute to achieving a higher level of autonomy.

Challenges and Requirements: This subsection addresses the challenges that must be overcome to achieve seamless autonomy in vehicles. It examines issues such as real-time data processing bottlenecks, communication latency, sensor accuracy, and the need for robust decision-making under varying conditions. The section also discusses the stringent requirements that autonomous vehicles demand, including high reliability, low-latency communication, and adaptive learning capabilities. By outlining these challenges and requirements, the chapter sets the context for the role of fog computing in addressing these issues and enhancing vehicle autonomy.

This chapter follows a logical progression of concepts, starting with autonomous vehicles, transitioning to intelligent systems, and concluding with fog computing integration challenges and requirements. It provides a comprehensive understanding of the context and motivations behind fog computing-based solutions for enabling autonomy in vehicles.

FOG COMPUTING: CONCEPTS AND PRINCIPLES

Understanding Fog Computing: In this section, the chapter delves into the fundamental concepts of fog computing. It explains how fog computing extends the capabilities of cloud computing by bringing data processing and storage closer to the edge of the network, where the data is generated. This section clarifies the concept of "fog nodes" or "edge nodes," which are devices situated between the cloud and end devices, capable of performing computation and data processing tasks. It discusses the motivation behind fog computing, emphasizing its ability to address latency, bandwidth, and real-time requirements, which are critical for applications like autonomous vehicles.

Fog vs. Cloud Computing: The chapter proceeds to differentiate between fog computing and traditional cloud computing. It highlights the distinctions in terms of data processing location, latency, and data volume. It explains how cloud computing involves centralized data centers, often leading to latency issues due to the distance between data sources and processing centers. In contrast, fog computing distributes processing tasks to edge nodes, reducing latency and enabling faster response times. This section clarifies how fog computing complements cloud computing rather than replacing it, and emphasizes the hybrid nature of these two paradigms in intelligent vehicle systems.

Advantages for Intelligent Vehicle Systems: Here, the advantages of adopting fog computing for intelligent vehicle systems are discussed in detail. The section highlights how fog computing addresses the unique challenges faced by autonomous vehicles, such as real-time decision-making, low-latency communication, and processing of massive sensor data. It explains how fog computing enhances vehicle autonomy by enabling on-the-fly data analysis, reducing dependence on centralized data centers, and optimizing resource utilization. The section also touches upon energy efficiency gains and the potential for enhanced security in a fog-based environment. By showcasing these advantages, the chapter lays the foundation for the subsequent exploration of fog-based solutions for intelligent vehicle systems.

This section establishes a strong understanding of fog computing's principles, its differentiation from cloud computing, and its applicability to addressing the specific requirements of intelligent vehicle systems. It provides a conceptual framework that prepares the reader for the more practical discussions regarding the integration of fog computing into autonomous vehicles.

FRAMEWORK FOR INTEGRATING FOG COMPUTING

The integration of fog computing into intelligent vehicle systems involves a dynamic and distributed architecture that enhances real-time data processing and decision-making. This architecture consists of three main components: fog nodes, vehicles, and cloud infrastructure. Fog nodes are strategically positioned edge devices that act as intermediaries between vehicles and the cloud (Al-Shareeda & Manickam, 2022; Palattella et al., 2019). They handle data processing, communication, and local decision-making. Vehicles generate a constant stream of sensor data, while the cloud infrastructure provides storage and advanced analytics capabilities.

Fog Node Placement and Distribution: The placement of fog nodes is critical to ensuring efficient data processing and minimizing latency. Fog nodes are strategically positioned in areas with high vehicle density or crucial traffic intersections. This placement optimizes data flow and minimizes the distance data needs to travel, reducing communication delays. Dynamic placement algorithms consider real-time traffic patterns and adjust fog node positions accordingly. By distributing processing closer to the data source, fog nodes minimize the need for data transmission to centralized cloud servers, enhancing response times and reducing network congestion.

Data Offloading and Resource Optimization: Data offloading refers to the process of transferring selected data and computational tasks from vehicles to fog nodes. Offloading decisions are made based on data significance, network conditions, and computational requirements. Advanced algorithms determine whether to process data locally or offload it for centralized processing. Offloading reduces the computational burden on vehicles, conserving energy and enhancing their computational capabilities. Resource optimization algorithms ensure that fog nodes distribute processing tasks efficiently, preventing overloads and ensuring optimal resource utilization.

Real-time Data Processing at the Edge: Real-time data processing at the edge is a core capability enabled by fog computing. Fog nodes process data as it is generated, using techniques such as stream processing and edge analytics. For instance, image recognition algorithms can identify pedestrians or road signs in real-time, enhancing vehicle safety. Fog nodes also facilitate sensor data fusion, combining inputs from multiple vehicles to create a more comprehensive view of the environment. This local processing reduces reliance on distant cloud servers, minimizing communication latency and ensuring rapid decision-making.

Case Study: Autonomous Intersection Management: To illustrate the framework's effectiveness, consider an autonomous intersection management scenario. As vehicles approach an intersection, they communicate with nearby fog nodes. These fog nodes analyze traffic conditions and make immediate decisions, such as adjusting traffic light timings or coordinating vehicle movements. This localized decision-making minimizes congestion and reduces the risk of collisions. Critical data, such as sudden braking events, can be offloaded to the cloud for in-depth analysis and long-term planning (Al-Shareeda & Manickam, 2022; Shahzad et al., 2022; Zhou et al., 2020).

Thus, the integration of fog computing into intelligent vehicle systems creates a powerful framework that enhances vehicle autonomy. By strategically placing fog nodes, optimizing resource utilization, and enabling real-time data processing at the edge, this framework addresses challenges related to latency, communication, and decision-making. Through its dynamic architecture, the integration of fog computing contributes to safer, more efficient, and more autonomous vehicles in our rapidly advancing world.

COMMUNICATION PROTOCOLS FOR FOG-BASED VEHICLE SYSTEMS

Figure 1 illustrates the communication protocols for fog-based vehicle systems and explained here.

Low-Latency Communication Requirements

In the realm of fog-based vehicle systems, achieving low-latency communication is of paramount importance. As vehicles become more autonomous, the need for real-time data exchange between vehicles, infrastructure, and fog nodes becomes critical for ensuring safe and efficient driving. Low-latency communication protocols play a pivotal role in enabling timely interactions and rapid decision-making within the intelligent vehicle ecosystem (Eftekhari et al., 2021; Lin et al., 2020).

Figure 1. Communication protocols for fog-based vehicle systems

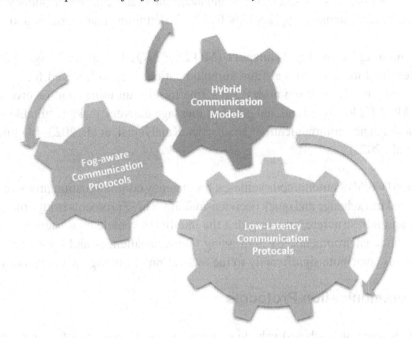

Understanding Low-Latency Communication: Low-latency communication refers to the minimal delay between the transmission of data and its receipt by the intended recipient. In the context of fog-based vehicle systems, low-latency communication protocols reduce the time it takes for data to travel from sensors on the vehicle to fog nodes or other vehicles, and vice versa. This rapid data exchange is vital for applications such as collision avoidance, cooperative driving, and real-time traffic management.

Challenges in Achieving Low Latency: Several challenges must be overcome to achieve low-latency communication in fog-based vehicle systems. One key challenge is the variability of communication channels. Wireless communication can be affected by signal interference, changing network conditions, and congestion. Moreover, the sheer volume of data generated by multiple vehicles can lead to network congestion, which may result in increased latency. Additionally, the need to process and transmit data in real time introduces computational challenges that can impact the overall latency of the system.

Protocols for Low-Latency Communication: To address these challenges, a variety of communication protocols have been developed to facilitate low-latency data exchange in fog-based vehicle systems (Eftekhari et al., 2021; Ma et al., 2019). These protocols prioritize rapid data transmission, efficient data compression, and reliable data delivery. Examples of such protocols include:

- **IEEE 802.11p (DSRC/WAVE):** This protocol, also known as Dedicated Short-Range Communication (DSRC) or Wireless Access in Vehicular Environments (WAVE), is designed specifically for vehicular communication. It operates in the 5.9 GHz frequency band and supports low-latency vehicle-to-vehicle (V2V) and vehicle-to-infrastructure (V2I) communication, making it suitable for applications like cooperative driving and intersection management.
- **Cellular V2X (C-V2X):** Cellular Vehicle-to-Everything (C-V2X) communication leverages cellular networks to enable V2V, V2I, and vehicle-to-network (V2N) communication. With the in-

troduction of 5G networks, C-V2X offers enhanced data rates and lower latency, making it suitable for bandwidth-intensive applications like high-definition map updates and real-time video streaming.

- **Message Queuing Telemetry Transport (MQTT):** MQTT is a lightweight publish-subscribe protocol designed for IoT and real-time communication. It is well-suited for fog-based systems where devices (vehicles and fog nodes) subscribe to relevant topics and receive updates in near real-time. MQTT's low overhead and efficient message delivery make it suitable for applications requiring low-latency communication (Koshariya, Kalaiyarasi, et al., 2023; Maguluri et al., 2023; Syamala et al., 2023).

In the fast-paced world of autonomous vehicles, low-latency communication protocols are crucial for enabling real-time data exchange and quick decision-making. These protocols bridge the gap between fog nodes, vehicles, and infrastructure, ensuring that the intelligent vehicle ecosystem operates seamlessly and safely in dynamic environments. By addressing latency challenges and optimizing data transmission, these protocols contribute significantly to the realization of intelligent and autonomous vehicles.

Fog-aware Communication Protocols

In the intricate landscape of fog-based vehicle systems, the development of fog-aware communication protocols has emerged as a strategic approach to enhancing efficiency and reliability. These protocols are specifically designed to harness the benefits of fog computing, optimizing data exchange between vehicles, fog nodes, and the cloud. By understanding the unique capabilities of fog nodes and their role in edge processing, fog-aware communication protocols enable seamless and intelligent data transmission within the intelligent vehicle ecosystem.

The Characteristics of Fog-Aware Communication Protocols

- **Dynamic Routing:** Fog-aware protocols intelligently route data based on factors such as proximity to fog nodes, computational load, and network conditions. This dynamic routing ensures that data is directed to the most suitable processing point, minimizing latency and maximizing resource utilization.
- **Edge Offloading:** These protocols enable data to be offloaded to fog nodes at the edge for immediate processing. Fog nodes can perform initial data filtering, aggregation, and lightweight analytics before transmitting relevant information to the cloud. This approach reduces the amount of data sent to the cloud, conserving bandwidth and improving response times.
- **Adaptive Data Compression:** Fog-aware protocols implement adaptive data compression techniques that adjust compression levels based on network congestion and available resources. This ensures efficient data transmission without compromising the quality of information.
- **Prioritized Messaging:** Fog-aware protocols enable the prioritization of messages based on their urgency and importance. Critical safety-related messages, such as collision warnings, can be assigned higher priority, ensuring they are delivered promptly and without delay.
- **Real-time Feedback Loop:** These protocols establish a real-time feedback loop between fog nodes and vehicles. Fog nodes can provide instantaneous acknowledgments and updates to vehicles, confirming successful data reception or requesting data retransmission if necessary.

Hybrid Communication Models

Hybrid communication models present a pragmatic approach to optimizing communication within fog-based vehicle systems. These models leverage the strengths of both cellular networks and dedicated vehicular communication technologies to achieve a balanced and effective data exchange framework (Sodhro et al., 2020; Yazdani et al., 2023).

Cellular-Vehicular Communication Integration (C-V2X): Hybrid models often integrate cellular technologies like 4G/5G with vehicle-to-everything (V2X) communication. This fusion combines the wide coverage and high data rates of cellular networks with the low-latency and direct communication of V2X, allowing vehicles to seamlessly switch between modes based on the application's requirements.

Application-Aware Communication: Hybrid models tailor communication choices based on the specific application. For latency-critical applications, such as collision warnings, direct V2X communication might be preferred. For less time-sensitive applications, cellular networks could be used, optimizing network utilization and ensuring efficient data transmission.

Challenges and Opportunities: While hybrid models offer versatility, they also introduce challenges related to protocol interoperability, network handovers, and dynamic decision-making. Addressing these challenges opens opportunities for adaptive algorithms that select the most suitable communication mode based on real-time conditions.

In conclusion, fog-aware communication protocols and hybrid communication models play pivotal roles in the success of fog-based vehicle systems. These approaches harness the benefits of fog computing and optimize data exchange, ensuring that intelligent vehicles operate seamlessly, safely, and efficiently in complex and evolving environments.

ENABLING VEHICLE AUTONOMY THROUGH FOG COMPUTING

The implementation of vehicle autonomy through fog computing is shown in Figure 2.

Figure 2. Enabling vehicle autonomy through fog computing

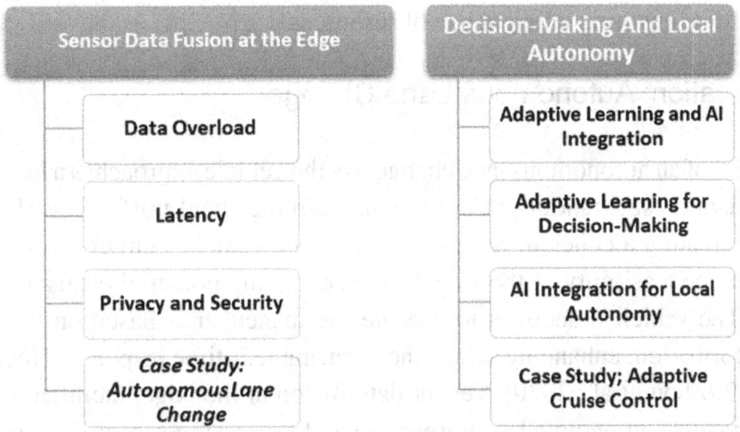

Sensor Data Fusion at the Edge

At the heart of enabling vehicle autonomy lies the crucial process of sensor data fusion at the edge. This process, empowered by fog computing, plays a pivotal role in enhancing vehicles' perception capabilities, enabling them to make informed decisions in real time. Sensor data fusion involves amalgamating data from various sensors—such as LiDAR, radar, cameras, and ultrasonic sensors—to create a comprehensive and accurate representation of the vehicle's surroundings (Neelakantam et al., 2020; Yang, Luo, Chu, & Zhou, 2020; Yang, Luo, Chu, Zhou, et al., 2020). When conducted at the edge through fog nodes, this fusion enhances vehicle autonomy, safety, and overall performance. Challenges Addressed by Sensor Data Fusion at the Edge is listed below.

- **Data Overload:** Vehicles generate an immense volume of sensor data, which can overwhelm communication networks and cloud servers. Edge-based sensor data fusion addresses this challenge by aggregating and preprocessing data locally, reducing the amount of data transmitted to the cloud.
- **Latency:** Real-time decision-making demands low-latency processing. By conducting sensor data fusion at the edge, fog nodes reduce communication delays, allowing vehicles to respond swiftly to changing conditions such as obstacles, pedestrians, or traffic signals.
- **Privacy and Security:** Transmitting raw sensor data to the cloud raises privacy and security concerns. Edge-based fusion enables vehicles to anonymize and aggregate data before transmission, mitigating privacy risks while still providing valuable insights.

Advantages of Edge-Based Sensor Data Fusion

- **Real-time Decision-Making:** Edge-based sensor data fusion enables vehicles to process and interpret sensor data on-the-fly. This empowers vehicles to make split-second decisions, crucial for scenarios like lane changes, emergency braking, and avoiding collisions.
- **Redundancy and Reliability:** By fusing data from multiple sensor modalities, edge nodes can enhance the reliability of perception systems. If one sensor modality is compromised, others can compensate, contributing to safer autonomous driving.
- **Bandwidth Efficiency:** Edge processing minimizes the amount of data sent to the cloud, conserving bandwidth and reducing communication congestion. Only relevant and high-priority information is transmitted, ensuring efficient use of resources.

Real-world Application: Autonomous Lane Change

Consider the scenario of an autonomous lane change. As the vehicle approaches a lane change maneuver, its sensors collect data on surrounding vehicles, lane markings, and traffic flow. Edge nodes conduct sensor data fusion, creating a coherent representation of the vehicle's environment. This fused data is then used to analyze the feasibility of the lane change, detecting potential obstacles and assessing safe gaps for merging. The vehicle's decision to execute the maneuver is based on the locally processed, high-confidence information, enhancing safety and ensuring real-time response (Hoermann et al., 2017; Shahzad et al., 2022; Zhou et al., 2020). Sensor data fusion at the edge, facilitated by fog computing, is a cornerstone of vehicle autonomy. By aggregating and processing sensor data locally, vehicles gain the ability to perceive and interpret their surroundings in real time, enabling safer and more efficient autonomous driving. Edge-based fusion addresses challenges related to data overload, latency, and security, ultimately contributing to the realization of intelligent and self-driving vehicles in our increasingly connected and automated world.

Decision-Making and Local Autonomy

Adaptive Learning and AI Integration: In the pursuit of enabling vehicle autonomy, the integration of adaptive learning and AI techniques plays a pivotal role in enhancing decision-making processes and local autonomy. By leveraging fog computing's capabilities, vehicles can dynamically learn from their interactions with the environment, adapt to changing conditions, and make informed decisions in real time. This integration empowers vehicles to navigate complex scenarios, predict outcomes, and ensure safe and efficient autonomous driving.

Adaptive Learning for Decision-Making: Adaptive learning involves the continuous refinement of algorithms based on real-world data and experiences. In fog-based vehicle systems, vehicles can learn from their interactions with other vehicles, pedestrians, and infrastructure. Adaptive learning algorithms can be employed to fine-tune decision-making processes such as lane changing, merging, and route planning. By analyzing historical data and incorporating feedback, vehicles become more adept at making contextually aware decisions.

AI Integration for Local Autonomy: Artificial intelligence, including machine learning and neural networks, can be seamlessly integrated into fog-based vehicle systems to enhance local autonomy. Fog nodes equipped with AI capabilities can process data from multiple sources, recognize patterns, and predict outcomes. For instance, fog nodes can identify aggressive driving behavior in nearby vehicles, enabling a proactive response to potential safety risks. By harnessing AI, vehicles can take actions that are aligned with safety protocols and regulatory standards, enhancing overall road safety.

Real-world Application: Adaptive Cruise Control: Consider the application of adaptive learning and AI integration in adaptive cruise control. As a vehicle navigates varying traffic conditions, its adaptive cruise control system learns from different scenarios, such as heavy traffic congestion, sudden lane changes, and merging vehicles. Fog nodes equipped with AI capabilities analyze data from multiple vehicles and predict potential disruptions in traffic flow. Based on these predictions, the vehicle adapts its speed and maintains a safe following distance, ensuring smoother traffic flow and minimizing abrupt braking.

Benefits of Adaptive Learning and AI Integration

- **Real-time Adaptation:** Adaptive learning and AI integration enable vehicles to adapt to dynamic and unpredictable situations in real time. This agility enhances overall road safety and efficiency.
- **Continuous Improvement:** Through continuous learning, vehicles refine their decision-making processes over time, becoming more adept at handling a wide range of scenarios and conditions.
- **Contextual Understanding:** AI-integrated fog nodes provide vehicles with a deeper understanding of their surroundings, enabling them to make decisions based on comprehensive data analysis.

The integration of adaptive learning and AI techniques into fog-based vehicle systems revolutionizes decision-making and local autonomy. By continuously learning from real-world interactions and leveraging AI capabilities, vehicles become more capable of navigating complex environments, predicting outcomes, and ensuring safe and efficient autonomous driving. This integration contributes significantly to the realization of truly intelligent and self-aware vehicles that can operate confidently and responsibly in a wide range of driving scenarios.

USE CASES AND APPLICATIONS

The use and applications of Fog-enabled advanced driver assistance systems are depicted in Figure 3.

Figure 3. Fog-enabled advanced driver assistance systems (ADAS)

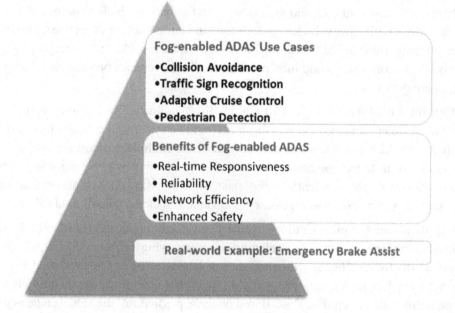

Fog-Enabled Advanced Driver Assistance Systems (ADAS)

Fog computing has a transformative impact on the development and effectiveness of Advanced Driver Assistance Systems (ADAS). By extending cloud capabilities to the edge, fog computing enhances the capabilities of ADAS, enabling real-time processing, quicker response times, and improved safety features (Bakhshi & Balador, 2019; Kadu & Singh, 2023; Zhong et al., 2021).

Fog-Enabled ADAS Use Cases

- **Collision Avoidance:** Fog-enabled ADAS can process data from multiple sensors, such as cameras and radar, in real time to detect potential collisions. Fog nodes analyze this data at the edge and provide immediate warnings or take control of vehicle systems to avoid accidents.
- **Traffic Sign Recognition:** Fog computing facilitates real-time processing of images captured by vehicle cameras. Fog nodes can identify and interpret traffic signs, ensuring that drivers receive accurate and timely information about speed limits, road conditions, and more.
- **Adaptive Cruise Control:** Fog-enabled ADAS enhances adaptive cruise control by integrating data from radar and LiDAR sensors. Fog nodes process this data at the edge to adjust the vehicle's speed and maintain a safe following distance in real-time traffic conditions.
- **Pedestrian Detection:** Fog computing empowers ADAS with improved pedestrian detection capabilities. Fog nodes process data from cameras and other sensors to identify pedestrians in real time, allowing the system to alert the driver or initiate emergency braking if necessary.

Benefits of Fog-Enabled ADAS

- **Real-time Responsiveness:** Fog-enabled ADAS significantly reduces communication latency, enabling immediate response to changing road conditions and potential hazards.
- **Reliability:** By processing data at the edge, fog nodes enhance the reliability of ADAS. Even in scenarios with limited or intermittent connectivity, fog-enabled systems can continue to function effectively.
- **Network Efficiency:** Fog computing minimizes the need to transmit large volumes of sensor data to the cloud, optimizing network bandwidth and reducing congestion.
- **Enhanced Safety:** The ability of fog-enabled ADAS to process data locally enhances safety by reducing the likelihood of accidents and providing drivers with timely alerts and interventions.

Real-World Example: Emergency Brake Assist

Imagine a fog-enabled ADAS application for emergency brake assist. As a vehicle approaches a sudden obstacle, fog nodes process data from sensors to determine if emergency braking is necessary. The fog nodes quickly calculate the required braking force and initiate the action, preventing a collision. The low-latency processing ensures that the decision and action occur in a fraction of a second, minimizing the risk of accidents.

Fog-enabled ADAS demonstrates the tangible benefits of integrating fog computing into intelligent vehicle systems. By bringing processing closer to the edge, fog-enabled systems provide real-time responsiveness, improved safety features, and efficient data processing for a range of ADAS applications. This integration paves the way for safer and more advanced driving experiences, moving us closer to the vision of fully autonomous vehicles.

Traffic Management and Cooperative Driving

Fog computing has the potential to revolutionize traffic management and cooperative driving, ushering in a new era of intelligent transportation systems that enhance efficiency, safety, and overall driving experience (Figure 4). By facilitating real-time communication and data sharing among vehicles, infrastructure, and cloud resources, fog-enabled solutions contribute to smoother traffic flow, reduced congestion, and collaborative driving practices.

Figure 4. Traffic management and cooperative driving

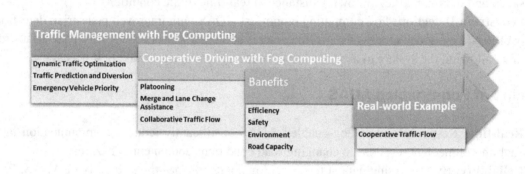

Traffic Management With Fog Computing

- **Dynamic Traffic Optimization:** Fog-enabled traffic management systems leverage data from various sources, including traffic cameras, sensors, and connected vehicles, to dynamically adjust traffic signal timings and manage traffic flow. Fog nodes situated at intersections analyze real-time data and optimize signal phases to alleviate congestion and reduce waiting times (Ijaz et al., 2021; Khalid et al., 2019; Neelakantam et al., 2020).
- **Traffic Prediction and Diversion:** Fog computing allows for the analysis of historical and real-time traffic data to predict congestion and traffic patterns. This information enables smart navigation systems to suggest alternate routes to drivers, diverting traffic away from congested areas and minimizing travel time.
- **Emergency Vehicle Priority:** Fog-enabled systems can detect the approach of emergency vehicles using GPS data and traffic sensors. Fog nodes communicate with traffic lights to create a green corridor, ensuring swift passage for emergency vehicles and minimizing response times.

Cooperative Driving With Fog Computing

- **Platooning:** Fog-enabled cooperative driving supports vehicle platooning, where groups of vehicles travel in close proximity, maintaining a safe distance through real-time communication. Fog nodes coordinate acceleration, braking, and steering, optimizing fuel efficiency and reducing congestion.

- **Merge and Lane Change Assistance:** Fog-enabled cooperative systems enable vehicles to communicate their intentions to neighboring vehicles during lane changes and merges. Fog nodes facilitate smoother transitions by coordinating speed adjustments and maintaining safe gaps.
- **Collaborative Traffic Flow:** Fog nodes facilitate communication between vehicles in traffic, allowing them to share information about speed, acceleration, and braking. This collaborative flow management reduces the risk of abrupt stops, improving traffic stability and safety.

Benefits of Fog-Enabled Traffic Management and Cooperative Driving

- **Efficiency:** Fog-enabled systems optimize traffic flow, reducing stop-and-go patterns and minimizing travel time, leading to fuel savings and reduced emissions.
- **Safety:** Real-time communication and coordination enhance safety by reducing the risk of collisions, preventing abrupt lane changes, and ensuring safe distances between vehicles.
- **Environment:** Fog-enabled cooperative driving contributes to eco-friendly practices by promoting fuel-efficient driving behaviors and reducing greenhouse gas emissions.
- **Road Capacity:** Improved traffic flow through fog-enabled traffic management and cooperative driving maximizes road capacity, delaying the need for costly infrastructure expansion.

Real-World Example: Cooperative Traffic Flow

Imagine a fog-enabled cooperative driving scenario where vehicles communicate their speeds and intentions to surrounding vehicles through fog nodes. In heavy traffic, fog nodes adjust vehicle speeds to maintain safe following distances and prevent traffic waves, resulting in smoother traffic flow and reduced congestion.

Fog-enabled traffic management and cooperative driving exemplify the transformative potential of fog computing in the realm of intelligent transportation. By fostering real-time communication, data exchange, and collaborative decision-making, fog-enabled solutions contribute to safer roads, efficient traffic flow, and a more sustainable transportation ecosystem. These applications pave the way for a future where vehicles and infrastructure work together seamlessly to create a smarter, safer, and more connected driving experience.

Vehicle-to-Infrastructure (V2I) Communication

Vehicle-to-Infrastructure (V2I) communication, a vital component of intelligent transportation systems, empowers vehicles to interact with infrastructure elements such as traffic signals, road signs, and toll booths. Fog computing enriches V2I communication by enabling real-time data exchange between vehicles and infrastructure, facilitating traffic management, safety enhancements, and improved overall driving experience (Damaj et al., 2022; Hakak et al., 2022; Shahzad et al., 2022).

V2I Communication Scenarios With Fog Computing

- **Traffic Signal Coordination:** Fog-enabled V2I communication allows vehicles to transmit real-time data to traffic signals. Fog nodes process this data and adjust signal timings based on traffic flow, optimizing signal phases to minimize congestion and improve traffic flow.

- **Real-time Road Information:** Fog nodes placed along roadways disseminate real-time road condition information, such as accidents, road closures, and weather conditions, to approaching vehicles. Drivers receive alerts and alternative route suggestions, ensuring safe and efficient navigation.
- **Parking Management:** Fog-enabled V2I systems assist in parking management by providing real-time data about available parking spaces in urban areas. Vehicles receive information about nearby parking availability, reducing search time and congestion.

Benefits of Fog-Enabled V2I Communication

- **Real-Time Information Exchange:** Fog computing enables instant data exchange between vehicles and infrastructure, facilitating timely decision-making and enhancing overall driving experience.
- **Efficient Traffic Flow:** By optimizing traffic signal timings and providing real-time road information, fog-enabled V2I communication contributes to smoother traffic flow and reduced congestion.
- **Safety Enhancements:** Fog-enabled V2I systems transmit alerts about hazards, construction zones, and emergency situations, enhancing driver awareness and road safety.
- **Eco-Friendly Driving:** Real-time data exchange helps drivers avoid congested areas and choose optimal routes, promoting fuel efficiency and reducing emissions.

Real-World Example: Adaptive Traffic Lights

Consider an intersection equipped with fog-enabled V2I communication. As a vehicle approaches the intersection, it communicates its position, speed, and intended direction to nearby fog nodes. The fog nodes process this data and adjust the traffic signal timings to allow a smooth passage for the approaching vehicle, reducing the need for unnecessary stops.

Fog-enabled Vehicle-to-Infrastructure (V2I) communication exemplifies the transformative potential of real-time data exchange between vehicles and infrastructure elements. By enabling seamless communication, fog computing enhances traffic management, safety, and overall efficiency in intelligent transportation systems. This integration not only improves the driving experience but also sets the stage for the realization of smarter, more connected, and more sustainable transportation ecosystems.

SECURITY CONSIDERATIONS IN FOG-BASED VEHICLE SYSTEMS

The various security considerations in fog-based vehicle systems are displayed in Figure 5.

Threats and Vulnerabilities in Fog Computing

The integration of fog computing into vehicle systems introduces new security challenges and vulnerabilities that must be addressed to ensure the safety and integrity of the intelligent transportation ecosystem. Some potential threats and vulnerabilities include data breaches, unauthorized access, malware attacks, and manipulation of sensor data. Fog nodes and edge devices, due to their distributed nature, can become potential attack vectors if not properly secured (Eftekhari et al., 2021; Padhy et al., 2023).

Figure 5. Security considerations in fog-based vehicle systems

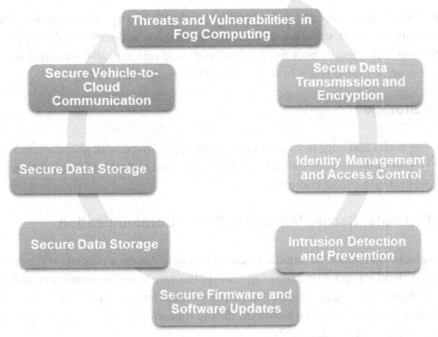

Secure Data Transmission and Encryption

Secure data transmission is paramount in fog-based vehicle systems to prevent interception and tampering of sensitive information. Encryption mechanisms, such as Transport Layer Security (TLS) protocols, can be employed to encrypt data exchanged between fog nodes, vehicles, and cloud servers. Fog nodes can also facilitate end-to-end encryption, ensuring that data remains secure even if intercepted during transmission.

Identity Management and Access Control

Robust identity management and access control mechanisms are essential to authenticate and authorize entities within fog-based vehicle systems. Fog nodes should employ secure authentication methods, such as digital certificates or biometric authentication, to verify the identity of vehicles and users. Access control policies should be implemented to limit data access based on user roles and responsibilities.

Intrusion Detection and Prevention

Intrusion detection and prevention systems (IDPS) can be deployed within fog-based vehicle systems to detect and mitigate unauthorized activities or attacks. These systems monitor network traffic and behavior patterns, raising alerts or taking corrective actions in real time. IDPS can help identify anomalies, such as unauthorized access attempts or abnormal data traffic, and respond proactively.

Secure Firmware and Software Updates

Fog nodes and edge devices require regular firmware and software updates to address vulnerabilities and improve security. However, updating these components must be done securely to prevent the introduction of malicious code. Secure boot mechanisms and code signing can ensure that only trusted and authorized updates are installed(Bakhshi & Balador, 2019; Kaur et al., 2020).

Secure Data Storage

Data stored on fog nodes and edge devices should be encrypted to prevent unauthorized access in case of device theft or compromise. Encryption keys should be managed securely, and access to stored data should be subject to strong access controls.

Real-world Example: Secure Vehicle-to-Cloud Communication

In a fog-based vehicle system, secure vehicle-to-cloud communication is crucial for transmitting vehicle data to centralized cloud servers. Fog nodes can encrypt the data before transmission, and cloud servers can use secure authentication mechanisms to validate fog nodes and vehicles. This ensures that sensitive vehicle data remains confidential and trustworthy during transmission and storage (Bakhshi & Balador, 2019; Khalid et al., 2019; Sun et al., 2019).

Security considerations are paramount in fog-based vehicle systems to safeguard against evolving cyber threats and vulnerabilities. By implementing secure data transmission, encryption, identity management, and intrusion detection, these systems can mitigate risks and ensure the integrity, confidentiality, and availability of data and services. As fog computing continues to play a pivotal role in intelligent transportation, robust security measures are essential to building a resilient and safe ecosystem for connected and autonomous vehicles.

PERFORMANCE EVALUATION AND CASE STUDIES

Metrics for Fog-Based Vehicle Systems

Evaluating the performance of fog-based vehicle systems involves measuring various metrics that reflect their effectiveness, efficiency, and impact on driving experience. Some key metrics include:

- **Latency:** The time it takes for data to travel between vehicles, fog nodes, and the cloud. Low latency is crucial for real-time decision-making.
- **Throughput:** The amount of data processed and transmitted within the system over a given period. High throughput is essential for handling the data generated by multiple vehicles.
- **Reliability:** The system's ability to maintain consistent and accurate data exchange, minimizing data loss or disruptions.
- **Scalability:** The system's capacity to handle a growing number of vehicles and devices without a significant drop in performance.

- **Energy Efficiency:** The energy consumption of fog nodes and vehicles in executing data processing and communication tasks.
- **Safety Impact:** The system's contribution to enhancing driving safety, such as reducing accidents, improving collision avoidance, and providing reliable traffic information.

Simulation and Experimentation

Simulation and experimentation are essential tools for assessing the performance of fog-based vehicle systems in controlled environments before real-world deployment. Simulators can model various scenarios, traffic patterns, and communication conditions to evaluate metrics such as latency, throughput, and system scalability. Experimentation on test tracks or closed environments allows researchers to validate the system's behavior, test algorithms, and fine-tune parameters under controlled conditions.

Case Studies: Real-World Implementations

- **Connected Vehicle Pilot Programs:** Various cities and regions worldwide have implemented connected vehicle pilot programs to assess the performance of fog-based vehicle systems. These programs involve equipping vehicles and infrastructure with communication devices to test real-time traffic management, safety improvements, and cooperative driving scenarios.
- **Smart Intersection Management:** Fog-enabled intelligent intersections have been deployed to enhance traffic flow and safety. These intersections use V2I communication to optimize traffic signal timings, manage pedestrian crossings, and mitigate congestion.
- **Cooperative Platooning:** Real-world tests of cooperative platooning involving trucks or buses demonstrate the benefits of fog-enabled communication in optimizing fuel efficiency, reducing emissions, and improving traffic flow.
- **Emergency Vehicle Priority:** Some regions have implemented fog-based systems that grant priority to emergency vehicles by controlling traffic signals and creating green corridors, ensuring swift passage during emergencies.

Performance evaluation and case studies are integral to understanding the effectiveness and impact of fog-based vehicle systems in real-world settings. By measuring key metrics, conducting simulations, and exploring case studies, researchers and practitioners can refine system design, algorithms, and deployment strategies. These efforts contribute to the continuous improvement of fog-enabled intelligent transportation systems, leading to safer, more efficient, and more connected roadways.

FUTURE TRENDS AND CHALLENGES

Emerging Technologies and Innovations

The future of fog-based vehicle systems holds promise as emerging technologies and innovations continue to shape the landscape of intelligent transportation (Damaj et al., 2022; Nkenyereye et al., 2021; Palattella et al., 2019; Shahzad et al., 2022). Some trends to watch for include:

- **5G Integration:** The integration of 5G networks will enhance data rates, lower latency, and provide higher network capacity, enabling more advanced applications and real-time communication in fog-based vehicle systems.
- **Edge AI:** Advances in edge artificial intelligence will empower fog nodes to perform complex data processing, pattern recognition, and decision-making at the edge, enabling vehicles to make more sophisticated autonomous decisions (Boopathi & Kanike, 2023; Koshariya, Khatoon, et al., 2023; Ramudu et al., 2023; Sengeni et al., 2023).
- **V2X Evolution:** The evolution of Vehicle-to-Everything (V2X) communication will extend beyond V2I and V2V, incorporating vehicle-to-pedestrian (V2P) and vehicle-to-grid (V2G) communication, enabling safer interactions with all road users and energy optimization.

Scalability and Network Management

As fog-based vehicle systems grow in scale and complexity, scalability and network management become critical challenges:

- **Data Overload:** With increasing vehicle density and data generation, fog nodes must efficiently manage and prioritize data to avoid congestion and ensure timely data processing.
- **Dynamic Scalability:** Fog systems need to dynamically scale resources to accommodate varying traffic conditions, adapt to changing road networks, and ensure consistent performance.
- **Interoperability:** Ensuring seamless communication and interoperability among various vehicle types, manufacturers, and infrastructure elements requires standardized protocols and interfaces.

Ethical and Regulatory Implications

The adoption of fog-based vehicle systems raises ethical and regulatory considerations:

- **Privacy Concerns:** The extensive data collection and communication within fog systems raise privacy concerns. Striking a balance between data sharing for safety and respecting user privacy becomes crucial.
- **Data Security:** Ensuring data security and preventing unauthorized access or tampering of sensitive vehicle data is a constant challenge.
- **Algorithmic Transparency:** As AI algorithms influence decision-making in autonomous vehicles, ensuring transparency and understanding in how these decisions are made becomes important for accountability and safety.
- **Regulations:** Establishing clear regulatory frameworks and standards for fog-based vehicle systems is necessary to ensure consistent safety, interoperability, and compliance across jurisdictions.

The future of fog-based vehicle systems holds both opportunities and challenges. Emerging technologies, improved communication infrastructure, and advancements in AI promise to enhance safety, efficiency, and convenience on the road (Anitha et al., 2023; Koshariya, Kalaiyarasi, et al., 2023; Koshariya, Khatoon, et al., 2023). However, managing scalability, addressing ethical concerns, and navigating complex regulatory landscapes remain vital tasks. As the intelligent transportation ecosystem evolves, collaboration among researchers, industry stakeholders, policymakers, and the public will be crucial to realize the full potential of fog-based vehicle systems while addressing the challenges that lie ahead.

CONCLUSION

The integration of fog computing into intelligent vehicle systems marks a significant milestone in the evolution of transportation. By extending cloud capabilities to the edge, fog computing transforms vehicles into intelligent, interconnected entities capable of real-time decision-making, enhanced safety, and improved efficiency. This paradigm shift paves the way for the realization of autonomous vehicles, cooperative driving, and more sustainable transportation ecosystems.

Fog computing's impact is profound, enhancing vehicle autonomy through sensor data fusion, enabling secure communication through V2I and V2V interactions, and optimizing traffic management with dynamic signal control. The framework for integrating fog computing provides a comprehensive blueprint for designing and deploying fog-enabled solutions, emphasizing architectural considerations, communication protocols, and adaptive learning.

As fog-based vehicle systems continue to advance, several challenges and opportunities emerge. Ensuring data security, managing scalability, and addressing ethical and regulatory concerns are pivotal to building trust and acceptance among users and stakeholders (Boopathi, 2023a; Boopathi et al., 2023; Boopathi & Kanike, 2023; Koshariya, Kalaiyarasi, et al., 2023). Emerging technologies like 5G and edge AI promise to further revolutionize the capabilities of fog-based systems, enhancing real-time communication, decision-making, and collaboration among vehicles and infrastructure (Venkateswaran et al., 2023).

In this dynamic landscape, collaboration between researchers, industry leaders, policymakers, and society at large is essential. The future of fog-based vehicle systems depends on collective efforts to overcome challenges, develop robust security measures, and shape ethical guidelines. As fog computing continues to reshape transportation, the vision of safer, more efficient, and interconnected roadways is within reach, promising a transformative impact on how we experience and navigate the world of intelligent transportation.

LIST OF ABBREVIATIONS

5G: Fifth Generation (Wireless Network)
ADAS: Advanced Driver Assistance Systems
AI: Artificial Intelligence
GPS: Global Positioning System
IDPS: Intrusion Detection and Prevention Systems
IoT: Internet of Things
LiDAR: Light Detection and Ranging
TLS: Transport Layer Security
V2G: Vehicle-to-Grid
V2I: Vehicle-to-Infrastructure
V2P: Vehicle-to-Pedestrian

V2V: Vehicle-to-Vehicle
V2X: Vehicle-to-Everything

REFERENCES

Ahangar, M. N., Ahmed, Q. Z., Khan, F. A., & Hafeez, M. (2021). A survey of autonomous vehicles: Enabling communication technologies and challenges. *Sensors (Basel)*, *21*(3), 706. doi:10.3390/s21030706 PMID:33494191

Al-Shareeda, M. A., & Manickam, S. (2022). COVID-19 vehicle based on an efficient mutual authentication scheme for 5G-enabled vehicular fog computing. *International Journal of Environmental Research and Public Health*, *19*(23), 15618. doi:10.3390/ijerph192315618 PMID:36497709

Alajali, W., Gao, S., & Alhusaynat, A. D. (2019). Fog computing based traffic and car parking intelligent system. *International Conference on Algorithms and Architectures for Parallel Processing*, 365–380.

Aljumah, A., Kaur, A., Bhatia, M., & Ahamed Ahanger, T. (2021). Internet of things-fog computing-based framework for smart disaster management. *Transactions on Emerging Telecommunications Technologies*, *32*(8), e4078. doi:10.1002/ett.4078

Anitha, C., Komala, C., Vivekanand, C. V., Lalitha, S., & Boopathi, S. (2023). Artificial Intelligence driven security model for Internet of Medical Things (IoMT). *IEEE Explore*, 1–7.

Bakhshi, Z., & Balador, A. (2019). An overview on security and privacy challenges and their solutions in fog-based vehicular application. *2019 IEEE 30th International Symposium on Personal, Indoor and Mobile Radio Communications (PIMRC Workshops)*, 1–7.

Barik, R. K., Priyadarshini, R., Dubey, H., Kumar, V., & Mankodiya, K. (2018). FogLearn: Leveraging fog-based machine learning for smart system big data analytics. *International Journal of Fog Computing*, *1*(1), 15–34. doi:10.4018/IJFC.2018010102

Boopathi, S. (2023a). Internet of Things-Integrated Remote Patient Monitoring System: Healthcare Application. In *Dynamics of Swarm Intelligence Health Analysis for the Next Generation* (pp. 137–161). IGI Global. doi:10.4018/978-1-6684-6894-4.ch008

Boopathi, S. (2023b). Securing Healthcare Systems Integrated With IoT: Fundamentals, Applications, and Future Trends. In Dynamics of Swarm Intelligence Health Analysis for the Next Generation (pp. 186–209). IGI Global.

Boopathi, S., & Kanike, U. K. (2023). Applications of Artificial Intelligent and Machine Learning Techniques in Image Processing. In *Handbook of Research on Thrust Technologies' Effect on Image Processing* (pp. 151–173). IGI Global. doi:10.4018/978-1-6684-8618-4.ch010

Boopathi, S., Pandey, B. K., & Pandey, D. (2023). Advances in Artificial Intelligence for Image Processing: Techniques, Applications, and Optimization. In Handbook of Research on Thrust Technologies' Effect on Image Processing (pp. 73–95). IGI Global.

Butt, F. A., Chattha, J. N., Ahmad, J., Zia, M. U., Rizwan, M., & Naqvi, I. H. (2022). On the integration of enabling wireless technologies and sensor fusion for next-generation connected and autonomous vehicles. *IEEE Access : Practical Innovations, Open Solutions*, *10*, 14643–14668. doi:10.1109/ACCESS.2022.3145972

Cui, Q., Wang, Y., Chen, K.-C., Ni, W., Lin, I.-C., Tao, X., & Zhang, P. (2018). Big data analytics and network calculus enabling intelligent management of autonomous vehicles in a smart city. *IEEE Internet of Things Journal*, *6*(2), 2021–2034. doi:10.1109/JIOT.2018.2872442

Damaj, I. W., Yousafzai, J. K., & Mouftah, H. T. (2022). Future trends in connected and autonomous vehicles: Enabling communications and processing technologies. *IEEE Access : Practical Innovations, Open Solutions*, *10*, 42334–42345. doi:10.1109/ACCESS.2022.3168320

Eftekhari, S. A., Nikooghadam, M., & Rafighi, M. (2021). Security-enhanced three-party pairwise secret key agreement protocol for fog-based vehicular ad-hoc communications. *Vehicular Communications*, *28*, 100306. doi:10.1016/j.vehcom.2020.100306

Fernandes, P., & Nunes, U. (2012). Platooning with DSRC-based IVC-enabled autonomous vehicles: Adding infrared communications for IVC reliability improvement. *2012 IEEE Intelligent Vehicles Symposium*, 517–522. 10.1109/IVS.2012.6232206

Hakak, S., Gadekallu, T. R., Maddikunta, P. K. R., Ramu, S. P., Parimala, M., De Alwis, C., & Liyanage, M. (2022). Autonomous Vehicles in 5G and beyond: A Survey. *Vehicular Communications*, 100551.

Hema, N., Krishnamoorthy, N., Chavan, S. M., Kumar, N., Sabarimuthu, M., & Boopathi, S. (2023). A Study on an Internet of Things (IoT)-Enabled Smart Solar Grid System. In *Handbook of Research on Deep Learning Techniques for Cloud-Based Industrial IoT* (pp. 290–308). IGI Global. doi:10.4018/978-1-6684-8098-4.ch017

Hoermann, S., Stumper, D., & Dietmayer, K. (2017). Probabilistic long-term prediction for autonomous vehicles. *2017 IEEE Intelligent Vehicles Symposium (IV)*, 237–243. 10.1109/IVS.2017.7995726

Ijaz, M., Li, G., Lin, L., Cheikhrouhou, O., Hamam, H., & Noor, A. (2021). Integration and applications of fog computing and cloud computing based on the internet of things for provision of healthcare services at home. *Electronics (Basel)*, *10*(9), 1077. doi:10.3390/electronics10091077

Kadu, A., & Singh, M. (2023). Fog-Enabled Framework for Patient Health-Monitoring Systems Using Internet of Things and Wireless Body Area Networks. In *Computational Intelligence: Select Proceedings of InCITe 2022* (pp. 607–616). Springer. doi:10.1007/978-981-19-7346-8_52

Karthik, S., Hemalatha, R., Aruna, R., Deivakani, M., Reddy, R. V. K., & Boopathi, S. (2023). Study on Healthcare Security System-Integrated Internet of Things (IoT). In Perspectives and Considerations on the Evolution of Smart Systems (pp. 342–362). IGI Global.

Kato, S., Tokunaga, S., Maruyama, Y., Maeda, S., Hirabayashi, M., Kitsukawa, Y., Monrroy, A., Ando, T., Fujii, Y., & Azumi, T. (2018). Autoware on board: Enabling autonomous vehicles with embedded systems. *2018 ACM/IEEE 9th International Conference on Cyber-Physical Systems (ICCPS)*, 287–296.

Kaur, J., Agrawal, A., & Khan, R. A. (2020). Security issues in fog environment: A systematic literature review. *International Journal of Wireless Information Networks*, 27(3), 467–483. doi:10.1007/s10776-020-00491-7

Khalid, T., Khan, A. N., Ali, M., Adeel, A., ur Rehman Khan, A., & Shuja, J. (2019). A fog-based security framework for intelligent traffic light control system. *Multimedia Tools and Applications*, 78(17), 24595–24615. doi:10.1007/s11042-018-7008-z

Kirsanova, A. A., Radchenko, G. I., & Tchernykh, A. N. (2021). Fog computing state of the art: Concept and classification of platforms to support distributed computing systems. *ArXiv Preprint ArXiv:2106.11726*.

Koshariya, A. K., Kalaiyarasi, D., Jovith, A. A., Sivakami, T., Hasan, D. S., & Boopathi, S. (2023). AI-Enabled IoT and WSN-Integrated Smart Agriculture System. In *Artificial Intelligence Tools and Technologies for Smart Farming and Agriculture Practices* (pp. 200–218). IGI Global. doi:10.4018/978-1-6684-8516-3.ch011

Koshariya, A. K., Khatoon, S., Marathe, A. M., Suba, G. M., Baral, D., & Boopathi, S. (2023). Agricultural Waste Management Systems Using Artificial Intelligence Techniques. In *AI-Enabled Social Robotics in Human Care Services* (pp. 236–258). IGI Global. doi:10.4018/978-1-6684-8171-4.ch009

Lin, S.-Y., Du, Y., Ko, P.-C., Wu, T.-J., Ho, P.-T., Sivakumar, V., & Subbareddy, R. (2020). Fog computing based hybrid deep learning framework in effective inspection system for smart manufacturing. *Computer Communications*, 160, 636–642. doi:10.1016/j.comcom.2020.05.044

Ma, M., He, D., Wang, H., Kumar, N., & Choo, K.-K. R. (2019). An efficient and provably secure authenticated key agreement protocol for fog-based vehicular ad-hoc networks. *IEEE Internet of Things Journal*, 6(5), 8065–8075. doi:10.1109/JIOT.2019.2902840

Maguluri, L. P., Ananth, J., Hariram, S., Geetha, C., Bhaskar, A., & Boopathi, S. (2023). Smart Vehicle-Emissions Monitoring System Using Internet of Things (IoT). In Handbook of Research on Safe Disposal Methods of Municipal Solid Wastes for a Sustainable Environment (pp. 191–211). IGI Global.

Neelakantam, G., Onthoni, D. D., & Sahoo, P. K. (2020). Reinforcement learning based passengers assistance system for crowded public transportation in fog enabled smart city. *Electronics (Basel)*, 9(9), 1501. doi:10.3390/electronics9091501

Nkenyereye, L., Islam, S. R., Bilal, M., Abdullah-Al-Wadud, M., Alamri, A., & Nayyar, A. (2021). Secure crowd-sensing protocol for fog-based vehicular cloud. *Future Generation Computer Systems*, 120, 61–75. doi:10.1016/j.future.2021.02.008

Padhy, S., Alowaidi, M., Dash, S., Alshehri, M., Malla, P. P., Routray, S., & Alhumyani, H. (2023). AgriSecure: A Fog Computing-Based Security Framework for Agriculture 4.0 via Blockchain. *Processes (Basel, Switzerland)*, 11(3), 757. doi:10.3390/pr11030757

Palattella, M. R., Soua, R., Khelil, A., & Engel, T. (2019). Fog computing as the key for seamless connectivity handover in future vehicular networks. *Proceedings of the 34th ACM/SIGAPP Symposium on Applied Computing*, 1996–2000. 10.1145/3297280.3297475

Pareek, K., Tiwari, P. K., & Bhatnagar, V. (2021). Fog computing in healthcare: A review. *IOP Conference Series. Materials Science and Engineering*, *1099*(1), 012025. doi:10.1088/1757-899X/1099/1/012025

Rahamathunnisa, U., Sudhakar, K., Murugan, T. K., Thivaharan, S., Rajkumar, M., & Boopathi, S. (2023). Cloud Computing Principles for Optimizing Robot Task Offloading Processes. In *AI-Enabled Social Robotics in Human Care Services* (pp. 188–211). IGI Global. doi:10.4018/978-1-6684-8171-4.ch007

Ramudu, K., Mohan, V. M., Jyothirmai, D., Prasad, D., Agrawal, R., & Boopathi, S. (2023). Machine Learning and Artificial Intelligence in Disease Prediction: Applications, Challenges, Limitations, Case Studies, and Future Directions. In *Contemporary Applications of Data Fusion for Advanced Healthcare Informatics* (pp. 297–318). IGI Global.

Rani, S., Srivastava, G., & ... (2023). Secure hierarchical fog computing-based architecture for industry 5.0 using an attribute-based encryption scheme. *Expert Systems with Applications*, 121180.

Reddy, M. A., Reddy, B. M., Mukund, C., Venneti, K., Preethi, D., & Boopathi, S. (2023). Social Health Protection During the COVID-Pandemic Using IoT. In *The COVID-19 Pandemic and the Digitalization of Diplomacy* (pp. 204–235). IGI Global. doi:10.4018/978-1-7998-8394-4.ch009

Samikannu, R., Koshariya, A. K., Poornima, E., Ramesh, S., Kumar, A., & Boopathi, S. (2022). Sustainable Development in Modern Aquaponics Cultivation Systems Using IoT Technologies. In *Human Agro-Energy Optimization for Business and Industry* (pp. 105–127). IGI Global.

Sengeni, D., Padmapriya, G., Imambi, S. S., Suganthi, D., Suri, A., & Boopathi, S. (2023). Biomedical Waste Handling Method Using Artificial Intelligence Techniques. In *Handbook of Research on Safe Disposal Methods of Municipal Solid Wastes for a Sustainable Environment* (pp. 306–323). IGI Global. doi:10.4018/978-1-6684-8117-2.ch022

Shahzad, A., Gherbi, A., & Zhang, K. (2022). Enabling fog–blockchain computing for autonomous-vehicle-parking system: A solution to reinforce iot–cloud platform for future smart parking. *Sensors (Basel)*, *22*(13), 4849. doi:10.3390/s22134849 PMID:35808345

Singh, G., & Singh, J. (2023). A Fog Computing based Agriculture-IoT Framework for Detection of Alert Conditions and Effective Crop Protection. *2023 5th International Conference on Smart Systems and Inventive Technology (ICSSIT)*, 537–543.

Sodhro, A. H., Sodhro, G. H., Guizani, M., Pirbhulal, S., & Boukerche, A. (2020). AI-enabled reliable channel modeling architecture for fog computing vehicular networks. *IEEE Wireless Communications*, *27*(2), 14–21. doi:10.1109/MWC.001.1900311

Sun, G., Sun, S., Sun, J., Yu, H., Du, X., & Guizani, M. (2019). Security and privacy preservation in fog-based crowd sensing on the internet of vehicles. *Journal of Network and Computer Applications*, *134*, 89–99. doi:10.1016/j.jnca.2019.02.018

Syamala, M., Komala, C., Pramila, P., Dash, S., Meenakshi, S., & Boopathi, S. (2023). Machine Learning-Integrated IoT-Based Smart Home Energy Management System. In *Handbook of Research on Deep Learning Techniques for Cloud-Based Industrial IoT* (pp. 219–235). IGI Global. doi:10.4018/978-1-6684-8098-4.ch013

Tripathy, S. S., Imoize, A. L., Rath, M., Tripathy, N., Bebortta, S., Lee, C.-C., Chen, T.-Y., Ojo, S., Isabona, J., & Pani, S. K. (2022). A novel edge-computing-based framework for an intelligent smart healthcare system in smart cities. *Sustainability (Basel)*, *15*(1), 735. doi:10.3390/su15010735

Venkateswaran, N., Vidhya, K., Ayyannan, M., Chavan, S. M., Sekar, K., & Boopathi, S. (2023). A Study on Smart Energy Management Framework Using Cloud Computing. In 5G, Artificial Intelligence, and Next Generation Internet of Things: Digital Innovation for Green and Sustainable Economies (pp. 189–212). IGI Global. doi:10.4018/978-1-6684-8634-4.ch009

Vilela, P. H., Rodrigues, J. J., Righi, R. da R., Kozlov, S., & Rodrigues, V. F. (2020). Looking at fog computing for e-health through the lens of deployment challenges and applications. *Sensors (Basel)*, *20*(9), 2553. doi:10.3390/s20092553 PMID:32365815

Wu, D., Liu, S., Zhang, L., Terpenny, J., Gao, R. X., Kurfess, T., & Guzzo, J. A. (2017). A fog computing-based framework for process monitoring and prognosis in cyber-manufacturing. *Journal of Manufacturing Systems*, *43*, 25–34. doi:10.1016/j.jmsy.2017.02.011

Yang, Y., Luo, X., Chu, X., & Zhou, M.-T. (2020). *Fog-enabled intelligent IoT systems*. Springer. doi:10.1007/978-3-030-23185-9

Yang, Y., Luo, X., Chu, X., Zhou, M.-T., Yang, Y., Luo, X., Chu, X., & Zhou, M.-T. (2020). Fog-enabled intelligent transportation system. *Fog-Enabled Intelligent IoT Systems*, 163–184.

Yazdani, A., Dashti, S. F., & Safdari, Y. (2023). A fog-assisted information model based on priority queue and clinical decision support systems. *Health Informatics Journal*, *29*(1). doi:10.1177/14604582231152792 PMID:36645733

Zhang, X., Zhong, H., Cui, J., Bolodurina, I., & Liu, L. (2022). Lbvp: A lightweight batch verification protocol for fog-based vehicular networks using self-certified public key cryptography. *IEEE Transactions on Vehicular Technology*, *71*(5), 5519–5533. doi:10.1109/TVT.2022.3157960

Zhong, H., Chen, L., Cui, J., Zhang, J., Bolodurina, I., & Liu, L. (2021). Secure and Lightweight Conditional Privacy-Preserving Authentication for Fog-Based Vehicular Ad Hoc Networks. *IEEE Internet of Things Journal*, *9*(11), 8485–8497. doi:10.1109/JIOT.2021.3116039

Zhou, Z., Liao, H., Wang, X., Mumtaz, S., & Rodriguez, J. (2020). When vehicular fog computing meets autonomous driving: Computational resource management and task offloading. *IEEE Network*, *34*(6), 70–76. doi:10.1109/MNET.001.1900527

Compilation of References

Abarca-Guerrero, L., Roa-Gutiérrez, F., & Rudín-Vega, V. (2018). WEEE resource management system in Costa Rica. *Resources*, *7*(1), 1–14. doi:10.3390/resources7010002

Abbas, S., Khan, M. A., Falcon-Morales, L. E., Rehman, A., Saeed, Y., Zareei, M., Zeb, A., & Mohamed, E. M. (2020). Modeling, simulation and optimization of power plant energy sustainability for IoT enabled smart cities empowered with deep extreme learning machine. *IEEE Access : Practical Innovations, Open Solutions*, *8*, 39982–39997. doi:10.1109/ACCESS.2020.2976452

Abd Wahab, M. H., Kadir, A. A., Tomari, M. R., & Jabbar, M. H. (2014). Smart Recycle Bin: A Conceptual Approach of Smart Waste Management with Integrated Web Based System. *2014 International Conference on IT Convergence and Security (ICITCS)*, 1-4. 10.1109/ICITCS.2014.7021812

Abu-Qdais, H., Shatnawi, N., & Esra'a, A. A. (2023). *Intelligent solid waste classification system using combination of image processing and machine learning models*. Academic Press.

Adedeji, O., & Wang, Z. (2019). Intelligent waste classification system using deep learning convolutional neural network. *Procedia Manufacturing*, *35*, 607–612. doi:10.1016/j.promfg.2019.05.086

Afolabi, H. A., & Aburas, A. (2021). An evaluation of machine learning classifiers for prediction of attacks to secure green IoT infrastructure. *International Journal (Toronto, Ont.)*, *9*(5).

Agarwal, R., & Sharma, D. K. (2021, February). Machine learning & Deep learning based Load Balancing Algorithms techniques in Cloud Computing. In *2021 International Conference on Innovative Practices in Technology and Management (ICIPTM)* (pp. 249-254). IEEE. 10.1109/ICIPTM52218.2021.9388349

Ahangar, M. N., Ahmed, Q. Z., Khan, F. A., & Hafeez, M. (2021). A survey of autonomous vehicles: Enabling communication technologies and challenges. *Sensors (Basel)*, *21*(3), 706. doi:10.3390/s21030706 PMID:33494191

Ahmad, R. a. (2019). *Green IoT—issues and challenges*. Academic Press.

Ahmed, M. B. (2021). A Decentralised Mechanism for Secure Cloud Computing Transactions. IEEE, 1-5.

Ahmed, K. I., Tabassum, H., & Hossain, E. (2019). Deep learning for radio resource allocation in multi-cell networks. *IEEE Network*, *33*(6), 188–195. doi:10.1109/MNET.2019.1900029

Aiswarya. (2023, August 2). *Green Data Centers: A Key to Sustainability*. https://www.analyticsinsight.net/green-data-centers-a-key-to-sustainability/

Akyurt, İ. Z., Kuvvetli, Y., & Deveci, M. (2020). Enterprise resource planning in the age of industry 4.0: A general overview. *Logistics*, *4*(0), 178–185.

Alajali, W., Gao, S., & Alhusaynat, A. D. (2019). Fog computing based traffic and car parking intelligent system. *International Conference on Algorithms and Architectures for Parallel Processing*, 365–380.

Al-Ansi, A., Al-Ansi, A. M., Muthanna, A., Elgendy, I. A., & Koucheryavy, A. (2021). Survey on intelligence edge computing in 6G: Characteristics, challenges, potential use cases, and market drivers. *Future Internet*, *13*(5), 118. doi:10.3390/fi13050118

Alavi, N., Shirmardi, M., Babaei, A., Takdastan, A., & Bagheri, N. (2015). Waste electrical andelectronic equipment (WEEE) estimation: A case study of Ahvaz City, Iran. *Journal of the Air & Waste Management Association*, *65*(3), 298–305. doi:10.1080/10962247.2014.976297 PMID:25947126

Alexander & Fontecchio. (2023, October 28). *Power usage effectiveness (PUE)*. https://www.techtarget.com/searchdatacenter/definition/power-usage-effectiveness-PUE

Alexander. (2023). *Data center infrastructure management (DCIM)*. https://www.techtarget.com/searchdatacenter/definition/data-center-infrastructure-management-DCIM

Alghamdi, R., Dahrouj, H., Al-Naffouri, T., & Alouini, M. S. (2023). *Toward Immersive Underwater Cloud-Enabled Networks: Prospects and Challenges*. IEEE BITS the Information Theory Magazine.

Ali, M. A., & PP, F. RSalama Abd Elminaam, D. (2022). A Feature Selection Based on Improved Artificial Hummingbird Algorithm Using Random Opposition-Based Learning for Solving Waste Classification Problem. *Mathematics*, *10*(15), 2675. doi:10.3390/math10152675

Ali, Z., Jiao, L., Baker, T., Abbas, G., Abbas, Z. H., & Khaf, S. (2019). A deep learning approach for energy efficient computational offloading in mobile edge computing. *IEEE Access : Practical Innovations, Open Solutions*, *7*, 149623–149633. doi:10.1109/ACCESS.2019.2947053

Aljumah, A., Kaur, A., Bhatia, M., & Ahamed Ahanger, T. (2021). Internet of things-fog computing-based framework for smart disaster management. *Transactions on Emerging Telecommunications Technologies*, *32*(8), e4078. doi:10.1002/ett.4078

Almalki, F. A. (2021). Green IoT for eco-friendly and sustainable smart cities: Future directions and opportunities. *Mobile Networks and Applications*, 1–25.

Alrayes, F. S., Asiri, M. M., Maashi, M. S., Nour, M. K., Rizwanullah, M., Osman, A. E., Drar, S., & Zamani, A. S. (2023). Waste classification using vision transformer based on multilayer hybrid convolution neural network. *Urban Climate*, *49*, 101483. doi:10.1016/j.uclim.2023.101483

Alsamhi, S. H., Ma, O., Ansari, M. S., & Meng, Q. (2019). Greening internet of things for greener and smarter cities: A survey and future prospects. *Telecommunication Systems*, *72*(4), 609–632. doi:10.1007/s11235-019-00597-1

Al-Shareeda, M. A., & Manickam, S. (2022). COVID-19 vehicle based on an efficient mutual authentication scheme for 5G-enabled vehicular fog computing. *International Journal of Environmental Research and Public Health*, *19*(23), 15618. doi:10.3390/ijerph192315618 PMID:36497709

Alsharif, M. H., Jahid, A., Kelechi, A. H., & Kannadasan, R. (2023). Green IoT: A review and future research directions. *Symmetry*, *15*(3), 757. doi:10.3390/sym15030757

Alsubai, S., Garg, H., & Alqahtani, A. (2023). A Novel Hybrid MSA-CSA Algorithm for Cloud Computing Task Scheduling Problems. *Symmetry*, *15*(10), 1931. doi:10.3390/sym15101931

Al-Turjman, F. S. (2019). *The green internet of things (g-iot)*. Hindawi. doi:10.1155/2019/6059343

Alvarez-de-los-Mozos, E., & Renteria, A. (2017). Collaborative robots in e-waste management. *Procedia Manufacturing*, *11*, 55–62. doi:10.1016/j.promfg.2017.07.133

Al-Zamil, A. S., & Saudagar, A. K. J. (2020). Drivers and challenges of applying green computing for sustainable agriculture: A case study. *Sustainable Computing : Informatics and Systems*, *28*, 100–264. doi:10.1016/j.suscom.2018.07.008

Andrae, A. S., & Edler, T. (2015). On global electricity usage of communication technology: Trends to 2030. *Challenges*, *6*(1), 117–157. doi:10.3390/challe6010117

Angus Loten. (2019, August 19). *Data-Center Market Is Booming Amid Shift to Cloud*. https://www.wsj.com/articles/data-center-market-is-booming-amid-shift-to-cloud-11566252481

Anitha, C., Komala, C., Vivekanand, C. V., Lalitha, S., & Boopathi, S. (2023). Artificial Intelligence driven security model for Internet of Medical Things (IoMT). *IEEE Explore*, 1–7.

Anley, M. B., & Awgichew, R. B. (2022, January). Machine Learning Approach for Green Usage of Computing Devices. *Proceeding of the 2 nd Deep Learning Indaba-X Ethiopia Conference 2021*.

Anthony, L. F. W., Kanding, B., & Selvan, R. (2020). Carbontracker: Tracking and predicting the carbon footprint of training deep learning models. *arXiv preprint arXiv:2007.03051*.

Anwar, M. (2013). Green computing and energy consumption issues in the modern age. *IOSR Journal of Computer Engineering*, *12*(6), 91–98. doi:10.9790/0661-1269198

Apple Inc. (2023, April 5). *Apple and global suppliers expand renewable energy to 13.7 gigawatts*. https://www.apple.com/in/newsroom/2023/04/apple-and-global-suppliers-expand-renewable-energy-to-13-point-7-gigawatts/

Aral, R. A., Keskin, S. R., & Kaya, M. (2018). Classification of trashnet dataset based on deep learning models. *Proceedings of the 2018 IEEE International Conference on Big Data (Big Data)*, 2058–206.

Araujo, M. G., Magrini, A., Mahler, C. F., & Bilitewski, B. (2012). A model for estimation of potential generation of waste electrical and electronic equipment in Brazil. *Waste Management (New York, N.Y.)*, *32*(2), 335–342. doi:10.1016/j.wasman.2011.09.020 PMID:22014584

Arikumar, K. S., & Natarajan, V. (2020). FIoT: a QoS-aware fog-IoT framework to minimize latency in IoT applications via fog offloading. In *Evolution in Computational Intelligence: Frontiers in Intelligent Computing: Theory and Applications (FICTA 2020)* (Vol. 1, pp. 551–559). Springer Singapore.

Arikumar, K. S., Prathiba, S. B., Alazab, M., Gadekallu, T. R., Pandya, S., Khan, J. M., & Moorthy, R. S. (2022). FL-PMI: Federated learning-based person movement identification through wearable devices in smart healthcare systems. *Sensors (Basel)*, *22*(4), 1377. doi:10.3390/s22041377 PMID:35214282

Ash, J., & Adams, R. P. (2020). On warm-starting neural network training. *Advances in Neural Information Processing Systems*, *33*, 3884–3894.

Asim, M., & Abd El-Latif, A. A. (2021). Intelligent computational methods for multi-unmanned aerial vehicle-enabled autonomous mobile edge computing systems. *ISA Transactions*. PMID:34933773

Aslanpour, M. S., Gill, S. S., & Toosi, A. N. (2020). Performance evaluation metrics for cloud, fog and edge computing: A review, taxonomy, benchmarks and standards for future research. *Internet of Things : Engineering Cyber Physical Human Systems*, *12*, 100273. doi:10.1016/j.iot.2020.100273

Aste, N., Buzzetti, M., Caputo, P., & Del Pero, C. (2018). Regional policies toward energy efficiency and renewable energy sources integration: Results of a wide monitoring campaign. *Sustainable Cities and Society, 36*, 215–224. doi:10.1016/j.scs.2017.10.005

Atiq, H. U., Ahmad, Z., Uz Zaman, S. K., Khan, M. A., Shaikh, A. A., & Al-Rasheed, A. (2023). Reliable resource allocation and management for IoT transportation using fog computing. *Electronics (Basel), 12*(6), 1452. doi:10.3390/electronics12061452

Austen & FitzGerald. (2016, August 26). *Rackspace to Go Private in $4.3 Billion Deal*. https://www.wsj.com/articles/rackspace-to-go-private-in-4-3-billion-deal-1472218264

Avasalcai, C., & Dustdar, S. (2018, October). Latency-aware decentralized resource management for IoT applications. In *Proceedings of the 8th International Conference on the Internet of Things* (pp. 1-4). 10.1145/3277593.3277637

Awasthi, A. K., & Li, J. (2017). Management of electrical and electronic waste: A comparative evaluation of China and India. *Renewable & Sustainable Energy Reviews, 76*, 434–447. doi:10.1016/j.rser.2017.02.067

Awasthi, A., & Goel, N. (2021). Phishing Website Prediction: A Comparison of Machine Learning Techniques. *Data Intelligence and Cognitive Informatics Proceedings of ICDICI, 2020*, 637–650.

Baccini, P., & Brunner, P. H. (2012). *Metabolism of the Anthroposphere: Analysis, Evaluation, Design*. MIT Press.

Bahers, J. B., & Kim, J. (2018). Regional approach of waste electrical and electronic equipment (WEEE) management in France. *Resources, Conservation and Recycling, 129*, 45–55. doi:10.1016/j.resconrec.2017.10.016

Bai, J., Lian, S., Liu, Z., Wang, K., & Liu, D. (2018). Deep learning based robot for automatically picking up garbage on the grass. *IEEE Transactions on Consumer Electronics, 64*(3), 382–389. doi:10.1109/TCE.2018.2859629

Bakhshi, Z., & Balador, A. (2019). An overview on security and privacy challenges and their solutions in fog-based vehicular application. *2019 IEEE 30th International Symposium on Personal, Indoor and Mobile Radio Communications (PIMRC Workshops)*, 1–7.

Balaji, D. (2017). *Smart trash can using internet of things*. Academic Press.

Baldwin, E. a. (2007). *Managing it innovation for business value: Practical strategies for it and business managers*. Academic Press.

Balicki, J. (2021). Many-objective quantum-inspired particle swarm optimization algorithm for placement of virtual machines in smart computing cloud. *Entropy (Basel, Switzerland), 24*(1), 58. doi:10.3390/e24010058 PMID:35052084

Baliga, J., Ayre, R. W., Hinton, K., & Tucker, R. S. (2010). Green cloud computing: Balancing energy in processing, storage, and transport. *Proceedings of the IEEE, 99*(1), 149–167. doi:10.1109/JPROC.2010.2060451

Bal, P. K., Mohapatra, S. K., Das, T. K., Srinivasan, K., & Hu, Y. C. (2022). A joint resource allocation, security with efficient task scheduling in cloud computing using hybrid machine learning techniques. *Sensors (Basel), 22*(3), 1242. doi:10.3390/s22031242 PMID:35161987

Barik, R. K., Priyadarshini, R., Dubey, H., Kumar, V., & Mankodiya, K. (2018). FogLearn: Leveraging fog-based machine learning for smart system big data analytics. *International Journal of Fog Computing, 1*(1), 15–34. doi:10.4018/IJFC.2018010102

Baron & Beran. (2022). *Achieving sustainable data Center growth*. https://carrier.huawei.com/~/media/CNBGV2/download/products/servies/Achieving-Sustainable-Data-Center-Growth.pdf

Baswaraju, S., Maheswari, V. U., Chennam, K. K., Thirumalraj, A., Kantipudi, M. P., & Aluvalu, R. (2023). Future Food Production Prediction Using AROA Based Hybrid Deep Learning Model in Agri-Sector. *Human-Centric Intelligent Systems*, 1-16.

Belkhir, L., & Elmeligi, A. (2018). Assessing ICT global emissions footprint: Trends to 2040 & recommendations. *Journal of Cleaner Production*, *177*, 448–463. doi:10.1016/j.jclepro.2017.12.239

Berral, J. L., Gavalda, R., & Torres, J. (2011, September). Adaptive scheduling on power-aware managed data-centers using machine learning. In *2011 IEEE/ACM 12th International Conference on Grid Computing* (pp. 66-73). IEEE. 10.1109/Grid.2011.18

Bharany, S., Badotra, S., Sharma, S., Rani, S., Alazab, M., Jhaveri, R. H., & Gadekallu, T. R. (2022). Energy efficient fault tolerance techniques in green cloud computing: A systematic survey and taxonomy. *Sustainable Energy Technologies and Assessments*, *53*, 102613. doi:10.1016/j.seta.2022.102613

Bhoi, S. K., Chakraborty, S., Verbrugge, B., Helsen, S., Robyns, S., Baghdadi, M. E., & Hegazy, O. (2022). Advanced edge computing framework for grid power quality monitoring of industrial motor drive applications. *2022 International Symposium on Power Electronics, Electrical Drives, Automation and Motion (SPEEDAM)*. 10.1109/SPEEDAM53979.2022.9841966

Birke, R., Chen, L. Y., & Smirni, E. (2012, June). Data centers in the cloud: A large scale performance study. In *2012 IEEE Fifth international conference on cloud computing* (pp. 336-343). IEEE. 10.1109/CLOUD.2012.87

Bleiwas, D. I., & Kelly, T. (2001). Obsolete Computers, "Gold Mine", or High-tech Trash? Resource Recovery From Recycling. Fact Sheet.

Bobulski, J., & Kubanek, M. (2019). Waste Classification System Using Image Processing and Convolutional Neural Networks. *International Conference on Artificial Neural Networks*, 350-361. 10.1007/978-3-030-20518-8_30

Boopathi, S. (2023b). Securing Healthcare Systems Integrated With IoT: Fundamentals, Applications, and Future Trends. In Dynamics of Swarm Intelligence Health Analysis for the Next Generation (pp. 186–209). IGI Global.

Boopathi, S., Pandey, B. K., & Pandey, D. (2023). Advances in Artificial Intelligence for Image Processing: Techniques, Applications, and Optimization. In Handbook of Research on Thrust Technologies' Effect on Image Processing (pp. 73–95). IGI Global.

Boopathi, S. (2023a). Internet of Things-Integrated Remote Patient Monitoring System: Healthcare Application. In *Dynamics of Swarm Intelligence Health Analysis for the Next Generation* (pp. 137–161). IGI Global. doi:10.4018/978-1-6684-6894-4.ch008

Boopathi, S., & Kanike, U. K. (2023). Applications of Artificial Intelligent and Machine Learning Techniques in Image Processing. In *Handbook of Research on Thrust Technologies' Effect on Image Processing* (pp. 151–173). IGI Global. doi:10.4018/978-1-6684-8618-4.ch010

Brien Posey. (2022, April 29). *Data center temperature and humidity guidelines*. https://www.techtarget.com/searchdatacenter/tip/Data-center-temperature-and-humidity-guidelines

Briscar, J. R. (2017). *Data Transmission and Energy Efficient Internet Data Centers. Am. UL Rev.*, *67*, 233.

Budde, F., & Volz, D. (2019). The next big thing? Quantum computing's potential impact on chemicals. *McKinsey*.

Bueno-Delgado, M.-V., Romero-G'azquez, J.-L., Jim'enez, P., & Pav'on-Mariño, P. (2019). Optimal path planning for selective waste collection in smart cities. *Sensors (Basel)*, *19*(9), 1973. doi:10.3390/s19091973 PMID:31035549 businessnorway.com. (2023, October 30). *Explore Norway's green data centre industry.* https://businessnorway.com/key-industries/data-centres

Butner, K. a. (2008). Mastering carbon management: Balancing trade-offs to optimize supply chain efficiencies. IBM Global Business Services. *IBM Institute for Business Value.*

Butt, S. A. (2020). Green Computing: Sustainable Design and Technologies. IEEE, 1-7.

Butt, F. A., Chattha, J. N., Ahmad, J., Zia, M. U., Rizwan, M., & Naqvi, I. H. (2022). On the integration of enabling wireless technologies and sensor fusion for next-generation connected and autonomous vehicles. *IEEE Access : Practical Innovations, Open Solutions, 10,* 14643–14668. doi:10.1109/ACCESS.2022.3145972

Buyya, R., Srirama, S. N., Casale, G., Calheiros, R., Simmhan, Y., Varghese, B., Gelenbe, E., Javadi, B., Vaquero, L. M., Netto, M. A. S., Toosi, A. N., Rodriguez, M. A., Llorente, I. M., Vimercati, S. D. C. D., Samarati, P., Milojicic, D., Varela, C., Bahsoon, R., Assuncao, M. D. D., ... Shen, H. (2018). A manifesto for future generation cloud computing: Research directions for the next decade. *ACM Computing Surveys, 51*(5), 1–38. doi:10.1145/3241737

Cao, E., Musa, S., Chen, M., Wei, T., Wei, X., Fu, X., & Qiu, M. (2023). Energy and reliability-aware task scheduling for cost optimization of DVFS-enabled cloud workflows. *IEEE Transactions on Cloud Computing, 11*(2), 2127–2143. doi:10.1109/TCC.2022.3188672

Carol Yan. (2023, April 26). *AWS Collaborates with WindEurope and Accenture to Streamline Wind Permitting in Europe.* https://aws.amazon.com/blogs/industries/aws-collaborates-with-windeurope-and-accenture-to-streamline-wind-permitting-in-europe/

Cassady Craighill. (2019, February 13). *Greenpeace Finds Amazon Breaking Commitment to Power Cloud with 100% Renewable Energy.* https://www.greenpeace.org/usa/news/greenpeace-finds-amazon-breaking-commitment-to-power-cloud-with-100-renewable-energy/

Catania, V., & Ventura, D. (2014). An Approach for Monitoring and Smart Planning of Urban Solid Waste Management Using Smart-M3 Platform. *Proceedings of the 15th Conference of Open Innovations Association FRUCT,* 24–31. 10.1109/FRUCT.2014.6872422

CBRE. (2023). *Global data center trends 2023.* Retrieved October 5, 2023, from https://www.cbre.com/insights/reports/global-data-center-trends-2023

Chakraborty, C., Kishor, A., & Rodrigues, J. J. (2022). Novel Enhanced-Grey Wolf Optimization hybrid machine learning technique for biomedical data computation. *Computers & Electrical Engineering, 99,* 107778. doi:10.1016/j.compeleceng.2022.107778

Chandrashekar, C., Krishnadoss, P., Kedalu Poornachary, V., Ananthakrishnan, B., & Rangasamy, K. (2023). HWACOA scheduler: Hybrid weighted ant colony optimization algorithm for task scheduling in cloud computing. *Applied Sciences (Basel, Switzerland), 13*(6), 3433. doi:10.3390/app13063433

Chatterjee, S., Nizamani, F. A., & Nürnberger, A. (n.d.). Classification of brain tumours in MR images using deep spatio-spatial models. *Scientific Reports, 12.*

Chavadi & Thangam. (2023). Global perspectives on social media usage within governments. *Global Perspectives on Social Media Usage Within Governments, 2023,* 1–353.

Chavadi, C., Manoj, G., Ganesan, S. K., Manoharan, S., & Thangam, D. (2023). Global Perspectives on Social Media Usage Within Governments. In Global Perspectives on Social Media Usage Within Governments (pp. 1-19). IGI Global.

Cheema, S., & Khan, G. N. (2020, October). Power and Performance Analysis of Deep Neural Networks for Energy-aware Heterogeneous Systems. In *2020 IEEE International Conference on Systems, Man, and Cybernetics (SMC)* (pp. 2184-2189). IEEE. 10.1109/SMC42975.2020.9283092

Chen, J., Wei, Z., Li, S., & Cao, B. (2020). Artificial intelligence aided joint bit rate selection and radio resource allocation for adaptive video streaming over F-RANs. *IEEE Wireless Communications, 27*(2), 36–43. doi:10.1109/MWC.001.1900351

Chen, M., Ma, Y., Li, Y., Wu, D., Zhang, Y., & Youn, C. H. (2017). Wearable 2.0: Enabling human-cloud integration in next generation healthcare systems. *IEEE Communications Magazine, 55*(1), 54–61. doi:10.1109/MCOM.2017.1600410CM

Chen, X., Pan, M., Li, X., & Zhang, K. (2022). Multi-mode operation and thermo-economic analyses of combined cooling and power systems for recovering waste heat from data centers. *Energy Conversion and Management, 266*, 115820. doi:10.1016/j.enconman.2022.115820

China Daily. (2021, December 9). *Green data centers in focus*. http://english.www.gov.cn/statecouncil/ministries/202112/09/content_WS61b13edac6d09c94e48a1f81.html

Chitra, S., & Jayalakshmi, V. (2021). A review of healthcare applications on Internet of Things. *Computer Networks, Big Data and IoT. Proceedings of ICCBI, 2020*, 227–237.

Chi, X., Streicher-Porte, M., Wang, M. Y. L., & Reuter, M. A. (2011). Informal electronic waste recycling: A sector review with special focus on China. *Waste Management (New York, N.Y.), 31*(4), 731–742. doi:10.1016/j.wasman.2010.11.006 PMID:21147524

Chi, X., Wang, M. Y. L., & Reuter, M. A. (2014). E-waste collection channels and household recycling behaviors in Taizhou of China. *Journal of Cleaner Production, 80*, 87–95. doi:10.1016/j.jclepro.2014.05.056

Chu, H., & Zhang, Y. (2023, August). A Green Granular Neural Network with Efficient Software-FPGA Co-designed Learning. In *IEEE 22nd International Conference on Cognitive Informatics and Cognitive Computing (ICCI* CC'2023)*. IEEE.

Coelho, C. N. Jr, Kuusela, A., Li, S., Zhuang, H., Ngadiuba, J., Aarrestad, T. K., Loncar, V., Pierini, M., Pol, A. A., & Summers, S. (2021). Automatic heterogeneous quantization of deep neural networks for low-latency inference on the edge for particle detectors. *Nature Machine Intelligence, 3*(8), 675–686. doi:10.1038/s42256-021-00356-5

conecomm.com. (2017). *2017 Cone Communications CSR Study*. https://conecomm.com/2017-csr-study/#download-the-research

Cossu, R., Salieri, V., & Bisinella, V. (2012). *Urban Mining: a Global Cycle Approach to Resource Recovery From Solid Waste*. CISA Publ.

Crowe, M., Elser, A., Gopfert, B., Mertins, L., Meyer, T., Schmid, J., Spillner, A. & Strobel, R. (2003). *Waste from Electrical and Electronic Equipment (WEEE) – Quantities, Dangerous Substances and Treatment Methods*. Academic Press.

Cui, J., & Forssberg, E. (2003). Mechanical recycling of waste electric and electronic equipment: A review. *Journal of Hazardous Materials, 99*(3), 243–263. doi:10.1016/S0304-3894(03)00061-X PMID:12758010

Cui, Q., Wang, Y., Chen, K.-C., Ni, W., Lin, I.-C., Tao, X., & Zhang, P. (2018). Big data analytics and network calculus enabling intelligent management of autonomous vehicles in a smart city. *IEEE Internet of Things Journal, 6*(2), 2021–2034. doi:10.1109/JIOT.2018.2872442

D'Oro, S., Bonati, L., Polese, M., & Melodia, T. (2022, May). Orchestran: Network automation through orchestrated intelligence in the open ran. In *IEEE INFOCOM 2022-IEEE Conference on Computer Communications* (pp. 270-279). IEEE.

Daigger, G. T. (2008, January). Aeesp Lecture: Evolving Urban Water and Residuals Management Paradigms: Water Reclamation and Reuse, Decentralization, Resource Recovery. In WEFTEC 2008 (pp. 1537-1565). Water Environment Federation.

Damaj, I. W., Yousafzai, J. K., & Mouftah, H. T. (2022). Future trends in connected and autonomous vehicles: Enabling communications and processing technologies. *IEEE Access : Practical Innovations, Open Solutions, 10*, 42334–42345. doi:10.1109/ACCESS.2022.3168320

Dan Thompson. (2022, August 2). *The rising importance of data center sustainability.* https://www.spglobal.com/marketintelligence/en/news-insights/research/the-rising-importance-of-data-center-sustainability

Daniel Hamad. (2019, April 10). *Data Centres: How Will the Rising Demand for Data be Powered?* https://www.arcadis.com/en-au/knowledge-hub/blog/australia/daniel-hamad/2019/data-centres-how-will-the-rising-demand-for-data-be-powered

Darwish, T. S., & Bakar, K. A. (2018). Fog based intelligent transportation big data analytics in the internet of vehicles environment: Motivations, architecture, challenges, and critical issues. *IEEE Access : Practical Innovations, Open Solutions, 6*, 15679–15701. doi:10.1109/ACCESS.2018.2815989

Das, K., Saha, S., Chowdhury, S., Reza, A. W., Paul, S., & Arefin, M. S. (2022, October). A Sustainable E-waste Management System and Recycling Trade for Bangladesh in Green IT. In *International Conference on Intelligent Computing & Optimization* (pp. 351-360). Cham: Springer International Publishing.

data4group.com. (2021, May 6). *What Is PUE?* https://www.data4group.com/es/diccionario-del-centro-de-datos/que-es-pue/

Datacenter Dynamics. (2019, September 13). *Greenpeace: China's data centers run mainly on coal, emit 99 million tons of CO2.* https://www.datacenterdynamics.com/es/noticias/greenpeace-los-centros-de-datos-de-china-funcionan-principalmente-con-carb%C3%B3n-emiten-99-millones-de-toneladas-de-co2/

DCS Content Team. (2023, July 7). *The Importance of Sustainable Data Centers: Why It Matters.* https://blog.datacentersystems.com/the-importance-of-sustainable-data-centers-why-it-matters-for-the-environment-and-your-organization

de Souza, R. G., Clímaco, J. C. N., Sant'Anna, A. P., Rocha, T. B., do Valle, R. de A.B., & Quelhas, O. L. G. (2016). Sustainability assessment and prioritisation of e-waste management options in Brazil. *Waste Management (New York, N.Y.), 57*, 46–56. doi:10.1016/j.wasman.2016.01.034 PMID:26852754

Debnath, B., Baidya, R., Biswas, N. T., Kundu, R., & Ghosh, S. K. (2015). E-waste recycling as criteria for green computing approach: analysis by QFD tool. In K. Maharatna, G. K. Dalapati, P. K. Banerjee, A. K. Mallick, & M. Mukherjee (Eds.), *Computational Advancement in Communication Circuits and Systems* (pp. 139–144). Springer. doi:10.1007/978-81-322-2274-3_17

Debnath, B., Roychoudhuri, R., & Ghosh, S. K. (2016). E-waste management—A potential route to green computing. *Procedia Environmental Sciences, 35*, 669–675. doi:10.1016/j.proenv.2016.07.063

Demestichas, K., & Daskalakis, E. (2020). Information and communication technology solutions for the circular economy. *Sustainability (Basel), 12*(18), 7272. doi:10.3390/su12187272

Demirci, M. (2015). A Survey of Machine Learning Applications for Energy-Efficient Resource Management in Cloud Computing Environments. In *2015 IEEE 14th International Conference on Machine Learning and Applications (ICMLA)* (pp. 205). 10.1109/ICMLA.2015.205

Demis, Suleyman, & Legg. (2017, January 3). *DeepMind's work in 2016: a round-up.* https://deepmind.google/discover/blog/deepminds-work-in-2016-a-round-up

device42.com. (2023, November 10). *Best Practices for Data Center Sustainability.* https://www.device42.com/data-center-infrastructure-management-guide/data-center-sustainability/

Dharfizi, A. D. (2018). The Energy Sector and the Internet of Things Sustainable Consumption and Enhanced Security through Industrial Revolution 4.0. *Journal of International Students*, 99–117.

Dhar, P. (2020). The carbon impact of artificial intelligence. *Nature Machine Intelligence*, 2(8), 423–425. doi:10.1038/s42256-020-0219-9

Dietterich, T. G. (2000, June). Ensemble methods in machine learning. In *International workshop on multiple classifier systems* (pp. 1-15). Springer Berlin Heidelberg.

Ding, Y., Zhang, S., Liu, B., Zheng, H., Chang, C., & Ekberg, C. (2019). Recovery of precious metals from electronic waste and spent catalysts: A review. *Resources, Conservation and Recycling*, 141, 284–298. doi:10.1016/j.resconrec.2018.10.041

Dittakavi, R. S. S. (2023). AI-Optimized Cost-Aware Design Strategies for Resource-Efficient Applications. *Journal of Science and Technology*, 4(1), 1–10.

Drew & Hufford. (2016, December 6). *Equinix to Buy Some Verizon Data Centers for $3.6 Billion.* https://www.wsj.com/articles/equinix-to-buy-some-verizon-data-centers-for-3-6-billion-1481034980?mod=article_inline

Driskell, D. (2022). *Strategies for Sustainable Data Centers.* Technology and Sustainability in Modern Society.

Dugdhe, S., Shelar, P., Jire, S., & Apte, A. (2016). Efficient waste collection system. *2016 International Conference on Internet of Things and Applications (IOTA)*, 143-147, 10.1109/IOTA.2016.7562711

EEA. (2003). Waste from electrical and electronics equipment (WEEE)-quantities, dangerous substances and treatment methods. EEA.

Eftekhari, S. A., Nikooghadam, M., & Rafighi, M. (2021). Security-enhanced three-party pairwise secret key agreement protocol for fog-based vehicular ad-hoc communications. *Vehicular Communications*, 28, 100306. doi:10.1016/j.vehcom.2020.100306

Elangovan, S., Sasikala, S., Kumar, S. A., Bharathi, M., Sangath, E. N., & Subashini, T. (2021, August). A deep learning based multiclass segregation of e-waste using hardware software co-simulation. *Journal of Physics: Conference Series*, 1997(1), 012039. doi:10.1088/1742-6596/1997/1/012039

El-Ela, A. A. A., El-Sehiemy, R. A., Shaheen, A. M., & Kotb, N. (2020). Optimal allocation of DGs with network reconfiguration using improved spotted hyena algorithm. *WSEAS Transactions on Power Systems*, 15, 60–67. doi:10.37394/232016.2020.15.7

Elgamal, A. S., Aletri, O. Z., Yosuf, B. A., Qidan, A. A., El-Gorashi, T., & Elmirghani, J. M. (2023). AI-Driven Resource Allocation in Optical Wireless Communication Systems. *arXiv preprint arXiv:2304.03880.* doi:10.1109/ICTON59386.2023.10207473

Elhadad, A., Alanazi, F., Taloba, A. I., & Abozeid, A. (2022). Fog computing service in the healthcare monitoring system for managing the real-time notification. *Journal of Healthcare Engineering*, 2022, 2022. doi:10.1155/2022/5337733 PMID:35340260

Elsken, T., Metzen, J. H., & Hutter, F. (2019). Neural architecture search: A survey. *Journal of Machine Learning Research*, 20(1), 1997–2017.

epa.gov. (2023, October 28). *Overview of Greenhouse Gases.* https://www.epa.gov/ghgemissions/overview-greenhouse-gases

europapress.es. (2021, September 28). *Energy sector leaders launch the 24/7 Carbonless Energy Pact (1).* https://www.europapress.es/comunicados/internacional-00907/noticia-comunicado-lideres-sector-energetico-lanzan-pacto-energia-carbono-24-20210928160229.html

Evangelidis, H., & Davies, R. (2023, November 29). *Are you aware of your digital carbon footprint?* Retrieved October 5, 2023, from https://www.capgemini.com/gb-en/insights/expert-perspectives/are-you-aware-of-your-digital-carbon-footprint/

Ewim, D. R. E., Ninduwezuor-Ehiobu, N., Orikpete, O. F., Egbokhaebho, B. A., Fawole, A. A., & Onunka, C. (2023). Impact of Data Centers on Climate Change: A Review of Energy Efficient Strategies. *The Journal of Engineering and Exact Sciences, 9*(6), 16397–01e. doi:10.18540/jcecvl9iss6pp16397-01e

expresscomputer.in. (2023, September 12). *Energy-Efficient Design Strategies for Sustainable Data Center Operations.* https://www.expresscomputer.in/data-center/energy-efficient-design-strategies-for-sustainable-data-center-operations/103415/#:~:text=Optimizing%20the%20supply%20of%20air,speed%20chillers%20and%20pumps%20and

Farjana, M., Fahad, A. B., Alam, S. E., & Islam, M. M. (2023). An IoT- and Cloud-Based E-Waste Management System for Resource Reclamation with a Data-Driven Decision-Making Process. *IoT., 4*(3), 202–220. doi:10.3390/iot4030011

Fei, F., Qu, L., Wen, Z., Xue, Y., & Zhang, H. (2016). How to integrate the informal recycling system into municipal solid waste management in developing countries: Based on a China's case in Suzhou urban area. *Resources, Conservation and Recycling, 110,* 74–86. doi:10.1016/j.resconrec.2016.03.019

Feng, B., Sun, K., Chen, M., & Gao, T. (2020). The impact of core technological capabilities of high-tech industry on sustainable competitive advantage. *Sustainability (Basel), 12*(7), 2980. doi:10.3390/su12072980

Fentis, A. a. (2019). *Short-term nonlinear autoregressive photovoltaic power forecasting using statistical learning approaches and in-situ observations.* Academic Press.

Fernandes, P., & Nunes, U. (2012). Platooning with DSRC-based IVC-enabled autonomous vehicles: Adding infrared communications for IVC reliability improvement. *2012 IEEE Intelligent Vehicles Symposium,* 517–522. 10.1109/IVS.2012.6232206

Florence, A. P., Shanthi, V., & Simon, C. B. (2016). Energy conservation using dynamic voltage frequency scaling for computational cloud. *TheScientificWorldJournal, 2016,* 2016. doi:10.1155/2016/9328070 PMID:27239551

Foley, J. (2007). *Google in Oregon: Mother Nature meets the data center.* InformationWeek's Google Weblog.

Forti, V., Balde, C. P., Kuehr, R., & Bel, G. (2020). The Global E-waste Monitor 2020: Quantities, Flows and the Circular Economy Potential. United Nations University.

Fraga-Lamas, P., Lopes, S. I., & Fernández-Caramés, T. M. (2021). Green IoT and edge AI as key technological enablers for a sustainable digital transition towards a smart circular economy: An industry 5.0 use case. *Sensors (Basel), 21*(17), 5745. doi:10.3390/s21175745 PMID:34502637

fsc.org. (2023, November 17). *The future of forests is in our hands.* https://fsc.org/en/how-to-be-a-forest-steward

Gaharwar, G., Pandya, J., & Pandya, S. (2022, July). Comprehensive Framework for Assessing Organization's Green Computing Adaptability. In *2022 1st International Conference on Sustainable Technology for Power and Energy Systems (STPES)* (pp. 1-6). IEEE. 10.1109/STPES54845.2022.10006640

Gai, K., Qiu, M., Zhao, H., & Sun, X. (2017). Resource management in sustainable cyber-physical systems using heterogeneous cloud computing. *IEEE Transactions on Sustainable Computing, 3*(2), 60–72. doi:10.1109/TSUSC.2017.2723954

Gallego, F., Martín, C., Díaz, M., & Garrido, D. (2023). Maintaining flexibility in smart grid consumption through deep learning and deep reinforcement learning. *Energy and AI, 13*, 100241. doi:10.1016/j.egyai.2023.100241

Gao, H., Huang, Z., Zhang, X., & Yang, H. (2023). Research and Design of a Decentralized Edge-Computing-Assisted LoRa Gateway. *Future Internet, 15*(6), 194. doi:10.3390/fi15060194

Gezer, V. W. (2021). Real-Time Edge Framework (RTEF): Decentralized Decision Making for Offloading. IEEE, 1-6.

Ghawi, R., & Pfeffer, J. (2019). Efficient hyperparameter tuning with grid search for text categorization using kNN approach with BM25 similarity. *Open Computer Science, 9*(1), 160–180. doi:10.1515/comp-2019-0011

Ghazal, T.M., Abbas, S., Munir, S., Khan, M.A., Ahmad, M., Issa, G.F., Zahra, S.B., Khan, M.A., & Hasan, M.K. (2022). *Alzheimer Disease Detection Empowered with Transfer Learning*. Academic Press.

Gholipour, N., Shoeibi, N., & Arianyan, E. (2021). An energy-aware dynamic resource management technique using deep q-learning algorithm and joint VM and container consolidation approach for green computing in cloud data centers. In *Distributed Computing and Artificial Intelligence, Special Sessions, 17th International Conference* (pp. 227-233). Springer International Publishing. 10.1007/978-3-030-53829-3_26

Ghosh, S., & Das, J. (2022). Dynamic Voltage and Frequency Scaling Approach for Processing Spatio-Temporal Queries in Mobile Environment. In *Green Mobile Cloud Computing* (pp. 185–199). Springer International Publishing. doi:10.1007/978-3-031-08038-8_9

Gill, S. S., Xu, M., Ottaviani, C., Patros, P., Bahsoon, R., Shaghaghi, A., Golec, M., Stankovski, V., Wu, H., Abraham, A., Singh, M., Mehta, H., Ghosh, S. K., Baker, T., Parlikad, A. K., Lutfiyya, H., Kanhere, S. S., Sakellariou, R., Dustdar, S., ... Uhlig, S. (2022). AI for next generation computing: Emerging trends and future directions. *Internet of Things : Engineering Cyber Physical Human Systems, 19*, 100514. doi:10.1016/j.iot.2022.100514

Girsang, A. S., Pratama, H., & Santo Agustinus, L. P. (2023). Classification Organic and Inorganic Waste with Convolutional Neural Network Using Deep Learning. *International Journal of Intelligent Systems and Applications in Engineering, 11*(2), 343–348.

Golilarz, N. A., Gao, H., Addeh, A., & Pirasteh, S. (2020, December). ORCA optimization algorithm: a new meta-heuristic tool for complex optimization problems. In *2020 17th International Computer Conference on Wavelet Active Media Technology and Information Processing (ICCWAMTIP)* (pp. 198-204). IEEE. 10.1109/ICCWAMTIP51612.2020.9317473

Gou, J., Yu, B., Maybank, S. J., & Tao, D. (2021). Knowledge distillation: A survey. *International Journal of Computer Vision, 129*(6), 1789–1819. doi:10.1007/s11263-021-01453-z

Goyal, V., & Garg, S. (n.d.). Green cloud computing for sustainable environment: A review. *IOSR Journal of Environmental Science, Toxicology and Food Technology*.

Grama Sachivalayam Andhra Pradesh. (2019). *Digital Assistant Training Handbook*. Author.

gray.com. (2023, October 17). *The Data Center Industry Is Booming*. https://www.gray.com/insights/the-data-center-industry-is-booming/

grcooling.com. (2020). *The Effects of Data Centers on the Environment*. https://www.grcooling.com/blog/the-effects-of-data-centers-on-the-environment/

Guazzone, M., Anglano, C., & Canonico, M. (2011, November). Energy-efficient resource management for cloud computing infrastructures. In *2011 IEEE Third International Conference on Cloud Computing Technology and Science* (pp. 424-431). IEEE. 10.1109/CloudCom.2011.63

Guitart, J. (2017). Toward sustainable data centers: A comprehensive energy management strategy. *Computing, 99*(6), 597–615. doi:10.1007/s00607-016-0501-1

Guo, C., Luo, F., Cai, Z., & Dong, Z. Y. (2021). Integrated energy systems of data centers and smart grids: State-of-the-art and future opportunities. *Applied Energy, 301*, 117474. doi:10.1016/j.apenergy.2021.117474

Guo, L. X., Xu, Liu, & Wang. (2019). Understanding firm performance on green sustainable practices through managers' ascribed responsibility and waste management: Green self-efficacy as moderator. *Sustainability (Basel), 11*(18), 49–76. doi:10.3390/su11184976

Gupta, H. (2020). *Trash Image Classification System using Machine Learning and Deep Learning Algorithms* [Doctoral dissertation]. National College of Ireland.

Gupta, C. (2022). E-Waste Management-Accelerating Green Computing. *International Journal of Early Childhood Special Education, 14*(5).

Gupta, T., Joshi, R., Mukhopadhyay, D., Sachdeva, K., Jain, N., Virmani, D., & Garcia-Hernandez, L. (2022). A deep learning approach based hardware solution to categorise garbage in environment. *Complex & Intelligent Systems, 8*(2), 1–24. doi:10.1007/s40747-021-00529-0

Hageluken, C. (2006). Improving metal returns and eco-efficiency in electronics recycling - a holistic approach for interface optimisation between pre-processing and integrated metals smelting and refining. *Proceedings of the 2006 IEEE International Symposium on Electronics and the Environment*, 218–223. 10.1109/ISEE.2006.1650064

Hakak, S., Gadekallu, T. R., Maddikunta, P. K. R., Ramu, S. P., Parimala, M., De Alwis, C., & Liyanage, M. (2022). Autonomous Vehicles in 5G and beyond: A Survey. *Vehicular Communications*, 100551.

Hameed, A., Khoshkbarforoushha, A., Ranjan, R., Jayaraman, P. P., Kolodziej, J., Balaji, P., Zeadally, S., Malluhi, Q. M., Tziritas, N., Vishnu, A., Khan, S. U., & Zomaya, A. (2016, July). A survey and taxonomy on energy efficient resource allocation techniques for cloud computing systems. *Computing, 98*(7), 751–774. doi:10.1007/s00607-014-0407-8

Han, J., & Orshansky, M. (2013, May). Approximate computing: An emerging paradigm for energy-efficient design. In *2013 18th IEEE European Test Symposium (ETS)* (pp. 1-6). IEEE.

Han, S., Mao, H., & Dally, W. J. (2015). Deep compression: Compressing deep neural networks with pruning, trained quantization and huffman coding. *arXiv preprint arXiv:1510.00149*.

Harding, A. C. (2015). *Improved methods for identifying, applying, and verifying industrial energy efficiency measures.* Academic Press.

Harmon, R. R. (2009). Sustainable I.T. services: Assessing the impact of green computing practices. IEEE, 1707-1717.

Hashem, I. A. T., Yaqoob, I., Anuar, N. B., Mokhtar, S., Gani, A., & Khan, S. U. (2015). The rise of "big data" on cloud computing: Review and open research issues. *Information Systems, 47*, 98–115. doi:10.1016/j.is.2014.07.006

Hassan, M. B., Saeed, R. A., Khalifa, O., Ali, E. S., Mokhtar, R. A., & Hashim, A. A. (2022, May). Green Machine Learning for Green Cloud Energy Efficiency. In *2022 IEEE 2nd International Maghreb Meeting of the Conference on Sciences and Techniques of Automatic Control and Computer Engineering (MI-STA)* (pp. 288-294). IEEE. 10.1109/MI-STA54861.2022.9837531

Hassan, N. G., Gillani, S., Ahmed, E., Yaqoob, I., & Imran, M. (2018). The role of edge computing in internet of things. *IEEE Communications Magazine, 56*(11), 110–115. doi:10.1109/MCOM.2018.1700906

HassanQ.SameenA. Z.SalmanH. M.Al-JibooryA.JaszczurM. (2023, May 25). The role of renewable energy and artificial intelligence towards environmental sustainability and net zero. Research Square. doi:10.21203/rs.3.rs-2970234/v1

Hatzivasilis, G., Ioannidis, S., Fysarakis, K., Spanoudakis, G., & Papadakis, N. (2021). The green blockchains of circular economy. *Electronics (Basel), 10*(16), 2008. doi:10.3390/electronics10162008

He, Y., Gu, Q., & Shi, M. (2020). *Trash Classification Using Convolutional Neural Networks Project Category: Computer Vision.* Academic Press.

Heeks, R., Subramanian, L., & Jones, C. (2015). Understanding e-Waste Management in Developing Countries: Strategies, Determinants, and Policy Implications in the Indian ICT Sector. *Information Technology for Development, 21*(4), 653–667. doi:10.1080/02681102.2014.886547

Helmreich, R. L., Merritt, A. C., & Wilhelm, J. A. (2017). The evolution of crew resource management training in commercial aviation. In *Human error in aviation* (pp. 275–288). Routledge. doi:10.4324/9781315092898-15

Hema, N., Krishnamoorthy, N., Chavan, S. M., Kumar, N., Sabarimuthu, M., & Boopathi, S. (2023). A Study on an Internet of Things (IoT)-Enabled Smart Solar Grid System. In *Handbook of Research on Deep Learning Techniques for Cloud-Based Industrial IoT* (pp. 290–308). IGI Global. doi:10.4018/978-1-6684-8098-4.ch017

Henderson, P., Hu, J., Romoff, J., Brunskill, E., Jurafsky, D., & Pineau, J. (2020). Towards the systematic reporting of the energy and carbon footprints of machine learning. *Journal of Machine Learning Research, 21*(1), 10039–10081.

Hewa, T., Gür, G., Kalla, A., Ylianttila, M., Bracken, A., & Liyanage, M. (2020). The role of blockchain in 6G: Challenges, opportunities and research directions. *2020 2nd 6G Wireless Summit (6G SUMMIT)*, 1-5.

He, X., Zhao, K., & Chu, X. (2021). AutoML: A survey of the state-of-the-art. *Knowledge-Based Systems, 212*, 106622. doi:10.1016/j.knosys.2020.106622

Hindrise. (2023). *E-Waste Management in India.* https://hindrise.org/resources/e-waste-management-in-india/

Hinton, K., Baliga, J., Feng, M., Ayre, R., & Tucker, R. S. (2011). Power consumption and energy efficiency in the internet. *IEEE Network, 25*(2), 6–12. doi:10.1109/MNET.2011.5730522

Hoermann, S., Stumper, D., & Dietmayer, K. (2017). Probabilistic long-term prediction for autonomous vehicles. *2017 IEEE Intelligent Vehicles Symposium (IV)*, 237–243. 10.1109/IVS.2017.7995726

Hossain, M., Al-Hamadani, S., & Rahman, R. (2015). E-waste: A Challenge for Sustainable Development. *Journal of Health & Pollution, 5.* PMID:30524771

Huang, J., Meng, Y., Gong, X., Liu, Y., & Duan, Q. (2014). A novel deployment scheme for green internet of things. *IEEE Internet of Things Journal, 1*(2), 196–205. doi:10.1109/JIOT.2014.2301819

Huiyu, L. (2019). Automatic Classifications and Recognition for Recycled Garbage by Utilizing Deep Learning Technology. *7th International Conference on Information Technology: IoT and Smart City*, 1–4.

Hussain, A., Draz, U., Ali, T., Tariq, S., Irfan, M., Glowacz, A., Antonino Daviu, J. A., Yasin, S., & Rahman, S. (2020). Waste Management and Prediction of Air Pollutants Using IoT and Machine Learning Approach. *Energies, 13*(15), 3930. doi:10.3390/en13153930

iea.org. (2023, October 22). *Data Centres and Data Transmission Networks.* https://www.iea.org/energy-system/buildings/data-centres-and-data-transmission-networks

iea.org. (2023, October 26). *Net Zero Emissions by 2050 Scenario (NZE).* https://www.iea.org/reports/global-energy-and-climate-model/net-zero-emissions-by-2050-scenario-nze

Ihnaini, B., Khan, M. A., Khan, T. A., Abbas, S., Daoud, M. S., Ahmad, M., & Khan, M. A. (2021). A Smart Healthcare Recommendation System for Multidisciplinary Diabetes Patients with Data Fusion Based on Deep Ensemble Learning. *Computational Intelligence and Neuroscience*, *2021*, 2021. doi:10.1155/2021/4243700 PMID:34567101

Ijaz, M., Li, G., Lin, L., Cheikhrouhou, O., Hamam, H., & Noor, A. (2021). Integration and applications of fog computing and cloud computing based on the internet of things for provision of healthcare services at home. *Electronics (Basel)*, *10*(9), 1077. doi:10.3390/electronics10091077

Ikhlayel, M. (2016). Differences of methods to estimate generation of waste electrical and electronic equipment for developing countries: Jordan as a case study. *Resources, Conservation and Recycling*, *108*, 134–139. doi:10.1016/j.resconrec.2016.01.015

International Energy Agency (IEA). (2023). *Data centres and data transmission networks*. Retrieved from https://www.iea.org/energy-system/buildings/data-centres-and-data-transmission-networks

International Energy Agency. (2023). *The World Energy Outlook 2023*. https://www.iea.org/reports/world-energy-outlook-2023

irena.org. (2019, February 15). *Innovation landscape for a renewable-powered future*. https://www.irena.org/publications/2019/Feb/Innovation-landscape-for-a-renewable-powered-future

Irtija, N., Anagnostopoulos, I., Zervakis, G., Tsiropoulou, E. E., Amrouch, H., & Henkel, J. (2021). Energy efficient edge computing enabled by satisfaction games and approximate computing. *IEEE Transactions on Green Communications and Networking*, *6*(1), 281–294. doi:10.1109/TGCN.2021.3122911

Islam, A., Ahmed, T., Awual, M. R., Rahman, A., Sultana, M., Aziz, A. A., Monir, M. U., Teo, S. H., & Hasan, M. (2020). Advances in sustainable approaches to recover metals from e-waste-A review. *Journal of Cleaner Production*, *244*, 118815. doi:10.1016/j.jclepro.2019.118815

Ismail, H., & Hanafiah, M. M. (2019). Discovering opportunities to meet the challenges of an effective waste electrical and electronic equipment recycling system in Malaysia. *Journal of Cleaner Production*, *238*, 117927. doi:10.1016/j.jclepro.2019.117927

Jacob Roundy. (2023, February 24). *A primer on hyperscale data centers*. https://www.techtarget.com/searchdatacenter/tip/A-primer-on-hyperscale-data-centers

Jain, A. a. (2013). Energy efficient computing-green cloud computing. IEEE, 978-982.

Jamil, B., Ijaz, H., Shojafar, M., Munir, K., & Buyya, R. (2022). Resource allocation and task scheduling in fog computing and internet of everything environments: A taxonomy, review, and future directions. *ACM Computing Surveys*, *54*(11s), 1–38. doi:10.1145/3513002

Jänick, M. (2012). Green growth: From a growing eco-industry to economic sustainability. *Energy Policy*, *48*, 13–21. doi:10.1016/j.enpol.2012.04.045

Jayalath, J. M. T. I., Chathumali, E. J. A. P. C., Kothalawala, K. R. M., & Kuruwitaarachchi, N. (2019, March). Green cloud computing: A review on adoption of green-computing attributes and vendor specific implementations. In *2019 International Research Conference on Smart Computing and Systems Engineering (SCSE)* (pp. 158-164). IEEE. 10.23919/SCSE.2019.8842817

Jim O'Donnell. (2023, March 17). *Data center, devices scrutinized for sustainability goals*. https://www.techtarget.com/sustainability/feature/Data-center-devices-scrutinized-for-sustainability-goals

Jin, S., Yang, Z., Królczykg, G., Liu, X., Gardoni, P., & Li, Z. (2023). Garbage detection and classification using a new deep learning-based machine vision system as a tool for sustainable waste recycling. *Waste Management (New York, N.Y.)*, *162*, 123–130. doi:10.1016/j.wasman.2023.02.014 PMID:36989995

Jonathan Bardelline. (2011, May 31). *Google Uses Sea Water to Cool Finland Data Center*. https://www.greenbiz.com/article/google-uses-sea-water-cool-finland-data-center#:~:text=Google's%20new%20data%20center%20in,by%20the%20Gulf%20of%20Finland

Julia Borgini. (2022, May 3). *Data center cooling systems and technologies and how they work*. https://www.techtarget.com/searchdatacenter/tip/Data-center-cooling-systems-and-technologies-and-how-they-work

Julia Borgini. (2023, May 22). *Navigate Energy Star data center standard and certification*. https://www.techtarget.com/searchdatacenter/tip/Navigate-Energy-Star-data-center-standard-and-certification

Kadu, A., & Singh, M. (2023). Fog-Enabled Framework for Patient Health-Monitoring Systems Using Internet of Things and Wireless Body Area Networks. In *Computational Intelligence: Select Proceedings of InCITe 2022* (pp. 607–616). Springer. doi:10.1007/978-981-19-7346-8_52

Kallapu, B., Dodmane, R., Thota, S., & Sahu, A. K. (2023). Enhancing Cloud Communication Security: A Blockchain-Powered Framework with Attribute-Aware Encryption. *Electronics (Basel)*, *12*(18), 3890. doi:10.3390/electronics12183890

Kamiya & Shakti Nagpal. (2015). *Green Cloud Computing Resource Managing Policies: A Survey*. Academic Press.

Kamiya, M. (2020). *The carbon footprint of streaming video on Netflix*. Retrieved October 5, 2023, from https://www.carbonbrief.org/factcheck-what-is-the-carbon-footprint-of-streaming-video-on-netflix/

Karkošková, S. (2023). Data governance model to enhance data quality in financial institutions. *Information Systems Management*, *40*(1), 90–110. doi:10.1080/10580530.2022.2042628

Karthiban, K., & Raj, J. S. (2020). An efficient green computing fair resource allocation in cloud computing using modified deep reinforcement learning algorithm. *Soft Computing*, *24*(19), 14933–14942. doi:10.1007/s00500-020-04846-3

Karthik, S., Hemalatha, R., Aruna, R., Deivakani, M., Reddy, R. V. K., & Boopathi, S. (2023). Study on Healthcare Security System-Integrated Internet of Things (IoT). In Perspectives and Considerations on the Evolution of Smart Systems (pp. 342–362). IGI Global.

Karumban, S., Sanyal, S., Laddunuri, M. M., Sivalinga, V. D., Shanmugam, V., Bose, V., . . . Murugan, S. P. (2023). Industrial Automation and Its Impact on Manufacturing Industries. In Revolutionizing Industrial Automation Through the Convergence of Artificial Intelligence and the Internet of Things (pp. 24-40). IGI Global.

Kashyap, P. K., Kumar, S., & Jaiswal, A. (2019, November). Deep learning based offloading scheme for IoT networks towards green computing. In *2019 IEEE International Conference on Industrial Internet (ICII)* (pp. 22-27). IEEE. 10.1109/ICII.2019.00015

Katal, A., Dahiya, S., & Choudhury, T. (2023). Energy efficiency in cloud computing data centers: A survey on software technologies. *Cluster Computing*, *26*(3), 1845–1875. doi:10.1007/s10586-022-03713-0 PMID:36060618

Kato, S., Tokunaga, S., Maruyama, Y., Maeda, S., Hirabayashi, M., Kitsukawa, Y., Monrroy, A., Ando, T., Fujii, Y., & Azumi, T. (2018). Autoware on board: Enabling autonomous vehicles with embedded systems. *2018 ACM/IEEE 9th International Conference on Cyber-Physical Systems (ICCPS)*, 287–296.

Kaur, J., Verma, R., Alharbe, N. R., Agrawal, A., & Khan, R. A. (2021). Importance of fog computing in healthcare 4.0. *Fog Computing for Healthcare 4.0 Environments: Technical, Societal, and Future Implications*, 79-101.

Kaur, A., Kaur, B., Singh, P., Devgan, M. S., & Toor, H. K. (2020). Load balancing optimization based on deep learning approach in cloud environment. *International Journal of Information Technology and Computer Science, 12*(3), 8–18. doi:10.5815/ijitcs.2020.03.02

Kaur, J., Agrawal, A., & Khan, R. A. (2020). Security issues in fog environment: A systematic literature review. *International Journal of Wireless Information Networks, 27*(3), 467–483. doi:10.1007/s10776-020-00491-7

Kaur, S. a. (2015). *Green Computing-Saving the environment with Intelligent use of computing.* Know Your CSI.

Kaza, S., Yao, L., Bhada-Tata, P., & Van Woerden, F. (n.d.). *What a Waste 2.0: A Global Snapshot of Solid Waste Management to 2050.* World Bank. https://openknowledge.worldbank.org/h andle/10986/30317

Keshavarzi, A., Ni, K., Van Den Hoek, W., Datta, S., & Raychowdhury, A. (2020). Ferroelectronics for edge intelligence. *IEEE Micro, 40*(6), 33–48. doi:10.1109/MM.2020.3026667

Kevin Wee. (2023, November 3). *The challenges in keeping data centers sustainable.* https://www.johnsoncontrols.com/en_id/insights/2021/in-the-news/the-challenges-in-keeping-data-centres-sustainable

Khalid, T., Khan, A. N., Ali, M., Adeel, A., ur Rehman Khan, A., & Shuja, J. (2019). A fog-based security framework for intelligent traffic light control system. *Multimedia Tools and Applications, 78*(17), 24595–24615. doi:10.1007/s11042-018-7008-z

Khan, A. H., Khan, M. A., Abbas, S., Siddiqui, S. Y., Saeed, M. A., Alfayad, M., & Elmitwally, N. S. (2021). Simulation, Modeling, and Optimization of Intelligent Kidney Disease Prediction Empowered with Computational Intelligence Approaches. Academic Press.

Khan, M.A., Abbas, S., Atta, A., Ditta, A., Alquhayz, H., Khan, M.F., & Naqvi, R.A. (2020). *Intelligent cloud based heart disease prediction system empowered with supervised machine learning.* Academic Press.

Khan, M.A., Abbas, S., Khan, K.M., Al Ghamdi, M.A., & Rehman, A. (2020). *Intelligent forecasting model of COVID-19 novel coronavirus outbreak empowered with deep extreme learning.* Academic Press.

Khan, T. AAbbas, SDitta, AKhan, M. AAlquhayz, HFatima, AKhan, M. F. (2020). IoMT-Based Smart Monitoring Hierarchical Fuzzy Inference System for Diagnosis of COVID-19 machine. *Computers, Materials & Continua, 65*(3), 1329–1342.

Khang, A. (Ed.). (2023). *AI and IoT-Based Technologies for Precision Medicine.* IGI Global. doi:10.4018/979-8-3693-0876-9

Khan, W. Z., Ahmed, E., Hakak, S., Yaqoob, I., & Ahmed, A. (2019). Edge computing: A survey. *Future Generation Computer Systems, 97*, 219–235. doi:10.1016/j.future.2019.02.050

Khatiwada, B., Jariyaboon, R., & Techato, K. (2023). E-waste management in Nepal: A case study overcoming challenges and opportunities. e-Prime-Advances in Electrical Engineering. *Electronics and Energy, 4*, 100155.

Khriji, S., Chéour, R., & Kanoun, O. (2022). Dynamic Voltage and Frequency Scaling and Duty-Cycling for Ultra Low-Power Wireless Sensor Nodes. *Electronics (Basel), 11*(24), 4071. doi:10.3390/electronics11244071

Kirsanova, A. A., Radchenko, G. I., & Tchernykh, A. N. (2021). Fog computing state of the art: Concept and classification of platforms to support distributed computing systems. *ArXiv Preprint ArXiv:2106.11726.*

Kishor, A., Chakraborty, C., & Jeberson, W. (2021). Intelligent healthcare data segregation using fog computing with internet of things and machine learning. *International Journal of Engineering Systems Modelling and Simulation, 12*(2-3), 188–194. doi:10.1504/IJESMS.2021.115533

Klinglmair, M., & Fellner, J. (2010). Urban mining in times of raw material shortage. *Journal of Industrial Ecology*, *14*(4), 666–679. doi:10.1111/j.1530-9290.2010.00257.x

Koshariya, A. K., Kalaiyarasi, D., Jovith, A. A., Sivakami, T., Hasan, D. S., & Boopathi, S. (2023). AI-Enabled IoT and WSN-Integrated Smart Agriculture System. In *Artificial Intelligence Tools and Technologies for Smart Farming and Agriculture Practices* (pp. 200–218). IGI Global. doi:10.4018/978-1-6684-8516-3.ch011

Koshariya, A. K., Khatoon, S., Marathe, A. M., Suba, G. M., Baral, D., & Boopathi, S. (2023). Agricultural Waste Management Systems Using Artificial Intelligence Techniques. In *AI-Enabled Social Robotics in Human Care Services* (pp. 236–258). IGI Global. doi:10.4018/978-1-6684-8171-4.ch009

Krishnan, S. R., Nallakaruppan, M. K., Chengoden, R., Koppu, S., Iyapparaja, M., Sadhasivam, J., & Sethuraman, S. (2022). Smart water resource management using Artificial Intelligence—A review. *Sustainability (Basel)*, *14*(20), 13384. doi:10.3390/su142013384

Krishna, R. (2015). *Study Paper On e-waste management, DDG(FA)*. TEC.

Krook, J., & Baas, L. (2013). Getting serious about mining the technosphere: A review of recent landfill mining and urban mining research. *Journal of Cleaner Production*, *55*, 1–9. doi:10.1016/j.jclepro.2013.04.043

Kumar, T. V. (2014). Green computing-an eco friendly approach for energy efficiency and minimizing e-waste. *International Journal of Engineering Research*, 356-359.

Kumar, M. R., Devi, B. R., Rangaswamy, K., Sangeetha, M., & Kumar, K. V. R. (2023, April). IoT-Edge Computing for Efficient and Effective Information Process on Industrial Automation. In *2023 International Conference on Networking and Communications (ICNWC)* (pp. 1-6). IEEE. 10.1109/ICNWC57852.2023.10127492

Kumar, R., Khatri, S. K., & Diván, M. J. (2022). Optimization of power consumption in data centers using machine learning based approaches: A review. *Iranian Journal of Electrical and Computer Engineering*, *12*(3), 3192. doi:10.11591/ijece.v12i3.pp3192-3203

Kumar, Y., Kaul, S., & Hu, Y. C. (2022). Machine learning for energy-resource allocation, workflow scheduling and live migration in cloud computing: State-of-the-art survey. *Sustainable Computing : Informatics and Systems*, *36*, 100780. doi:10.1016/j.suscom.2022.100780

Lau, W. K., Chung, S. S., & Zhang, C. (2013). A material flow analysis on current electrical and electronic waste disposal from Hong Kong households. *Waste Management (New York, N.Y.)*, *33*(3), 714–721. doi:10.1016/j.wasman.2012.09.007 PMID:23046876

Lavanya, R. (2020). Green Scrum Model: Implementation of Scrum in Green and Sustainable Software Engineering. *International Research Journal of Engineering and Technology*, 1583-1587.

Lepakshi, V. A., & Prashanth, C. S. R. (2020, March). Efficient resource allocation with score for reliable task scheduling in cloud computing systems. In *2020 2nd International Conference on Innovative Mechanisms for Industry Applications (ICIMIA)* (pp. 6-12). IEEE. 10.1109/ICIMIA48430.2020.9074914

Leung, A., Cai, Z. W., & Wong, M. H. (2006). Environmental contamination from electronic waste recycling at Guiyu, southeast China. *Journal of Material Cycles and Waste Management*, *8*(1), 21–33. doi:10.1007/s10163-005-0141-6

Liang, L., Wang, W., Jia, Y., & Fu, S. (2016). A cluster-based energy-efficient resource management scheme for ultra-dense networks. *IEEE Access : Practical Innovations, Open Solutions*, *4*, 6823–6832. doi:10.1109/ACCESS.2016.2614517

Liang, T., Glossner, J., Wang, L., Shi, S., & Zhang, X. (2021). Pruning and quantization for deep neural network acceleration: A survey. *Neurocomputing*, *461*, 370–403. doi:10.1016/j.neucom.2021.07.045

Li, D., Zhao, D., Zhang, Q., & Chen, Y. (2019). Reinforcement learning and deep learning based lateral control for autonomous driving. *IEEE Computational Intelligence Magazine, 14*(2), 83–98. doi:10.1109/MCI.2019.2901089

Light, J. (2020). Green networking: A simulation of energy efficient methods. *Procedia Computer Science, 171*, 1489–1497. doi:10.1016/j.procs.2020.04.159

Li, H., Jin, Z., & Krishnamoorthy, S. (2021). E-waste management using machine learning. In *Proceedings of the 6th International Conference on Big Data and Computing* (pp. 30-35). 10.1145/3469968.3469973

Li, J., Lu, H., Guo, J., Xu, Z., & Zhou, Y. (2007). Recycle technology for recovering resourcesand products from waste printed circuit boards. *Environmental Science & Technology, 41*(6), 1995–2000. doi:10.1021/es0618245 PMID:17410796

Li, L., Fan, Y., Tse, M., & Lin, K. Y. (2020). A review of applications in federated learning. *Computers & Industrial Engineering, 149*, 106854. doi:10.1016/j.cie.2020.106854

Lilhore, U. K., Simaiya, S., Dalal, S., & Damaševičius, R. (2023). A smart waste classification model using hybrid CNN-LSTM with transfer learning for sustainable environment. *Multimedia Tools and Applications*, 1–25. doi:10.1007/s11042-023-16677-z

Lin, S.-Y., Du, Y., Ko, P.-C., Wu, T.-J., Ho, P.-T., Sivakumar, V., & Subbareddy, R. (2020). Fog computing based hybrid deep learning framework in effective inspection system for smart manufacturing. *Computer Communications, 160*, 636–642. doi:10.1016/j.comcom.2020.05.044

Liu, G., Ying, Z., Zhao, L., Yuan, X., & Chen, Z. (2018, July). A New Deep Transfer Learning Model for Judicial Data Classification. In *2018 IEEE International Conference on Internet of Things (iThings) and IEEE Green Computing and Communications (GreenCom) and IEEE Cyber, Physical and Social Computing (CPSCom) and IEEE Smart Data (SmartData)* (pp. 126-131). IEEE. 10.1109/Cybermatics_2018.2018.00053

Liu, C., Sharan, L., Adelson, E. H., & Rosenholtz, R. (2010). Exploring features in a bayesian framework for material recognition. In *Computer Vision and Pattern Recognition (CVPR)*. IEEE. 10.1109/CVPR.2010.5540207

Liu, H. (2022). Research on cloud computing adaptive task scheduling based on ant colony algorithm. *Optik (Stuttgart), 258*, 168677. doi:10.1016/j.ijleo.2022.168677

Liu, L., Zhang, Q., Zhai, Z. J., Yue, C., & Ma, X. (2020). State-of-the-art on thermal energy storage technologies in data center. *Energy and Building, 226*, 110345. doi:10.1016/j.enbuild.2020.110345

Li, W. Y., Yang, T., Delicato, F. C., Pires, P. F., Tari, Z., Khan, S. U., & Zomaya, A. Y. (2018). On enabling sustainable edge computing with renewable energy resources. *IEEE Communications Magazine, 56*(5), 94–101. doi:10.1109/MCOM.2018.1700888

Li, X. (2023). CNN-GRU model based on attention mechanism for large-scale energy storage optimization in smart grid. *Frontiers in Energy Research, 11*, 1228256. Advance online publication. doi:10.3389/fenrg.2023.1228256

Li, Y., & Liu, W. (2023). Deep learning-based garbage image recognition algorithm. *Applied Nanoscience, 13*(2), 1415–1424. doi:10.1007/s13204-021-02068-z

Lucivero, F. (2019). Big Data, Big Waste? A Reflection on the Environmental Sustainability of Big Data Initiatives. *Science and Engineering Ethics, 26*(2), 1009–1030. doi:10.1007/s11948-019-00171-7 PMID:31893331

Luo, J., & Hu, D. (2023). An Image Classification Method Based on Adaptive Attention Mechanism and Feature Extraction Network. *Computational Intelligence and Neuroscience, 2023*, 2023. doi:10.1155/2023/4305594 PMID:36844695

Luo, T., Wong, W. F., Goh, R. S. M., Do, A. T., Chen, Z., Li, H., Jiang, W., & Yau, W. (2023). Achieving Green AI with Energy-Efficient Deep Learning Using Neuromorphic Computing. *Communications of the ACM, 66*(7), 52–57. doi:10.1145/3588591

Lu, Z., Whalen, I., Boddeti, V., Dhebar, Y., Deb, K., Goodman, E., & Banzhaf, W. (2019, July). Nsga-net: neural architecture search using multi-objective genetic algorithm. In *Proceedings of the genetic and evolutionary computation conference* (pp. 419-427). 10.1145/3321707.3321729

Lv, T., & Ai, Q. (2016). Interactive energy management of networked microgrids-based active distribution system considering large-scale integration of renewable energy resources. *Applied Energy, 163*, 408–422. doi:10.1016/j.apenergy.2015.10.179

Ma, X., Wang, Z., Zhou, S., Wen, H., & Zhang, Y. (2018, June). Intelligent healthcare systems assisted by data analytics and mobile computing. In *2018 14th International Wireless Communications & Mobile Computing Conference (IWCMC)* (pp. 1317-1322). IEEE. doi:10.1109/IWCMC.2018.8450377

Magnier, B., & Hayat, K. (2023). Revisiting Mehrotra and Nichani's Corner Detection Method for Improvement with Truncated Anisotropic Gaussian Filtering. *Sensors (Basel), 23*(20), 8653. doi:10.3390/s23208653 PMID:37896745

Magoulas, G. D., & Prentza, A. (1999). Machine learning in medical applications. In *Advanced course on artificial intelligence* (pp. 300–307). Springer Berlin Heidelberg.

Maguluri, L. P., Ananth, J., Hariram, S., Geetha, C., Bhaskar, A., & Boopathi, S. (2023). Smart Vehicle-Emissions Monitoring System Using Internet of Things (IoT). In Handbook of Research on Safe Disposal Methods of Municipal Solid Wastes for a Sustainable Environment (pp. 191–211). IGI Global.

Maksimovic, M. (2017). The role of green internet of things (G-IoT) and big data in making cities smarter, safer and more sustainable. *International Journal of Computing and Digital Systems*, 175-184.

Ma, L., Chen, Y., Sun, Y., & Wu, Q. (2012). Virtualization maturity reference model for green software. *2012 International Conference on Control Engineering and Communication Technology*. 10.1109/ICCECT.2012.230

Malapur, B. S., & Pattanshetti, V. R. (2017). IoT based waste management: An application to smart city. *IEEE International Conference on Energy, Communication, Data Analytics and Soft Computing (ICECDS)*, 2476-2486. 10.1109/ICECDS.2017.8389897

Malathi, K., & Priyadarsini, K. (2023). Hybrid lion–GA optimization algorithm-based task scheduling approach in cloud computing. *Applied Nanoscience, 13*(3), 2601–2610. doi:10.1007/s13204-021-02336-y

Ma, M., He, D., Wang, H., Kumar, N., & Choo, K.-K. R. (2019). An efficient and provably secure authenticated key agreement protocol for fog-based vehicular ad-hoc networks. *IEEE Internet of Things Journal, 6*(5), 8065–8075. doi:10.1109/JIOT.2019.2902840

Manganelli, M., Soldati, A., Martirano, L., & Ramakrishna, S. (2021). Strategies for improving the sustainability of data centers via energy mix, energy conservation, and circular energy. *Sustainability (Basel), 13*(11), 6114. doi:10.3390/su13116114

Manikandan, N., Divya, P., & Janani, S. (2022). BWFSO: Hybrid Black-widow and Fish swarm optimization Algorithm for resource allocation and task scheduling in cloud computing. *Materials Today: Proceedings, 62*, 4903–4908. doi:10.1016/j.matpr.2022.03.535

Manikandan, N., Gobalakrishnan, N., & Pradeep, K. (2022). Bee optimization based random double adaptive whale optimization model for task scheduling in cloud computing environment. *Computer Communications, 187*, 35–44. doi:10.1016/j.comcom.2022.01.016

Manogaran, G., Thota, C., Lopez, D., & Sundarasekar, R. (2017). Big data security intelligence for healthcare industry 4.0. *Cybersecurity for Industry 4.0: Analysis for Design and Manufacturing*, 103-126.

Mao, J., Bhattacharya, T., Peng, X., Cao, T., & Qin, X. (2020). Modelling energy consumption of virtual machines in DVFS-enabled cloud data centres. *2020 IEEE 39th International Performance Computing and Communications Conference (IPCCC)*.

Mao, W. L., Chen, W. C., Wang, C. T., & Lin, Y. H. (2021). Recycling waste classification using optimized convolutional neural network. *Resources, Conservation and Recycling, 164*, 105132. doi:10.1016/j.resconrec.2020.105132

Market Trends. (2020, July 8). *Delivering World-Class Insights with Data-Driven Technologies*. https://www.analyticsinsight.net/delivering-world-class-insights-with-data-driven-technologies/

Martinez, I., Hafid, A. S., & Jarray, A. (2020). Design, resource management, and evaluation of fog computing systems: A survey. *IEEE Internet of Things Journal, 8*(4), 2494–2516. doi:10.1109/JIOT.2020.3022699

Mary Riddle. (2023, February 7). *How Meta Sources 100 Percent Renewable Energy Around the World*. https://www.triplepundit.com/story/2023/meta-renewable-energy/765686

Maryam Arbabzadeh. (2022). *Clouding the issue: Are Amazon, Google, and Microsoft really helping companies go green?* https://www.climatiq.io/blog/cloud-computing-amazon-google-microsoft-helping-companies-go-green

Masand, A., Chauhan, S., Jangid, M., Kumar, R., & Roy, S. (2021). Scrapnet: An efficient approach to trash classification. *IEEE Access : Practical Innovations, Open Solutions, 9*, 130947–130958. doi:10.1109/ACCESS.2021.3111230

Masanet, E., Shehabi, A., Lei, N., Smith, S., & Koomey, J. (2020). Recalibrating global data center energy-use estimates. *Science, 367*(6481), 984–986. doi:10.1126/science.aba3758 PMID:32108103

Masoudi, J., Barzegar, B., & Motameni, H. (2022). Energy-aware virtual machine allocation in DVFS-enabled cloud data centres. *IEEE Access : Practical Innovations, Open Solutions, 10*, 3617–3630. doi:10.1109/ACCESS.2021.3136827

Masud, M. H., Akram, W., Ahmed, A., Ananno, A. A., Mourshed, M., Hasan, M., & Joardder, M. U. H. (2019). Towards the effective E-waste management in Bangladesh: A review. *Environmental Science and Pollution Research International, 26*(2), 1250–1276. doi:10.1007/s11356-018-3626-2 PMID:30456610

Matsveichuk, N. M., &Sotskov, Y. N. (2023). *Digital Technologies, Internet of Things and Cloud Computations Used in Agriculture: Surveys and Literature in Russian*. Academic Press.

Matthews, S., Francis, C., McMichael, C., Hendrickson, T., & Hart, D.J. (1997). Disposition and End-of-Life Options for Personal Computers: Green Design Initiative Technical Report. Carnegie Mellon University.

mckinsey.com. (2022, June 26). *How to navigate rising energy costs and inflation*. https://www.mckinsey.com/featured-insights/themes/how-to-navigate-rising-energy-costs-and-inflation

Meenu. (2021, February 9). *Green Data Centers: Enhancing Digital Transformation by Conserving Environment*. https://www.analyticsinsight.net/green-data-centers-enhancing-digital-transformation-by-conserving-environment/

Meghan Rimol. (2021, May 3). *Your Data Center is Old. Now What?* https://www.gartner.com/smarterwithgartner/your-data-center-is-old-now-what

Mehlin, V., Schacht, S., & Lanquillon, C. (2023). Towards energy-efficient Deep Learning: An overview of energy-efficient approaches along the Deep Learning Lifecycle. *arXiv preprint arXiv:2303.01980.*

Meskers, C., Hagelüken, C., Salhofer, S., & Spitzbart, M. (2009). *Impact of Pre-processing Routes on Precious Metal Recovery From PCs.* Academic Press.

Miraz, M. H., Ali, M., Excell, P. S., & Picking, R. (2015). A review on Internet of Things (IoT), Internet of everything (IoE) and Internet of nano things (IoNT). *2015 Internet Technologies and Applications (ITA),* 219-224.

Mirjalili, S., Mirjalili, S. M., & Lewis, A. (2014). Grey wolf optimizer. *Advances in Engineering Software, 69,* 46–61. doi:10.1016/j.advengsoft.2013.12.007

Misra, D., Das, G., Chakrabortty, T., & Das, D. (2018). An IoT-based waste management system monitored by cloud. *Journal of Material Cycles and Waste Management, 20*(3), 1–9. doi:10.1007/s10163-018-0720-y

Mmeah, S. a. (2018). Assessing the Influence of Green Computing Practices on Sustainable I.T. Services. *International Journal of Computer Applications Technology and Research,* 390-397.

Mmeah, S. a. (2018). Assessing the Influence of Green Computing Practices on Sustainable IT Services. *International Journal of Computer Applications Technology and Research,* 390-397.

Modius. (2023, March 10). *The Role of Data Center Infrastructure Management (DCIM) Software in Running a More Profitable Data Center.* https://www.linkedin.com/pulse/role-data-center-infrastructure-management-dcim-software-running/

Moedjahedy, J. H., & Taroreh, M. (2019). Green data centre analysis and design for energy efficiency using clustered and virtualization method. *2019 1st International Conference on Cybernetics and Intelligent System (ICORIS).*

Moorthy, R. S., & Pabitha, P. (2020). Optimal detection of phising attack using SCA based K-NN. *Procedia Computer Science, 171,* 1716–1725. doi:10.1016/j.procs.2020.04.184

Moorthy, R. S., & Pabitha, P. (2021). Design of Wireless Sensor Networks Using Fog Computing for the Optimal Provisioning of Analytics as a Service. In *Machine Learning and Deep Learning Techniques in Wireless and Mobile Networking Systems* (pp. 153–173). CRC Press. doi:10.1201/9781003107477-9

Moorthy, R. S., & Pabitha, P. (2021). Prediction of Parkinson's disease using improved radial basis function neural network. *Computers, Materials & Continua, 68*(3), 3101–3119. doi:10.32604/cmc.2021.016489

Morley, J., Widdicks, K., & Hazas, M. (2018). Digitalisation, energy and data demand: The impact of Internet traffic on overall and peak electricity consumption. *Energy Research & Social Science, 38,* 128–137. doi:10.1016/j.erss.2018.01.018

Morozov, V., Kalnichenko, O., & Mezentseva, O. O. M. (2020). The method of interaction modeling on basis of deep learning the neural networks in complex IT-projects. *International Journal of Computing, 19*(1), 88–96. doi:10.47839/ijc.19.1.1697

Mory-Alvarado, A. M. (2023). *Green IT in small and medium-sized enterprises: A systematic literature review.* Elsevier.

Mukta, T. A. (2021). *E-Waste Management Strategies for Implementing Green Computing.* Academic Press.

Mukta, T. A. (2020). Review on E-waste management strategies for implementing green computing. *Int. J. Comput. Appl,* 45–52.

Mulay, M. R., Chauhan, A., Patel, S., Balakrishnan, V., Halder, A., & Vaish, R. (2019). Candle soot: Journey from a pollutant to a functional material. *Carbon, 144,* 684–712. doi:10.1016/j.carbon.2018.12.083

Multi objective Ant colony Optimization Algorithm for Resource Allocation in Cloud Computing Prasad Devarasetty. (n.d.). https://www.researchgate.net/publication/277477534

Munir, H., Pervaiz, H., Hassan, S. A., Musavian, L., Ni, Q., Imran, M. A., & Tafazolli, R. (2018). Computationally intelligent techniques for resource management in mmwave small cell networks. *IEEE Wireless Communications*, *25*(4), 32–39. doi:10.1109/MWC.2018.1700400

Murugesan, S. (2008). *Harnessing green I.T.: Principles and practices*. IEEE.

Murugesan, S. (2008). *Harnessing green IT: Principles and practices*. IEEE.

Murugesan, S. (2008). Harnessing green IT: Principles and practices. *IT Professional*, *10*(1), 24–33. doi:10.1109/MITP.2008.10

Muthugala, M. V. J., Samarakoon, S. B. P., & Elara, M. R. (2020). Tradeoff between Area Coverage and Energy Usage of a Self-Reconfigurable Floor Cleaning Robot based on User Preference. *IEEE Access : Practical Innovations, Open Solutions*, *8*, 76267–76275. doi:10.1109/ACCESS.2020.2988977

MyClimate. (2023). *What is a digital carbon footprint?* Retrieved October 5, 2023, from https://www.myclimate.org/en/information/faq/faq-detail/what-is-a-digital-carbon-footprint/: https://www.myclimate.org/en/information/faq/faq-detail/what-is-a-digital-carbon-footprint/

Mythili, T., & Anbarasi, A. (2022). A concatenation of deep and texture features for medicinal trash image classification using EnSegNet-DNN-based transfer learning. *Materials Today: Proceedings*, *62*, 4691–4698. doi:10.1016/j.matpr.2022.03.129

Mytton, D., & Ashtine, M. (2022). Sources of data center energy estimates: A comprehensive review. *Joule*, *6*(9), 2032–2056. doi:10.1016/j.joule.2022.07.011

N. P. M K., Rastogi, & A. K. (2023). Demand Forecasting in Supply Chain Management using CNN-LSTM Hybrid Model. In *14th International Conference on Computing Communication and Networking Technologies (ICCCNT)* (pp. 1-5). 10.1109/ICCCNT56998.2023.10307665

Nabeeh, N. A., Smarandache, F., Abdel-Basset, M., El-Ghareeb, H. A., & Aboelfetouh, A. (2019). An integrated neutrosophic-topsis approach and its application to personnel selection: A new trend in brain processing and analysis. *IEEE Access : Practical Innovations, Open Solutions*, *7*, 29734–29744. doi:10.1109/ACCESS.2019.2899841

Nabil Taha. (2023). *The path to data center decarbonization starts now*. https://www.datacenterdynamics.com/en/opinions/the-path-to-data-center-decarbonization-starts-now/

Nagrath, P., Jain, R., Madan, A., Arora, R., Kataria, P., & Hemanth, J. (2021). SSDMNV2: A real time DNN-based face mask detection system using single shot multibox detector and MobileNetV2. *Sustainable Cities and Society*, *66*, 102692. doi:10.1016/j.scs.2020.102692 PMID:33425664

Naim, A. (2021). Green Information Technologies in Business Operations. *Periodica Journal of Modern Philosophy, Social Sciences and Humanities*, 36-49.

Nalla, K., & Pothabathula, S. V. (2021). Green IoT and Machine Learning for Agricultural Applications. *Green Internet of Things and Machine Learning: Towards a Smart Sustainable World*, 189-214.

Narciso, D. A., & Martins, F. G. (2020). Application of machine learning tools for energy efficiency in industry: A review. *Energy Reports*, *6*, 1181–1199. doi:10.1016/j.egyr.2020.04.035

Needhidasan, S., Samuel, M., & Chidambaram, R. (2014). Electronic waste – an emerging threat to the environment of urban India. *Journal of Environmental Health Science & Engineering, 12*(1), 12. doi:10.1186/2052-336X-12-36 PMID:24444377

Neelakantam, G., Onthoni, D. D., & Sahoo, P. K. (2020). Reinforcement learning based passengers assistance system for crowded public transportation in fog enabled smart city. *Electronics (Basel), 9*(9), 1501. doi:10.3390/electronics9091501

Negi, P. S., Negi, V., & Pandey, A. C. (2011). Impact of information technology on learning, teaching and human resource management in educational sector. *International Journal of Computer Science and Telecommunications, 2*(4), 66–72.

Nelson, B., Zytner, R. G., Dulac, Y., & Cabral, A. R. (2022). Mitigating fugitive methane emissions from closed landfills: A pilot-scale field study. *The Science of the Total Environment, 851*, 158351. doi:10.1016/j.scitotenv.2022.158351 PMID:36049680

Nicole Loher. (2023, March 15). *What Does it Mean to Be 'Water Positive'?* https://sustainability.fb.com/blog/2023/03/15/what-does-it-mean-to-be-water-positive/

Ning, Z., Dong, P., Wang, X., Guo, L., Rodrigues, J. J., Kong, X., Huang, J., & Kwok, R. Y. (2019). Deep reinforcement learning for intelligent internet of vehicles: An energy-efficient computational offloading scheme. *IEEE Transactions on Cognitive Communications and Networking, 5*(4), 1060–1072. doi:10.1109/TCCN.2019.2930521

Nkenyereye, L., Islam, S. R., Bilal, M., Abdullah-Al-Wadud, M., Alamri, A., & Nayyar, A. (2021). Secure crowd-sensing protocol for fog-based vehicular cloud. *Future Generation Computer Systems, 120*, 61–75. doi:10.1016/j.future.2021.02.008

Nnorom, I. C., & Osibanjo, O. (2008). Overview of electronic waste (e-waste) management practices and legislations, and their poor applications in the developing countries. *Resources, Conservation and Recycling, 52*(6), 843–858. doi:10.1016/j.resconrec.2008.01.004

Nokia. (2023). *Energy efficiency.* Retrieved October 5, 2023, from https://www.nokia.com/networks/bss-oss/ava/energy-efficiency/?did=D000000007BR&gad_source=1&gclid=CjwKCAiAvdCrBhBREiwAX6-6Uu8DkwDjjBpM81xs6ol-ZKW-poLxSFWvpyW6fg5ISSpj6nsBsyQn9UhoCgMoQAvD_BwE

NSWMA. (2013). *Waste characterization and per capita generation rate report.* NSWMA.

NVIDIA. (2022). *What is green computing?* Retrieved from https://blogs.nvidia.com/blog/what-is-green-computing/

Obermeyer, Z., & Emanuel, E. J. (2016). Predicting the future—Big data, machine learning, and clinical medicine. *The New England Journal of Medicine, 375*(13), 1216–1219. doi:10.1056/NEJMp1606181 PMID:27682033

OECD. (2023). *Data usage per mobile broadband user.* Retrieved October 5, 2023, from https://www.oecd.org/digital/broadband/broadband-statistics-update.htm

Ojo, A. O., Raman, M., & Downe, A. G. (2019). Toward green computing practices: A Malaysian study of green belief and attitude among Information Technology professionals. *Journal of Cleaner Production, 224*, 246–255. doi:10.1016/j.jclepro.2019.03.237

Olujobi, O. J., Okorie, U. E., Olarinde, E. S., & Aina-Pelemo, A. D. (2023). Legal responses to energy security and sustainability in Nigeria's power sector amidst fossil fuel disruptions and low carbon energy transition. *Heliyon, 9*(7), e17912. doi:10.1016/j.heliyon.2023.e17912 PMID:37483776

Omitaomu, O. A., & Niu, H. (2021). Artificial Intelligence Techniques in Smart Grid: A Survey. *Smart Cities, 4*(2), 548–568. doi:10.3390/smartcities4020029

Orlins, S., & Guan, D. (2016). China's toxic informal e-waste recycling: Local approaches to a global environmental problem. *Journal of Cleaner Production, 114*, 71–80. doi:10.1016/j.jclepro.2015.05.090

Osibanjo, O., & Nnorom, I. C. (2007). The challenge of electronic waste (e-waste) management in developing countries. *Waste Management & Research, 25*(6), 489–501. doi:10.1177/0734242X07082028 PMID:18229743

Ounifi, H. A., Gherbi, A., & Kara, N. (2022). Deep machine learning-based power usage effectiveness prediction for sustainable cloud infrastructures. *Sustainable Energy Technologies and Assessments, 52*, 101967. doi:10.1016/j.seta.2022.101967

Ozkaya, U., & Seyfi, L. (2019). Fine-tuning models comparisons on garbage classification for recyclability. *arXiv preprint arXiv:1908.04393*.

Öztürk, E., Ferreira, F., Jomaa, H., Schmidt-Thieme, L., Grabocka, J., & Hutter, F. (2022, June). Zero-Shot AutoML with Pretrained Models. In *International Conference on Machine Learning* (pp. 17138-17155). PMLR.

Padhy, S., Alowaidi, M., Dash, S., Alshehri, M., Malla, P. P., Routray, S., & Alhumyani, H. (2023). AgriSecure: A Fog Computing-Based Security Framework for Agriculture 4.0 via Blockchain. *Processes (Basel, Switzerland), 11*(3), 757. doi:10.3390/pr11030757

Palattella, M. R., Soua, R., Khelil, A., & Engel, T. (2019). Fog computing as the key for seamless connectivity handover in future vehicular networks. *Proceedings of the 34th ACM/SIGAPP Symposium on Applied Computing*, 1996–2000. 10.1145/3297280.3297475

Pandi, K. M., & Somasundaram, K. (2016). Energy efficient in virtual infrastructure and green cloud computing: A review. *Indian Journal of Science and Technology, 9*(11), 1–8. doi:10.17485/ijst/2016/v9i11/89399

Papagianni, C., Mangues-Bafalluy, J., Bermudez, P., Barmpounakis, S., De Vleeschauwer, D., Brenes, J., ... Pepe, T. (2020, June). 5Growth: AI-driven 5G for Automation in Vertical Industries. In *2020 European Conference on Networks and Communications (EuCNC)* (pp. 17-22). IEEE. 10.1109/EuCNC48522.2020.9200919

Parajuly, K., Thapa, K. B., Cimpan, C., & Wenzel, H. (2017). Electronic waste and informal recycling in Kathmandu, Nepal: Challenges and opportunities. *Journal of Material Cycles and Waste Management*.

Pareek, K., Tiwari, P. K., & Bhatnagar, V. (2021). Fog computing in healthcare: A review. *IOP Conference Series. Materials Science and Engineering, 1099*(1), 012025. doi:10.1088/1757-899X/1099/1/012025

Park, S., & Seo, J. (2018). Analysis of Air-side economizers in terms of cooling-energy performance in a data center considering Exhaust air recirculation. *Energies, 11*(2), 444. doi:10.3390/en11020444

Parmentola, A., Petrillo, A., Tutore, I., & De Felice, F. (2021). Is blockchain able to enhance environmental sustainability? A systematic review and research agenda from the perspective of Sustainable Development Goals (SDGs). *Business Strategy and the Environment*. Advance online publication. doi:10.1002/bse.2882

Patil, A., & Patil, D. R. (2019). An analysis report on green cloud computing current trends and future research challenges. SSRN *Electronic Journal*. doi:10.2139/ssrn.3355151

Paul Evans. (2018, April 19). *Data center HVAC cooling systems*. https://theengineeringmindset.com/data-center-hvac-cooling-systems/

Paul, S. G., Saha, A., Arefin, M. S., Bhuiyan, T., Biswas, A. A., Reza, A. W., ... Moni, M. A. (2023). A Comprehensive Review of Green Computing: Past, Present, and Future Research. *IEEE Access*.

Paul, A., Pinjari, H., Hong, W. H., Seo, H. C., & Rho, S. (2018). Fog computing-based IoT for health monitoring system. *Journal of Sensors, 2018*, 2018. doi:10.1155/2018/1386470

Pazowski, P. a. (2015). *Green computing: latest practices and technologies for ICT sustainability*. Academic Press.

Pearce, F. (2018). Energy hogs: can world's huge data centers be made more efficient. *Yale Environment, 360*(3).

Pedrycz, W. (2018). Granular computing for data analytics: a manifesto of human-centric computing. *IEEE/CAA Journal of Automatica Sinica, 5*(6), 1025-1034.

Peng, Z., Barzegar, B., Yarahmadi, M., Motameni, H., & Pirouzmand, P. (2020). Energy-aware scheduling of workflow using a heuristic method on green cloud. *Scientific Programming, 2020*, 1–14. doi:10.1155/2020/8898059

Perez-Belis, V., Bovea, M. D., & Ibanez-Fores, V. (2015). An in-depth literature review of the waste electrical and electronic equipment context: Trends and evolution. *Waste Management & Research, 33*(1), 3–29. doi:10.1177/0734242X14557382 PMID:25406121

Peri, G. L., Licciardi, G. R., Matera, N., Mazzeo, D., Cirrincione, L., & Scaccianoce, G. (2022). Disposal of green roofs: A contribution to identifying an "Allowed by legislation" end—of—life scenario and facilitating their environmental analysis. *Building and Environment, 226*, 109–739. doi:10.1016/j.buildenv.2022.109739

Peter Judge. (2022, March 22). *SEC proposals could force disclosure of Scope 1, 2, and 3 emissions*. https://www.data-centerdynamics.com/en/news/sec-proposals-could-force-disclosure-of-scope-1-2-and-3-emissions/

Petrosyan, A. (2023). *Worldwide digital population 2023*. Retrieved October 5, 2023, from https://www.statista.com/statistics/1044012/us-digital-audience/

Phiri, M., Mulenga, M., Zimba, A., & Eke, C. I. (2023). Deep learning techniques for solar tracking systems: A systematic literature review, research challenges, and open research directions. *Solar Energy, 262*, 111803. doi:10.1016/j.solener.2023.111803

Podder, S. K. (2022). *Green computing practice in ICT-based methods: innovation in web-based learning and teaching technologies*. IGI Global.

Posey, B. (2022). *What Is the Akida Event Domain Neural Processor?* Academic Press.

Potluri, S., & Rao, K. S. (2020). Optimization model for QoS based task scheduling in cloud computing environment. *Indonesian Journal of Electrical Engineering and Computer Science, 18*(2), 1081–1088. doi:10.11591/ijeecs.v18.i2.pp1081-1088

Prasad, K. D., Murthy, P. K., Gireesh, C. H., Prasad, M., & Sravani, K. (2021). Prioritization of e-waste management strategies towards green computing using AHP-QFD approach. *Proc. Eng, 3*, 33–40.

Praveenchandar, J., & Tamilarasi, A. (2021). Dynamic resource allocation with optimized task scheduling and improved power management in cloud computing. *Journal of Ambient Intelligence and Humanized Computing, 12*(3), 4147–4159. doi:10.1007/s12652-020-01794-6

prnewswire.com. (2021, March 23). *Data Center Market Size to Cross over $ 519 Billion by 2025*. https://www.prnewswire.com/news-releases/data-center-market-size-to-cross-over--519-billion-by-2025---technavio-301253991.html

Puppala Ramya, V. (2023, February 10). Optimized Deep Learning-Based E-Waste Management in IoT Application via Energy-Aware Routing. *Cybernetics and Systems*, 1–30. Advance online publication. doi:10.1080/01969722.2023.2175119

Purnomo, A., Anam, F., Afia, N., Septianto, A., & Mufliq, A. (2021, August). Four decades of the green computing study: a bibliometric overview. In *2021 International Conference on Information Management and Technology (ICIMTech)* (Vol. 1, pp. 795-800). IEEE. 10.1109/ICIMTech53080.2021.9535069

PVcase. (2023). *The environmental impact of computing problems and possible solutions.* Retrieved from https://pvcase.com/blog/the-environmental-impact-of-computing-problems-and-possible-solutions/#:~:text=Cloud%20computing%20is%20only%20as,of%20e%2Dwaste%20is%20recycled

Rahamathunnisa, U., Sudhakar, K., Murugan, T. K., Thivaharan, S., Rajkumar, M., & Boopathi, S. (2023). Cloud Computing Principles for Optimizing Robot Task Offloading Processes. In *AI-Enabled Social Robotics in Human Care Services* (pp. 188–211). IGI Global. doi:10.4018/978-1-6684-8171-4.ch007

Rahman, M. W., Islam, R., Hasan, A., Bithi, N. I., Hasan, M. M., & Rahman, M. M. (2022). Intelligent waste management system using deep learning with IoT. *Journal of King Saud University. Computer and Information Sciences, 34*(5), 2072–2087. doi:10.1016/j.jksuci.2020.08.016

Raja, S. P. (2021). Green computing and carbon footprint management in the IT sectors. *IEEE Transactions on Computational Social Systems, 8*(5), 1172–1177. doi:10.1109/TCSS.2021.3076461

Rajkumar, K., & Dhanakoti, V. (2020, December). Methodological Methods to Improve the Efficiency of Cloud Storage by applying De-duplication Techniques in Cloud Computing. In *2020 2nd International Conference on Advances in Computing, Communication Control and Networking (ICACCCN)* (pp. 876-884). IEEE. 10.1109/ICACCCN51052.2020.9362940

Raju, K. B., Dara, S., Vidyarthi, A., Gupta, V. M., & Khan, B. (2022). Smart heart disease prediction system with IoT and fog computing sectors enabled by cascaded deep learning model. *Computational Intelligence and Neuroscience, 2022,* 2022. doi:10.1155/2022/1070697 PMID:35047027

Ramudu, K., Mohan, V. M., Jyothirmai, D., Prasad, D., Agrawal, R., & Boopathi, S. (2023). Machine Learning and Artificial Intelligence in Disease Prediction: Applications, Challenges, Limitations, Case Studies, and Future Directions. In Contemporary Applications of Data Fusion for Advanced Healthcare Informatics (pp. 297–318). IGI Global.

Ramya, P., & Ramya, V. (2023). E-waste management using hybrid optimization-enabled deep learning in IoT-cloud platform. *Advances in Engineering Software, 176,* 103353. doi:10.1016/j.advengsoft.2022.103353

Rani, S., Srivastava, G., & ... (2023). Secure hierarchical fog computing-based architecture for industry 5.0 using an attribute-based encryption scheme. *Expert Systems with Applications, 121180.*

Ranjani, M. (2012). Green computing - maturity model for virtualization. *International Journal of Data Mining Techniques and Applications, 1*(2), 29–35. doi:10.20894/IJDMTA.102.001.002.002

Rao, H. (n.d.). *Machine Learning endorsing "Green Computing" by optimizing energy efficiency of data centers.* https://www.linkedin.com/pulse/machine-learning-endorsing-green-computing-optimizing-himanshu-rao

Rashmi Singh. (2023, February 28). *Green building regulations give impetus to sustainable data centers in India.* https://india.mongabay.com/2023/02/green-building-regulations-give-impetus-to-sustainable-data-centers-in-india/

Rautela, R., Arya, S., Vishwakarma, S., Lee, J., Kim, K. H., & Kumar, S. (2021). E-waste management and its effects on the environment and human health. *The Science of the Total Environment, 773,* 145623. doi:10.1016/j.scitotenv.2021.145623 PMID:33592459

Ravi, G. S., Smith, K. N., Gokhale, P., & Chong, F. T. (2021, November). Quantum Computing in the Cloud: Analyzing job and machine characteristics. In *2021 IEEE International Symposium on Workload Characterization (IISWC)* (pp. 39-50). IEEE. 10.1109/IISWC53511.2021.00015

Reddy, K. H. K., Goswami, R. S., & Roy, D. S. (2023). A futuristic green service computing approach for smart city: A fog layered intelligent service management model for smart transport system. *Computer Communications, 212*, 151–160. doi:10.1016/j.comcom.2023.08.001

Reddy, M. A., Reddy, B. M., Mukund, C., Venneti, K., Preethi, D., & Boopathi, S. (2023). Social Health Protection During the COVID-Pandemic Using IoT. In *The COVID-19 Pandemic and the Digitalization of Diplomacy* (pp. 204–235). IGI Global. doi:10.4018/978-1-7998-8394-4.ch009

Research and Markets. (2020, February 13). *Comprehensive Data Center Market Outlook and Forecast 2020-2025.* https://www.globenewswire.com/news-release/2020/02/13/1984742/0/en/Comprehensive-Data-Center-Market-Outlook-and-Forecast-2020-2025.html

Reuter, M., Hudson, C., Hagelüken, C., Heiskanen, K., Meskers, C., & Schaik, A. (2013). *Metal Recycling - Opportunities, Limits, Infrastructure.* United Nations Environment Programme.

Rich Miller. (2022, February 12). *Cloud Titans Were the Largest Buyers of Renewable Energy in 2021.* https://www.datacenterfrontier.com/featured/article/11427604/cloud-titans-were-the-largest-buyers-of-renewable-energy-in-2021

Ritcey, G. M. (2006). Solvent extraction in hydrometallurgy: Present and future. *Tsinghua Science and Technology, 11*(2), 137–152. doi:10.1016/S1007-0214(06)70168-7

Rjoub, G., Bentahar, J., Abdel Wahab, O., & Saleh Bataineh, A. (2021). Deep and reinforcement learning for automated task scheduling in large-scale cloud computing systems. *Concurrency and Computation, 33*(23), e5919. doi:10.1002/cpe.5919

Robert McFarlane. (2021, September 27). *Considerations for sustainable data center design.* https://www.techtarget.com/searchdatacenter/tip/Considerations-for-sustainable-data-center-design

Rodoshi, R. T., Kim, T., & Choi, W. (2020). Resource management in cloud radio access network: Conventional and new approaches. *Sensors (Basel), 20*(9), 2708. doi:10.3390/s20092708 PMID:32397540

Roy, S., & Gupta, S. (2014). The green cloud effective framework: An environment friendly approach reducing CO2 level. *2014 1st International Conference on Non Conventional Energy (ICONCE 2014).*

Rugwiro, U., Gu, C., & Ding, W. (2019). Task scheduling and resource allocation based on ant-colony optimization and deep reinforcement learning. *Journal of Internet Technology, 20*(5), 1463–1475.

Ryder, C. (2008). *Improving energy efficiency through application of infrastructure virtualization: introducing IBM WebSphere Virtual Enterprise.* The Sageza Group.

S&P Global. (2019). *Avoiding garbage in machine learning shell.* Retrieved from https://www.spglobal.com/en/research-insights/articles/avoiding-garbage-in-machine-learning-shell

Sadegh Safarzadeh, M., Bafghi, M. S., Moradkhani, D., & Ojaghi Ilkhchi, M. (2007). A review on hydrometallurgical extraction and recovery of cadmium from various resources. *Minerals Engineering, 20*(3), 211–220. doi:10.1016/j.mineng.2006.07.001

Sagar, S. a. (2021). A review: Recent trends in green computing. *Green Computing in Smart Cities: Simulation and Techniques*, 19-34.

Saidi, K., & Bardou, D. (2023). Task scheduling and VM placement to resource allocation in Cloud computing: Challenges and opportunities. *Cluster Computing, 26*(5), 3069–3087. doi:10.1007/s10586-023-04098-4

Saleem, Y., Crespi, N., Rehmani, M. H., & Copeland, R. (2019). Internet of things-aided smart grid: Technologies, architectures, applications, prototypes, and future research directions. *IEEE Access : Practical Innovations, Open Solutions*, 7, 62962–63003. doi:10.1109/ACCESS.2019.2913984

Sam Mackilligin. (2023, November 16). *Decarbonising Data Centers*. https://infrastructure.aecom.com/2020/decarbonising-data-centres

Samann, F. E. (2017). The design and implementation of smart trash bins. *Academic Journal of Nawroz University*, 6(3), 141–148. doi:10.25007/ajnu.v6n3a103

Samarasinghe, K. R., & Medis, A. (2020). Artificial intelligence based strategic human resource management (AISHRM) for industry 4.0. *Global Journal of Management and Business Research*, 20(2), 7–13. doi:10.34257/GJMBRGVOl20IS2PG7

Samikannu, R., Koshariya, A. K., Poornima, E., Ramesh, S., Kumar, A., & Boopathi, S. (2022). Sustainable Development in Modern Aquaponics Cultivation Systems Using IoT Technologies. In *Human Agro-Energy Optimization for Business and Industry* (pp. 105–127). IGI Global.

Säntti, R. (2022). *Five ways to battle data waste*. Capgemini. Retrieved from https://www.capgemini.com/insights/expert-perspectives/five-ways-to-battle-data-waste/

Sanyal, S., Kalimuthu, M., Arumugam, T., Aruna, R., Balaji, J., Savarimuthu, A., & Patil, S. (2023). Internet of Things and Its Relevance to Digital Marketing. In *Opportunities and Challenges of Industrial IoT in 5G and 6G Networks* (pp. 138–154). IGI Global. doi:10.4018/978-1-7998-9266-3.ch007

Sarala, B., Sumathy, G., Kalpana, A. V., & Jasmine Hephzipah, J. (2023). Glioma brain tumor detection using dual convolutional neural networks and histogram density segmentation algorithm. *Biomedical Signal Processing and Control*, 85, 104859. doi:10.1016/j.bspc.2023.104859

Sarkar, N. I., & Gul, S. (2021). Green computing and internet of things for smart cities: technologies, challenges, and implementation. *Green Computing in Smart Cities: Simulation and Techniques*, 35-50.

Sathiyamoorthi, V., Keerthika, P., Suresh, P., Zhang, Z. J., Rao, A. P., & Logeswaran, K. (2021). Adaptive fault tolerant resource allocation scheme for cloud computing environments. *Journal of Organizational and End User Computing*, 33(5), 135–152. doi:10.4018/JOEUC.20210901.oa7

Savitz, A. (2013). *The triple bottom line: how today's best-run companies are achieving economic, social and environmental success-and how you can too*. John Wiley & Sons.

Sayed, K., & Gabbar, H. (2018). *Building Energy Management Systems (BEMS)*. doi:10.1002/9781119422099.ch2

Schluep, M., Müller, E., & Rochat, D. (2012). *E-Waste Assessment Methodology Training & Reference Manual*. Empa - Swiss Federal Laboratories for Materials Science and Technology.

Schumacher, K. A., & Agbemabiese, L. (2019). Towards comprehensive e-waste legislation inthe United States: Design considerations based on quantitative and qualitative assessments. *Resources, Conservation and Recycling*, 149, 605–621. doi:10.1016/j.resconrec.2019.06.033

Scogland, T. R. W., Lin, H., & Feng, W. C. (2010). A first look at integrated GPUs for green high-performance computing. *Computer Science (Berlin, Germany)*, 25(3-4), 125–134. doi:10.1007/s00450-010-0128-y

Sean Ratka and Francisco Boshell. (2020, June 26). *The Nexus between Data Centres, Efficiency And Renewables: A Role Model For The Energy Transition*. https://energypost.eu/the-nexus-between-data-centres-efficiency-and-renewables-a-role-model-for-the-energy-transition/

Sebastian-Coleman, L. (2022). *The Culture Challenge: Organizational Accountability for Data. Meeting the Challenges of Data Quality Management.* Academic Press. doi:10.1016/B978-0-12-821737-5.00008-0

Seitz, J. (2014). *Analysis of Existing E-Waste Practices in MENA Countries. The Regional Solid Waste Exchange of Information and Expertise Network in Mashreq and Maghreb Countries.* SWEEP-Net.

Selvaraj, S., Prabhu Kavin, B., Kavitha, C., & Lai, W. C. (2022). A Multiclass Fault Diagnosis Framework Using Context-Based Multilayered Bayesian Method for Centrifugal Pumps. *Electronics (Basel)*, *11*(23), 4014. doi:10.3390/electronics11234014

Sengeni, D., Padmapriya, G., Imambi, S. S., Suganthi, D., Suri, A., & Boopathi, S. (2023). Biomedical Waste Handling Method Using Artificial Intelligence Techniques. In *Handbook of Research on Safe Disposal Methods of Municipal Solid Wastes for a Sustainable Environment* (pp. 306–323). IGI Global. doi:10.4018/978-1-6684-8117-2.ch022

Senge, P. (2008). The necessary revolution: How individuals and organisations are working together to create a sustainable world. *Management Today*, 54–57.

Senthil Kumar, R., Saravanan, S., Pandiyan, P., Suresh, K. P., & Leninpugalhanthi, P. (2022). Green Energy Using Machine and Deep Learning. *Machine Learning Algorithms for Signal and Image Processing*, 429-444.

SEPA. (2011). Recycling and disposal of electronic waste: Health hazards and environmental impacts. SEPA.

Serrano, R., Duran, C., Sarmiento, M., Pham, C. K., & Hoang, T. T. (2022). ChaCha20–Poly1305 Authenticated Encryption with Additional Data for Transport Layer Security 1.3. *Cryptography*, *6*(2), 30. doi:10.3390/cryptography6020030

Seth, J. K., & Chandra, S. (2018). MIDS: Metaheuristic based intrusion detection system for cloud using k-NN and MGWO. In *Advances in Computing and Data Sciences: Second International Conference, ICACDS 2018, Dehradun, India, April 20-21, 2018, Revised Selected Papers, Part I 2* (pp. 411-420). Springer Singapore.

Shahzad, A., Gherbi, A., & Zhang, K. (2022). Enabling fog–blockchain computing for autonomous-vehicle-parking system: A solution to reinforce iot–cloud platform for future smart parking. *Sensors (Basel)*, *22*(13), 4849. doi:10.3390/s22134849 PMID:35808345

Shaikh, P. H., Nor, N. B. M., Nallagownden, P., Elamvazuthi, I., & Ibrahim, T. (2014). A review on optimized control systems for building energy and comfort management of smart sustainable buildings. *Renewable & Sustainable Energy Reviews*, *34*, 409–429. doi:10.1016/j.rser.2014.03.027

Sharanya, S., Venkataraman, R., & Murali, G. (2022). Predicting remaining useful life of turbofan engines using degradation signal based echo state network. *International Journal of Turbo & Jet-Engines*, (0).

Sharma, N., Singha, N., & Dutta, T. (2015). Smart Bin Implementation for Smart Cities. *International Journal of Scientific and Engineering Research*, *6*(9), 787–791.

Sharma, R., Bala, A., & Singh, A. (2022). Virtual machine migration for green cloud computing. *2022 IEEE International Conference on Distributed Computing and Electrical Circuits and Electronics (ICDCECE).* 10.1109/ICDCECE53908.2022.9793067

Shayeghi, H., & Shahryari, E. (2017). Integration and management technique of renewable energy resources in microgrid. *Energy Harvesting and Energy Efficiency: Technology, Methods, and Applications*, 393-421.

Shreyas Madhav, A. V., Rajaraman, R., Harini, S., & Kiliroor, C. C. (2022). Application of artificial intelligence to enhance collection of E-waste: A potential solution for household WEEE collection and segregation in India. *Waste Management & Research*, *40*(7), 1047–1053. doi:10.1177/0734242X211052846 PMID:34726090

Shu, W., Cai, K., & Xiong, N. N. (2021). Research on strong agile response task scheduling optimization enhancement with optimal resource usage in green cloud computing. *Future Generation Computer Systems*, *124*, 12–20. doi:10.1016/j.future.2021.05.012

Silva, G., Schulze, B., & Ferro, M. (2021). *Performance and energy efficiency analysis of machine learning algorithms towards green ai: a case study of decision tree algorithms* [Master's thesis]. National Lab. for Scientific Computing.

Singh, G., & Singh, J. (2023). A Fog Computing based Agriculture-IoT Framework for Detection of Alert Conditions and Effective Crop Protection. *2023 5th International Conference on Smart Systems and Inventive Technology (ICSSIT)*, 537–543.

Singh, N., Duan, H., & Tang, Y. (2020). Toxicity evaluation of E-waste plastics and potential repercussions for human health. *Environment International*, *137*, 105–559. doi:10.1016/j.envint.2020.105559 PMID:32062437

Singh, S. (2015). Green computing strategies & challenges. *2015 International Conference on Green Computing and Internet of Things (ICGCIoT)*. 10.1109/ICGCIoT.2015.7380564

Škare, M., & Soriano, D. R. (2021). A dynamic panel study on digitalization and firm's agility: What drives agility in advanced economies 2009–2018. *Technological Forecasting and Social Change*, *163*, 120418. doi:10.1016/j.techfore.2020.120418

Sneha, T.V., Singh, P., & Pandey, P. (2023). Green cloud computing: Goals, techniques, architectures, and research challenges. *2023 International Conference on Advancement in Computation & Computer Technologies (InCACCT)*. 10.1109/InCACCT57535.2023.10141845

Sodhro, A. H., Sodhro, G. H., Guizani, M., Pirbhulal, S., & Boukerche, A. (2020). AI-enabled reliable channel modeling architecture for fog computing vehicular networks. *IEEE Wireless Communications*, *27*(2), 14–21. doi:10.1109/MWC.001.1900311

Soesanto, H., Maarif, M. S., Anwar, S., & Yurianto, Y. (2023). Current Trend, Future Direction, and Enablers of e-Waste Management: Bibliometric Analysis and Literature Review. *Polish Journal of Environmental Studies*, *32*(4), 3455–3465. doi:10.15244/pjoes/163607

Sofia, A. S., & Kumar, P. G. (2015). Implementation of energy efficient green computing in cloud computing. *International Journal of Enterprise Network Management*, *6*(3), 222–237. doi:10.1504/IJENM.2015.071135

Songhorabadi, M., Rahimi, M., MoghadamFarid, A. M., & Haghi Kashani, M. (2023). Fog computing approaches in IoT-enabled smart cities. *Journal of Network and Computer Applications*, *211*, 103557. doi:10.1016/j.jnca.2022.103557

Song, Q., & Li, J. (2015). A review on human health consequences of metals exposure to ewaste in China. *Environmental Pollution*, *196*, 450–461. doi:10.1016/j.envpol.2014.11.004 PMID:25468213

Sreelakshmi, K., Akarsh, S., Vinayakumar, R., & Soman, K.P. (2019). *Capsule Networks and Visualization for Segregation of Plastic and Non-Plastic Wastes*. IEEE.

SRF. (2020). *Co2-Fussabdruck im Internet: Surfe ich das Klima kaputt?* Retrieved October 5, 2023, from https://www.srf.ch/kultur/gesellschaft-religion/co2-fussabdruck-im-internet-surfe-ich-das-klima-kaputt

Srikanth. (2023, September 12). *The concept of data center sustainability is revolutionizing the IT industry: Mohammed Atif, Director, Park Place Technologies*. https://www.expresscomputer.in/features/the-concept-of-data-center-sustainability-is-revolutionizing-the-it-industry-mohammed-atif-director-park-place-technologies/103376/

Sriram, V. P., Sanyal, S., Laddunuri, M. M., Subramanian, M., Bose, V., Booshan, B., . . . Thangam, D. (2023). Enhancing Cybersecurity Through Blockchain Technology. In Handbook of Research on Cybersecurity Issues and Challenges for Business and FinTech Applications (pp. 208-224). IGI Global.

Statista. (2023, October 23). *Number of data centers worldwide in 2023, by country.* https://www.statista.com/statistics/1228433/data-centers-worldwide-by-country/

Suja, F., Abdul Rahman, R., Yusof, A., & Masdar, M. S. (2014). E-waste management scenarios in Malaysia. *Journal of Waste Management, 2014*, 1–7. doi:10.1155/2014/609169

Sun, P., Wang, J., Xu, Y., & Wang, L. (2020). Research and application on power generation safety monitoring and cloud platform. *2020 IEEE 1st China International Youth Conference on Electrical Engineering (CIYCEE).*

Sun, G., Sun, S., Sun, J., Yu, H., Du, X., & Guizani, M. (2019). Security and privacy preservation in fog-based crowd sensing on the internet of vehicles. *Journal of Network and Computer Applications, 134*, 89–99. doi:10.1016/j.jnca.2019.02.018

Sun, Y., Ochiai, H., & Esaki, H. (2021). Decentralized deep learning for multi-access edge computing: A survey on communication efficiency and trustworthiness. *IEEE Transactions on Artificial Intelligence, 3*(6), 963–972. doi:10.1109/TAI.2021.3133819

Syamala, M., Komala, C., Pramila, P., Dash, S., Meenakshi, S., & Boopathi, S. (2023). Machine Learning-Integrated IoT-Based Smart Home Energy Management System. In *Handbook of Research on Deep Learning Techniques for Cloud-Based Industrial IoT* (pp. 219–235). IGI Global. doi:10.4018/978-1-6684-8098-4.ch013

Tabaa, M., Monteiro, F., Bensag, H., & Dandache, A. (2020). Green Industrial Internet of Things from a smart industry perspectives. *Energy Reports, 6*, 430–446. doi:10.1016/j.egyr.2020.09.022

Talwani, S., Singla, J., Mathur, G., Malik, N., Jhanjhi, N. Z., Masud, M., & Aljahdali, S. (2022). Machine-Learning-Based Approach for Virtual Machine Allocation and Migration. *Electronics (Basel), 11*(19), 3249. doi:10.3390/electronics11193249

Tambe, A., & Shrawankar, U. (2014). Virtual batching approach for green computing. *International Conference for Convergence for Technology*-2014.

Taylor, R. (2023). *Worldwide data created 2023*. Retrieved October 5, 2023, from https://www.statista.com/statistics/871513/worldwide-data-created/: https://www.statista.com/statistics/871513/worldwide-data-created/

Terazono, A., Murakami, S., Abe, N., Inanc, B., Moriguchi, Y., Sakai, S.-i., Kojima, M., Yoshida, A., Li, J., Yang, J., Wong, M. H., Jain, A., Kim, I.-S., Peralta, G. L., Lin, C.-C., Mungcharoen, T., & Williams, E. (2006). Current status and research on E-waste issues in Asia. *Journal of Material Cycles and Waste Management, 8*(1), 1–12. doi:10.1007/s10163-005-0147-0

Tewari, A. a. (2020). *Security, privacy and trust of different layers in Internet-of-Things (IoTs) framework.* Elsevier. doi:10.1016/j.future.2018.04.027

Thangam, D., Malali, A. B., Subramaniyan, S. G., Mariappan, S., Mohan, S., & Park, J. Y. (2021). Blockchain Technology and Its Brunt on Digital Marketing. In Blockchain Technology and Applications for Digital Marketing (pp. 1-15). IGI Global. doi:10.4018/978-1-7998-8081-3.ch001

Thangam, D., Malali, A. B., Subramanian, G., Mohan, S., & Park, J. Y. (2022). Internet of things: a smart technology for healthcare industries. In *Healthcare Systems and Health Informatics* (pp. 3–15). CRC Press.

The E-Waste Guide. (2017). *ewaste guide.info: a knowledge base for the sustainable recycling of e-Waste-Image gallery.* Available: https://ewasteguide.info/images_galleries

The Global Renewables Outlook. (2020, April 24). *Global Renewables Outlook: Energy transformation 2050*. https://www.irena.org/publications/2020/Apr/Global-Renewables-Outlook-2020

The Shift Project. (2019). *Lean ICT Report*. Retrieved October 5, 2023, from https://theshiftproject.org/en/home/

Thirumalraj, A., & Balasubramanian, P. K. (n.d.). *Designing a Modified Grey Wolf Optimizer Based Cyclegan Model for Eeg Mi Classification in Bci*. Academic Press.

Thirumalraj, A., & Rajesh, T. (2023). *An Improved ARO Model for Task Offloading in Vehicular Cloud Computing in VANET*. Academic Press.

Thirumalraj, A., Asha, V., & Kavin, B. P. (2023). An Improved Hunter-Prey Optimizer-Based DenseNet Model for Classification of Hyper-Spectral Images. In *AI and IoT-Based Technologies for Precision Medicine* (pp. 76–96). IGI Global. doi:10.4018/979-8-3693-0876-9.ch005

Thite, M., Kavanagh, M. J., & Johnson, R. D. (2012). Evolution of human resource management and human resource information systems. *Introduction To Human Resource Management*, 2-34.

timesofindia.com. (2023, September 25). *Sustainable Data Centers: Paving the Path for a Greener Tomorrow*. https://timesofindia.indiatimes.com/gadgets-news/sustainable-data-centers-paving-the-path-for-a-greener-tomorrow/articleshow/103925381.cms

Tiyajamorn, P., Lorprasertkul, P., Assabumrungrat, R., Poomarin, W., & Chancharoen, R. (2019, November). Automatic trash classification using convolutional neural network machine learning. In *2019 IEEE International Conference on Cybernetics and Intelligent Systems (CIS) and IEEE Conference on Robotics, Automation and Mechatronics (RAM)* (pp. 71-76). IEEE. 10.1109/CIS-RAM47153.2019.9095775

Townend, P., Clement, S., Burdett, D., Yang, R., Shaw, J., Slater, B., & Xu, J. (2019, April). Improving data center efficiency through holistic scheduling in kubernetes. In *2019 IEEE International Conference on Service-Oriented System Engineering (SOSE)* (pp. 156-15610). IEEE. 10.1109/SOSE.2019.00030

Tripathy, S. S., Imoize, A. L., Rath, M., Tripathy, N., Bebortta, S., Lee, C.-C., Chen, T.-Y., Ojo, S., Isabona, J., & Pani, S. K. (2022). A novel edge-computing-based framework for an intelligent smart healthcare system in smart cities. *Sustainability (Basel)*, *15*(1), 735. doi:10.3390/su15010735

Tuli, S., Basumatary, N., Gill, S. S., Kahani, M., Arya, R. C., Wander, G. S., & Buyya, R. (2020). HealthFog: An ensemble deep learning based Smart Healthcare System for Automatic Diagnosis of Heart Diseases in integrated IoT and fog computing environments. *Future Generation Computer Systems*, *104*, 187–200. doi:10.1016/j.future.2019.10.043

Tuli, S., Gill, S. S., Xu, M., Garraghan, P., Bahsoon, R., Dustdar, S., Sakellariou, R., Rana, O., Buyya, R., Casale, G., & Jennings, N. R. (2022). HUNTER: AI based holistic resource management for sustainable cloud computing. *Journal of Systems and Software*, *184*, 111124. doi:10.1016/j.jss.2021.111124

Turner, C. J., Emmanouilidis, C., Tomiyama, T., Tiwari, A., & Roy, R. (2019). Intelligent decision support for maintenance: An overview and future trends. *International Journal of Computer Integrated Manufacturing*, *32*(10), 936–959. doi:10.1080/0951192X.2019.1667033

Tuysuz, M. F., & Trestian, R. (2020). From serendipity to sustainable green IoT: Technical, industrial and political perspective. *Computer Networks*, *182*, 107–469. doi:10.1016/j.comnet.2020.107469

Txture. (2022). *Txture Cloud Transformation YouTube channel*. Retrieved from https://www.youtube.com/@txturecloudtransformation

U.S. Environmental Protection Agency. (2023, May 17). *U.S. EPA's Energy Star Program Develops Energy-Saving Guidance for Co-Location Data Centers in Collaboration with Equinix and Iron Mountain.* https://www.epa.gov/newsreleases/us-epas-energy-star-program-develops-energy-saving-guidance-co-location-data-centers

United Nations Development Programme (UNDP). (2023). *Three ways digital transformation accelerates sustainable and inclusive development.* Retrieved from https://www.undp.org/blog/three-ways-digital-transformation-accelerates-sustainable-and-inclusive-development

uptimeinstitute.com. (2021, October 22). *2021 Data Center Industry Survey Results.* https://uptimeinstitute.com/2021-data-center-industry-survey-results

usgbc.org. (2023, November 14). *LEED-certified green buildings are better buildings.* https://www.usgbc.org/leed

Utilities One. (2023). *Demystifying the impact of AI on energy management.* Retrieved October 5, 2023, from https://utilitiesone.com/demystifying-the-impact-of-ai-on-energy-management: https://utilitiesone.com/demystifying-the-impact-of-ai-on-energy-management

Vailshery, A. (2023). *Internet of Things (IoT) Connected Devices Worldwide 2022-2030.* Retrieved October 5, 2023, from https://www.statista.com/statistics/1183457/iot-connected-devices-worldwide/

Van Heddeghem, W., Lambert, S., Lannoo, B., Colle, D., Pickavet, M., & Demeester, P. (2014). Trends in worldwide ICT electricity consumption from 2007 to 2012. *Computer Communications, 50,* 64–76. doi:10.1016/j.comcom.2014.02.008

Veerendra Mulay. (2018, June 5). *StatePoint Liquid Cooling system: A new, more efficient way to cool a data center.* https://engineering.fb.com/2018/06/05/data-center-engineering/statepoint-liquid-cooling/

Velis, C. A., Wilson, D. C., Rocca, O., Smith, S. R., Mavropoulos, A., & Cheeseman, C. R. (2012). An analytical framework and tool ('InteRa') for integrating the informal recycling sector in waste and resource management systems in developing countries. *Waste Management & Research, 30*(9_suppl), 43–66. doi:10.1177/0734242X12454934 PMID:22993135

Venkateswaran, N., Vidhya, K., Ayyannan, M., Chavan, S. M., Sekar, K., & Boopathi, S. (2023). A Study on Smart Energy Management Framework Using Cloud Computing. In 5G, Artificial Intelligence, and Next Generation Internet of Things: Digital Innovation for Green and Sustainable Economies (pp. 189–212). IGI Global. doi:10.4018/978-1-6684-8634-4.ch009

Verma, P., Tiwari, R., Hong, W. C., Upadhyay, S., & Yeh, Y. H. (2022). FETCH: A deep learning-based fog computing and IoT integrated environment for healthcare monitoring and diagnosis. *IEEE Access : Practical Innovations, Open Solutions, 10,* 12548–12563. doi:10.1109/ACCESS.2022.3143793

Verwiebe, P. A., Seim, S., Burges, S., Schulz, L., & Müller-Kirchenbauer, J. (2021). Modeling Energy Demand—A Systematic Literature Review. *Energies, 14*(23), 7859. doi:10.3390/en14237859

Vhora, F., & Gandhi, J. (2020, March). A comprehensive survey on mobile edge computing: challenges, tools, applications. In *2020 fourth international conference on computing methodologies and communication (ICCMC)* (pp. 49-55). IEEE. 10.1109/ICCMC48092.2020.ICCMC-0009

Vilela, P. H., Rodrigues, J. J., Righi, R. da R., Kozlov, S., & Rodrigues, V. F. (2020). Looking at fog computing for e-health through the lens of deployment challenges and applications. *Sensors (Basel), 20*(9), 2553. doi:10.3390/s20092553 PMID:32365815

Vo, A. H., Vo, M. T., & Le, T. (2019). A novel framework for trash classification using deep transfer learning. *IEEE Access : Practical Innovations, Open Solutions, 7,* 178631–178639. doi:10.1109/ACCESS.2019.2959033

Voskergian, D., & Ishaq, I. (2023). *Smart E-waste Management System Utilizing Internet of Things and Deep Learning Approaches*. Academic Press.

Wadhwa, B., & Verma, A. (2014). Energy saving approaches for Green Cloud Computing. *Recent Advances in Engineering and Computational Sciences*.

Wang, X., Goyal, V., Yu, J., Bertacco, V., Boutros, A., Nurvitadhi, E., . . . Das, R. (2021, May). Compute-capable block RAMs for efficient deep learning acceleration on FPGAs. In *2021 IEEE 29th Annual International Symposium on Field-Programmable Custom Computing Machines (FCCM)* (pp. 88-96). IEEE. 10.1109/FCCM51124.2021.00018

Wang, D., Zhong, D., & Souri, A. (2021). Energy management solutions in the Internet of Things applications: Technical analysis and new research directions. *Cognitive Systems Research*, *67*, 33–49. doi:10.1016/j.cogsys.2020.12.009

Wang, F., Huisman, J., Stevels, A., & Balde, C. P. (2013). Enhancing e-waste estimates: Improving data quality by multivariate Input-Output Analysis. *Waste Management (New York, N.Y.)*, *33*(11), 2397–2407. doi:10.1016/j.wasman.2013.07.005 PMID:23899476

Wang, W., Tian, Y., Zhu, Q., & Zhong, Y. (2017). Barriers for household e-waste collection in China: Perspectives from formal collecting enterprises in Liaoning Province. *Journal of Cleaner Production*, *153*, 299–308. doi:10.1016/j.jclepro.2017.03.202

Wang, X., Han, Y., Leung, V. C., Niyato, D., Yan, X., & Chen, X. (2020). Convergence of edge computing and deep learning: A comprehensive survey. *IEEE Communications Surveys and Tutorials*, *22*(2), 869–904. doi:10.1109/COMST.2020.2970550

Wang, Z., Zhang, B., & Guan, D. (2016). Take responsibility for electronic-waste disposal. *Nature*, *536*(7614), 23–25. doi:10.1038/536023a PMID:27488785

Weiss, K., Khoshgoftaar, T. M., & Wang, D. (2016). A survey of transfer learning. *Journal of Big Data*, *3*(1), 1–40. doi:10.1186/s40537-016-0043-6

Whitney Pipkin. (2022, October 21). *Data center decisions could have big land use impacts in Virginia's Prince William County*. https://www.bayjournal.com/news/growth_conservation/data-center-decisions-could-have-big-land-use-impacts-in-virginia-s-prince-william-county/article_51ef20a2-5166-11ed-9409-b386158a70c3.html

Widmer, R., Oswald-Krapf, H., Sinha-Khetriwal, D., Schnellmann, M., & Böni, H. (2005). Global perspectives on e-waste. *Environmental Impact Assessment Review*, *25*(5), 436–458. doi:10.1016/j.eiar.2005.04.001

Wilson, D. C., Acun, F., Jana, S., Ardanaz, F., Eastep, J. M., Paschalidis, I. C., & Coskun, A. K. (2023, November). An End-to-End HPC Framework for Dynamic Power Objectives. In *Proceedings of the SC'23 Workshops of The International Conference on High Performance Computing, Network, Storage, and Analysis* (pp. 1801-1811). 10.1145/3624062.3624262

Wong, M. H., Wu, S. C., Deng, W. J., Yu, X. Z., Luo, Q., Leung, A. O. W., Wong, C. S. C., Luksemburg, W. J., & Wong, A. S. (2007). Export of toxic chemicals – a review of the case of uncontrolled electronic-waste recycling. *Environmental Pollution*, *149*(2), 131–140. doi:10.1016/j.envpol.2007.01.044 PMID:17412468

Wong, T. H., Rogers, B. C., & Brown, R. R. (2020). Transforming cities through water-sensitive principles and practices. *One Earth*, *3*(4), 436–447. doi:10.1016/j.oneear.2020.09.012

World Bank. (2012). *What a Waste: A Global Review of Solid Waste Management Urban Development Series Knowledge Papers*. Academic Press.

World Health Organization (WHO). (2023). *Electronic waste (e-waste) fact sheet*. Retrieved from https://www.who.int/news-room/fact-sheets/detail/electronic-waste-(e-waste)

Wu, C. J., Raghavendra, R., Gupta, U., Acun, B., Ardalani, N., Maeng, K., ... Hazelwood, K. (2022). Sustainable ai: Environmental implications, challenges and opportunities. *Proceedings of Machine Learning and Systems*, *4*, 795–813.

Wu, D., Liu, S., Zhang, L., Terpenny, J., Gao, R. X., Kurfess, T., & Guzzo, J. A. (2017). A fog computing-based framework for process monitoring and prognosis in cyber-manufacturing. *Journal of Manufacturing Systems*, *43*, 25–34. doi:10.1016/j.jmsy.2017.02.011

Xu, M. a. (2019). Optimized renewable energy use in green cloud data centers. Springer, 314330.

Xu, L. D., Xu, E. L., & Li, L. (2018). Industry 4.0: State of the art and future trends. *International Journal of Production Research*, *56*(8), 2941–2962. doi:10.1080/00207543.2018.1444806

Xu, M. T. (2020). A self-adaptive approach for managing applications and harnessing renewable energy for sustainable cloud computing. *IEEE Transactions on Sustainable Computing*, 544–558.

Yabanci, O. (2019). From human resource management to intelligent human resource management: A conceptual perspective. *Human-Intelligent Systems Integration*, *1*(2-4), 101–109. doi:10.1007/s42454-020-00007-x

Yadav, M., & Mishra, A. (2023). An enhanced ordinal optimization with lower scheduling overhead based novel approach for task scheduling in cloud computing environment. *Journal of Cloud Computing (Heidelberg, Germany)*, *12*(1), 1–14. doi:10.1186/s13677-023-00392-z

Yamini, B., & Vetri Selvi, D. (2010). Cloud virtualization: A potential way to reduce global warming. *Recent Advances in Space Technology Services and Climate Change 2010 (RSTS & CC-2010)*.

Yang, Y., Luo, X., Chu, X., Zhou, M.-T., Yang, Y., Luo, X., Chu, X., & Zhou, M.-T. (2020). Fog-enabled intelligent transportation system. *Fog-Enabled Intelligent IoT Systems*, 163–184.

Yang, G., Li, P., Xiao, K., He, Y., Xu, G., Wang, C., & Chen, X. (2023). An Efficient Attribute-Based Encryption Scheme with Data Security Classification in the Multi-Cloud Environment. *Electronics (Basel)*, *12*(20), 4237. doi:10.3390/electronics12204237

Yang, H. Q., Lu, T., Guo, W. J., Chang, S., & Song, J. F. (2020, January). RecycleTrashNet: Strengthening Training Efficiency for Trash Classification Via Composite Pooling. In *Proceedings of the 2020 2nd International Conference on Advanced Control, Automation and Artificial Intelligence (ACAAI 2020), Wuhan, China* (pp. 11-13). 10.12783/dtetr/acaai2020/34190

Yang, Y., Luo, X., Chu, X., & Zhou, M.-T. (2020). *Fog-enabled intelligent IoT systems*. Springer. doi:10.1007/978-3-030-23185-9

Yazdani, A., Dashti, S. F., & Safdari, Y. (2023). A fog-assisted information model based on priority queue and clinical decision support systems. *Health Informatics Journal*, *29*(1). doi:10.1177/14604582231152792 PMID:36645733

Yevgeniy Sverdlik. (2020, February 27). *Study: Data Centers Responsible for 1 Percent of All Electricity Consumed Worldwide*. https://www.datacenterknowledge.com/energy/study-data-centers-responsible-1-percent-all-electricity-consumed-worldwide#close-modal

Yuan, L., He, Q., Tan, S., Li, B., Yu, J., Chen, F., ... Yang, Y. (2021, April). Coopedge: A decentralized blockchain-based platform for cooperative edge computing. In *Proceedings of the Web Conference 2021* (pp. 2245-2257). 10.1145/3442381.3449994

Yu, P., Zhou, F., Zhang, X., Qiu, X., Kadoch, M., & Cheriet, M. (2020). Deep learning-based resource allocation for 5G broadband TV service. *IEEE Transactions on Broadcasting*, *66*(4), 800–813. doi:10.1109/TBC.2020.2968730

Yu, Y. (2020, August). A computer vision based detection system for trash bins identification during trash classification. *Journal of Physics: Conference Series, 1617*(1), 012015. doi:10.1088/1742-6596/1617/1/012015

Zahra, S.B., Khan, M.A., Abbas, S., Khan, K.M., Al-Ghamdi, M.A., & Almotiri, S.H. (2021). *Marker-Based and Marker-Less Motion Capturing Video Data: Person and Activity Identification Comparison Based on Machine Learning Approaches.* Academic Press.

Zeng, X., Duan, H., Wang, F., & Li, J. (2017). Examining environmental management of ewaste: China's experience and lessons. *Renewable & Sustainable Energy Reviews, 72*, 1076–1082. doi:10.1016/j.rser.2016.10.015

Zeng, X., Mathews, J. A., & Li, J. (2018). *Urban Mining of E-Waste Is Becoming More Cost-Effective Than Virgin Mining.* Academic Press.

Zhang, X., Wang, Y., Lu, S., Liu, L., & Shi, W. (2019, July). OpenEI: An open framework for edge intelligence. In *2019 IEEE 39th International Conference on Distributed Computing Systems (ICDCS)* (pp. 1840-1851). IEEE.

Zhang, A. N., Chu, S. C., Song, P. C., Wang, H., & Pan, J. S. (2022). Task scheduling in cloud computing environment using advanced phasmatodea population evolution algorithms. *Electronics (Basel), 11*(9), 1451. doi:10.3390/electronics11091451

Zhang, H., Tian, Y., Tian, C., & Zhai, Z. (2023). Effect of key structure and working condition parameters on a compact flat-evaporator loop heat pipe for chip cooling of data centers. *Energy, 284*, 128658. doi:10.1016/j.energy.2023.128658

Zhang, Q. a. (2021). A survey on data center cooling systems: Technology, power consumption modeling and control strategy optimization. *Journal of Systems Architecture*, 102253.

Zhang, Q., Meng, Z., Hong, X., Zhan, Y., Liu, J., Dong, J., Bai, T., Niu, J., & Deen, M. J. (2021). A survey on data center cooling systems: Technology, power consumption modeling and control strategy optimization. *Journal of Systems Architecture, 119*, 102253. doi:10.1016/j.sysarc.2021.102253

Zhang, Q., Yang, Q., Zhang, X., Bao, Q., Su, J., & Liu, X. (2021). Waste image classification based on transfer learning and convolutional neural network. *Waste Management (New York, N.Y.), 135*, 150–157. doi:10.1016/j.wasman.2021.08.038 PMID:34509053

Zhang, Q., Zhang, X., Mu, X., Wang, Z., Tian, R., Wang, X., & Liu, X. (2021). Recyclable waste image recognition based on deep learning. *Resources, Conservation and Recycling, 171*, 105636. doi:10.1016/j.resconrec.2021.105636

Zhang, S., & Forssberg, E. (1997). Mechanical separation-oriented characterization of electronic scrap. *Resources, Conservation and Recycling, 21*(4), 247–269. doi:10.1016/S0921-3449(97)00039-6

Zhang, T., Lei, C., Zhang, Z., Meng, X. B., & Chen, C. P. (2021). AS-NAS: Adaptive scalable neural architecture search with reinforced evolutionary algorithm for deep learning. *IEEE Transactions on Evolutionary Computation, 25*(5), 830–841. doi:10.1109/TEVC.2021.3061466

Zhang, X., Zhong, H., Cui, J., Bolodurina, I., & Liu, L. (2022). Lbvp: A lightweight batch verification protocol for fog-based vehicular networks using self-certified public key cryptography. *IEEE Transactions on Vehicular Technology, 71*(5), 5519–5533. doi:10.1109/TVT.2022.3157960

Zhong, H., Chen, L., Cui, J., Zhang, J., Bolodurina, I., & Liu, L. (2021). Secure and Lightweight Conditional Privacy-Preserving Authentication for Fog-Based Vehicular Ad Hoc Networks. *IEEE Internet of Things Journal, 9*(11), 8485–8497. doi:10.1109/JIOT.2021.3116039

Zhou, E. P. (2022). Machine Learning For The Classification And Separation Of E-Waste. In *2022 IEEE MIT Undergraduate Research Technology Conference (URTC)* (pp. 1-5). IEEE. 10.1109/URTC56832.2022.10002242

Zhou, H., Jiang, K., Liu, X., Li, X., & Leung, V. C. (2021). Deep reinforcement learning for energy-efficient computation offloading in mobile-edge computing. *IEEE Internet of Things Journal, 9*(2), 1517–1530. doi:10.1109/JIOT.2021.3091142

Zhou, X., Wang, R., Wen, Y., & Tan, R. (2021). Joint IT-facility optimization for green data centers via deep reinforcement learning. *IEEE Network, 35*(6), 255–262. doi:10.1109/MNET.011.2100101

Zhou, Z., Liao, H., Wang, X., Mumtaz, S., & Rodriguez, J. (2020). When vehicular fog computing meets autonomous driving: Computational resource management and task offloading. *IEEE Network, 34*(6), 70–76. doi:10.1109/MNET.001.1900527

Zong, Z. (2020). An Improvement of Task Scheduling Algorithms for Green Cloud Computing. In *2020 15th International Conference on Computer Science & Education (ICCSE)* (pp. 654-657). 10.1109/ICCSE49874.2020.9201785

About the Contributors

K. Dinesh Kumar is working as an Assistant Professor in the School of Computing, Amrita Vishwa Vidyapeetham University. Ph.D. degree received from the School of Computer Science and Engineering, VIT University. Published international publications and participated in various international and national conferences. Active review member for various reputed leading research journals like IEEE transactions, Springer, Wiley, IGI Global, Taylor & Francis Online, Hindawi etc. and also acted as review member for international conferences like MBDAS-2021, Shenzhen, China and ICMEMSCE-2018, Xiamen, China. Co-author of three textbooks. Certified AWS Solution Architect (Associate). Research area is Cloud computing and Machine learning also have interests in Fog computing and Computer networks.

Vijayakumar Varadarajan is currently a Program Leader fir Engineering in Ajeenkya DY Patil University India and also an Adjunct Professor in School of Computer Science and Engineering, University of New South Wales, Sydney, Australia. He is also a Visiting Postdoc Scientist in Centro de Tecnologia, Federal University of Piauí, Brazil. He was a Professor and Associate Dean for School of Computing Science and Engineering at VIT University, Chennai, India. He has more than 21 years of experience including industrial and institutional. He also served as a Team Lead in industries like Satyam, Mahindra Satyam and Tech Mahindra for several years. He has completed Diploma with First Class Honors. He has completed BE CSE and MBA HRD with First Class. He has also completed ME CSE with First Rank Award. He has completed his PhD from Anna University in 2012. He has published many articles in national and international level journals/conferences/books. He is a reviewer in IEEE Transactions, Inderscience and Springer Journals. He has initiated a number of international research collaborations with universities in Europe, Australia, Africa, Malaysia, Singapore and North & South America. He had also initiated joint research collaboration between VIT University and various industries. He is also the Lead Guest Editor for few journals in Inderscience, Springer, Elsevier, IOS, UM and IGI Global. He also organized several international conferences and special sessions in USA, Vietnam, Africa, Malaysia and India including ARCI, IEEE, ACSAT, ISRC, ISBCC, ICBCC etc. His research interests include computational areas covering grid computing, cloud computing, computer networks, cyber security and big data. He received his university-level Best Faculty Award for 2015–2016. He is also a member of several national and international professional bodies including IFSA, EAI, BIS, ISTE, IAENG, CSTA, IEA, etc.

Nidal Nasser received his B.Sc. and M.Sc. degrees with Honors in Computer Engineering from Kuwait University, State of Kuwait, in 1996 and 1999, respectively. He completed his Ph.D. in the School of Computing at Queen's University, Kingston, Ontario, Canada, in 2004. He is currently an Associate Professor and Chairman of Electrical and Computer Engineering Department at Alfaisal University, Saudi Arabia. He worked in the School of Computer Science at University of Guelph, Guelph, Ontario,

Canada (2004-2011). Dr. Nasser was the founder and Director of the Wireless Networking and Mobile Computing Research Lab @ Guelph (WiNG:). He has authored 129 journal publications, refereed conference publications and book chapters in the area of wireless communication networks and systems. He has also given tutorials in major international conferences. Dr. Nasser is currently serving as an associate editor of Wiley's International Journal of Wireless Communications and Mobile Computing, Wiley's International Journal on Communication Systems, Wiley's Security and Communication Networks Journal and International Journal of Ad Hoc & Sensor Wireless Networks. He has been a member of the technical program and organizing committees of several international IEEE conferences and workshops. Dr. Nasser is a member of several IEEE technical committees. He received Fund for Scholarly and Professional Development Award in 2004 from Queen's University. He received the Computing Faculty Appreciation Award from the University of Guelph-Humber. He received the Best Research Paper Award at the ACS/IEEE International Conference on Computer Systems and Applications (AICCSA'08), at the International Wireless Communications and Mobile Computing Conference (IWCMC'09), at the International Wireless Communications and Mobile Computing Conference (IWCMC'11), and at the International Conference on Computing, Management and Telecommunications (ComManTel'13).

Ravi Kumar Poluru received M.Tech (Software Engineering) from JNTU Anantapur, In 2014. He completed his Ph.D. degree (Internet of Things) in the School of Computer Science and Engineering from the Vellore Institute of Technology (Deemed by the university), Vellore, In 2021. He has 10 years of teaching and research experience as a Junior Research Fellow and Senior Research Fellow from the Ministry of Tribal Affairs, Government of India. Currently, he is an Associate Professor in the Department of Information Technology, Institute of Aeronautical Engineering, Hyderabad. His current research interests include the Internet of Things, Machine Learning, Wireless Sensor Networks, Optimization Techniques, and Nature Inspired Optimization Techniques. He has 20+ publications, including 13-International Journal research papers (SCI and IEEE Transactions) and conferences (International) -2, Book Chapters-6 (Scopus), Patents- 2 Granted and textbooks - 2. He is the reviewer for the E-Learning and Digital Media (SAGE) and WILEY Publications, Conducted Guest Lecture in various Engineering Colleges in India. He received Award for this contribution towards the number of publications in the year 2018.

* * *

A. V. Kalpana is an accomplished academician and researcher, currently serving as an Assistant Professor in the Department of Data Science and Business Systems at the School of Computing, SRM Institute of Science & Technology, Kattankulathur, Chennai. She completed her Bachelor's degree in Computer Science and Engineering in 2004 from the University of Madras, showcasing her foundational understanding of computer science. Further enhancing her expertise, she pursued a Master's degree in Computer Science and Engineering from Anna University. She holds a Ph.D. degree from Anna University, Chennai, reflecting her dedication to advancing knowledge in her field. Her research contributions are evident through numerous publications in reputable journals and international conference proceedings. Driven by a passion for knowledge, her research interests span across Machine Learning, Deep Learning, Wireless Sensor Networks, and the Internet of Things (IoT).

Sahaya Beni Prathiba B. received Bachelor's and Master's degrees in Computer Science and Engineering from Anna University, Chennai. I have secured 23rd rank among 2581 candidates in Master of Engineering. Currently, she is working as an Assistant Professor at the Centre for Cyber Physical Systems, School of Computer Science and Engineering, Vellore Institute of Technology, Chennai. She was also a recipient of the Anna Centenary Research Fellowship. Her research interest includes 5G/6G, Vehicle-to-Everything, Software Defined Networking, Autonomous Vehicular Networks, Industry 5.0, and Metaverse. She is serving as a reviewer for IEEE TRANSACTIONS and reputed Elsevier journals.

Syed Muzamil Basha is a distinguished professor at the REVA University, Bangalore, and Karnataka, India. He completed his postgraduate degree in Computer Science from the Vellore Institute of Technology at Vellore in 2011 and his Ph.D. in Sentiment Analysis from the Vellore Institute of Technology at Vellore in 2020. He has published more than 85 papers in refereed conferences and journals, and has applied for or been granted more than 10 patents. He is author or editor of 20 book chapters & books, including textbooks on Artificial Intelligence, Pattern Recognition, Machine Learning and Deep Learning. He is currently acting as Industry based Sponsored Researcher, University of Ha'il, KSA. Because of the quality publications made as full time research scholar, he has twice been designated as best researcher at Vellore Institute of Technology. Recently, achieved young researcher award by Computer Vision, High Performance Computing, Smart Devices and Networks (CHSN-2022). Conducted Guest Lecture in various Engineering Colleges in India. Aside from serving as program or general chair of many major conferences in Machine Learning and Computation Intelligence. He is an Executive editor of the journal computational Learning & Intelligence. He is a senior member of IEEE and Life time Member of ISTE.

Sampath Boopathi is an accomplished individual with a strong academic background and extensive research experience. He completed his undergraduate studies in Mechanical Engineering and pursued his postgraduate studies in the field of Computer-Aided Design. Dr. Boopathi obtained his Ph.D. from Anna University, focusing his research on Manufacturing and optimization. Throughout his career, Dr. Boopathi has made significant contributions to the field of engineering. He has authored and published over 160 research articles in internationally peer-reviewed journals, highlighting his expertise and dedication to advancing knowledge in his area of specialization. His research output demonstrates his commitment to conducting rigorous and impactful research. In addition to his research publications, Dr. Boopathi has also been granted one patent and has three published patents to his name. This indicates his innovative thinking and ability to develop practical solutions to real-world engineering challenges. With 17 years of academic and research experience, Dr. Boopathi has enriched the engineering community through his teaching and mentorship roles. He has served in various engineering colleges in Tamilnadu, India, where he has imparted knowledge, guided students, and contributed to the overall academic development of the institutions. Dr. Sampath Boopathi's diverse background, ranging from mechanical engineering to computer-aided design, along with his specialization in manufacturing and optimization, positions him as a valuable asset in the field of engineering. His research contributions, patents, and extensive teaching experience exemplify his expertise and dedication to advancing engineering knowledge and fostering innovation.

Chandan Chavadi is the Dean & Professor of the Presidency Business School, Presidency College, Bengaluru. A PhD from Karnatak University, Dharwad, and an MBA (Mktg.) He holds a primary degree in B.E. (E&C). He has two years of corporate experience before his moving to academics. He has been in academics for the last 21 years. He has 32 papers & 2 book reviews to his credit, published in reputed journals and magazines such as ABDC-C journals, Web of Science, Scopus, UGC Care list and other indexed journals. Under Google scholar indices, he has total citations of 90, h-index of 6 & i10-index of 4. His paper has been accepted for publication in the IIM –A Vikalpa. Five of his research papers were recently published in the IIM Kozhikode Society And Management Review journal, IIM-Shillong Journal of Management Science, Business Perspective & Research journal of K J Somaiya Institute of Management, the MDI journal "Vision", and IIM-Lucknow journal "Metamorphosis". He is the recipient of Labdhi Bhandari Best Paper Award for the 7th IIM-A International Marketing Conference held on 11th to 13th Jan 2017. He is a recognized PhD Guide in Management for Bangalore City University.

Vinod D. has around 18 years of experience in teaching and research in various Institutes under the department of Computer Science and Engineering. Completed his Ph.D. in the Domain of Information Security in the year 2018, with highly commendable reviews at Sathyabama Institute of Science and Technology. Completed his Masters of Engineering, in the department of Computer Science and Engineering at GKM Engineering College, Anna University in 2009. Completed the Bachelor of Technology in Information Technology in Adhiyamaan Engineering College, Anna University. He is currently acting as Scientific Advisor in Ja-Assure Singapore. He has published around 3 Science indexed and 30 Scopus Indexed Journals and more than 25 International Conference presentations across prestigious Universities of India and other Countries. Has an h-index of 3, Reviewer for 6 Scopus indexed journals, and 3 Science indexed journal. Presently working as Assistant Professor in Department of Computing Technologies, School of Computing, College of Engineering and Technology, SRM Institute of Science and Technology, SRM Nagar, Kattankulathur, Chennai, TN, India.

Premalatha G. received her Bachelor's of Engineering degree in Computer Science and Engineering from Anna University 2006, Master's of Engineering degree in Computer Science and Engineering from Anna University 2010.She obtained her doctorate degree from Anna University 2021.She is currently working as an Assistant Professor in the Department of Data Science and Business Systems, SRMIST, Kattankulathur, Chennai. Her area of specialization includes image processing, Video Surveillance, Machine Learning.

Shivnath Ghosh is a professor at Department of Computer Science & Engineering at Brainware University.

Sam Goundar is an Editor-in-Chief of the International Journal of Blockchains and Cryptocurrencies (IJFC) – Inderscience Publishers, Editor-in-Chief of the International Journal of Fog Computing (IJFC) – IGI Publishers, Section Editor of the Journal of Education and Information Technologies (EAIT) – Springer and Editor-in-Chief (Emeritus) of the International Journal of Cloud Applications and Computing (IJCAC) – IGI Publishers. He is also with more than 20 high impact factor journals. As a researcher, apart from Blockchains, Cryptocurrencies, Fog Computing, Mobile Cloud Computing and Cloud Computing, Dr. Sam Goundar also researches in Educational Technology, Artificial Intelligence,

ICT in Climate Change, ICT Devices in the Classroom, Using Mobile Devices in Education, and e-Government. He has published on all these topics. He was a Research Fellow with the United Nations University. He is a Senior Lecturer in CS at British University Vietnam, Adjunct Senior Lecturer at The University of the South Pacific, Affiliate Professor of Information Technology at Pontificia Universidad Catolica Del Peru, and Adjunct Professor at The University of Fiji.

Arthi K., a distinguished academician, earned an M.Tech in 2005, laying the foundation for a prolific academic journey. In 2015, they earned a Ph.D., solidifying expertise in their research areas. With over 15 years of teaching and research experience in the dynamic realms of IoT and Artificial Intelligence she has imparted knowledge and guidance to aspiring minds. She has accountability of more than 30 reputed publications and 1 design grant. She is a active member in ACM and ISTE professional societies.

Arikumar K. S. is currently working as an Assistant Professor (Senior Grade) in the School of Computer Science and Engineering (SCOPE) at VIT-AP University, Amaravati, Andhra Pradesh, India. He completed his Bachelor's, Master's degree in Computer Science and Engineering and Ph.D. under the Faculty of Information and Communication Engineering from Anna University, Chennai. His current research interests include Wireless Sensor Networks, Health Care, Autonomous Vehicle and IoT. He has published more than 30 Conferences and Journals papers. He has 12 years of teaching and 10 years of research experience.

Gudivada Lokesh is an experienced academician with a focus in the field of Computer Science & Engineering. With 11 years of experience, he has proven his expertise in guiding students, Undergraduate and Post Graduate levels in their academics, research projects and research papers. Lokesh is currently working as an Associate Professor at Vemu Institute of Technology (Affiliated to JNTUA) Chittoor. Lokesh's extensive academic background includes a Master's degree in Computer Science & Engineering from Sathyabama University Chennai. Presently he is pursuing Ph.D in Computer Science & Engineering from Mohan Babu University, Tirupathi, and Under the Supervision of Dr K K Baseer. Lokesh's teaching and research interests include Networks, Cloud Computing Environment, Grid Computing, Mobile Computing, and Computer Organization. Lokesh is committed to providing quality education to his students and his passion for teaching has earned his recognition in the field of academics.

Haritha M. is having 13years of teaching and research experience in commerce and management at various institutions. She also performed various administrative roles and responsibilities such as Program Coordinator, Examination in charge and IQAC Coordinator. She has authored 12 books in commerce and management and committee member for various research associations.

M. Ramprasath is an accomplished academician and researcher, currently serving as an Assistant Professor in the Department of Data Science and Business Systems at the School of Computing, SRM Institute of Science & Technology, Kattankulathur, Chennai. She completed her bachelor's degree in computer science and engineering in 2006 from anna University of chenni, showcasing her foundational understanding of computer science. Further enhancing her expertise, she pursued a Master's degree in information Technology from Sathyabama University. She holds a Ph.D. degree from Sathyabama University, Chennai, reflecting her dedication to advancing knowledge in her field. Her research contributions are evident through numerous publications in reputable journals and international conference proceedings. Driven by a passion for knowledge, her research interests span across Question Answering system, Semantic web, Machine Learning, Deep Learning, Wireless Sensor Networks, and the Internet of Things (IoT).

Thirupathi Manickam, M.Com, M.Phil, B.Ed, TN-SET, KSET, Ph.D., is presently working as an Assistant professor in the Department of Professional Studies at Christ University, Bangalore. It is one of the leading institutions in Bangalore, Karnataka, and the institution is Accredited by NIRF, NBA and NAAC Accredited university. He has more than seven years of teaching and Research experience. He has 66 citations and four h-Index. He has published 27 research papers in Scopus, Web of Science, UGC-CARE, and UGC-approved and leading international journals, and 15 presented papers at national and international conferences. He has also participated in over 50 seminars, conferences, FDP & workshops at the National and International Levels. His areas of expertise are Financial Accounting, Corporate Accounting, Financial Management, Management Accounting, Taxation, Digital Marketing and Technology Management.

Badrinath N. is working as Associate Professor in Department of Computer Science and Engineering, Vellore Institute of Technology, Vellore. He has 19 years of opulent experience in Teaching, research and Industrial Exposure with Research activities. He has published several papers in international & national journals and conferences and delivered guest lectures in many national & international seminars. He has completed BE (CSE) from Bharathidasan University, M. Tech (Advanced Computing) from SASTRA university and PhD in soft computing from Bharathidasan University.

Manikandan N. obtained his Bachelors of Engineering degree in Computer Science Engineering from Anna University 2005. Then he obtained his Master's of engineering degree in in Computer Science Engineering from Anna University 2011. He completed his PhD in Sathyabama Institute of Science and Technology 2020, He is currently working an Assistant Professor Department of Data science and Business Systems, SRMIST, Katankulathur, Chennai. His specializations include cloud computing, Operating systems, Big data.

Ramakrishna Narasimhaiah is an assistant professor at Jain University, a prestigious institution, boasting a wealth of knowledge and expertise acquired over a span of fourteen years in the realm of education, encompassing roles as a lecturer, researcher, and avid learner. Drawing from a background in economics, particularly in the domains of finance, insurance, and banking, this analysis aims to elucidate the key facets pertaining to the development of young individuals.

Pabitha Parameshwaran received her Bachelor of Engineering from University of Madras and Masters in Engineering from Anna University in Computer Science and Engineering. She received her Doctor of Philosophy from Anna University, Chennai. She is presently working as an Assistant professor in the Department of Computer Technology at Madras Institute of Technology campus, Anna University, Chennai. She has nearly twenty years of Academic and Industry experience. She is an active researcher in the field of Computer Science especially Web technologies, Machine Intelligence, and Software Engineering. She has several research publications in various International Journals and Conferences which are indexed in IEEE.

Anto Arockia Rosaline R., Assistant Professor in the Department of Computing Technologies, SRM Institute of Science and Technology, Kattankulathur Campus, has more than 19 years of experience in teaching and research. She received PhD in Computer Science from VIT University. Her areas of interests are Data Analytics, Data Science, Machine Learning, Text Analytics. She has guided a number of U.G. and P.G. projects. She has presented papers in various international conferences and published number of papers in refereed International Journals. She has delivered invited talk in various forums. She is a member of IEI and Computer society of India. She has received the Active participation Women award. She has also received three Best paper awards for the Oral presentation in various conferences.

Sankar Ganesh R. served in various reputed institutions. Currently employed as an Associate Professor at Vel Tech Rangarajan Dr. Sagunthala R&D Institute of Science and Technology, Chennai. His areas of specialisation are marketing and human resource management. He has 11 years of teaching experience and 4 years of industry experience. He has published Two design patent with IPR India. In addition to acting as a resource person in various conferences and FDPs, he has presented papers in many forums, participated in various national and international seminars and conferences, attended various faculty development programs, and published various research papers in Scopus, Web of Science, and UGC Care listed journals. He also guided Ph.D. scholars and postgraduate students in their research work.

Piyal Roy is an Assistant Professor at department of Computer Science & Engineering at Brainware University.

B. Rupa Devi, Associate Professor in Dept of CSE and at present, heading the Department of Master of Computer Applications in AITS, Tirupati, is pursuing her PhD in Dept of CSE, SVUCE, SVU, Tirupati under the guidance of Dr Ch D V Subba Rao. Her field of specialization is Data Mining, Machine Learning and Data Analytics. She has published 15 International Journals and has co-authored two books. She has 20 years of teaching experience.

Suchitra S. received her B.E. degree from the University of Madras, Chennai, India in 1997. M.E., degree from Anna University, Chennai, India in 2009 and received her Ph.D., from Anna University, Chennai, India in 2017. She has 26 years of teaching experience and currently she is working as an Associate Professor in the Department of Data Science and Business Systems, SRM Institute of Science and Technology, Kattankulathur, Chengalpattu District, Tamil Nadu, India. She has published more than 50 research papers in reputed international journals and conferences. Her research interest is Computer Vision, Big Data, Artificial Intelligence, Machine Learning, Deep Learning, and IoT.

Anil Sharma is an educator and researcher with over 15 years of experience in teaching management courses and guiding research. He holds a Ph.D. in Management from Kavayitri Bahinabai Chaudhari North Maharashtra University. His doctoral research focused on studying the impact of organized retailing on consumer buying behavior. Education - Ph.D., Management - Kavayitri Bahinabai Chaudhari North Maharashtra University - Post Graduate Diploma in Applied Statistics - IGNOU - MBA, Marketing Management - BE, Computer Engineering Teaching Expertise - Core Areas: Operations Management, Sales Management, Business Research Methods, Quantitative Techniques, Entrepreneurship - Secondary Areas: Marketing Management, Services Marketing, Global Marketing Research Interests - Consumer Behavior and Retail Management - Technology Adoption and Innovation Diffusion - Entrepreneurship and Small Business Growth Publications - 2 Scopus indexed papers, 12 UGC CARE listed journals, 20 peer-reviewed journals - 6 Edited books and book chapters on innovation, sustainability, rural development Academic Experience - Currently Assistant Professor, Parul Institute of Management and Research, Vadodara - Previously taught at institutes like Sinhgad College, G.H. Raisoni Institute and SKIPS University - PhD Evaluator and Paper Setter for North Maharashtra University - Developed teaching methodologies, mentored students, contributed to NAAC accreditation Academic Achievements - Received best paper awards at various international conferences - Awarded minor research project grant of Rs. 50,000 by North Maharashtra University - Appreciated for faculty development and student counselling initiatives.

Rajalakshmi Shenbaga Moorthy received her B.Tech under the stream of Information Technology from Mookambigai College of Engineering, Anna University in 2010. She completed her Master of Engineering (M.E) in the stream of Computer Science and Engineering from Madras Institute of Technology, Anna University in 2013. She is the Gold Medalist in her B.Tech and M.E programme. She completed Ph.D in the field of Analytics at Department of Computer Technology, Madras Institute of Technology, Anna University and also she is working as an Assistant Professor in Sri Ramachandra Institute of Higher Education and Research, Chennai. Her research area includes Analytics, Analysis of Algorithms, Machine learning. She is the recipient of Chancellor Shri. V. R. Venkataachalam Award for best publication. She has published various International research papers, which is indexed in IEEE explorer and Elsevier publications.

J. Shobana received Ph.D. in Computer Science and Engineering. Working as an Assistant Professor in the Department of Computer Science and Engineering, SRM Institute of Science and Technology has been serving the Education Profession for the past 18 years. Her area of interest is Text Mining, Natural Language Processing, Artificial Intelligence and Machine Learning.

Nancy Sundar has completed her Masters in Computer Science and Engineering in Bannari Amman Institue of Technology, Sathyamangalam. She was awarded doctorate degree from Anna University in 2019. She is an alumna of PSG College of Technology too. She has more than 14 years of teaching experience in various engineering colleges. Her areas of interest include Data Mining, Big Data, Social Network Analysis and Oncogenic studies. She has coordinated many international Conferences. She has published more than 25 research papers in various reputed and refereed journals and attended conferences both in India and Abroad. She has received funding from department of science and technology for her work in Social Network Analysis. Her research contribution facilitated the mankind with the identification of new rescuable cancer mutants. She has guided and guiding students in both UG and PG level. She is an active participant in accreditation process and also in teaching learning process.

Ravikumar Tammineni obtained his Ph.D. in Computer Science and Engineering from GITAM-A Deemed-to be University, Visakhapatnam, India, 2020. He received his M. Tech in Computer Science and Engineering from JNT University, Hyderabad in 2008. He is having 15 years of teaching experience in various Engineering Colleges. Presently he is working as an Associate Professor in the Department of CSE, at Aditya Institute of Technology and Management. He is having more than 20 publications in various reputed international journals and conferences . His current research interest includes Machine Learning, IoT, Software Engineering and Artificial intelligence.

Dhanabalan Thangam is presently working as Associate Professor at Presidency Business School, Bengaluru, India. Earlier he was worked as Post-Doctoral researcher in Konkuk School of Business, Konkuk University, Seoul, Korea South. He received his Ph.D. degree in Management from Alagappa University, Tamilnadu, India. His current research interests are marketing, small business management and artificial intelligence in management fields. He has authored several books, research articles, and proceedings presented at many professional conferences and venues. He has received Two funding support from Indian Council of Social Science Research.

Index

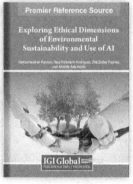

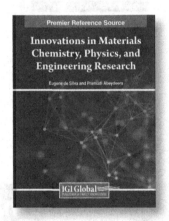
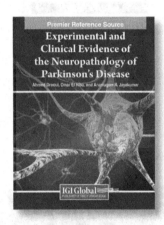
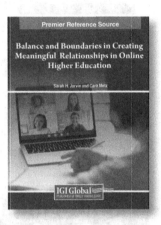

Printed in the United States
by Baker & Taylor Publisher Services